Neurochemical Systems and Signaling

The human brain is made up of billions of neurons that communicate with each other through chemical messengers, which are referred to as neuroactive substances. These neuroactive substances include neurotransmitters, neuromodulators, and neurohormones. Some neurotransmitters also act as neuromodulators and neurohormones. It is unlikely that there would ever be a consensus about the meanings of these neuroactive substances, including neurotransmitters, since the term 'neurotransmitter' has traditionally been used very loosely indeed, to include neurotransmitters, neurohormones, and neuromodulators. Any alterations in the functioning of these neuroactive substances can cause diseases.

The brain is the ultimate center that regulates all neurological and behavioral aspects of the body through neuronal communications mediated via various neurochemicals. Thus, neurological and psychiatric disorders are, in most cases, the result of disturbed neurochemical balance. Besides the multifaceted involvement of billions of neuronal cells, the central nervous system is a complex organization with a diverse number of neurotransmitter systems, as compared to the autonomic nervous system, in which the parasympathetic system works on the 'rest and digest' phenomenon, and the sympathetic system works on the 'fight or flight' phenomenon. There are more than 20 neurotransmitter systems and multiple receptors for each neurotransmitter. Any alterations in neurochemical balance are expressed in the form of neurological or psychiatric disorders such as epilepsy, Parkinson's disease, Alzheimer's disease, psychosis, depression, etc. Acetylcholine, noradrenaline, dopamine, and 5-HT are of the utmost importance among neurotransmitters for their profound role in the pathogenesis of various neurological and psychiatric disorders in humans. Yet the involvement of various proteins and peptides, such as neurotrophic factors, growth factors, and endogenous chemical compounds, cannot be ignored.

Day by day, the suffering of people due to an imbalance of neurotransmitters is increasing. Various factors, for example stress, diet, genetics, and toxins such as alcohol and nicotine, contribute to this imbalance. This imbalance may lead to mental health complaints. The main purpose of this book is to give a comprehensive overview of the neurological diseases associated with neurochemical imbalances. This book will help readers gain a comprehensive understanding of neuronal signaling and related neurological disorders, as well as status and future opportunities and challenges. It will provide a brief account of neurotransmission, as either a study or high-yield revision aid.

Neurochemical Systems and Signaling

From Molecules to Networks

Edited by
Ghulam Md Ashraf
Department of Medical Laboratory Sciences, College of Health Sciences, and Research Institute for Medical and Health Sciences, University of Sharjah, Sharjah, United Arab Emirates

CRC Press
Taylor & Francis Group
Boca Raton London New York

CRC Press is an imprint of the
Taylor & Francis Group, an **informa** business

First edition published 2023
by CRC Press
6000 Broken Sound Parkway NW, Suite 300, Boca Raton, FL 33487-2742

and by CRC Press
4 Park Square, Milton Park, Abingdon, Oxon, OX14 4RN

CRC Press is an imprint of Taylor & Francis Group, LLC

Library of Congress Cataloging-in-Publication Data
Names: Ashraf, Ghulam Md, 1984– editor.
Title: Neurochemical systems and signaling : from molecules to networks / edited by Ghulam Md Ashraf.
Description: First edition. | Boca Raton : CRC Press, 2023. |
Includes bibliographical references and index.
Identifiers: LCCN 2022041672 (print) | LCCN 2022041673 (ebook) |
ISBN 9780367210625 (hbk) | ISBN 9781032439273 (pbk) | ISBN 9780429265198 (ebk)
Subjects: MESH: Synaptic Transmission | Neurotransmitter Agents–physiology | Mental Disorders–physiopathology |
Emotions–physiology | Nerve Net–chemistry
Classification: LCC QP360 (print) | LCC QP360 (ebook) | NLM WL 102.8 | DDC 612.8–dc23/eng/20230119
LC record available at https://lccn.loc.gov/2022041672
LC ebook record available at https://lccn.loc.gov/2022041673

ISBN: 9780367210625 (hbk)
ISBN: 9781032439273 (pbk)
ISBN: 9780429265198 (ebk)

DOI: 10.1201/9780429265198

Typeset in Times
by Newgen Publishing UK

Contents

SECTION III Typical Neurochemical Processes

Preface

Neurochemistry allows the brain to function by utilizing molecules called neurotransmitters. Owing to variations in the neurotransmitters, neurochemistry varies from individual to individual. Different environmental factors, such as diet, medications, and various pharmaceuticals, significantly impact the levels of neurotransmitters and their receptors in the brain. In species ranging from microscopic invertebrates to highly evolved mammals, neurotransmission regulates both excitatory and inhibitory processes in the central nervous system, underpins sensory processing, and regulates autonomic and motor outputs. Neurotransmitters are essential regulators of message transmission across synapses and to receptors. Neurotransmitters can have an inhibitory or excitatory impact, which slows down or speeds up the brain.

As highly specialized human brain processes, learning and memory entail complicated interactions between neurotransmitters and cellular events. Neurophysiological, psychological, and pharmaceutical views have all contributed to our understanding of these processes over time. For different neurotransmitters, there are different types of receptors. Depending on the situation, a neurotransmitter might have either an excitatory or inhibitory impact. Studies suggest that neurotransmitter deficiency, imbalances, and malfunctions are exceedingly frequent in modern society. Such alteration of neurotransmitters is responsible for various diseases, because when neurotransmitters aren't functioning correctly, the mind and body can't communicate accurately.

The primary purpose of this book is to give a comprehensive overview of the neurological diseases associated with neurochemical imbalances. This book contains 16 chapters divided into three sections. It is expected that this book will be very effective in giving readers a complete idea about neuronal signaling and linked neurological disorders, as well as the current status and future opportunities and challenges.

Ghulam Md Ashraf

Editor Biography

Dr. Ghulam Md Ashraf is working as Associate Professor in Department of Medical Laboratory Sciences, College of Health Sciences, and Research Institute for Medical and Health Sciences, University of Sharjah, United Arab Emirates. He has a Ph.D. in Biochemistry from Aligarh Muslim University with over 12 years of teaching and research experience in various disciplines of biological and medical sciences. Dr. Ashraf has taught various subjects, such as Advanced Clinical Chemistry, Biochemistry, Biology, Clinical Biochemistry, Cell Biology, Genetics, Immunology, Intellectual Property Rights, Introduction to Research, Molecular Biology, Neurology, Recombinant DNA Technology and Research Methods at graduate and postgraduate levels.

His primary fields of research are biochemistry and neurology, currently focusing on understanding the molecular and behavioral mechanism of effects of antidiabetic drugs in psychotic and dementia conditions. Dr. Ashraf's lab is currently investigating the effect of antidiabetic drugs in the possible attenuation/reversal of antipsychotic drug-induced weight gain in psychotic conditions. He is also investigating molecular and behavioral mechanisms of novel therapeutic combinations in multiple sclerosis and Alzheimer's disease. Dr. Ashraf has published 391 research articles (citations: 9334; H-index: 52; i10 index: 202) and have been involved in 23 research grants.

Dr. Ashraf is also involved in editorial roles in renowned scientific journals such as the *Journal of Advanced Research* (Associate Editor), *Scientific Reports* (Senior EBM), *PLoS One* (Academic Editor), and *Current Neuropharmacology* (Section Editor). He is professionally associated with the Royal Society of Medicine (Fellow), Royal Society of Biology (Member), American Society for Biochemistry and Molecular Biology (Member), and Canadian Association of Neuroscience (Member). Dr. Ashraf has 301 Publons-verified editor records and 596 Publons-verified reviewer records.

For his scientific contributions, Dr. Ashraf has been recognized by various international organizations. He was approved as a subject of biographical record in *Marquis Who's Who in the World* (2020). Dr. Ashraf has been recognized as Expertscape World Expert in Alzheimer's disease and Nervous System Diseases. Most recently, he has been listed in the World's Top 2% of Scientists for consecutive years, announced by Stanford University-Elsevier.

Contributors

Nana Akua Acheampong
Kennesaw State University
Atlanta, GA, USA

Jawaid Ahsan
Department of Biotechnology, Central University of South Bihar
Gaya, Bihar, India

Md. Sayeed Akhtar
Department of Clinical Pharmacy, College of Pharmacy, King Khalid University
Abha, Saudi Arabia

Raushanara Akter
Department of Pharmacy, Brac University
Dhaka, Bangladesh

Md. Zeeshan Ali
Department of Biotechnology, Central University of South Bihar
Gaya, Bihar, India

Ashwani Arya
Department of Pharmaceutical Education and Research, BPS Women University
Sonipat, Haryana, India

Anushree
Department of Biotechnology, Central University of South Bihar
Gaya, Bihar, India

Faisal Ashraf Bhat
Department of Pharmacology, School of Pharmaceutical Education and Research, Jamia Hamdard
New Delhi, India

Shradha Bisht
Sardar Bhagwan Singh University
Dehradun, Uttarakhand, India

S. B. Chandrasekar
Hi-Tech Lab, Drug Testing Laboratory, Drug Control Department
Bangalore, India

Manish Dhall
Swami Dayanand Post Graduate Institute of Pharmaceutical Sciences, Sharma University of Health Sciences
Rohtak, Haryana, India

Faizana Fayaz
Department of Pharmaceutical Chemistry, Delhi Institute of Pharmaceutical Sciences and Research
Delhi, India

Neha Goyal
Department of Biotechnology, Jaypee Institute of Information Technology
Noida, Uttar Pradesh, India

Vivek Kumar Gupta
Department of Biochemistry, University of Allahabad
Allahabad, India

Hemavathi G
Hi-Tech Lab, Drugs Testing Laboratory, Drugs Control Department
Bangalore, India

Bakr Ahmed Hameed
Department of Pharmaceutics, School of Pharmaceutical Education and Research, Jamia Hamdard
New Delhi, India

Ashif Iqubal
Department of Pharmacology, School of Pharmaceutical Education and Research, Jamia Hamdard
New Delhi, India

Aadil Javed
Department of Biotechnology, Graduate School of Natural and Applied Sciences, Ege University
Bornova, Izmir, Turkey

Chinmay Jogdeo
School of Pharmacy, Dr. Vishwanath Karad MIT World Peace University
Pune, India

Md. Tanvir Kabir
Department of Pharmacy, Brac University
Dhaka, Bangladesh

Renu Kadian
Ram Gopal College of Pharmacy
Sultanpur, Farruknagar, Gurugram, Haryana, India

Ozaifa Kareem
Department of Pharmaceutical Sciences, University of Kashmir
Hazratbal, Srinagar, Jammu and Kashmir, India

Simrat Kaur
Department of Biotechnology, Jaypee Institute of Information Technology
Noida, Uttar Pradesh, India

Boshra Khaled
Department of Clinical Pharmacy and Pharmacology, RAK College of Pharmaceutical Sciences, RAK Medical and Health Sciences University
Ras Al Khaimah, UAE

Huda Khaled
Department of Clinical Pharmacy and Pharmacology, RAK College of Pharmaceutical Sciences, RAK Medical and Health Sciences University
Ras Al Khaimah, UAE

Fazlullah Khan
Department of Toxicology and Pharmacology, Faculty of Pharmacy, The Institute of Pharmaceutical Sciences (TIPS), Tehran University of Medical Sciences
Tehran, Iran

Moteb Khobrani
Department of Clinical Pharmacy, College of Pharmacy, King Khalid University
Abha, Saudi Arabia

Eswar Kumar Kilari
A.U. College of Pharmaceutical Sciences, Andhra University
Visakhapatnam, India

Asiya Mahtab
Department of Pharmaceutics, School of Pharmaceutical Sciences and Research, Jamia Hamdard
New Delhi, India

Mohd Mazhar
Department of Pharmacology, Delhi Pharmaceutical Sciences and Research University
Delhi, India

Kamal Niaz
Department of Pharmacology and Toxicology, Faculty of Bio-Sciences, Cholistan University of Veterinary and Animal Sciences
Bahawalpur, Pakistan

Newman Osafo
Kwame Nkrumah University of Science and Technology
Kumasi, Ghana

Joseph Owusu-Sarfo
Kwame Nkrumah University of Science and Technology
Kumasi, Ghana

Anil T. Pawar
School of Pharmacy, Dr. Vishwanath Karad MIT World Peace University
Pune, India

Faheem Hyder Pottoo
Department of Pharmacology, College of Clinical Pharmacy, Imam Abdulrahman Bin Faisal University
Dammam, Saudi Arabia

Lokesh Prasad
Hi-Tech Lab, Drug Testing Laboratory, Drug Control Department
Bangalore, India

Swati Putta
Department of Pharmacology, A.U. College of Pharmaceutical Sciences, Andhra University
Visakhapatnam, India

Syed Arman Rabbani
Department of Clinical Pharmacy and Pharmacology, RAK College of Pharmaceutical Sciences, RAK Medical and Health Sciences University
Ras Al Khaimah, UAE

Rachana
Department of Biotechnology, Jaypee Institute of Information Technology
Noida, Uttar Pradesh, India

Jawad ur Rahman
Department of Microbiology, College of Medicine, Imam Abdulrahman Bin Faisal University
Dammam, Saudi Arabia

Manoj K Sarangi
Sardar Bhagwan Singh University
Dehradun, Uttarakhand, India

Sadat Shafi
Pharmaceutical Medicine, Department of Pharmacology, School of Pharmaceutical Education and Research, Jamia Hamdard
New Delhi, India

Bechan Sharma
Department of Biochemistry, University of Allahabad
Allahabad, India

Shrestha Sharma
Department of Pharmaceutical Sciences, School of Medical & Allied Sciences, K.R. Mangalam University
Gurgaon, India

Mamta F. Singh
Sardar Bhagwan Singh University
Dehradun, Uttarakhand, India

Vinod Singh
Gurukul Kangri University
Haridwar, Uttarakhand, India

Aman Upaganlawar
SNJB's Shriman Suresh Dada Jain College of Pharmacy
Nashik, India

Oduro Kofi Yeboah
Kwame Nkrumah University of Science and Technology
Kumasi, Ghana

Section I

Transmitters and Systems

1 Adrenergic Neurotransmission

Md. Tanvir Kabir and Raushanara Akter
School of Pharmacy, Brac University, Dhaka, Bangladesh

CONTENTS

1.1 INTRODUCTION

The adrenergic transmitter system consists of adrenergic receptors (ARs; a group of G protein-coupled receptors) that are activated via catecholamines (CAs), particularly epinephrine (adrenaline) and norepinephrine (noradrenaline; Magnusson and Brim 2014). There are two main groups of ARs: α and β (with multiple subtypes for each type; Freundt 1965). The only source of norepinephrine (NE) is the neurons of the ascending dorsal noradrenergic bundle of the locus coeruleus (Descarries and Droz 1970). Furthermore, the ascending dorsal noradrenergic bundle of the locus coeruleus system is associated with various cognitive functions, including selective attention, learning, and memory (Sara 2009). The locus coeruleus is innervated by the prefrontal cortex projections (Jodo et al. 1998). ARs of the prefrontal cortex make significant contributions to both selective memory and attention. It is known that α_2-AR of the prefrontal cortex is predominantly significant for attention, whereas the β subtypes contribute significantly to memory (Sara 2009).

In the case of normal aging, both memory and attention have been found to decline. Attention deficit during aging can be ameliorated by using specific agonists of α_2-ARs in nonhuman primates, which further indicates that there is a loss of receptor function during aging (Arnsten and Contant 1992). In a similar manner, memory deficits could be ameliorated with a NE or β_2-AR-specific agonist in the case of normal aging (Sternberg et al. 1985; Ramos et al. 2008). Degeneration of the neurons of the ascending dorsal noradrenergic bundle of the locus coeruleus is one of the renowned characteristics of Alzheimer's disease (AD; Grudzien et al. 2007). Nonetheless, AD treatment with β- or α_2-AR agonists was found to be unsuccessful and rather worsened pathology, further indicating that cognitive deficit cannot be solely credited to a loss of function in NE or AR (Riekkinen et al. 1999; Ni et al. 2006). Most chromaffin cells are found in the adrenal medulla and these cells are derived embryologically from neuroectoderm. Furthermore, sustentacular and ganglion cells are also located in medulla. Chromaffin cells (which store CAs in their secretory vesicles, also called as chromaffin granules) exist as clusters and in trabeculae, where the ganglion cells are located alone or in clusters interspersed among chromaffin cells or along with nerve fibers. The sustentacular cells are found at the periphery of nests or clusters of chromaffin cells (Fung et al. 2008; Eiden and Jiang 2018).

At the center of the adrenal gland, the precursor chromaffin cells differentiate by responding to cortisol (a glucocorticoid). A small number of these cells also travel to generate paraganglia,

DOI: 10.1201/9780429265198-2

groups of chromaffin cells on both sides of the aorta, most of which are mainly located at the origin of the inferior mesenteric artery or at the bifurcation of the aorta, called the organ of Zuckerkandl (Tosun et al. 2017). Indeed, CAs are discharged by the chromaffin granules and sympathetic axons via exocytosis, where all soluble granule components (such as bioactive peptides, chromogranins, and enzymes) are co-released into the extracellular space and ultimately gain access to the circulation (Goumon et al. 2000; Day and Gorr 2003). CAs are mostly removed from the synaptic clefts via neuronal uptake (reuptake); however, the organic cation transporter family might mediate some non-neuronal uptake. Following neuronal uptake, cytosolic CAs can then be either re-transferred into the storage vesicles or deaminated and metabolized via oxidation (by the monoamine oxidases) or *O*-methylation (by catechol *O*-methyltransferase, or COMT), after neuronal uptake. Liver is essential for the full degradation of CAs to vanillylmandelic acid. Interestingly, CAs are one of the shortest-lived (only 1–2 minutes) signaling molecules in plasma. Neuronal uptake largely clears them from the circulation; however, they are also subjected to sulfoconjugation of a ring hydroxyl group in the alimentary canal or direct renal excretion (Kuhar et al. 1999; Eisenhofer 2001; Fung et al. 2008; Goldman and Schafer 2012).

In this chapter, we summarize adrenergic neurotransmission; mechanisms associated with synthesis, storage, release, and termination of action of CAs; and signal transduction pathways associated with ARs. In addition, we also focus on the effects mediated by ARs in various effector organs and the possible roles of CAs in various diseases.

1.2 INTERCELLULAR COMMUNICATION IN THE NERVOUS SYSTEM

Two diverging neuronal communication theories emerged in the late nineteenth century from the early reports of neuronal cell structure by Ramón y Cajal and Camillo Golgi (Glickstein 2006). It was proposed by Golgi that in order to communicate directly with each other, neurons generate an interconnected reticulum (similar to cardiac muscle cells) and use openings in their membranes. In contrast, Cajal claimed that neurons are individual cells containing adjoining cell membranes, and that communication between individual neurons occurs chemically at the contact sites. Although these theories are opposing in nature, both have been reported to be correct and both processes play a role in features of neuronal communication in the brains of mammals. Gap junctions play a role in direct communication between neurons (called 'electrical synapses'). Neurotransmitters are used for the communication across cell membranes at the regions where two neurons communicate with each other, where the site of contact is called a 'chemical synapse' (Glickstein 2006; Shoykhet and Clark 2011).

1.2.1 Electrical Synapses

The cell membranes of adjacent neurons are firmly bound together in a gap-junction plaque at the electrical synapse (Martin et al. 2020). Indeed, each plaque possesses many channels composed of connexin proteins. Twenty-one connexin genes have been identified in humans. Every channel is made of two hemichannels, with one on each cell membrane. It was revealed that two hemichannels link together to generate a functional gap junction amid two neurons, which eventually allows intercellular diffusion of ions and small molecules, including glucose, adenosine triphosphate (ATP), and cyclic adenosine monophosphate (cAMP). Therefore, gap junctions permit neurons to share information regarding their excitable and metabolic states, providing a process for large-scale neuronal network dynamics and regulation of energy demands (Mammano 2013). In addition, gap junction channels can be closed owing to increased Ca^{2+} or decreased intracellular pH levels; as both processes take place in damaged cells, at the gap junction paired hemichannels might play role in isolating healthy neurons from those injured during trauma or ischemia. It has been confirmed that unpaired hemichannels located outside of gap junction plaques might play a role in ischemic neuronal cell death (Nielsen et al. 2012; Belousov et al. 2017; Laird and Lampe 2018; Totland et al. 2020). Like neurons, glia are also linked via gap junctions. For example, brain astrocytes create an interlinked cellular network, which permits long-distance Ca^{2+} signal propagation across numerous cells. Furthermore, myelin layers are produced via oligodendrocytes in the central nervous system (CNS) and via Schwann cells in the peripheral nervous system are connected via gap junctions. Myelin sheath obtains structural stability by myelin gap junctions, which allows fast diffusion of nutrients and other components across the sheath to the underlying axon (Brenner et al. 1986; Fields and Stevens-Graham 2002; Nualart-Marti et al. 2013; Fields et al. 2015).

1.2.2 Chemical Synapses

1.2.2.1 Neuromuscular Junction

In the nervous system, the neuromuscular junction (NMJ) is an example of chemical synaptic transmission that has been extensively studied. At the presynaptic neuron, an action potential results in neurotransmitter release that leads to the activation of various ion channels on the muscle cell membrane, which eventually leads to a postsynaptic action potential (Holz and Fisher 1999; Lodish et al. 2000a). However, every stage in this mechanism is elegantly modulated and regulated in health and may be disturbed in the case of disease. Growing knowledge regarding NMJs provides a critical understanding of pathophysiology and synaptic function. The NMJ is composed of three different anatomic constituents, including the postsynaptic muscle fiber, the synaptic cleft comprising the basement membrane, and the presynaptic nerve terminal. In the spinal cord, presynaptic nerve terminal generates from the myelinated axon of a motoneuron (Naguib et al. 2002; Jimsheleishvili and Sherman 2019).

In the ventral gray matter of the spinal cord, lower motoneurons send their myelinated axons via the ventral root to peripheral muscle targets. By approaching the muscle fibers, axon loses its myelin sheath and branches into a fine network of terminals, where the diameter of each one is around 2 μm. Interestingly, each branch contains multiple swellings in its course ('presynaptic boutons'), wherein the nerve builds synaptic contact by means of muscle fiber. Terminal Schwann cells cover the presynaptic boutons, which recycle neurotransmitters, supply growth factors, and may play role in recovery following muscle and nerve damage (Song et al. 1999; Murray 2008). Presynaptic boutons are located over particular areas of the cell membrane of muscles, termed end-plate areas. Moreover, underlying each bouton is a particular area of the cell membrane that possesses an extremely high level of voltage-gated Na^+ channels and nicotinic acetylcholine (nACh) receptors (Sanes and Lichtman 1999; Gooch 2003). The end-plate areas and presynaptic boutons are separated via a 100-nm-wide synaptic cleft comprising the extracellular matrix and basement membrane. Indeed, the basement membrane anchors numerous proteins, such as acetylcholinesterase, an enzyme that plays role in rapid acetylcholine (ACh) hydrolysis in the synaptic cleft.

At the NMJ, multiple different functional events take place during synaptic transmission. First, in the presynaptic boutons, the action potential that arrives from the motor axon results in depolarization of the membrane, which results in the entry of Ca^{2+} through voltage-gated Ca^{2+} channels on the presynaptic membrane. In addition, the entry of Ca^{2+} leads to the fusion of ACh-containing synaptic vesicles with the presynaptic cell membrane and ACh release into the synaptic cleft. Then, ACh quickly diffuses to the postsynaptic membrane and binds with the nACh receptor (nAChR); two ACh molecules are essential for activation of the nAChR. After activation, the nAChR opens and permits both K^+ and Na^+ to flow via the ion pore (Toutant and Massoulié 1988; Hughes et al. 2006). Nonetheless, inward Na^+ current rules over the outward K^+ current, which further leads to net depolarization of the muscle membrane at the end-plate area. Before encountering voltage-gated Na^+ channels, this alleged end-plate potential propagates a short distance. Interestingly, these Na^+ channels open when the membrane potential increases to a critical threshold value, which only allows Na^+ ions to flow into the cell and generates an all-or-nothing muscle action potential. In the synaptic cleft, acetylcholinesterases stop the depolarizing activity of ACh at the postsynaptic membrane by quickly hydrolyzing ACh into choline and acetate (Lodish et al. 2000b).

1.2.2.2 Chemical Synapses in the Central Nervous System

In the CNS, chemical synapses function based on similar basic principles to those governing synaptic transmission at the NMJ, even though the cadre of postsynaptic receptors and neurotransmitters is considerably more diverse in the CNS. Furthermore, in the CNS, a neuron might produce and store more than a single neurotransmitter; however, it releases the same group of neurotransmitters at all of its synapses. Based on their appearance under electron microscopy, CNS synapses are classified into asymmetric (Gray type I) and symmetric (Gray type II) synapses. These synapses are also known as inhibitory and excitatory synapses, respectively. Every single neuron generates its own complement of neurotransmitters that are transported to all synaptic contact sites in the axon and packaged into synaptic vesicles (Holz and Fisher 1999).

Ca^{2+} currents induce synaptic vesicles to fuse with the cell membrane when an action potential reaches the axon, which further triggers the release of neurotransmitters into the synaptic cleft. Subsequently, on the postsynaptic membrane, neurotransmitters play a role in their corresponding metabotropic and ionotropic receptors. Activation of ionotropic receptor results in a hyperpolarizing (inhibitory) current or depolarizing (excitatory) current. In addition, these subthreshold currents are known as inhibitory and excitatory postsynaptic potentials (IPSPs and EPSPs, respectively). Occasionally, temporal and spatial summation of EPSPs in the dendritic tree of the postsynaptic neuron depolarizes the somatic membrane adequately enough to cross the threshold and trigger an action potential (Rekling et al. 2000; Purves et al. 2001).

In comparison with the NMJ, a distinctive characteristic of chemical synaptic transmission in the CNS is its deficiency of dependability on a single cell level, where an action potential in the presynaptic neuron does not essentially induce a postsynaptic neuron to generate its own action potential. Interestingly, such a deficiency in determinism is expected to allow individual differences in terms of responses to the same stimuli. Similar to ACh hydrolysis at the NMJ, following interaction with the receptor, the neurotransmitter is removed from the synapse via active reuptake into the terminal, diffusion, or enzymatic degradation (von Gersdorff and Hull 2003; Lawes 2004; Gooch 2014).

1.3 ADRENERGIC NEUROTRANSMISSION

Adrenergic transmission is restricted to the sympathetic division of the autonomic nervous system. Three closely associated endogenous CAs play roles in this neurotransmission. CA neurotransmitters include epinephrine (EPI), NE, and dopamine (DA; Figure 1.1).

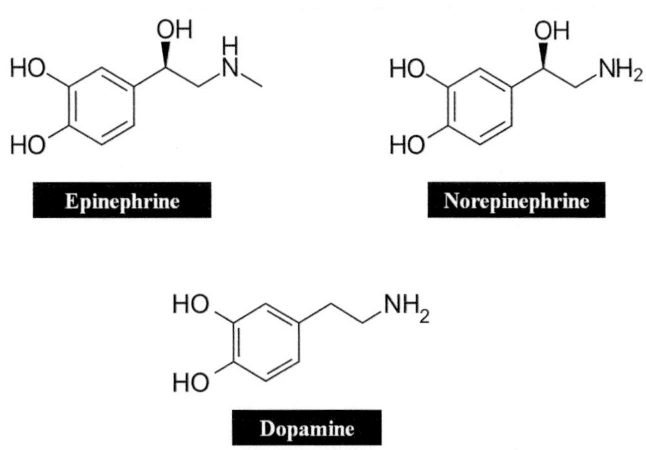

FIGURE 1.1 Chemical structures of catecholamines.

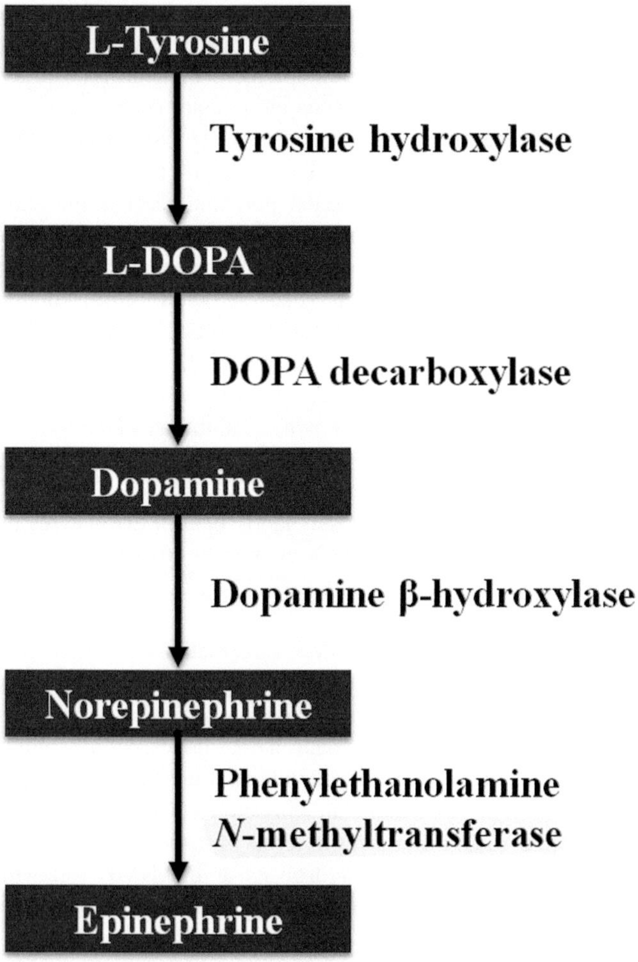

FIGURE 1.2 Biosynthesis of catecholamines.

1.3.1 Synthesis and Storage of Catecholamines

NE is the major neurotransmitter generated in sympathetic nerves from L-tyrosine (an amino acid). In this process, multiple enzymes expressed in these neurons are involved (Hussain and Maani 2019) (Figure 1.2). In CA biosynthesis, tyrosine hydroxylase is the rate-limiting enzyme that converts L-tyrosine to L-DOPA in the biosynthetic pathway. Reactions play a role in the cytoplasm of the cells expressing these enzymes, up to the point at which DA is generated. Phenylethanolamine *N*-methyltransferase is abundantly found in the adrenal medulla, where it methylates NE to generate EPI. Following EPI synthesis, it is taken up into storage granules where it remains stored until its release (Flatmark 2000; Marisa et al. 2012; Berends et al. 2019).

1.3.2 Release of Catecholamines

When the nerves containing the storage granules are depolarized, NE is released into the synaptic cleft. Even though the full mechanism is yet to be fully revealed, the process includes increased levels of intracellular calcium, which takes place along with depolarization (Bikle 2000; Katra and Laurita 2005; Blanke and VanDongen 2008; Trebichalská and Holubcová 2020). This mechanism is tightly controlled via various neurotransmitter and hormone receptors that are expressed on sympathetic neurons. Interestingly, these nerve endings usually have a receptor that is activated via NE, which plays a role in automatically suppressing NE release following neuronal firings (α_2-ARs). Amphetamine (a CNS stimulant) plays a major role in mediating NE release from these nerve endings, which is independent of sympathetic nerve firing (Starke 2001; Molderings et al. 2002; Bermingham and Blakely 2016; Pei et al. 2016; Hökfelt et al. 2018).

1.3.3 Termination of the Action of Catecholamines

Reuptake of NE into sympathetic nerve endings is the major process that ends the activity of NE on postsynaptic cells. Indeed, a specific transporter located on these neurons transport NE back into these cells. Cocaine blocks the action of this transporter (Zhou 2004; Sitte and Freissmuth 2010). Therefore, the main psychotropic activities of cocaine include suppression of the action of DA transporters in brain centers, while numerous cardiovascular activities take place due to induction of NE activities in the periphery. In addition, three more mechanisms can terminate NE activity, including transformation to inactive metabolites, uptake into other tissues, and diffusion out of the synaptic cleft into the bloodstream. Indeed, monoamine oxidase (MAO) makes a significant contribution to inactivating CAs. There are two subtypes of this enzyme: MAOA and MAOB. MAO is mainly located in the liver and nerve endings, but this enzyme is also found in various other tissues. It has been reported that mitochondrial-associated MAO nerve terminals can cause the inactivation of NE. In the liver, MAO blocks various substances, including tyramine (an amino acid found in in protein-containing foods), from gaining access to the systemic circulation in an intact condition by deactivating tyramine before it reaches the systemic circulation (Sivasubramaniam et al. 2003; Kaludercic et al. 2014). As tyramine is highly effective in triggering NE release from nerve endings, there is an increased risk of adverse reactions, including significantly elevated blood pressure, due to the consumption of tyramine-containing foods by individuals who are administered an MAO inhibitor to suppress activity of liver enzyme. COMT is another enzyme that plays vital roles in the metabolism of CAs. It has been reported that CAs that gain access to systemic circulation are widely methylated via COMT (Fiedorowicz and Swartz 2004; Garcia and Santos 2018; Sub Laban and Saadabadi 2020).

1.4 ADRENERGIC RECEPTORS AND SIGNAL TRANSDUCTION PATHWAYS

Three major groups are differentiated by their signal transduction pathways and by their pharmacology. Indeed, α_1-ARs are coupled to Gq/G11, which leads to phospholipase

TABLE 1.1
Features of Human α and β-Adrenergic Receptor Subtypes

Adrenergic Receptor Subtypes	Location	Chromosome	Amino Acids	Signal Transduction Pathways	References
α_{1A}	Cerebral cortex, cerebellum, blood vessels, heart, liver	8	430–476	Increased release of Ca^{2+} and protein kinase C (PKC) activation	(Bylund 1992; Hieble et al. 1995; Michel et al. 1995; Graham et al. 1996)
α_{1B}	Blood vessels, kidney, spleen, fetal brain	5	515	Elevated Ca^{2+} release and PKC activation	(Bylund 1992; Hieble et al. 1995; Michel et al. 1995; Graham et al. 1996)
α_{1D}	Cerebral cortex, aorta	20	560	Increased release of Ca^{2+} and PKC activation	(Bylund 1992; Hieble et al. 1995; Michel et al. 1995; Graham et al. 1996)
α_{2A}	Hippocampus, pancreas, locus coeruleus, small intestine	10	450	Reduced level of cyclic adenosine monophosphate (cAMP) level	(Regan and Cotecchia 1992; Bylund 1992; Bylund and Chacko 1999; Giovannitti et al. 2015)
α_{2B}	Thalamus, liver	2	450	Reduced cAMP level	(Regan and Cotecchia 1992; Bylund 1992; Bylund and Chacko 1999; Giovannitti et al. 2015)
α_{2C}	Hippocampus, lung, heart, olfactory bulb, aorta	4	461	Reduced cAMP level	(Regan and Cotecchia 1992; Bylund 1992; Bylund and Chacko 1999; Giovannitti et al. 2015)
β_1	Hypothalamus, heart, cerebral cortex, kidney	10	477	Increased activity of intracellular cAMP	(Minneman et al. 1981; Molinoff 1984; Hoffmann et al. 2004; Gnegy 2012)
β_2	Cerebral cortex, olfactory bulb, cerebellum, liver, lung, smooth muscle, hippocampus	5	413	Increased activity of intracellular cAMP	(Minneman et al. 1981; Molinoff 1984; Hoffmann et al. 2004; Gnegy 2012)
β_3	White and brown adipocytes	8	402/408	Increased activity of intracellular cAMP	(Minneman et al. 1981; Molinoff 1984; Hoffmann et al. 2004; Gnegy 2012; Patel 2012)

C activation, which further results in elevated release of Ca^{2+} and protein kinase C (PKC) activation. In addition, α_1-ARs activate mitogen-activated protein (MAP) kinases (MAPKs) and phospholipase D (Table 1.1). However, α_2-ARs are coupled to Gi/Go proteins, which lead to stimulation of phospholipase A2 and the inhibition of adenylyl cyclase activities (Table 1.1). Similar to D2 dopamine receptors, activation of α_2-ARs results in the release of G protein βγ (Gβγ) subunits, which further leads to suppression of Ca^{2+} channels and activation of K^+ channels. Interestingly, all three subtypes of the β-AR are coupled to G proteins and trigger adenylyl cyclase function (Figure 1.3). G protein-independent MAPK activation via activation of β-AR has also been revealed. The functional implication of this activity has not yet been fully revealed, and it takes place only with increased concentrations of agonists. The crystal structures of β_1- and β_2-ARs have been elucidated (Cherezov et al. 2007; Rasmussen et al. 2007; Warne et al. 2008), which will shed light on issues related to ligand and G protein functional selectivity.

Even though all subtypes within a grouping use the same effector pathways, their activities are not unnecessary, as confirmed in specific knockout mouse models. At the synapse, all AR subtypes have postsynaptic locations; however, the α_2-AR is also found presynaptically, which can play role as an autoreceptor to suppress NE release and as a presynaptic controller of the secretion of other neurotransmitters. It has been suggested that the α_2-AR has an autoreceptor action via releasing Gβγ from $G_{i/o}$, which subsequently binds with and suppresses P/Q- and N-type Ca^{2+} channels. Furthermore, α_{2A} and α_{2C}-ARs act as autoreceptors. Numerous β receptors are located in glia, wherein they exert several actions, including controlling inflammatory cascades, controlling the availability of glucose, and decreasing glutamate uptake. Similar to DA receptors, ARs are also subjected to several post-translational modifications. Furthermore, they are phosphorylated, glycosylated, and palmitoylated on several residues. Processes through which phosphorylation controls G protein-coupled receptor (GPCR) action were first explained by utilizing rhodopsin and β-AR (Lefkowitz 2004).

Protein kinase A (PKA), as well as GPCR kinase (GRK) can cause phosphorylation and downregulation of β-ARs. PKA facilitates heterologous desensitization or agonist nonspecific desensitization. Nonetheless, agonist binding with the receptor exposes areas that can be phosphorylated via GRK-2. Binding of arrestin (a protein) with the phosphorylated receptor prevents coupling to G proteins, which leads to receptor desensitization. In the CNS, NE has a significant role in vigilance, arousal, and attention. In the medial preoptic region, activation of α_1-ARs can result in wakefulness, whereas blocking these receptors results in sedation. In the hypothalamus, α_1-, α_2-, and β-ARs are significant constituents of the ascending arousal pathway (Mitchell and Weinshenker 2010). In the hippocampus and amygdala, NE induces long-term memory consolidation via activities at α_1- and β-ARs. Moreover, α_2-AR agonists increase

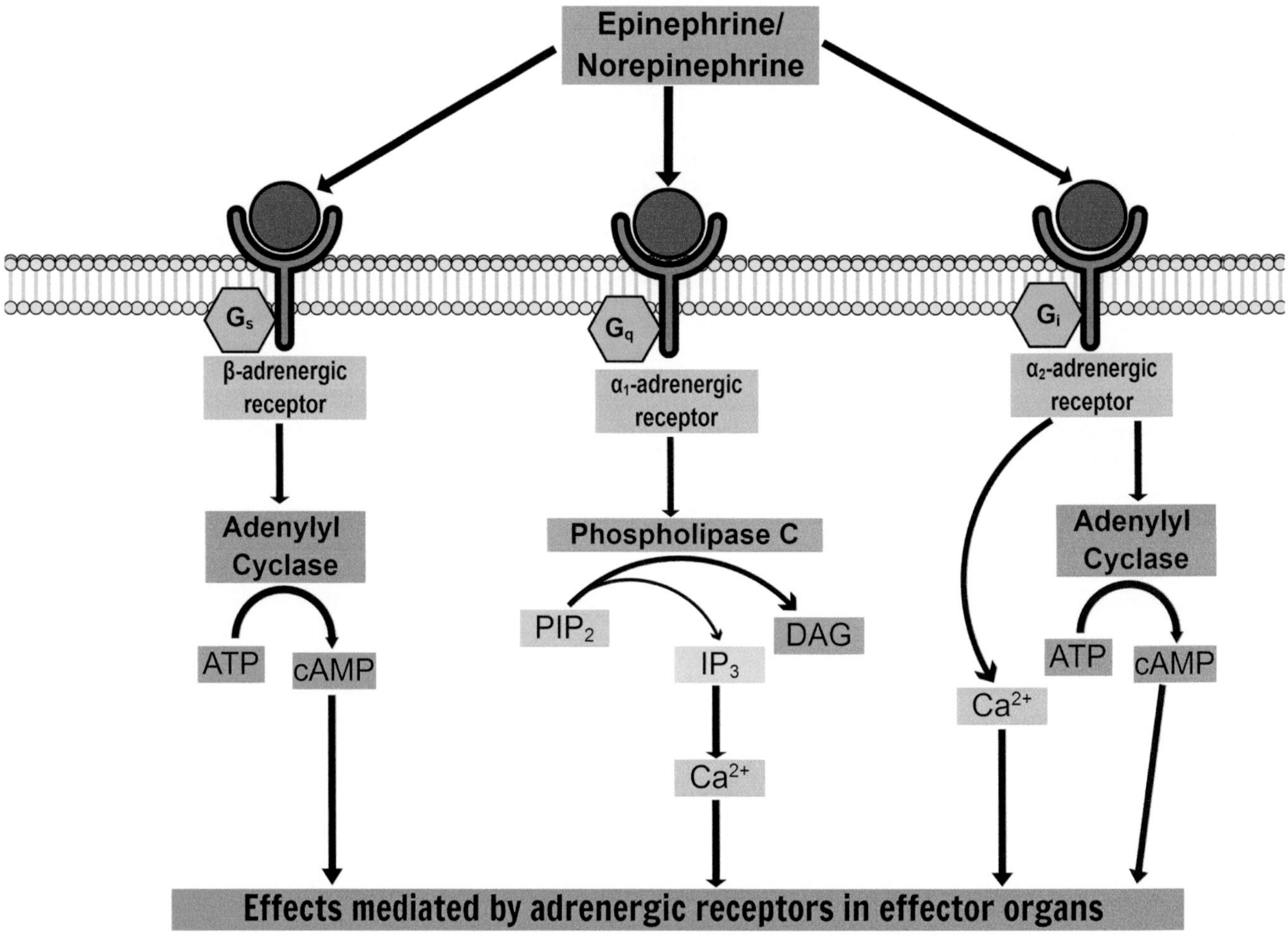

FIGURE 1.3 Signal transduction pathways of adrenergic receptors. Epinephrine or norepinephrine can bind with α_1, α_2, or β-adrenergic receptors. Typically, an α_1-adrenergic receptor couples to G_q, which leads to increased level of intracellular Ca^{2+} and subsequent effects, while α_2-adrenergic receptors couple to G_i and results in decreased cyclic adenosine monophosphate (cAMP) action and subsequent responses. The coupling of β-adrenergic receptors to G_s results in increased intracellular cAMP action and various effects. ATP, adenosine triphosphate; DAG, diacylglycerol; IP_3, inositol trisphosphate; PIP_2, phosphatidylinositol 4,5-bisphosphate.

TABLE 1.2
Effects Mediated by Adrenergic Receptors in Effector Organs

Effector Organs	Receptor	Effects	References
Heart			
Sinoatrial node	β_1, β_2	Increased heart rate	(Gordan et al. 2015; MacDonald et al. 2020)
Atria	β_1, β_2	Increased conduction velocity and contractility	(Gordan et al. 2015; Issa et al. 2019)
Atrioventricular node	β_1, β_2	Increased conduction velocity and automaticity	(Antzelevitch and Burashnikov 2011; Kester et al. 2012; Temple et al. 2016)
His-Purkinje system	β_1, β_2	Increased conduction velocity and automaticity	(Bartos et al. 2015; Schwarzwald 2018)
Ventricles	β_1, β_2	Increased rate of idioventricular pacemakers, automaticity, conduction velocity, and contractility	(López-Sendó et al. 2004; Bauer 2017; Gangwani and Nagalli 2020)
Eye			
Radial muscle, iris	α_1	Contraction (mydriasis)	(Pintor 2009; Kordasz et al. 2014)
Ciliary muscle	β_2	Relaxation for far vision	(McDougal and Gamlin 2015; Forrester et al. 2016)
Veins (systemic)	α_1, α_2; β_2	Constriction; dilations	(Pacak 2007, 2011)
Arterioles			
Coronary	α_1, α_2; β_2	Constriction; dilations	(Vatner et al. 1991; Sun et al. 2002; Brown 2018)
Skin and mucosa	α_1, α_2	Constriction	(Pacak 2011; Taylor and Cassagnol 2020)

TABLE 1.2 (Continued)
Effects Mediated by Adrenergic Receptors in Effector Organs

Effector Organs	Receptor	Effects	References
Skeletal muscle	α; β_2	Constriction; dilations	(Sun et al. 2002; Roatta and Passatore 2008; Sarelius and Pohl 2010; Korthuis 2011)
Cerebral	α_1	Constriction	(Fraser et al. 1971; Purkayastha and Raven 2011)
Pulmonary	α_1, β_2	Constriction; dilations	(Sun et al. 2002; Delong and Sharma 2019)
Abdominal viscera	α_1, β_2	Constriction; dilations	(Pacak 2007; Fang et al. 2020)
Salivary glands	α_1, α_2	Constriction	(Scully 2003; Tobin 2008)
Renal	α_1, α_2; β_1, β_2	Constriction; dilations	(Wang et al. 2006; Fang et al. 2020)
Stomach			
Motility and tone	α_1, α_2; β_2	Reduce	(Gullikson et al. 1991; Seiler et al. 2008)
Sphincters	α_1	Contraction	(Sahyoun et al. 1982)
Intestine			
Motility and tone	α_1, α_2; β_1, β_2	Reduce	(Seiler et al. 2008)
Sphincters	α_1	Contraction	(Biancani et al. 1997; Luckensmeyer and Keast 1998)
Secretion	α_2	Suppression	(Nakaki et al. 1982; Sagrada et al. 1987; Cox and Cuthbert 1989; Tanila et al. 1993)
Gallbladder and ducts	β_2	Relaxation	(Laukkarinen et al. 2012)
Lung			
Tracheal and bronchial muscle	β_2	Relaxation	(Barnes 1993; Proskocil and Fryer 2005; An et al. 2012)
Bronchial glands	α_1, β_2	Reduced secretion; elevated secretion	(Bai 1992; Bara et al. 2010; Fahy and Dickey 2010; Widdicombe and Wine 2015)
Kidney			
Renin secretion	α_1, β_1	Reduce; elevate	(Karlberg 1983; Hesse and Johns 1985; Kurtz 2012)
Uterus	α_1	Pregnant: contraction (α_1)	(Hajagos-Tóth et al. 2016, 2017; Chen et al. 2018)
	β_2	Pregnant: relaxation (β_2); nonpregnant: relaxation (β_2)	(Hajagos-Tóth et al. 2016, 2017; Chen et al. 2018)
Pancreas			
Acini	α	Reduced secretion	(Yamamura et al. 2012; Fujii et al. 2013)
Islets (β cells)	α_2	Reduced secretion	(Davani et al. 2004; Prentki and Nolan 2006; Hughes et al. 2018)
	β_2	Elevated secretion	(Lacey et al. 1991; Ceasrine et al. 2018)
Liver	α_1, β_2	Glycogenolysis and gluconeogenesis	(Exton 1987; Paredes-Flores and Mohiuddin 2020)
Fat cells	α_2, β_1	Lipolysis	(Langin et al. 1995; Barbe et al. 1997; Lafontan et al. 1997; Carey 1998)
	β_3	Thermogenesis	(Cypess et al. 2015; Peng et al. 2015; Lim et al. 2019)
Skeletal muscle	β_2	Increased contractility; glycogenolysis; K^+ uptake	(Yamamoto et al. 2007; Dinuzzo et al. 2013; Hostrup et al. 2014)
Urinary bladder			
Detrusor	β_2	Relaxation	(Andersson and Arner 2004; Badawi et al. 2007; Propping et al. 2016)
Trigone and sphincter	α_1	Contraction	(Michel and Vrydag 2006; Hill 2015)
Sex organs, male skin	α_1	Ejaculation	(Rowland et al. 2010; McMahon 2016; Clement and Giuliano 2016)
Pilomotor muscles	α_1	Contraction	(Wehrwein et al. 2016)
Sweat glands	α_1	Localized secretion	(Cui and Schlessinger 2015; Baker 2019)

prefrontal cortex cognitive activity, such as working memory. In the prefrontal cortex, release of increased NE levels can act on α_1-ARs, which further weakens prefrontal cortex activity. Numerous antipsychotic agents block α1-ARs and might improve some prefrontal impairment via this process. Following NE injection into the hypothalamic paraventricular nucleus, it activates postsynaptic α_2-ARs to exert a strong eating response. Numerous therapeutically active antidepressants have been found to elevate NE levels, which act by binding to ARs (Ventura et al. 2003; Ramos and Arnsten 2007; Mitchell and Weinshenker 2010; Berridge and Spencer 2016; Maletic et al. 2017).

1.5 CATECHOLAMINES AND ADRENOCEPTORS

It is crucial to understand the organ-specific activity and functions of ARs in elucidating different organ responses to CAs. In different vascular beds, ARs were initially classified into α and β types, based on the rank order of potency of different CAs (Ahlquist 1948). ARs have since been classified into α_1, α_2, β_1, β_2, β_3, and dopamine D_{1-5} receptors based on ligand-binding analyses and responses to synthetic antagonists and agonists (Lefkowitz 1979; Hoffman et al. 1980; Motulsky and Insel 1982; Arch 2001; Missale et al. 2010; Beaulieu and Gainetdinov 2011). It is known that α_1-ARs are postsynaptic in nature and expressed on effector tissues, including vascular smooth muscle (Pacak 2011). It has been reported that stimulation can lead to vasoconstriction and raised blood pressure, whereas stimulation of other α_1-ARs can result in uterine contraction, intestinal relaxation, and pupillary dilation (Table 1.2).

Several α_2-ARs are found presynaptically and suppress the secretion of NE after stimulation, whereas others on vascular smooth muscle are extrasynaptic and postsynaptic, and result in vasoconstriction after stimulation (Langer et al. 1980; Yamaguchi and Kopin 1980; Pacak 2011) (Table 1.2). Guanabenz, methyldopa, and clonidine are the examples of classic α_2-agonists. In the brain, the main role of an α_2-agonist is to inhibit sympathetic outflow and therefore it decreases blood pressure in individuals with pseudopheochromocytoma. Indeed, the clonidine suppression test is crucial to rule out or prove the presence of pheochromocytoma (a rare tumor that develops in adrenal gland tissue) in individuals with insignificantly increased levels of plasma normetanephrine or NE (Eisenhofer et al. 2003). β_1-ARs are involved with various activities and stimulation of these receptors in the heart results in positive chronotropic and inotropic activities, whereas stimulation of β_1-ARs somewhere else results in lipolysis in fat cells and elevated secretion of renin in the kidney (Table 1.2) (Pacak 2011). β_1-AR antagonists are commonly used drugs in individuals with tachyarrhythmia secondary to CA excess in pheochromocytoma. Stimulation of β_2-ARs results in an increased level of NE release from sympathetic nerves, relaxation of uterine and intestinal smooth muscles, glycogenolysis, vasodilation of some blood vessels (particularly those found in skeletal muscle), and bronchodilation.

Typical β_2-ARs agonists involve isoetharine, terbutaline, albuterol, and metaproterenol. Antagonists of both β_1- and β_2-ARs include timolol, nadolol, alprenolol, and propranolol, and these agents are infrequently used in individuals with pheochromocytoma because of their non-selectivity, which leads to a more intense effect on brain activity (e.g., sleepiness and tiredness). DA exerts very weak agonist activity at α- and β-ARs. D_1 receptors are mostly found in mesenteric, renal, coronary, and cerebral vascular beds. It has been observed that stimulation of these receptors can lead to vasodilation (Goldberg 1975; Creese et al. 1984). In the kidney, stimulation of D_1 receptors results in natriuresis and diuresis. If DA doses are elevated, it can stimulate both α- and β-ARs, which can result in vasoconstriction and raised blood pressure. At presynaptic terminals, D_2 receptors are located on sympathetic nerve endings, and stimulation of these receptors can lead to the suppression of NE release. While other D_2 receptors are located in sympathetic ganglia and their stimulation suppresses ganglionic transmission, others are found in the brain and are involved in emesis and suppression of the release of prolactin (Willems et al. 1985). D_{3-5} receptors are also involved in several brain activities (Beaulieu and Gainetdinov 2011). EPI exerts more strong effects on β_2-ARs than NE, in terms of a precise CA effect on ARs. In comparison with NE, it has been demonstrated that EPI shows higher or equal affinity toward α_1- and α_2-ARs, and equivalent potency toward β_1-ARs.

Because of these differences, EPI has activities on different populations of ARs, as compared to NE. EPI exerts strong activity on β_2-ARs of the skeletal muscle vasculature, which can lead to vasodilation. The metabolic activities of EPI include hypokalemia, increased oxygen consumption, thermogenesis, hyperlipidemia, and hyperglycemia. An EPI-mediated hyperglycemic response takes place as a result of various actions, including the stimulation of glucagon release, suppression of insulin secretion, stimulation of gluconeogenesis, suppression of glycogen synthase, and hepatic glycogen phosphorylase stimulation (Baker 2005; Pacak 2011; Giovannitti et al. 2015). NE released locally within the vasculature results in vasoconstriction induced by α_1-ARs. Interestingly, this and the inotropic and chronotropic activities of neurally released NE facilitated by means of cardiac β_1-ARs, show a primary and important activity of the sympathoneural system in cardiovascular regulation, mainly in blood pressure maintenance.

The differences in EPI and NE activities and ARs might be considered in the clinical setting. EPI-secreting tumors are most frequently involved with tachyarrhythmias, whereas NE-secreting pheochromocytomas are most frequently linked with hypertension, often along with hypertensive crises if NE is abruptly secreted. Nearly all organs and tissues in the body express ARs (some of them in a certain pattern). CAs tend to reduce intestinal tone and motility, and suppress intestinal secretion. The graded adrenergic response induces the secretion of renin in the kidney, and gallbladder relaxation, along with relaxation of the detrusor muscle that surrounds the urinary bladder. It has been reported that α_2-AR increases platelet aggregation and shunts blood toward the cardiopulmonary area with the induction of low-pressure cardiac baroreceptors (Goldenberg et al. 1948, 1950). In addition, sweating of palms of the hands and certain other regions (usually known as adrenergic sweating) takes place because of the induction of apocrine glands by α-ARs. Gut motility can be suppressed by CAs; in some individuals this can result in pseudo-obstruction or hypodynamic ileus. Indeed, most of the endocrine organs are influenced by CAs. Therefore, stimulation of α-ARs can reduce the release of most preformed hormones, while stimulation of β-ARs can induce the release of preformed hormones. This involves the release of antidiuretic hormone from the posterior pituitary, melatonin from the pineal gland, and insulin from acinar cells of the pancreas.

1.6 POSSIBLE ROLES OF CATECHOLAMINES IN VARIOUS DISEASES

1.6.1 Pheochromocytoma

Nearly all paragangliomas and pheochromocytomas generate CAs. There is an increased level of CAs in pheochromocytoma tissues, which creates a volcano that may erupt at any time. Marked eruptions lead to CA storms known as ‘spells’ or ‘attacks’. Indeed, acute CA crises can attack unpredictably and may result in a life-threatening condition. There is a well-identified genotype–biochemical phenotype relationship, which can provide cost-effective and proper genetic testing of individuals with these tumors. Nowadays, the generation of EPI and NE is optimally analyzed by measuring their *O*-methylated metabolites, metanephrine or normetanephrine, respectively. DA is a minor constituent, but some paragangliomas generate only DA or DA along with NE. For these tumours, methoxytyramine (an *O*-methylated metabolite of DA) is the best-known biochemical marker. However, an adjusted clonidine suppression test along with measurement of plasma normetanephrine level has also been introduced for individuals with ambiguous biochemical findings. Along with the differences in expression of CA enzymes, the existence of either regulated or constitutive secretory pathways can lead to the production and release of an exclusive mutation-dependent CA, which can eventually lead to several clinical presentations (Pacak 2011).

1.6.2 Alzheimer's Disease

AD is a neurodegenerative disease. The characteristics of this disorder include attention impairments, such as agnosia, chronic amnesia, aphasia, and apraxia (Soldan et al. 2016). Worldwide, the occurrence of this devastating disorder has increased significantly in recent years (Chibnik et al. 2017). Indeed, early detection of biomarkers might be reliable and consistent, and can help in implementing timely interferences that can lead to an improved diagnosis and decreased disease burden (Fliessbach and Schneider 2018). Among the neurotransmitter impairments studied in the case of AD, CAs including EPI, NE, and DA have been extensively analyzed in terms of their activities in neurotransmitter efflux, which involve facilitating various cognition functions, including spatial memory, working memory, and recognition memory (Yates et al. 1983; Liao et al. 2002; Nelson et al. 2011; Bensmann et al. 2019). Moreover, neuropathological findings indicate a relationship between molecular processes, functional alterations that are taking place in the CAs, and AD neuropathology (Youdim 2018). DA contributes significantly in the case of both the formation of amyloid beta and the progression of cognitive deficit (Kumari et al. 2019). Inconsistent findings have also reported a link between CA levels and AD (Fitzgerald 2010; Chalermpalanupap et al. 2013). Nonetheless, some studies have confirmed that there are increased levels of DA in individuals with AD versus controls (Yates et al. 1983). It has also been indicated that the levels of EPI and NE might be increased in some patients with AD (Fitzgerald 2010).

1.6.3 Parkinson's Disease

Parkinson's disease (PD) is another common neurodegenerative disease. The characteristics of this disease include mesencephalic nigral dopaminergic neuronal degeneration along with Lewy bodies, which are intracytoplasmic neuronal inclusions containing insoluble aggregated alpha-synuclein (Forno 1996; Lees et al. 2009; Halliday et al. 2011). This disease usually takes place in individuals who are ≥60 years of age and its occurrence is rising in the elderly populations of Europe (De Rijk et al. 1997). Cardinal motor features include resting tremor, muscular rigidity, and bradykinesia, mediated by the depletion of DA in the striatum due to a loss of projections from mesencephalic dopaminergic neurons. Laboratory tests (such as brain magnetic resonance imaging and blood tests) did not find any disease-specific results and thus a clinical diagnosis was created as per the neurological examination, history, and response towards DA replacement therapy (Daniel and Lees 1993).

Along with motor deficits, signs and symptoms of PD-associated nonmotor disturbances include hallucinations, cognitive deficit, anhedonia, sleep disturbance, and orthostatic hypotension (Chaudhuri and Martinez-Martin 2008). Hallucinations or cognitive deficit usually take place in advanced stages of the disease, and various nonmotor symptoms, including hyposmia, constipation, or disturbance in rapid eye movement sleep, have been found to occur before the appearance of motor symptoms (Tan et al. 1996; Ross et al. 2012). It has been reported that clinical motor symptoms appear when the striatal dopamine level is depleted by 60–70%. Dopaminergic neuronal degeneration (which lasts for years) takes place before the appearance of motor symptoms in the preclinical stage (Savica et al. 2010). Thus, it is believed that dopaminergic neurons have already begun to moderately degenerate at the time of a PD diagnosis. Multiple clinical studies have evaluated drug actions in modifying the progression of PD; however, the findings are debated (Olanow et al. 2009).

1.6.4 Autoimmune Diseases

Adrenergic immunomodulation plays a vital role in autoimmune diseases (Frieri 2003). Numerous studies in animals and humans have confirmed the association of CAs and the sympathetic nervous system in the pathogenesis of autoimmune diseases (Kavelaars et al. 1998; Bijlsma et al. 1999; Wahle et al. 2002; Wilder 2002). An impairment of autonomous nervous system has been observed in individuals with autoimmune diseases, including rheumatoid arthritis (Wahle et al. 2001). In adult rats, the intensity of an experimental allergic encephalomyelitis was found to be increased as a result of a chemical peripheral sympathectomy. Moreover, this activity was reversed via the administration of a β-AR agonist (Lorton et al. 2005). Interestingly, the administration of β_2-ARs agonists decreased the intensity in an experimental allergic neuritis model that stimulated inflammation of peripheral nerves, similar to Guillain-Barré disease (a rare neurological disorder; Lorton et al. 2005). In an animal model of arthritis, systemic NE depletion reduced joint destruction and inflammatory activity. Furthermore, it was supposed that, other than the adrenergic activities on vascular muscles, this might take place because of the interaction of immune cells and CAs, which eventually modulates inflammation in rheumatic disease (Lorton et al. 2005).

1.6.5 Familial Dysautonomia

Familial dysautonomia (FD) is a hereditary sensory and autonomic neuropathy. Furthermore, FD is a developmental and chronic neurodegenerative condition that takes place because of an autosomal recessive mutation in *ELP1* (formerly known as *IKBKAP*). FD not only affects the peripheral nervous system, but also disrupts the CNS, which mainly affects the optic nerve and retina (Lefcort et al. 2017). The characteristics of FD include consistent neuropathological irregularities of the sensory and autonomic nervous systems (Pearson et al. 1974), along with severe depletion of number of sympathetic neurons (Pearson and Pytel 1978). Both orthostatic hypotension and hypertensive crises without compensatory tachycardia can also occur in FD. Clinical observations in FD involve aberrant heart rate and blood pressure responses to postural alteration (Pearson et al. 1974; Ziegler et al. 1976; Axelrod et al. 1993), prolonged QT interval, and reduced sympathetic and excessive parasympathetic responses, as revealed by power spectral analysis of heart rate variability (Maayan et al. 1987). Other studies of plasma levels of CAs in the case of FD (Ziegler et al. 1976; Axelrod et al. 1994) did not find a consistent irregularity, except for the failure of NE levels to increase during orthostasis. In the case of patients with FD, this defect is not specific and can occur as a result of impairment in any of multiple sites that participate in the release, uptake, or generation of NA.

1.7 FUTURE DIRECTIONS IN CATECHOLAMINE RESEARCH

Over the last 50 years, the speed of CA research has been hindered to some extent. Nonetheless, important understanding and knowledge regarding these neurotransmitter systems may arise from the results being obtained from three associated fields, all of which are gaining the required speed. The first field is functional interactions between CAs and their co-transmitters, particularly amino acids and peptides. The second field involves regulating heterodimerization between CA receptors with those for other neuromodulators/neurotransmitters and an enhanced understanding of how this influences the activity of downstream targets (such as gene transcription). The third field includes how the transmission of CA affects, and is affected by, glial cells. Indeed, these fields are promising and can revive the pharmacology of CAs (Stanford and Heal 2019). In terms of therapeutics, early successes with β_2-ARs agonists (e.g., antiasthmatic drugs) and β_1-ARs antagonists (e.g., antihypertensive agents) enhanced drug development towards ever-increased selectivity toward receptors. Nonetheless, this technique did not constantly generate the projected developments in increasing efficacy or decreasing side effects.

In the case of therapeutic indications, numerous pharmacological activities provide an optimum balance of safety and efficacy. Most of the CA transmission-associated molecular targets have been explored, both separately and in combination with additive pharmacological processes, which has significantly improved research and development in the pharmaceutical sector in the identification of possible therapeutic uses for novel compounds via pharmacological phenotyping by utilizing 'black box' behavioral protocols. *In vitro* pharmacological profiling of various compounds to detect their mechanism(s) of action via high-throughput screening approaches has also been carried out retrospectively. However, there is no robust presumption regarding which type of approach is essential for therapeutic efficacy. More studies are required to anticipate whether this technique will be successful or not (Stanford and Heal 2019).

1.8 CONCLUSION

CAs, including NE, EPI, and DA, make a significant contribution to physiological homeostasis, especially in the adaptation to chronic and acute physical and mental stressful stimuli. In spite of structural similarities, CAs vary in their affinities for particular ARs, therefore stimulating a wide range of physiological activities. Numerous powerful new technologies have been introduced in CA research over the last few years. In addition, many neurotransmitters have also been identified. These developments are helping researchers to conduct stronger and more complex studies to elucidate the role of CAs in normal and pathologic conditions.

REFERENCES

Ahlquist RP (1948) A study of the adrenotropic receptors. Am J Physiol 153:586–600. https://doi.org/10.1152/ajplegacy.1948.153.3.586

An SS, Wang WCH, Koziol-White CJ, et al (2012) TAS2R activation promotes airway smooth muscle relaxation despite β 2-adrenergic receptor tachyphylaxis. Am J Physiol Lung Cell Mol Physiol 303:L304. https://doi.org/10.1152/ajplung.00126.2012

Andersson K-E, Arner A (2004) Urinary bladder contraction and relaxation: physiology and pathophysiology. Physiol Rev 84:935–986. https://doi.org/10.1152/physrev.00038.2003

Antzelevitch C, Burashnikov A (2011) Overview of basic mechanisms of cardiac arrhythmia. Card Electrophysiol Clin 3:23–45. https://doi.org/10.1016/j.ccep.2010.10.012

Arch JRS (2001) The β3-adrenergic system and β3-adrenergic agonists. Rev Endocr Metab Disord 2:385–393. https://doi.org/10.1023/A:1011852500209

Arnsten AFT, Contant TA (1992) Alpha-2 adrenergic agonists decrease distractibility in aged monkeys performing the delayed response task. Psychopharmacology (Berl) 108:159–169. https://doi.org/10.1007/BF02245302

Axelrod FB, Glickstein JS, Weider J, et al (1993) The effects of postural change and exercise on renal haemodynamics in familial dysautonomia. Clin Auton Res 3:195–200. https://doi.org/10.1007/BF01826233

Axelrod FB, Krey L, Glickstein JS, et al (1994) Atrial natriuretic peptide response to postural change and medication in familial dysautonomia. Clin Auton Res 4:311–318. https://doi.org/10.1007/BF01821531

Badawi JK, Seja T, Uecelehan H, et al (2007) Relaxation of human detrusor muscle by selective beta-2 and beta-3 agonists and endogenous catecholamines. Urology 69:785–790. https://doi.org/10.1016/j.urology.2007.01.059

Bai TR (1992) Beta2 adrenergic receptors in asthma: a current perspective. Lung 170:125–141. https://doi.org/10.1007/BF00174316

Baker JG (2005) The selectivity of β-adrenoceptor antagonists at the human β1, β2 and β3 adrenoceptors. Br J Pharmacol 144:317–322. https://doi.org/10.1038/sj.bjp.0706048

Baker LB (2019) Physiology of sweat gland function: the roles of sweating and sweat composition in human health. Temperature 6:211–259. https://doi.org/10.1080/23328940.2019.1632145

Bara I, Ozier A, Tunon De Lara JM, et al (2010) Pathophysiology of bronchial smooth muscle remodelling in asthma. Eur Respir J 36:1174–1184. https://doi.org/10.1183/09031936.00019810

Barbe P, Stich V, Galitzky J, et al (1997) In vivo increase in β-adrenergic lipolytic response in subcutaneous adipose tissue of obese subjects submitted to a hypocaloric diet. J Clin Endocrinol Metab 82:63–69. https://doi.org/10.1210/jcem.82.1.3686

Barnes PJ (1993) β-adrenoceptors on smooth muscle, nerves and inflammatory cells. Life Sci 52:2101–2109. https://doi.org/10.1016/0024-3205(93)90725-I

Bartos DC, Grandi E, Ripplinger CM (2015) Ion channels in the heart. Compr Physiol 5:1423–1464. https://doi.org/10.1002/cphy.c140069

Bauer M (2017) Cardiovascular anatomy and pharmacology. In: Farag E, Argalious M, Tetzlaff JE, Sharma D (eds). Basic Sciences in Anesthesia. Cham: Springer International Publishing, pp. 195–228.

Beaulieu JM, Gainetdinov RR (2011) The physiology, signaling, and pharmacology of dopamine receptors. Pharmacol Rev 63:182–217.

Belousov AB, Fontes JD, Freitas-Andrade M, Naus CC (2017) Gap junctions and hemichannels: communicating cell death in neurodevelopment and disease. BMC Cell Biol 18(Suppl.1):4.

Bensmann W, Zink N, Roessner V, et al (2019) Catecholaminergic effects on inhibitory control depend on the interplay of prior task experience and working memory demands. J Psychopharmacol 33:678–687. https://doi.org/10.1177/0269881119827815

Berends AMA, Eisenhofer G, Fishbein L, et al (2019) Intricacies of the molecular machinery of catecholamine biosynthesis and secretion by chromaffin cells of the normal adrenal medulla and in pheochromocytoma and paraganglioma. Cancers (Basel) 11:1121.

Bermingham DP, Blakely RD (2016) Kinase-dependent regulation of monoamine neurotransmitter transporters. Pharmacol Rev 68:888–953. https://doi.org/10.1124/pr.115.012260

Berridge CW, Spencer RC (2016) Differential cognitive actions of norepinephrine a2 and a1 receptor signaling in the prefrontal cortex. Brain Res 1641:189–196.

Biancani P, Sohn UD, Rich HG, et al (1997) Signal transduction pathways in esophageal and lower esophageal sphincter circular muscle. Am J Med 103:23S–28S. https://doi.org/10.1016/S0002-9343(97)00316-1

Bijlsma JWJ, Cutolo M, Masi AT, Chikanza IC (1999) The neuroendocrine immune basis of rheumatic diseases. Immunol Today 20:298–301. https://doi.org/10.1016/S0167-5699(98)01422-4

Bikle D (2000) Vitamin D: production, metabolism, and mechanisms of action. www.endotext.org.

Blanke ML, VanDongen AMJ (2008) Activation mechanisms of the NMDA receptor. In: Van Dongen, AM. (ed.). Biology of the NMDA Receptor. Boca Raton, FL: CRC Press, pp. 283–312.

Brenner T, Lisak RP, Rostami A, et al (1986) Astrocytes, oligodendrocytes, and Schwann cells share a common antigenic determinant that cross-reacts with myelin basic protein: identification with monoclonal antibody. J Neurosci 6:1925–1933. https://doi.org/10.1523/jneurosci.06-07-01925.1986

Brown DL (2018) Coronary physiology and pathophysiology. In: Brown, DL. (ed.). Cardiac Intensive Care. Philadelphia, PA: Elsevier, pp. 60–67.

Bylund DB (1992) Subtypes of α1- and α2-adrenergic receptors. FASEB J 6:832–839. https://doi.org/https://doi.org/10.1096/fasebj.6.3.1346768

Bylund DB, Chacko DM (1999) Characterization of α2 adrenergic receptor subtypes in human ocular tissue homogenates. Invest Ophthalmol Vis Sci 40:2299–2306.

Carey GB (1998) Mechanisms regulating adipocyte lipolysis. Adv Exp Med Biol 441:157–170. https://doi.org/10.1007/978-1-4899-1928-1_15

Ceasrine AM, Lin EE, Lumelsky DN, et al (2018) Adrb2 controls glucose homeostasis by developmental regulation of pancreatic islet vasculature. Elife 7. https://doi.org/10.7554/eLife.39689

Chalermpalanupap T, Kinkead B, Hu WT, et al (2013) Targeting norepinephrine in mild cognitive impairment and Alzheimer's disease. Alzheimer's Res Ther 5:21.

Chaudhuri KR, Martinez-Martin P (2008) Quantitation of non-motor symptoms in Parkinson's disease. Eur J Neurol 15:2–8. https://doi.org/10.1111/j.1468-1331.2008.02212.x

Chen X, Merouch M, Mazur G, et al (2018) Phenylephrine, a common cold remedy active ingredient, suppresses uterine contractions through cAMP signalling. Sci Rep 8:11666. https://doi.org/10.1038/s41598-018-30094-5

Cherezov V, Rosenbaum DM, Hanson MA, et al (2007) High-resolution crystal structure of an engineered human beta2-adrenergic G protein-coupled receptor. Science 318:1258–1265. https://doi.org/10.1126/science.1150577

Chibnik LB, Wolters FJ, Bäckman K, et al (2017) Trends in the incidence of dementia: design and methods in the Alzheimer Cohorts Consortium. Eur J Epidemiol 32:931–938. https://doi.org/10.1007/s10654-017-0320-5

Clement P, Giuliano F (2016) Physiology and pharmacology of ejaculation. Basic Clin Pharmacol Toxicol 119:18–25. https://doi.org/10.1111/bcpt.12546

Cox HM, Cuthbert AW (1989) Antisecretory activity of the alpha2-adrenoceptor agonist, xylazine in rat jejunal epithelium. Naunyn Schmiedebergs Arch Pharmacol 339:669–674. https://doi.org/10.1007/BF00168660

Creese I, Sibley DR, Leff SE (1984) Agonist interactions with dopamine receptors: focus on radioligand-binding studies. Fed Proc 43:2779–2784. https://doi.org/10.1111/j.1749-6632.1984.tb14499.x

Cui CY, Schlessinger D (2015) Eccrine sweat gland development and sweat secretion. Exp Dermatol 24:644–650. https://doi.org/10.1111/exd.12773

Cypess AM, Weiner LS, Roberts-Toler C, et al (2015) Activation of human brown adipose tissue by a β3-adrenergic receptor agonist. Cell Metab 21:33–38. https://doi.org/10.1016/j.cmet.2014.12.009

Daniel SE, Lees AJ (1993) Parkinson's Disease Society Brain Bank, London: overview and research. J Neural Transm Suppl 39:165–172.

Davani B, Portwood N, Bryzgalova G, et al (2004) Aged transgenic mice with increased glucocorticoid sensitivity in pancreatic β-cells develop diabetes. Diabetes 39(Suppl.1): S51–S59.

Day R, Gorr SU (2003) Secretory granule biogenesis and chromogranin A: master gene, on/off switch or assembly factor? Trends Endocrinol Metab 14:10–13. https://doi.org/10.1016/S1043-2760(02)00011-5

De Rijk MC, Tzourio C, Breteler MMB, et al (1997) Prevalence of parkinsonism and Parkinson's disease in Europe: The EUROPARKINSON collaborative study. J Neurol Neurosurg Psychiatry 62:10–15. https://doi.org/10.1136/jnnp.62.1.10

Delong C, Sharma S (2019) Physiology, Peripheral Vascular Resistance. Treasure Island, FL: StatPearls Publishing.

Descarries L, Droz B (1970) Intraneural distribution of exogenous norepinephrine in the central nervous system of the rat. J Cell Biol 44:385–399. https://doi.org/10.1083/jcb.44.2.385

Dinuzzo M, Mangia S, Maraviglia B, Giove F (2013) Regulatory mechanisms for glycogenolysis and K^+ uptake in brain astrocytes. Neurochem Int 63:458–464. https://doi.org/10.1016/j.neuint.2013.08.004

Eiden LE, Jiang SZ (2018) What's new in endocrinology: the chromaffin cell. Front Endocrinol (Lausanne) 9:711. https://doi.org/10.3389/fendo.2018.00711

Eisenhofer G (2001) The role of neuronal and extraneuronal plasma membrane transporters in the inactivation of peripheral catecholamines. Pharmacol Ther 91:35–62.

Eisenhofer G, Goldstein DS, Walther MM, et al (2003) Biochemical diagnosis of pheochromocytoma: how to distinguish true- from false-positive test results. J Clin Endocrinol Metab 88:2656–2666. https://doi.org/10.1210/jc.2002-030005

Exton JH (1987) Mechanisms of hormonal regulation of hepatic glucose metabolism. Diabetes Metab Rev 3 :163–183. https://doi.org/10.1002/dmr.5610030108

Fahy JV, Dickey BF (2010) Airway mucus function and dysfunction. N Engl J Med 363:2233–2247. https://doi.org/10.1056/nejmra0910061

Fang F, Ding L, He Q, Liu M (2020) Preoperative management of pheochromocytoma and paraganglioma. Front Endocrinol (Lausanne) 11:586795 https://doi.org/10.3389/fendo.2020.586795

Fiedorowicz JG, Swartz KL (2004) The role of monoamine oxidase inhibitors in current psychiatric practice. J Psychiatr Pract 10:239–248.

Fields RD, Stevens-Graham B (2002) Neuroscience: new insights into neuron-glia communication. Science 298:556–562.

Fields RD, Woo DH, Basser PJ (2015) Glial regulation of the neuronal connectome through local and long-distant communication. Neuron 86:374–386.

Fitzgerald PJ (2010) Is elevated norepinephrine an etiological factor in some cases of Alzheimers disease? Curr Alzheimer Res 7:506–516. https://doi.org/10.2174/156720510792231775

Flatmark T (2000) Catecholamine biosynthesis and physiological regulation in neuroendocrine cells. Acta Physiol Scand 168:1–17.

Fliessbach K, Schneider A (2018) [Biomarker-based diagnostics of Alzheimer's disease: concept of suspected non-Alzheimer pathology]. Nervenarzt 89:345–358. https://doi.org/10.1007/s00115-018-0488-2

Forno LS (1996) Neuropathology of Parkinson's disease. J Neuropathol Exp Neurol 55:259–272.

Forrester JV, Dick AD, McMenamin PG, et al (2016) General and ocular pharmacology. In: Forrester JV, Dick AD, McMenamin PG, et al. (eds). The Eye. Philadelphia, PA: Elsevier, pp. 338–369.

Fraser RAR, Stein BM, Pool JL (1971) Adrenergic blockade of hypocapnic cerebral arterial constriction. Stroke 2:219–231. https://doi.org/10.1161/01.STR.2.3.219

Freundt KJ (1965) Adrenergic alpha- and beta-receptors in the mouse iris. Nature 206:725–726. https://doi.org/10.1038/206725b0

Frieri M (2003) Neuroimmunology and inflammation: implications for therapy of allergic and autoimmune diseases. Ann Allergy Asthma Immunol 90:34–40. https://doi.org/10.1016/s1081-1206(10)61658-4

Fujii A, Morimoto-Tochigi A, Walkup RD, et al (2013) Lacritin-induced secretion of tear proteins from cultured monkey lacrimal acinar cells. Investig Ophthalmol Vis Sci 54:2533–2540. https://doi.org/10.1167/iovs.12-10394

Fung MM, Viveros OH, O'Connor DT (2008) Diseases of the adrenal medulla. Acta Physiol (Oxf) 192:325–335.

Gangwani MK, Nagalli S (2020) Idioventricular Rhythm. Treasure Island, FL: StatPearls Publishing.

Garcia E, Santos C (2018) Toxicity, Monoamine Oxidase Inhibitor. Treasure Island, FL: StatPearls Publishing.

Giovannitti JA, Thoms SM, Crawford JJ (2015) Alpha-2 adrenergic receptor agonists: a review of current clinical applications. Anesth Prog 62:31–38. https://doi.org/10.2344/0003-3006-62.1.31

Glickstein M (2006) Golgi and Cajal: the neuron doctrine and the 100th anniversary of the 1906 Nobel Prize. Curr Biol 16:R147–R151.

Gnegy ME (2012) Catecholamines. In: Brady S, Siegel G, Albers RW, Price DL (eds). Basic Neurochemistry. Philadelphia, PA: Elsevier, pp. 283–299.

Goldberg LI (1975) The dopamine vascular receptor. Biochem Pharmacol 24:651–653.

Goldenberg M, Pines KL, de F. Baldwin E, et al (1948) The hemodynamic response of man to nor-epinephrine and epinephrine and its relation to the problem of hypertension. Am J Med 5:792–806. https://doi.org/10.1016/0002-9343(48)90158-2

Goldenberg M, Aranow H, Smith AA, Faber M (1950) Pheochromocytoma and essential hypertensive vascular disease. Arch Intern Med 86:823–836. https://doi.org/10.1001/archinte.1950.00230180028003

Goldman L, Schafer AI (2012) Goldman's Cecil Medicine: Twenty Fourth Edition. Philadelphia, PA: Elsevier.

Gooch CL (2003) Neuromuscular junction, normal. In: Aminoff M, Daroff R (eds). Encyclopedia of the Neurological Sciences. Philadelphia, PA: Elsevier, pp. 505–507.

Gooch CL (2014) Neuromuscular junction, normal. In: Daroff R, Aminoff M (eds). Encyclopedia of the Neurological Sciences. Philadelphia, PA: Elsevier, pp. 423–424.

Gordan R, Gwathmey JK, Xie L-H (2015) Autonomic and endocrine control of cardiovascular function. World J Cardiol 7:204. https://doi.org/10.4330/wjc.v7.i4.204

Goumon Y, Lugardon K, Gadroy P, et al (2000) Processing of proenkephalin-A in bovine chromaffin cells: identification of natural derived fragments by N-terminal sequencing and matrix-assisted laser desorption ionization-time of flight mass spectrometry. J Biol Chem 275:38355–38362. https://doi.org/10.1074/jbc.M007557200

Graham RM, Perez DM, Hwa J, Piascik MT (1996) α1-Adrenergic receptor subtypes. Circ Res 78:737–749. https://doi.org/10.1161/01.RES.78.5.737

Grudzien A, Shaw P, Weintraub S, et al (2007) Locus coeruleus neurofibrillary degeneration in aging, mild cognitive impairment and early Alzheimer's disease. Neurobiol Aging 28:327 335. https://doi.org/10.1016/j.neurobiolaging.2006.02.007

Gullikson GW, Virina MA, Loeffler R, Erwin WD (1991) Alpha 2-adrenergic model of gastroparesis: validation with renzapride, a stimulator of motility. Am J Physiol - Gastrointest Liver Physiol 261:G426–G432. https://doi.org/10.1152/ajpgi.1991.261.3.g426

Hajagos-Tóth J, Bóta J, Ducza E, et al (2016) The effects of progesterone on the alpha2-adrenergic receptor subtypes in late-pregnant uterine contractions in vitro. Reprod Biol Endocrinol 14:33. https://doi.org/10.1186/s12958-016-0166-9

Hajagos-Tóth J, Ducza E, Samavati R, et al (2017) Obesity in pregnancy: a novel concept on the roles of adipokines in uterine contractility. Croat Med J 58:96–104. https://doi.org/10.3325/cmj.2017.58.96

Halliday G, Lees A, Stern M (2011) Milestones in Parkinson's disease – clinical and pathologic features. Mov Disord 26:1015–1021.

Hesse IF, Johns EJ (1985) The role of alpha-adrenoceptors in the regulation of renal tubular sodium reabsorption and renin secretion in the rabbit. Br J Pharmacol 84:715–724. https://doi.org/10.1111/j.1476-5381.1985.tb16154.x

Hieble JP, Bylund DB, Clarke DE, et al (1995) International Union of Pharmacology. X. Recommendation for nomenclature of alpha 1-adrenoceptors: consensus update. Pharmacol Rev 47:267–270.

Hill WG (2015) Control of urinary drainage and voiding. Clin J Am Soc Nephrol 10:480–492. https://doi.org/10.2215/CJN.04520413

Hoffman BB, Lefkowitz RJ, Hoffman BB (1980) Alpha-adrenergic receptor subtypes. N Engl J Med 302:1390–1396.

Hoffmann C, Leitz MR, Oberdorf-Maass S, et al (2004) Comparative pharmacology of human β-adrenergic receptor subtypes – characterization of stably transfected receptors in CHO cells. Naunyn Schmiedebergs Arch Pharmacol 369:151–159. https://doi.org/10.1007/s00210-003-0860-y

Hökfelt T, Barde S, Xu ZQD, et al (2018) Neuropeptide and small transmitter coexistence: fundamental studies and relevance to mental illness. Front Neural Circuits 12:106.

Holz RW, Fisher SK (1999) Synaptic transmission. In: Siegel GJ, Agranoff BW, Albers RW, et al., (eds). Basic Neurochemistry: Molecular, Cellular and Medical Aspects. 6th edition. Philadelphia: Lippincott-Raven

Hostrup M, Kalsen A, Ørtenblad N, et al (2014) β2-Adrenergic stimulation enhances Ca^{2+} release and contractile properties of skeletal muscles, and counteracts exercise-induced reductions in Na^+-K^+-ATPase Vmax in trained men. J Physiol 592:5445–5459. https://doi.org/10.1113/jphysiol.2014.277095

Hughes BW, Kusner LL, Kaminski HJ (2006) Molecular architecture of the neuromuscular junction. Muscle Nerve 33:445–461.

Hughes JW, Ustione A, Lavagnino Z, Piston DW (2018) Regulation of islet glucagon secretion: beyond calcium. Diabetes Obes Metab 20:127–136. https://doi.org/10.1111/dom.13381

Hussain LS, Maani CV (2019) Physiology, Noradrenergic Synapse. Treasure Island, FL: StatPearls Publishing.

Issa ZF, Miller JM, Zipes DP (2019) Atrial fibrillation. In: Issa ZF, Miller JM, Zipes DP (eds). Clinical Arrhythmology and Electrophysiology. Treasure Island, FL: Elsevier, pp. 421–548.

Jimsheleishvili S, Marwaha K, Sherman AI. (2019) Physiology, Neuromuscular Transmission. Treasure Island, FL: StatPearls Publishing.

Jodo E, Chiang C, Aston-Jones G (1998) Potent excitatory influence of prefrontal cortex activity on noradrenergic locus coeruleus neurons. Neuroscience 83:63–79. https://doi.org/10.1016/S0306-4522(97)00372-2

Kaludercic N, Mialet-Perez J, Paolocci N, et al (2014) Monoamine oxidases as sources of oxidants in the heart. J Mol Cell Cardiol 73:34–42.

Karlberg BE (1983) Adrenergic regulation of renin release and effects on angiotensin and aldosterone. Acta Med Scand 213:33–40. https://doi.org/10.1111/j.0954-6820.1983.tb01611.x

Katra RP, Laurita KR (2005) Cellular mechanism of calcium-mediated triggered activity in the heart. Circ Res 96:535–542. https://doi.org/10.1161/01.RES.0000159387.00749.3c

Kavelaars A, Van De Steenwijk TJDEV, et al. (1998) The reactivity of the cardiovascular system and immunomodulation by catecholamines in juvenile chronic arthritis. Ann N Y Acad Sci 840:698–704. https://doi.org/10.1111/j.1749-6632.1998.tb09608.x

Kester M, Karpa KD, Vrana KE (2012) Cardiovascular system. In: Kester M, Karpa KD, Vrana KE (eds). Elsevier's Integrated Review Pharmacology. Philadelphia, PA: Elsevier, pp. 125–151.

Kordasz ML, Manicam C, Steege A, et al (2014) Role of α1-adrenoceptor subtypes in pupil dilation studied with gene-targeted mice. Invest Ophthalmol Vis Sci 55:8295–8301. https://doi.org/10.1167/iovs.14-15706

Korthuis RJ (2011) Regulation of vascular tone in skeletal muscle. In: Granger DN, Granger J, (eds). Skeletal Muscle Circulation. Morgan & Claypool Life Sciences; San Rafael, CA: pp. 734.

Kuhar MJ, Couceyro PR, Lambert PD (1999) Storage and Release of Catecholamines. In: Siegel GJ, Agranoff BW, Albers RW, Fisher SK, Uhler MD, (eds). Basic Neurochemistry: Molecular, Cellular and Medical Aspects, 6th edition. Philadelphia: Lippincott-Raven.

Kumari S, Mukherjee A, Mukhopadhyay CK (2019) Dopamine promotes cathepsin B-mediated amyloid precursor protein degradation by reactive oxygen species-sensitive mechanism in neuronal cell. Mol Cell Biochem 454:153–163. https://doi.org/10.1007/s11010-018-3460-3

Kurtz A (2012) Control of renin synthesis and secretion. Am J Hypertens 25:839–847. https://doi.org/10.1038/ajh.2011.246

Lacey RJ, Berrow NS, Scarpello JHB, Morgan NG (1991) Selective stimulation of glucagon secretion by β2-adrenoceptors in isolated islets of Langerhans of the rat. Br J Pharmacol 103:1824–1828. https://doi.org/10.1111/j.1476-5381.1991.tb09870.x

Lafontan M, Betuing S, Saulnier-Blache JS, et al (1997) Regulation of fat-cell function by α2-adrenergic receptors. Adv Pharmacol 42:496–498. https://doi.org/10.1016/S1054-3589(08)60797-8

Laird DW, Lampe PD (2018) Therapeutic strategies targeting connexins. Nat Rev Drug Discov 17:905–921.

Langer SZ, Massingham R, Shepperson NB (1980) Presence of postsynaptic α2-adrenoreceptors of predominantly extrasynaptic location in the vascular smooth muscle of the dog hind limb. Clin Sci (Lond) 59(Suppl. 6):225s–228s. https://doi.org/10.1042/cs059225s

Langin D, Tavernier G, Lafontan M (1995) Regulation of beta3-adrenoceptor expression in white fat cells. Fundam Clin Pharmacol 9:97–106. https://doi.org/https://doi.org/10.1111/j.1472-8206.1995.tb00268.x

Laukkarinen J, Sand J, Nordback I (2012) The underlying mechanisms: how hypothyroidism affects the formation of common bile duct stonesa review. HPB Surg 2012:102825. https://doi.org/10.1155/2012/102825

Lawes N (2004) Neuroplasticity. In: Stokes M (ed.). Physical Management in Neurological Rehabilitation. St. Louis, MO: Elsevier, pp. 57–72.

Lees AJ, Hardy J, Revesz T (2009) Parkinson's disease. Lancet 373:2055–2066.

Lefcort F, Mergy M, Ohlen SB, et al (2017) Animal and cellular models of familial dysautonomia. Clin Auton Res 27:235–243.

Lefkowitz RJ (1979) Direct binding studies of adrenergic receptors: biochemical, physiologic, and clinical implications. Ann Intern Med 91:450–458.

Lefkowitz RJ (2004) Historical review: a brief history and personal retrospective of seven-transmembrane receptors. Trends Pharmacol Sci 25:413–422. https://doi.org/10.1016/j.tips.2004.06.006

Liao R, Lai W-S, Lin J (2002) The role of catecholamines in retention performance of a partially baited radial eight-arm maze for rats. Chin J Physiol 45:177–185.

Lim S, Park J, Um J-Y (2019) Ginsenoside Rb1 induces beta 3 adrenergic receptor-dependent lipolysis and thermogenesis in 3T3-L1 adipocytes and db/db mice. Front. Pharmacol 10:1154.

Lodish H, Berk A, Zipursky SL, et al (2000a) Molecular Cell Biology: Section 21.4 Neurotransmitters, Synapses, and Impulse Transmission. 4th edition. New York: W. H. Freeman.

Lodish H, Berk A, Zipursky SL, et al (2000b) The Action Potential and Conduction of Electric Impulses. New York: W.H. Freeman.

López-Sendó J, Swedberg K, McMurray J, et al (2004) Expert consensus document on β-adrenergic receptor blockers: The Task Force on Beta-Blockers of the European Society of Cardiology. Eur Heart J 25:1341–1362. https://doi.org/10.1016/j.ehj.2004.06.002

Lorton D, Lubahn C, Bellinger D (2005) Potential use of drugs that target neural-immune pathways in the treatment of rheumatoid arthritis and other autoimmune diseases. Curr Drug Target Inflammation Allergy 2:1–30. https://doi.org/10.2174/1568010033344499

Luckensmeyer GB, Keast JR (1998) Activation of α- and β-adrenoceptors by sympathetic nerve stimulation in the large intestine of the rat. J Physiol 510:549–561. https://doi.org/10.1111/j.1469-7793.1998.549bk.x

Maayan C, Axelrod FB, Akselrod S, et al (1987) Evaluation of autonomic dysfunction in familial dysautonomia by power spectral analysis. J Auton Nerv Syst 21:51–58. https://doi.org/10.1016/0165-1838(87)90091-9

MacDonald EA, Rose RA, Quinn TA (2020) Neurohumoral control of sinoatrial node activity and heart rate: insight from experimental models and findings from humans. Front Physiol 11:170. https://doi.org/10.3389/fphys.2020.00170

McDougal DH, Gamlin PD (2015) Autonomic control of the eye. Compr Physiol 5:439–473. https://doi.org/10.1002/cphy.c140014

McMahon CG (2016) Emerging and investigational drugs for premature ejaculation. Transl Androl Urol 5:487–501. https://doi.org/10.21037/tau.2016.04.02

Magnusson KR, Brim BL (2014) The aging brain. In: Reference Module in Biomedical Sciences. Elsevier.

Maletic V, Eramo A, Gwin K, et al (2017) The role of norepinephrine and its α-adrenergic receptors in the pathophysiology and treatment of major depressive disorder and schizophrenia: a systematic review. Front Psychiatry 8:1.

Mammano F (2013) Gap junctions: cell–cell channels in animals. Available at: www.ncbi.nlm.nih.gov/books/NBK6455/

Marisa V, Carvalho F, Lourdes M, et al (2012) Adrenaline and Noradrenaline: Partners and Actors in the Same Play. In Contreras CM (Editor). Neuroscience – Dealing With Frontiers. London: InTech.

Martin EA, Lasseigne AM, Miller AC (2020) Understanding the molecular and cell biological mechanisms of electrical synapse formation. Front Neuroanat 14:12.

Michel MC, Kenny B, Schwinn DA (1995) Classification of α1-adrenoceptor subtypes. Naunyn Schmiedebergs Arch Pharmacol 352:1–10. https://doi.org/10.1007/BF00169183

Michel MC, Vrydag W (2006) α1-, α2- and β-adrenoceptors in the urinary bladder, urethra and prostate. Br J Pharmacol 147(Suppl. 2):S88–S119.

Minneman KP, Pittman RN, Molinoff PB (1981) Beta-adrenergic receptor subtypes: properties, distribution, and regulation. Annu Rev Neurosci 4:419–461. https://doi.org/10.1146/annurev.ne.04.030181.002223

Missale C, Fiorentini C, Collo G, Spano P (2010) The neurobiology of dopamine receptors: evolution from the dual concept to heterodimer complexes. J Recept Signal Transduct 30:347–354.

Mitchell HA, Weinshenker D (2010) Good night and good luck: norepinephrine in sleep pharmacology. Biochem Pharmacol 79:801–809.

Molderings GJ, Bönisch H, Hammermann R, et al (2002) Noradrenaline release-inhibiting receptors on PC12 cells devoid of α2- and CB1 receptors: similarities to presynaptic imidazoline and EDG receptors. Neurochem Int 40:157–167. https://doi.org/10.1016/S0197-0186(01)00076-6

Molinoff PB (1984) Alpha- and beta-adrenergic receptor subtypes properties, distribution and regulation . Drugs 28:1–15. https://doi.org/10.2165/00003495-198400282-00002

Motulsky HJ, Insel PA (1982) Adrenergic receptors in man: direct identification, physiologic regulation, and clinical alterations. N Engl J Med 307:18–29.

Murray M (2008) Organization of the spinal cord. In: Conn PM (ed.). Neuroscience in Medicine: Third Edition. New York: Humana Press, pp 201–215.

Naguib M, Flood P, McArdle JJ, Brenner HR (2002) Advances in neurobiology of the neuromuscular junction: implications for the anesthesiologist. Anesthesiology 96:202–231.

Nakaki T, Nakadate T, Yamamoto S, Kato R (1982) Alpha-2 adrenergic inhibition of intestinal secretion induced by prostaglandin E1, vasoactive intestinal peptide and dibutyryl cyclic AMP in rat jejunum. J Pharmacol Exp Ther 220:637 LP–641.

Nelson AJD, Cooper MT, Thur KE, et al (2011) The effect of catecholaminergic depletion within the prelimbic and infralimbic medial prefrontal cortex on recognition memory for recency, location, and objects. Behav Neurosci 125:396–403. https://doi.org/10.1037/a0023337

Ni Y, Zhao X, Bao G, et al (2006) Activation of β2-adrenergic receptor stimulates γ-secretase activity and accelerates amyloid plaque formation. Nat Med 12:1390–1396. https://doi.org/10.1038/nm1485

Nielsen MS, Axelsen LN, Sorgen PL, et al (2012) Gap junctions. Compr Physiol 2:1981–2035. https://doi.org/10.1002/cphy.c110051

Nualart-Marti A, Solsona C, Fields RD (2013) Gap junction communication in myelinating glia. Biochim Biophys Acta Biomembr 1828:69–78.

Olanow CW, Rascol O, Hauser R, et al (2009) A double-blind, delayed-start trial of rasagiline in Parkinson's disease. N

Engl J Med 361:1268–1278. https://doi.org/10.1056/NEJMoa0809335
Pacak K (2007) Preoperative management of the pheochromocytoma patient. J Clin Endocrinol Metab 92:4069–4079. https://doi.org/10.1210/jc.2007-1720
Pacak K (2011) Phaeochromocytoma: a catecholamine and oxidative stress disorder. Endocr Regul 45:65–90. https://doi.org/10.4149/endo_2011_02_65
Pacak K (2011b) Phaeochromocytoma: a catecholamine and oxidative stress disorder. Endocr Regul 45:65–90. https://doi.org/10.4149/endo_2011_02_65
Pacak K (2011c) Phaeochromocytoma: a catecholamine and oxidative stress disorder. Endocr Regul 45:65–90. https://doi.org/10.4149/endo_2011_02_65
Paredes-Flores MA, Mohiuddin SS (2020) Biochemistry, Glycogenolysis. Treasure Island, FL: StatPearls Publishing.
Patel RN (2012) Industrial applications of biocatalytic hydrolysis (esters, amides, epoxides, nitriles) and biocatalytic dynamic kinetic resolution. In: Yamamoto H, Carreira E (eds). Comprehensive Chirality. New York: Elsevier, pp. 288–317.
Pearson J, Pytel BA (1978) Quantitative studies of sympathetic ganglia and spinal cord intermedio-lateral gray columns in familial dysautonomia. J Neurol Sci 39:47–59. https://doi.org/10.1016/0022-510X(78)90187-9
Pearson J, Axelrod F, Dancis J (1974) Current concepts of dysautonomia: neuropathological defects. Ann N Y Acad Sci 228:288–300. https://doi.org/10.1111/j.1749-6632.1974.tb20517.x
Pei Y, Asif-Malik A, Canales JJ (2016) Trace amines and the trace amine-associated receptor 1: pharmacology, neurochemistry, and clinical implications. Front Neurosci 10:148.
Peng XR, Gennemark P, O'Mahony G, Bartesaghi S (2015) Unlock the thermogenic potential of adipose tissue: pharmacological modulation and implications for treatment of diabetes and obesity. Front Endocrinol (Lausanne) 6:174. https://doi.org/10.3389/fendo.2015.00174
Pintor J (2009) Autonomic nervous system: ophthalmic control. In: Squire L (ed.). Encyclopedia of Neuroscience. New York: Elsevier, pp. 967–974.
Prentki M, Nolan CJ (2006) Islet β cell failure in type 2 diabetes. J Clin Invest 116:1802–1812. https://doi.org/10.1172/JCI29103
Propping S, Lorenz K, Michel MC, et al (2016) β-adrenoceptor-mediated relaxation of urinary bladder muscle in β2-adrenoceptor knockout mice. Front Pharmacol 7:118. https://doi.org/10.3389/fphar.2016.00118
Proskocil BJ, Fryer AD (2005) β2-agonist and anticholinergic drugs in the treatment of lung disease. Proc Am Thorac Soc 2:305–310.
Purkayastha S, Raven PB (2011) The functional role of the alpha-1 adrenergic receptors in cerebral blood flow regulation. Indian J Pharmacol 43:502–506. https://doi.org/10.4103/0253-7613.84950
Purves D, Augustine GJ, Fitzpatrick D, et al (Editors) (2001) Neuroscience. 2nd edition. Summation of Synaptic Potentials. Sunderland (MA): Sinauer Associates.
Ramos BP, Arnsten AFT (2007) Adrenergic pharmacology and cognition: focus on the prefrontal cortex. Pharmacol Ther 113:523–536.
Ramos BP, Colgan LA, Nou E, Arnsten AFT (2008) β2 adrenergic agonist, clenbuterol, enhances working memory performance in aging animals. Neurobiol Aging 29:1060–1069. https://doi.org/10.1016/j.neurobiolaging.2007.02.003
Rasmussen SGF, Choi H-J, Rosenbaum DM, et al (2007) Crystal structure of the human β2 adrenergic G-protein-coupled receptor. Nature 450:383–387. https://doi.org/10.1038/nature06325
Regan JW, Cotecchia S (1992) The α-adrenergic receptors: new subtypes, pharmacology, and coupling mechanisms. In: Brann M (ed.). Molecular Biology of G-Protein-Coupled Receptors. Boston, MA: Birkhäuser, pp. 76–112.
Rekling JC, Funk GD, Bayliss DA, et al (2000) Synaptic control of motoneuronal excitability. Physiol Rev 80:767–852.
Riekkinen M, Laakso MP, Jäkälä P, Riekkinen P (1999) Clonidine impairs sustained attention and memory in Alzheimer's disease. Neuroscience 92:975–982. https://doi.org/10.1016/S0306-4522(99)00070-6
Roatta S, Passatore M (2008) Autonomic effects on skeletal muscle. In: Binder MD, Hirokawa N, Windhorst U (eds). Encyclopedia of Neuroscience. Berlin: Springer, pp. 250–253.
Ross GW, Abbott RD, Petrovitch H, et al (2012) Pre-motor features of Parkinson's disease: the Honolulu-Asia Aging study experience. Park Relat Disord 18(Suppl. 1):S199–S202. https://doi.org/10.1016/s1353-8020(11)70062-1
Rowland D, McMahon CG, Abdo C, et al (2010) Disorders of orgasm and ejaculation in men. J Sex Med 7:1668–1686. https://doi.org/10.1111/j.1743-6109.2010.01782.x
Sagrada A, Fargeas MJ, Bueno L (1987) Involvement of alpha-1 and alpha-2 adrenoceptors in the postlaparotomy intestinal motor disturbances in the rat. Gut 28:955–959. https://doi.org/10.1136/gut.28.8.955
Sahyoun HA, Costall B, Naylor RJ (1982) Catecholamine-induced relaxation and contraction of the lower oesophageal and pyloric sphincters of guinea-pig stomach: modification by domperidone. J Pharm Pharmacol 34:318–324. https://doi.org/10.1111/j.2042-7158.1982.tb04715.x
Sanes JR, Lichtman JW (1999) Development of the vertebrate neuromuscular junction. Annu Rev Neurosci 22:389–442. https://doi.org/10.1146/annurev.neuro.22.1.389
Sara SJ (2009) The locus coeruleus and noradrenergic modulation of cognition. Nat Rev Neurosci 10:211–223. https://doi.org/10.1038/nrn2573
Sarelius I, Pohl U (2010) Control of muscle blood flow during exercise: local factors and integrative mechanisms. Acta Physiol 199:349–365. https://doi.org/10.1111/j.1748-1716.2010.02129.x
Savica R, Rocca WA, Ahlskog JE (2010) When does Parkinson disease start? Arch Neurol 67:798–801. https://doi.org/10.1001/archneurol.2010.135
Schwarzwald CC (2018) Disorders of the cardiovascular system. In: Reed SM, Bayly WM, Sellon DC (eds). Equine Internal Medicine: Fourth Edition. St. Louis, MO: Elsevier, pp. 387–541.
Scully C (2003) Drug effects on salivary glands: dry mouth. Oral Dis 9:165–176. https://doi.org/10.1034/j.1601-0825.2003.03967.x
Seiler R, Rickenbacher A, Shaw S, et al (2008) Role of selective α and β adrenergic receptor mechanisms in rat jejunal longitudinal muscle contractility. J Gastrointest Surg 12:1087–1093. https://doi.org/10.1007/s11605-007-0327-4
Shoykhet M, Clark RSB (2011) Structure, function, and development of the nervous system. In: Fuhrman BP (ed.). Pediatric Critical Care. Philadelphia, PA: Elsevier, pp. 783–804.
Sitte HH, Freissmuth M (2010) The reverse operation of $Na^{+}Cl^{-}$-coupled neurotransmitter transporters – why amphetamines take two to tango. J Neurochem 112:340–355.

Sivasubramaniam SD, Finch CC, Rodriguez MJ, et al (2003) A comparative study of the expression of monoamine oxidase-A and -B mRNA and protein in non-CNS human tissues. Cell Tissue Res 313:291–300. https://doi.org/10.1007/s00441-003-0765-6

Soldan A, Pettigrew C, Cai Q, et al (2016) Hypothetical preclinical Alzheimer disease groups and longitudinal cognitive change. JAMA Neurol 73:698–705. https://doi.org/10.1001/jamaneurol.2016.0194

Song A, Tracey DJ, Ashwell KWS (1999) Development of the rat phrenic nerve and the terminal distribution of phrenic afferents in the cervical cord. Anat Embryol (Berl) 200:625–643. https://doi.org/10.1007/s004290050310

Stanford SC, Heal DJ (2019) Catecholamines: knowledge and understanding in the 1960s, now, and in the future. Brain Neurosci Adv 3:239821281881068. https://doi.org/10.1177/2398212818810682

Starke K (2001) Presynaptic autoreceptors in the third decade: focus on α2-adrenoceptors. J. Neurochem. 78:685–693.

Sternberg DB, Martinez JL, Gold PE, McGaugh JL (1985) Age-related memory deficits in rats and mice: enhancement with peripheral injections of epinephrine. Behav Neural Biol 44:213–220. https://doi.org/https://doi.org/10.1016/S0163-1047(85)90212-2

Sub Laban T, Saadabadi A (2020) Monoamine Oxidase Inhibitors (MAOI). Treasure Island, FL: StatPearls Publishing

Sun D, Huang A, Mital S, et al (2002) Norepinephrine elicits β2-receptor-mediated dilation of isolated human coronary arterioles. Circulation 106:550–555. https://doi.org/10.1161/01.CIR.0000023896.70583.9F

Tan A, Salgado M, Fahn S (1996) Rapid eye movement sleep behavior disorder preceding Parkinson's disease with therapeutic response to levodopa. Mov Disord 11:214–216. https://doi.org/10.1002/mds.870110216

Tanila H, Kauppila T, Taira T (1993) Inhibition of intestinal motility and reversal of postlaparotomy ileus by selective α2-adrenergic drugs in the rat. Gastroenterology 104:819–824. https://doi.org/10.1016/0016-5085(93)91018-D

Taylor BN, Cassagnol M (2020) Alpha Adrenergic Receptors. Treasure Island, FL: StatPearls Publishing.

Temple IP, Logantha SJRJ, Absi M, et al (2016) Atrioventricular node dysfunction and ion channel transcriptome in pulmonary hypertension. Circ Arrhythm Electrophysiol 9:e003432. https://doi.org/10.1161/CIRCEP.115.003432

Tobin G (2008) Salivary secretion control. In: Binder MD, Hirokawa N, Windhorst U (eds). Encyclopedia of Neuroscience. Berlin: Springer, pp. 3565–3571.

Tosun M, Karabekir HS, Durmaz MO, et al (2017) Anatomic origin and molecular genetics in neuroblastoma. In: Gowda C (ed.). Neuroblastoma – Current State and Recent Updates. Vienna: InTech.

Totland MZ, Rasmussen NL, Knudsen LM, Leithe E (2020) Regulation of gap junction intercellular communication by connexin ubiquitination: physiological and pathophysiological implications. Cell Mol Life Sci 77:573–591.

Toutant J-P, Massoulié J (1988) Cholinesterases: tissue and cellular distribution of molecular forms and their physiological regulation. Berlin: Springer, pp 225–265.

Trebichalská Z, Holubcová Z (2020) Perfect date – the review of current research into molecular bases of mammalian fertilization. J Assist Reprod Genet 37:243–256.

Vatner SF, Young MA, Vatner DE (1991) Large coronary artery regulation by α- and β-adrenergic receptors in conscious calves. In: Nakamura M, Vanhoutte PM (eds). Coronary Circulation in Physiological and Pathophysiological States. Tokyo: Springer Japan, pp. 29–41.

Ventura R, Cabib S, Alcaro A, et al (2003) Norepinephrine in the prefrontal cortex is critical for amphetamine-induced reward and mesoaccumbens dopamine release. J Neurosci 23:1879–1885. https://doi.org/10.1523/jneurosci.23-05-01879.2003

von Gersdorff H, Hull C (2003) Synaptic transmission. In: Conn PM (ed.). Neuroscience in Medicine. Totowa, NJ: Humana Press, pp. 73–87.

Wahle M, Stachetzki U, Krause A, et al (2001) Regulation of beta2-adrenergic receptors on CD4 and CD8 positive lymphocytes by cytokines in vitro. Cytokine 16:205–209. https://doi.org/10.1006/cyto.2001.0965

Wahle M, Krause A, Pierer M, et al (2002) Immunopathogenesis of rheumatic diseases in the context of neuroendocrine interactions. Ann N Y Acad Sci 966:355–364. https://doi.org/10.1111/j.1749-6632.2002.tb04235.x

Wang D, Jose P, Wilcox CS (2006) beta(1) Receptors protect the renal afferent arteriole of angiotensin-infused rabbits from norepinephrine-induced oxidative stress . J Am Soc Nephrol 17:3347–3354. https://doi.org/10.1681/ASN.2006030212

Warne T, Serrano-Vega MJ, Baker JG, et al (2008) Structure of a β1-adrenergic G-protein-coupled receptor. Nature 454:486–491. https://doi.org/10.1038/nature07101

Wehrwein EA, Orer HS, Barman SM (2016) Overview of the anatomy, physiology, and pharmacology of the autonomic nervous system. Compr Physiol 6:1239–1278.

Widdicombe JH, Wine JJ (2015) Airway gland structure and function. Physiol Rev 95:1241–1319. https://doi.org/10.1152/physrev.00039.2014

Wilder RL (2002) Neuroimmunoendocrinology of the rheumatic diseases: past, present, and future. Ann N Y Acad Sci 966:13–19. https://doi.org/10.1111/j.1749-6632.2002.tb04197.x

Willems JL, Buylaert WA, Lefebvre RA, Bogaert MG (1985) Neuronal dopamine receptors on autonomic ganglia and sympathetic nerves and dopamine receptors in the gastrointestinal system. Pharmacol Rev 37:165–216.

Yamaguchi I, Kopin IJ (1980) Differential inhibiton of alpha-1 and alpha-2 adrenoceptor-mediated pressor responses in pithed rats. J Pharmacol Exp Ther 214:275–281.

Yamamoto DL, Hutchinson DS, Bengtsson T (2007) β2-adrenergic activation increases glycogen synthesis in L6 skeletal muscle cells through a signalling pathway independent of cyclic AMP. Diabetologia 50:158–167. https://doi.org/10.1007/s00125-006-0484-0

Yamamura Y, Motegi K, Kani K, et al (2012) TNF-α inhibits aquaporin 5 expression in human salivary gland acinar cells via suppression of histone H4 acetylation. J Cell Mol Med 16:1766–1775. https://doi.org/10.1111/j.1582-4934.2011.01456.x

Yates CM, Simpson J, Gordon A, et al (1983) Catecholamines and cholinergic enzymes in pre-senile and senile Alzheimer-type dementia and down's syndrome. Brain Res 280:119–126. https://doi.org/10.1016/0006-8993(83)91179-4

Youdim MBH (2018) Monoamine oxidase inhibitors, and iron chelators in depressive illness and neurodegenerative diseases. J Neural Transm 125:1719–1733.

Zhou J (2004) Norepinephrine transporter inhibitors and their therapeutic potential. Drugs Future 29:1235–1244.

Ziegler MG, Lake CR, Kopin IJ (1976) Deficient sympathetic nervous response in familial dysautonomia. N Engl J Med 294:630–633. https://doi.org/10.1056/NEJM197603182941202

2 Cholinergic Neurotransmission

Anil T. Pawar and Chinmay Jogdeo
School of Pharmacy, Dr. Vishwanath Karad MIT World Peace University, Pune, India

Aman Upaganlawar
SNJB's Shriman Suresh Dada Jain College of Pharmacy, Nashik, India

S.B. Chandrasekar
Hi-Tech Lab, Drug Testing Laboratory, Drug Control Department, Bangalore, India

Swati Putta
Department of Pharmacology, A.U. College of Pharmaceutical Sciences, Andhra University, Visakhapatnam, India

CONTENTS

DOI: 10.1201/9780429265198-3

2.1 INTRODUCTION

The human nervous system consists of the central (CNS) and peripheral (PNS) nervous systems. Brain and spinal cord are components of the CNS. The CNS relays and processes signals of the PNS. This CNS processing responds to the peripheral division of the nervous system (Waugh and Grant, 2006). Important functions such as sensory perception, auditory and visual processing, wakefulness, language, and consciousness are the responsibility of the CNS. The PNS consists of nerves (cranial and spinal) and their branches, ganglions, and sensory receptors (Tortora and Derrickson, 2009). It includes any nerves that leave or enter the CNS. The two main divisions of the PNS are afferent and efferent. The afferent division carries signals to the CNS, while the efferent division conducts signals away from the CNS. The efferent portion of the PNS is further divided into the somatic and autonomic nervous systems (Howland et al., 2006). The somatic nervous system carries out voluntary functions such as movement of voluntary muscles. In contrast, involuntary functions such as the functioning of smooth muscles and glands are regulated by the autonomic nervous system (ANS) (Howland et al., 2006).

The ANS has two divisions, parasympathetic and sympathetic (Waugh and Grant, 2006). The parasympathetic and sympathetic divisions are also referred to as the craniosacral and thoracolumbar divisions, respectively, to reflect their anatomic basis (Tortora and Derrickson, 2009). Most organs are supplied by both divisions of the ANS, which produce opposite effects, with a few exceptions (Rang et al., 2007). The ANS is of paramount importance in two situations: in high-stress situations that require us to 'fight' or take 'flight'; and in non-urgent situations that permit us to 'rest' and 'digest'. The ANS and the endocrine system are linked to each other and with sensory input and information from higher centers of the CNS, such as the cerebral cortex at the level of the midbrain and the medulla (Tortora and Derrickson, 2009). These links led early researchers to call the PNS a trophotropic system (i.e., a system leading to growth) and the sympathetic system an ergotropic one (i.e., leading to energy expenditure) (Katzung and Trevor, 2015).

Autonomic nerve fibers connect the CNS with their target organs by a two-neuron pathway. These two neurons are the preganglionic and the postganglionic neurons. The preganglionic neuron originates in the CNS and synapses outside the spinal cord. The postganglionic neuron connects the preganglionic neuron to the target organs (Tortora and Derrickson, 2009). The synapse is a place where two neurons, or a neuron and an effector cell, interact with the help of chemical messengers known as neurotransmitters (Waugh and Grant, 2006). The terms sympathetic and parasympathetic are not based on the type of neurotransmitter released from the nerve endings or on the effect produced by nerve activity (excitatory or inhibitory); they are just structural designations (Katzung and Trevor, 2015).

Small-molecule neurotransmitters and neuropeptides are two important classes of neurotransmitters based on their size (Tortora and Derrickson, 2009). Acetylcholine (ACh) is one of the small-molecule neurotransmitters. Neurons that release ACh are called cholinergic neurons. The basal ganglia and the cortex reticular-activating system are two CNS locations where ACh acts as a neurotransmitter (Bhandari, 2018). ACh is released by all preganglionic neurons and postganglionic parasympathetic fibers. In addition, postganglionic sympathetic fibers of sweat glands also release ACh (Craig and Stitzel, 2004). Alterations in cholinergic neurotransmission are involved in the pathogenesis of various diseases. The objectives of the present chapter are to summarize the workings of the cholinergic system, its role in the pathophysiology of diseases, and the use of drugs targeting cholinergic systems.

2.2 BASIC STEPS IN NEUROCHEMICAL TRANSMISSION

The main processes that occur during neurochemical transmission for many transmitters such as ACh, monoamines, amino acids, and adenosine triphosphate (ATP) involve following steps:

(a) Uptake of precursors: In this step, the precursors needed for the biosynthesis of neurotransmitters are transported into neurons.
(b) Synthesis of neurotransmitter: The enzymes present in the neurons utilize the precursors and synthesize the transmitter.

(c) Uptake/transport of neurotransmitter into vesicles: The synthesized transmitter then enters into membrane-bound structures known as vesicles and is stored at the nerve terminals. The vesicles protect the synthesized neurotransmitter from degradation by cytoplasmic enzymes.
(d) Degradation of surplus neurotransmitters: The surplus neurotransmitters are acted upon by cytoplasmic enzymes which cause their degradation.
(e) Depolarization by propagated action potential: The action potential generated at the cell body propagates along the neuronal axon and reaches the nerve terminal.
(f) Influx of Ca^{2+} in response to depolarization: There are voltage-sensitive Ca^{2+} channels in the membrane of nerve terminals. These ion channels open during the depolarizing phase of the action potential, resulting in an influx of Ca^{2+} into the presynaptic neuron as Ca^{2+} is more concentrated in the extraneuronal fluids.
(g) Release of transmitter by exocytosis: The increased concentration of Ca^{2+} inside presynaptic neurons triggers exocytosis of synaptic vesicles and the subsequent release of the neurotransmitters within those vesicles into the synaptic cleft.
(h) Interaction with pre- and postsynaptic receptors: The released neurotransmitter binds to specific receptors located on the plasma membrane of pre- and postsynaptic neurons.
(i) Inactivation of neurotransmitter: The released neurotransmitter is metabolized and thereby inactivated by the enzymes present in the synaptic cleft to form degradation products.
(j) Reuptake of neurotransmitter and degradation products by nerve terminals: The neurotransmitter and its degradation products are again taken into nerve terminals.

2.3 STEPS IN CHOLINERGIC NEUROTRANSMISSION

There are five major steps of cholinergic neurotransmission: synthesis, storage, release, termination of action, and receptor effects of ACh (Figure 2.1). These steps provide potential targets for drug therapy (Katzung and Trevor, 2015)

2.3.1 Biosynthesis of Acetylcholine

The biosynthesis of ACh is summarized in Figure 2.2. ACh is synthesized in the cytoplasm of cholinergic neurons as follows:

(a) Choline in the extracellular fluid is actively transported into the neuronal cytoplasm by the axonal membrane transport system known as Na^+ choline co-transporter. The transport of choline can be inhibited by a drug called hemicholinium (Tripathi, 2010). The choline portion of ACh has a quaternary ammonium group that imparts positive charge on ACh making it extremely hydrophilic and membrane-impermeable (Barar, 2000).
(b) The intraneuronal choline is acetylated by the choline acetyltransferase (ChAT) in the axonal cytoplasm to form ACh (Tripathi, 2010). This reaction requires ATP and acetyl coenzyme A (acetyl-CoA). The mitochondria supply acetyl-CoA. The acetyl-CoA is synthesized from pyruvate, which is derived from glucose. Glucose enters the neuron by facilitated transport (Craig and Stitzel, 2004). The action of ChAT is regulated by neuronal depolarization, calcium influx, and enzyme phosphorylation by a number of protein kinases (Bruce and Hersh, 1989). However, ChAT action is not the limiting step for ACh synthesis, as ChAT is present in excess in the neuronal cytoplasm (Collier and Katz, 1974).

2.3.2 Storage of Acetylcholine

Synthesized ACh is transported into the synaptic vesicles by the vesicular ACh transporter (VAChT; Tripathi, 2010). ACh concentration in the vesicle is about 100-fold higher than in the cytoplasm (Parsons et al., 1993). Two vesicular protons are exchanged for one molecule of ACh. The transporter can be blocked by vesamicol, a non-competitive inhibitor (Bahr and Parsons, 1986). The fully developed synaptic vesicle contains ACh, as well as peptides, ATP, and proteoglycan (Tripathi, 2010). The synaptic vesicles also contain a co-transmitter that modifies the effect of ACh (Howland et al., 2006). A VAChT knockdown mouse model shows neuromuscular problems, highlighting the importance of this transport in the nervous system (Ferreira-Vieira et al., 2016).

2.3.3 Release of Acetylcholine

The arrival of an action potential at nerve ending opens voltage-sensitive calcium channels in the membrane of synaptic end bulbs (Katzung and Trevor, 2015). This leads to an influx of Ca^{2+} into the presynaptic neuron as Ca^{2+} is more concentrated in the extraneuronal fluids. The raised Ca^{2+} levels in neuronal cytoplasm promote the release of vesicular contents into the synaptic cleft by exocytosis (Tripathi, 2010).

Calcium causes exocytosis by binding to the vesicle-associated membrane protein synaptotagmin, which favors an association between a second vesicle-associated membrane protein – synaptobrevin – and a related synaptosomal nerve-associated protein – synaptotaxin – on the inside of the plasma membrane (Katzung and Trevor, 2015). This association brings the vesicle membrane close to the plasma membrane, causing an integration of the vesicle membrane with the terminal membrane and opening of the pore into the synapse (Tripathi, 2010). This pore formation causes the influx of cations in the synapse and thereby liberation of ACh into the synaptic cleft (Katzung and Trevor, 2015). All the proteins involved in the exocytosis of ACh are collectively known as SNARE proteins (soluble *N*-ethylmaleimide-sensitive factor attachment receptor; Südhof and Rizo, 2011). Botulinum

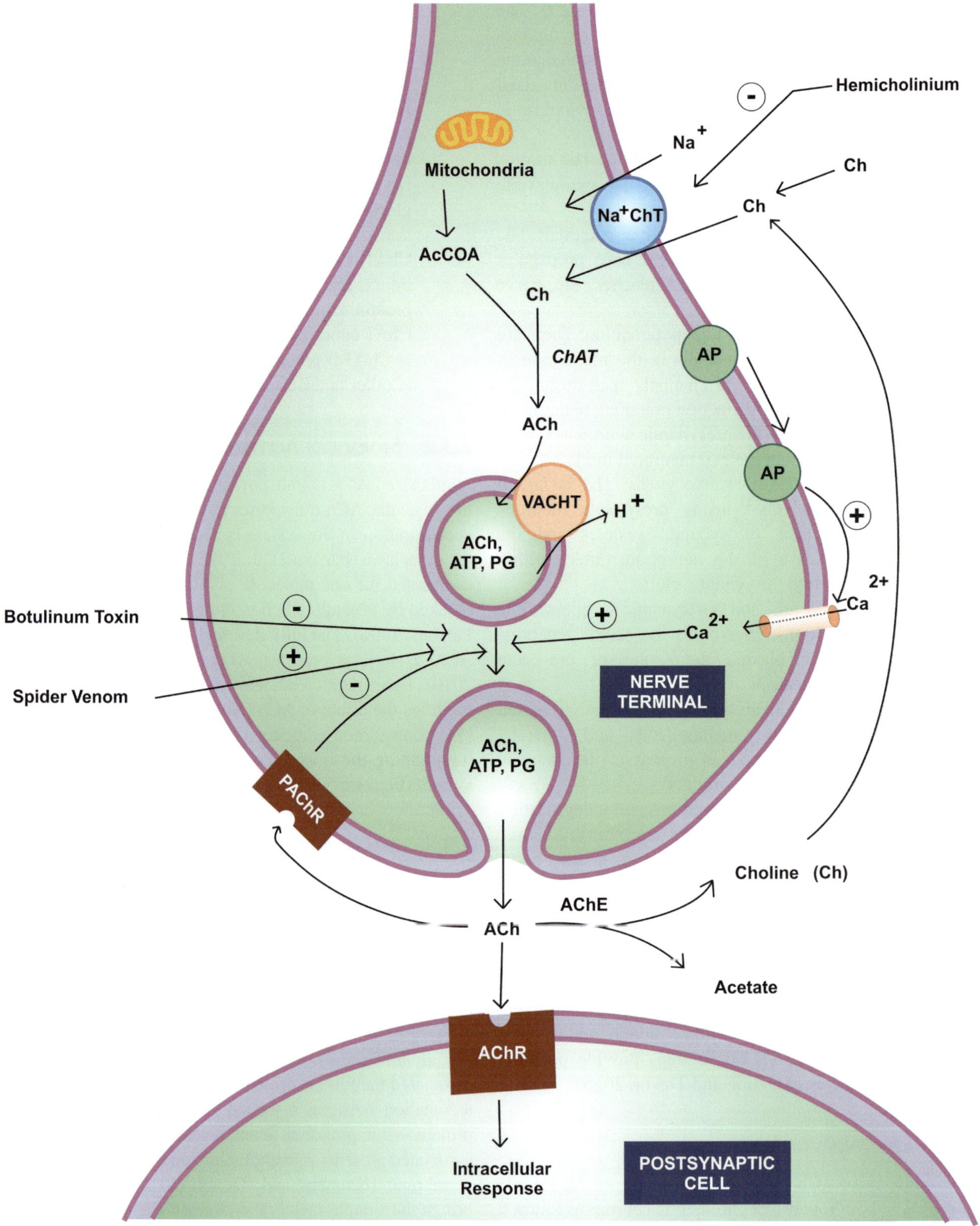

FIGURE 2.1 Steps in cholinergic neurotransmission. AcCOA, acetyl coenzyme A; ACh, acetylcholine; AChE, acetylcholinesterase; AChR, acetylcholine receptor; AP, action potential; ATP, adenosine triphosphate; Ca^{2+}, calcium ion; ChAT, choline acetyltransferase; Ch, choline; H^+, hydrogen ion; Na^+ChT, Na^+ choline co-transporter; Na^+, sodium ion; PAChR, presynaptic acetylcholine receptor; PG, proteoglycan; VACHT, vesicular acetylcholine transporter.

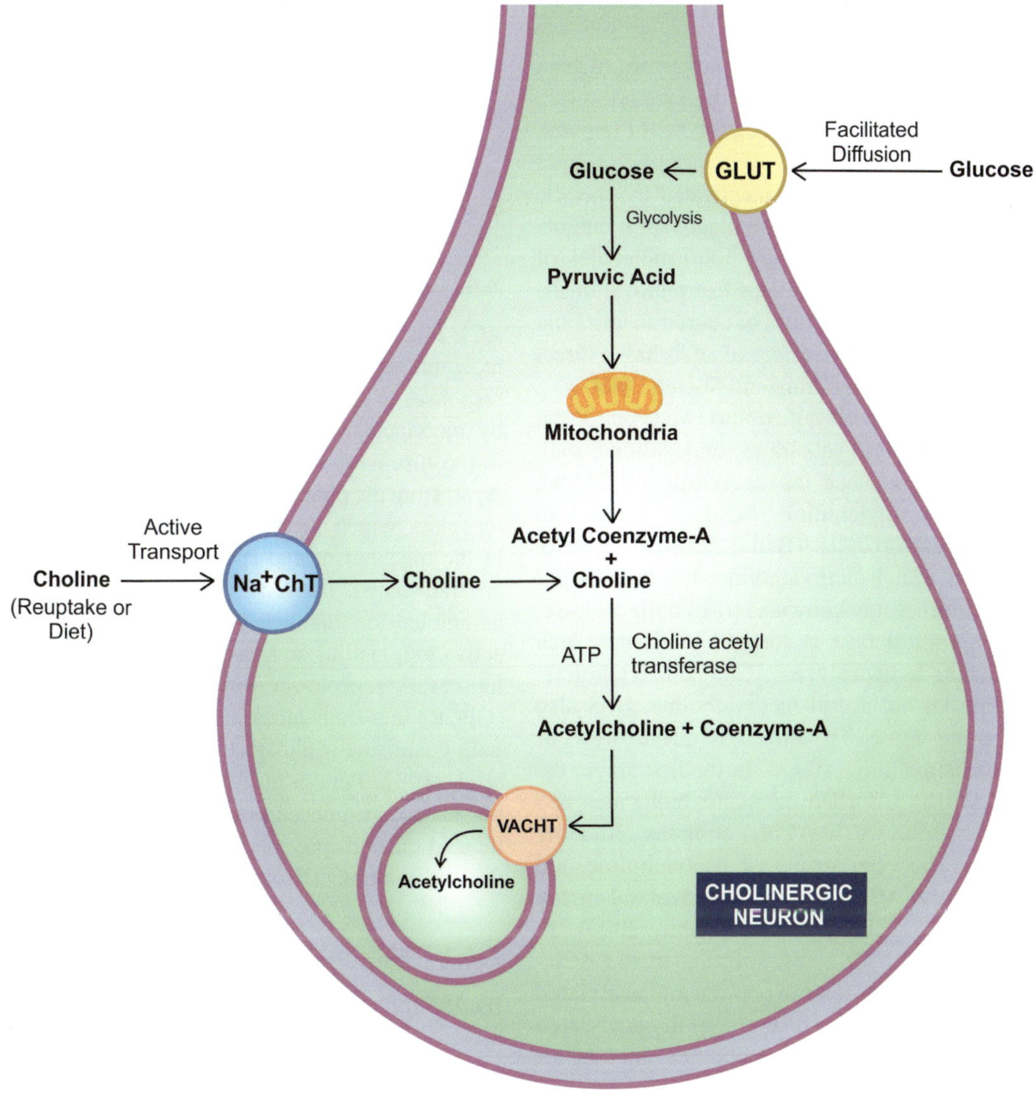

FIGURE 2.2 Biosynthesis of acetylcholine. ATP, adenosine triphosphate, GLUT, glucose transporter; Na^+ChT, Na^+ choline co-transporter; VACHT, vesicular acetylcholine transporter.

neurotoxins block ACh release by proteolytic cleavage of SNARE proteins. Contrary to this, the black widow spider venom causes ACh release (Rang et al., 2007).

After exocytosis, the empty vesicle is recaptured by endocytosis and returns to the interior of the nerve terminal, where it fuses with the larger endosomal membrane. Three alternative pathways involved in recycling of synaptic vesicles are local refilling with neurotransmitters without undocking; local recycling with undocking; and full recycling of vesicles with passage through an endosomal intermediate (Südhof, 2004). The endosome forms new vesicles, which take up transmitter from the cytosol utilizing specific transport proteins and are again docked on the presynaptic membrane. This sequence, which typically takes several minutes, is controlled by various trafficking proteins associated with the plasma membrane and the vesicles, as well as cytosolic proteins (Katzung and Trevor, 2015).

Modulators, including ACh itself, modulate the release of ACh by acting on presynaptic receptors. These presynaptic receptors regulate neurotransmitter release primarily by modifying Ca^{2+} entry into the nerve terminal by inhibiting opening of calcium channels (Katzung and Trevor, 2015). Inhibitory receptors on postganglionic parasympathetic nerve endings are involved in the autoinhibition of ACh release. Other modulators, such as noradrenaline, also inhibit the release of ACh (Tripathi, 2010).

2.3.4 Receptor Binding

The released ACh is then diffused across the synaptic space and attaches to the specific receptors. These receptors are known as ACh receptors or cholinoceptors. ACh binding to cholinoceptors produces a wide variety of biological responses within the cell (Tripathi, 2010).

2.3.5 Degradation of Acetylcholine

The enzyme that degrades ACh is acetylcholinesterase (AChE). It splits ACh into choline and acetate in the synaptic cleft, leading to rapid termination of the effects of ACh (Tripathi, 2010). Neurotransmitters other than ACh usually re-enter the presynaptic neurons and are then inactivated by specific enzymes, making this mechanism of ACh inactivation unique. Each molecule of AChE can hydrolyze 5000 molecules of ACh per second (Potter et al., 1984). This fast removal of the neurotransmitter is crucial to the control of neurotransmission. The intensity and duration of the effect of ACh has a direct relation to the rate of its release (Craig and Stitzel, 2004).

Cholinesterases have been classified into two types based on their affinities for different substrates: endogenous, such as ACh; and exogenous, such as acetyl-methylcholine, butyrylcholine, and benzoylcholine. AChE is located in nervous tissues and erythrocytes (Barar, 2000). It rapidly hydrolyzes ACh and acetyl-methylcholine, but it does not hydrolyze butyrylcholine. It is known as true cholinesterase. However, butyrylcholinesterase is located in tissues such as heart and plasma. It hydrolyzes ACh, benzoylcholine, and butyrylcholine but not acetyl-methylcholine. It is also called pseudo-cholinesterase (Tripathi, 2010). Hydrolysis of ACh by AChE consists of three stages. In the first stage, the cationic head of ACh is adsorbed on AChE molecules and forms a substrate enzyme (ACh–AChE) complex (Katzung and Trevor, 2015). The ester group of acetylcholine and the esteratic site of the AChE are mainly involved in the formation of the ACh–AChE complex (Barar, 2000). In the second stage of ACh hydrolysis, splitting of the ACh–AChE complex takes place to release choline and acetylated enzymes (Tripathi, 2010). In the final stage, the acetylated enzyme reacts with water to release acetic acid and AChE (Barar, 2000).

2.3.6 Recycling of Choline

Choline released by the hydrolysis of ACh is continuously transported back into the presynaptic neuron by a high-affinity active transport system (Ferreira-Vieira et al., 2016). Two choline transporters have been identified: a low-affinity sodium-independent transporter which can only be inhibited by high concentrations of hemicholinium; and a high- affinity sodium-dependent hemicholinium sensitive transporter (CHT1; Kuhar and Murrin, 1978). CHT1 is mainly found in cholinergic neurons and supplies choline for ACh synthesis. Choline reuptake is usually the rate-limiting step in ACh synthesis. It has also been shown that an increase in neuronal firing leads to an increase in choline uptake and a subsequent increase in ACh synthesis (Kobayashi et al., 2002; Vickroy et al., 1985).

2.4 ACETYLCHOLINE RECEPTORS

ACh binds to the specific receptors known as cholinergic receptors or cholinoceptors. In 1914, Dale noted two types of responses to ACh, one mimicked by muscarine and the other by nicotine (Dale, 1914). These findings subsequently led to the discovery of muscarinic and nicotinic ACh receptors. Apart from the pharmacological differences between these two receptors, which are quite substantial, there is also a difference in the mechanism and speed by which their cellular signals are transduced. The nicotinic receptor is a pentameric subunit assembled ion channel, which is opened in response to ACh activation, leading to a fast response (Noda et al., 1982). The muscarinic receptor, however, is a G protein-coupled receptor (GPCR) that transduces its cellular response via interactions with guanosine triphosphate (GTP)-binding proteins (Fukuda et al., 1987; Peralta et al., 1987). Due to these interactions, muscarinic responses are much slower. These receptors are located on both pre- and postsynaptic membranes, and have diverse functions (Broadley and Kelly, 2001).

2.4.1 Nicotinic Receptors

By 1980, the approximate structure of the nicotinic receptor had been elucidated by protein chemistry using the electric organ of the electric eel *Torpedo* as a source of nicotinic receptor (Raftery et al., 1980).

Nicotinic receptors recognize nicotine, besides binding to ACh. These receptors are initially stimulated and then blocked by nicotine or ACh (Barar, 2000). Based on location, nicotinic receptors are divided into three major categories, namely the muscle, ganglionic, and CNS receptors (Table 2.1). Muscle receptors are mainly located at the skeletal neuromuscular junction; ganglionic receptors are in control of transmission at sympathetic and parasympathetic ganglia; and CNS receptors are prevalent in the brain. The nicotinic receptors of autonomic ganglia are different from those of the neuromuscular junction

TABLE 2.1
Subtypes of Nicotinic Receptor

Receptor Subtype	Muscle Type	Ganglion Type		CNS Type
Main Location	Postsynaptic receptor on skeletal neuromuscular junction	Postganglionic receptor in autonomic ganglia	Pre- and postsynaptic receptors in many brain regions	Pre- and postsynaptic receptors in many brain regions
Response	Excitation by increasing cation permeability (mainly Na^+ and K^+)	Excitation by increasing cation permeability (mainly Na^+ and K^+)	Excitation by increasing cation permeability (mainly Na^+ and K^+)	Excitation by increasing Ca^{2+} permeability

TABLE 2.2
Subtypes of Muscarinic Receptor

Receptor subtype	M_1 (neural)	M_2 (Cardiac)	M_3 (Glandular/Smooth Muscle)	M_4	M_5
Main Locations	Autonomic ganglia, glands, and cerebral cortex	Heart and central nervous system (CNS)	Exocrine glands, smooth muscle, eye, endothelium, and CNS	CNS	CNS (localized in substantia nigra), salivary glands, iris/ciliary muscle
Response	Excitation	Inhibition	Stimulation	Inhibition	Excitation

type of receptors. All nicotinic receptors act as ligand-gated sodium ion channels (Barar, 2000). Five subunits (namely, α, β, γ, δ, and ε) form the receptor–channel complex. Each subunit has four membrane-spanning helical domains. One of the helices of each subunit constitutes the central pore of the ion channel. There are two binding sites for ACh in the receptor–channel complex (Rang and Dale, 2007). These ACh binding sites are located at the interface between the extracellular domain of each of the α-subunits and its neighbor. Binding to both these sites by ACh molecules is essential for the ion channel to open, causing a sodium ion influx (Howland et al., 2006).

2.4.2 Muscarinic Receptors

The first muscarinic receptor was cloned from a pig brain cDNA library. The amino acid sequence that was predicted displayed a similar homology to the beta-adrenergic receptor and the visual pigment rhodopsin (Dixon et al., 1986).We now know that the muscarinic ACh receptor belongs to the superfamily of heterotrimeric GPCRs (Ishii and Kurachi, 2006). Muscarinic signals usually involve secondary messengers activated via the G proteins; however, responses independent of secondary messengers are also seen. Initial evidence from pharmacological studies pointed to the possibility of at least three subtypes of the muscarinic receptor. Subsequently, muscarinic receptors in the heart, salivary glands, and cerebral cortex were found to have low, medium, and high affinities for the antagonist pirenzepine (Hammer et al., 1980). Further studies and cloning experiments led to the differentiation of the muscarinic receptors into five subclasses (M1–M5) (Table 2.2). These five subtypes have differential coupling to G proteins, with the odd-numbered receptors (M1, M3, and M5) coupling to the G_q/G_{11} family triggering the inositol triphosphate pathway and the even-numbered receptors (M2 and M4) to the G_i/G_o family inhibiting adenylyl cyclase activity (Bonner et al., 1987; Fukuda et al., 1987).

2.5 EFFECTS OF CHOLINERGIC NERVOUS STIMULATION ON EFFECTOR ORGANS

ACh-induced activation of cholinergic receptors causes biological responses within the body. On the basis of involvement of types of cholinergic receptors, cholinergic effects are characterized as having either muscarinic or nicotinic effects.

2.5.1 Muscarinic Effects

Muscarinic receptors are broadly distributed throughout the human body and mediate a range of physiological functions depending on the location and the receptor subtype. To date, the exact locations and roles of all five subtypes of the muscarinic receptors have not been completely elucidated (Abrams et al., 2006). These receptor subtypes have different roles within the same body system, with a lot of interplay between the subtypes. A few muscarinic receptors-mediated effects of ACh are discussed in the following subsections.

2.5.1.1 Cardiovascular Effects

Muscarinic receptor stimulation of the M2 subtype in the heart modulates pacemaker activity and atrioventricular (AV) conduction (Barar, 2000). Sinoatrial nodal cells are hyperpolarized by ACh and thereby decrease the rate of diastolic depolarization of the cell. This causes a decrease in the rate of impulse generation, which may lead to bradycardia or even cardiac arrest. The refractory period is increased at the AV node, Purkinje fibers, and bundle of His, slowing down the conduction. The overall effect is a decrease in the force of atrial contraction and decrease in the refractory period of atrial fibers (Tripathi, 2010).In the heart, ACh is supplied by both neuronal and non-neuronal sources. The non-neuronal source is ACh synthesized and secreted by heart muscles (cardiomyocytes) which bind to cholinergic receptors on the outer cell membrane of cardiomyocytes in an autocrine or paracrine manner (Rocha-Resende et al., 2012). Non-neuronal ACh regulates some of the important functions of the heart such as heart rate, offsetting hypertrophic signals, action potential propagation, and energy metabolism (Saw et al., 2018). There is an amplification system of neuronal and non-neuronal sources of ACh as a small amount of neuronal ACh stimulates the synthesis and secretion of ACh by cardiomyocytes (Lewartowski and Mackiewicz, 2015). Intravenous administration of small doses of ACh stimulates the muscarinic receptors on vascular endothelial cells and causes relaxation of arterial smooth muscle cells. This causes a fall in blood pressure (Craig and Stitzel, 2004). The release of an endothelium-dependent relaxing factor (nitric oxide) primarily controls vasodilation (Tripathi, 2010).

2.5.1.2 The Eye

Immunoprecipitation studies have shown that all five subtypes of muscarinic receptor are present in the human eye; the M3 receptor is the most prominent in the ciliary muscles and the iris sphincter (Gil et al., 1997). ACh causes smooth muscle contraction in two structures of the eye, the ciliary muscles and the iris sphincter. Iris sphincter contraction causes miosis, while contraction of ciliary muscles decreases the tension on suspensory ligaments and thereby provides accommodation for near vision (Barar, 2000). These effects are accompanied by an increased humor outflow, leading to a reduction in intraocular tension (Craig and Stitzel, 2004).

2.5.1.3 Gastrointestinal Tract

The M2 and M3 receptor subtypes are considered to be the most functionally important in humans, with the M2 receptor population outnumbering the M3 population by 4:1 (Kerr et al., 1995). Muscarinic stimulation leads to salivation, secretion of gastric acid, improved intestinal tone, peristalsis, and relaxation of the anal sphincter (Barar, 2000). Smooth muscle membrane depolarization and increased calcium influx stimulate contraction in the smooth muscles of the gastrointestinal tract (Tripathi, 2010).

2.5.1.4 Respiratory System

In healthy lungs, the action of ACh on muscarinic receptors controls smooth muscle tone, mucus secretion, vasodilation, bronchoconstriction, and the stimulation of respiratory secretions (Barar, 2000). Even though all five receptor subtypes exist in the lungs, there is only evidence for a functional role of M1, M2, and M3 receptors, and antagonists targeting these receptors have been used to treat lung diseases (Alagha et al., 2014). Bronchoconstriction via smooth muscle contraction is mainly considered to be the action on M3 receptors. In addition, M3 receptors are also responsible for mucus secretion (Buels and Fryer, 2012).

2.5.1.5 Urinary System

The M2 and M3 receptor subtypes are predominant in the detrusor muscles, with the M2 subtype outnumbering the M3 subtype (Matsumoto et al., 2010). However, it has been demonstrated that it is this minority population of M3 receptors that mediates the detrusor muscle contraction in vitro (Fetscher et al., 2002; Wang et al., 1995). ACh stimulates detrusor muscle of the bladder and promotes urination. This effect is facilitated by the relaxation of trigone muscle and the external sphincter muscles (Tripathi, 2010).

2.5.1.6 Glands

Sweating, lacrimation, salivation, tracheobronchial secretions, nasopharyngeal, and gastric secretions – basically secretions from all parasympathetically innervated glands – are increased by ACh (Barar, 2000). However, the intestinal and pancreatic glands are not much affected. Milk and bile secretion is unaffected (Tripathi, 2010).

2.5.2 Nicotinic Actions

ACh produces the following effects through activation of nicotinic receptors.

2.5.2.1 Cardiovascular System

Nicotinic receptor activation leads to positive chronotropic and positive inotropic effects on myocardium (Tripathi, 2010). Cardiac output and blood pressure are also increased (Barar, 2000). Experiments on an isolated perfused rat heart demonstrated that nicotine administration caused a brief decrease followed by a much larger increase in heart rate, which slowly returned to baseline within 10–15 minutes. Administration of antagonists specific to different receptor subunits showed that the $\alpha 7$ subunits participate in the initial nicotine-induced heart rate decrease, whereas, $\beta 4$ subunits are involved in mediating the subsequent nicotine-induced rise in heart rate (Ji et al., 2002).

2.5.2.2 Respiratory System

At smaller doses, ACh stimulates respiration through activation of chemoreceptors in the aortic arch and carotid bodies. However, respiratory centers are directly stimulated by high doses (Barar, 2000). In toxic doses, the respiratory centers in the brain stem are inhibited, causing respiratory depression. At such high doses, it also acts at the receptors on the respiratory muscles, causing respiratory depression (Tripathi, 2010).

2.5.2.3 Central Nervous System

In the human brain, there are nine different nicotinic receptor subunits ($\alpha 2$–7 and $\beta 2$–4), which combine as homo- or heteromeric complexes into different pentameric receptors with diverse functions. Of these, the $\alpha 7$-containing subtypes are the most functionally predominant (Dani and Bertrand, 2007). Nicotinic receptor activation causes a combination of CNS stimulation and depression. Common effects are convulsions, tremors, respiratory depression or stimulation, and the release of antidiuretic hormone by the pituitary gland (Tripathi, 2010).

2.5.2.4 Digestive System

Nicotinic receptor activation causes an increase in gastric acid secretions. There is also an improvement in motility and tone of the gastrointestinal tract (Craig and Stitzel, 2004).

2.5.2.5 Skeletal Muscles

Using iontophoresis, contraction of the muscle fiber is seen upon the application of ACh to the muscle endplate. Intra-arterial injection of high doses of ACh can lead to twitching and fasciculation. However, intravenous injections are usually ineffective due to rapid hydrolysis of ACh (Tripathi, 2010).

2.6 DRUGS AFFECTING CHOLINERGIC NEUROTRANSMISSION

Cholinergic agonists and cholinergic antagonists are the categories of drugs that affect cholinergic neurotransmission.

2.6.1 Cholinergic Agonists

Cholinergic agonists are drugs that facilitate cholinergic neurotransmission. They are also called parasympathomimetics or cholinomimetics as they mimic cholinergic transmission (Barar, 2000). Depending on the mechanism of action, they are classified as direct-acting and indirectly acting cholinergic agonists (Katzung and Trevor, 2015). Directly acting agonists bind to cholinoceptors directly, whereas indirectly acting agonists prevents the degradation of ACh by inhibiting cholinesterase activity in the synaptic cleft, thereby facilitating cholinergic neurotransmission. Indirectly acting agonists are also known as anticholinesterases or cholinesterase inhibitors, and they produce either reversible or irreversible inhibition of cholinesterase (Katzung and Trevor, 2015; Tripathi, 2010).

2.6.2 Cholinergic Antagonists

'Cholinergic antagonist' is a broad term for compounds that inhibit the effects of ACh and other agonists by binding to muscarinic or nicotinic cholinergic receptors (Katzung and Trevor, 2015). The drugs that selectively block muscarinic receptors are known as anticholinergic agents, antimuscarinics, or parasympatholytics (Barar, 2000).These agents act as competitive inhibitors of ACh at muscarinic receptors and are clinically useful agents. They also block a few unusual sympathetic cholinergic neurons, like those connected to the salivary and sweat glands (Craig and Stitzel, 2004). The representative drugs of this class include atropine, scopolamine, and other belladonna alkaloids (Tripathi, 2010).

The second category of compounds is the ganglion blocking drugs which are selective for nicotinic receptors (Barar, 2000). Ganglion blockers competitively inhibit ACh and other agonists at the neuronal receptors of sympathetic and parasympathetic autonomic ganglia (Charles et al., 2004). Some agents also block the ion channels gated by nicotinic cholinoceptors. These drugs are of great importance in research due to their ability to block all autonomic outflow; however, a wide array of adverse effects arising out of their non-selectivity limits their use (Katzung and Trevor, 2015).

The third group of compounds which are considered nicotinic antagonists are the neuromuscular blockers, which intervene in the transmission of efferent impulses to skeletal muscles (Barar, 2000).The neuromuscular blockers are administered along with anesthetics to provide optimum muscle relaxation at low doses of anesthetics, consequently leading to a much more rapid recovery (Tripathi, 2010).

2.7 ROLE OF CHOLINERGIC TRANSMISSION IN PATHOPHYSIOLOGY AND DISEASE MANAGEMENT

ACh plays several important physiological roles in brain and muscle function. Alterations in cholinergic transmission are linked to several pathological conditions, as summarized below.

2.7.1 Myasthenia Gravis

Myasthenia gravis is an autoimmune disease in which the normal contact between the muscles and the nerves is interrupted at the neuromuscular junction (Barar, 2000). Antibodies recognize and destroy or block the cholinoceptors at the neuromuscular junction, subsequently reducing the sensitivity of muscles to ACh (Katzung and Trevor, 2015). Antibodies block or impede nicotinic receptor function by (1) cross-linking receptors, which triggers their internalization and degradation; (2) causing a breakdown of the postsynaptic membrane; and (3) binding to the nicotinic receptor and blocking the action of ACh. Overall effects are muscle weakness and fatigue, ptosis, diplopia, and difficulty in speaking and swallowing (Tripathi, 2010).

2.7.2 Alzheimer's Disease

Alzheimer's disease (AD) is an irreversible, progressive neurological disorder marked by loss of memory, cognitive function, and changes in behavior and personality (Blennow et al., 2006; Liu et al., 2019). In recent years, research has provided a number of different hypotheses to explain the cause of AD, of which the cholinergic hypothesis is at the forefront. The cholinergic hypothesis was proposed by Peter Davies and A.J.F. Maloney in 1976 (Davies and Maloney, 1976). The theory states that the development of AD is due to structural alterations in the cholinergic synapses, loss of specific ACh receptors, the death of ACh-generating neurons, and, subsequently, deterioration of cholinergic neurotransmission, all of which lead to an accumulation of the ACh hydrolyzing enzyme AChE. Based on this hypothesis, AChE inhibitors were found to reduce cognitive impairment in patients with AD by blocking degradation of ACh. These agents have been used for more than 20 years, since the first drug for the treatment of AD – tacrine – was approved by the U.S. Food and Drugs Administration (Davis and Powchik, 1995). These compounds offer only a temporary stabilization of symptoms and improvement in quality of life for a short period, usually 1–3 years. They cannot stop disease progression (Stanciu et al., 2019).

2.7.3 Parkinson's Disease

Parkinson's disease is an extrapyramidal motor disorder with distinguishing symptoms, namely rigidity, tremor, and hypokinesia (Barar, 2000).The pathology of Parkinsonism includes neuronal degradation in the substantia nigra, pars compacta, and nigrostriatal tract. This causes imbalance between dopaminergic and cholinergic systems, leading to motor defects (Tripathi, 2010). Even though the cholinergic system is not affected directly, its suppression by anticholinergic drugs restores balance, which is clinically useful in the management of disease symptoms (Barar, 2000). Muscarinic antagonists are commonly used in the early stages of the disease or as an adjunct to the available first-line therapeutic drug – levodopa (Craig and Stitzel, 2004).

2.7.4 Epilepsy

Studies conducted on patients and animal models of seizures have demonstrated altered nicotinic receptor activity in certain forms of epilepsy, including juvenile myoclonic epilepsy and autosomal dominant nocturnal frontal lobe epilepsy (Ghasemi and Hadipour-Niktarash, 2015). Mutations in muscarinic receptors in several different brain areas are believed to be associated with several types of epilepsy, whereas that of nicotinic receptors result in specific types of epilepsy such as autosomal dominant nocturnal frontal lobe epilepsy. Epileptogenesis is proposed to be determined by cholinergic projection neurons and their interaction with other neurons (Wang et al., 2017).

2.7.5 Cholinergic Anti-Inflammatory Pathway

The inflammatory reflex consists of sensory afferent and motor efferent vagus nerves. The afferent vagus nerve brings inflammatory signals from peripheral parts of the body to the brain (Goehler et al., 2000; Watkins et al., 1995). In response to this, the efferent vagus signals suppress cytokine production via release of ACh. This pathway is known as the cholinergic anti-inflammatory pathway (Pavlov and Tracey, 2005; Tracey, 2002).

2.7.6 Huntington's Disease

Stratum is one of the regions affected majorly by neurodegeneration in Huntington's disease. The striatal cholinergic system is reported to be affected in patients with the disease (Smith et al., 2006). Studies have suggested that alteration of cholinergic transmission in the brain significantly contributes to the behavioral symptoms and neuropathology in Huntington's disease (D'Souza and Waldvogel, 2016). There is clinical evidence to support the beneficial effects of cholinergic drugs in patients with Huntington's disease (Smith et al., 2006).

2.7.7 Schizophrenia

It is proposed that the dysregulation of the central muscarinic cholinergic system contributes to the progression of schizophrenia (Berman et al., 2007). This muscarinic hypothesis of schizophrenia is validated by preclinical and clinical studies. Antipsychotic drugs are reported to cause cholinergic alterations in schizophrenic patients. Many studies have been conducted to optimize the use of cholinergic compounds for the treatment of schizophrenia (Terry, 2008).

2.7.8 Motion Sickness

Enhanced cholinergic activity plays a vital role in manifestations of motion sickness (Eisenman, 2009). Pharmacotherapy of motion sickness primarily involves anticholinergics and antihistamines (Murdin et al., 2011). Hyoscine, an anticholinergic drug, is the most effective prophylactic drug for motion sickness (Barar, 2000). Scopolamine is the widely used medication for motion sickness, which acts by central anticholinergic effects (Leung and Hon, 2019). Many antihistamines also have anticholinergic properties, which is suggested as a reason for its usefulness against motion sickness (Karrim et al., 2017).

2.7.9 Glaucoma

Glaucoma is a disorder caused by an unnaturally high intraocular pressure which leads to optic nerve damage. It is one of the top causes of blindness in geriatrics. Muscarinic agonists and cholinesterase inhibitors reduce the raised intraocular pressure by causing contraction of the ciliary body, leading to an efflux of aqueous humor and reducing its secretion (Tripathi, 2010). Pilocarpine is the most commonly used cholinergic for glaucoma (Gupta et al., 2008). A combination of clonidine 0.125% and pilocarpine 1% has been reported to show positive effects in patients with bilateral primary open-angle glaucoma. Therefore, this combination can be used in patients for whom beta blockers are contraindicated (Sihota et al. 2002).

2.7.10 Bradycardia

Atropine is useful for symptomatic sinus and AV node bradycardia (Tripathi, 2010). Anticholinergics can also be used for prevention of bradycardia during laparoscopic urologic and gynecologic surgery (Aghamohammadi et al., 2009; Steer et al., 2019).

2.7.11 Peptic Ulcers

Secretion of gastric acid is mediated by the acetylcholine through M3 muscarinic receptors located on parietal cells (Xie et al., 2005). Accordingly, antimuscarinic drugs decrease gastric secretion and offer symptomatic relief in gastric and duodenal ulcers (Katzung and Trevor, 2015).

2.8 RECENT DEVELOPMENTS AND CHALLENGES

Macrophages are an important component of the immune system against microbial infection and cancer. They have a central role in the phagocytosis of microorganisms and programmed cell removal of cancer cells (Aderem and Underhill, 1999). The non-neuronal cholinergic system regulates phagocytosis in macrophages (Kawashima et al., 2015). It was recently proposed that pharmacological modulation of the non-neuronal cholinergic system is a new strategy for the prevention and/or therapy of infectious diseases and cancer (Reichrath et al., 2016).

It is now known that a non-neuronal cholinergic system that plays a role in immunity works in tandem with the neuronal cholinergic system. A number of immune cell types can directly produce ACh and respond to ACh signals. Immune cells that can synthesize ACh include lymphocytes, natural killer cells, and B lymphocytes (Cox et al., 2020). ACh signaling plays a vital role in immune responses to pathologies such as sepsis, viral infections, arthritis, multiple sclerosis, and even diabetes. The role of immune cell-derived ACh was first identified during the response caused by vagus nerve stimulation to sepsis. It was demonstrated that mice lacking CD4+ T cells could not produce ACh in the spleen, and hence the cholinergic anti-inflammatory pathway could not protect the mice (Rosas-Ballina et al., 2011). Vasodilation is an important marker of inflammation and helps slows down blood flow to allow immune cell migration to infection sites. It is known that vascular endothelial cells expressing the M3 receptor are stimulated to produce nitric oxide, which then causes vasodilation. However, interestingly, nerves that innervate the arteries are mostly adrenergic, and hence the source of ACh was unknown. It was demonstrated that choline acetyltransferase-expressing T cells are responsible for the vasodilation (Olofsson et al., 2016). This further supported the earlier finding that CD4+ T lymphocytes expressing choline acetyltransferase release ACh in response to vagus nerve stimulation (Rosas-Ballina et al., 2011). In addition to immune cells, the vesicular ACh transporter is also expressed by pancreatic alpha cells, which release ACh (Rodriguez-Diaz et al., 2011). This pancreatic ACh stimulates the insulin-secreting β-cells (M3 and M3 receptors) and somatostatin-secreting δ-cell by acting on M1 receptors (Molina et al., 2014). Activation of the α7 nicotinic ACh receptor was shown to improve streptozotocin-induced hyperglycemia and restores functional β-cells in mice (Gupta et al., 2018). In addition, it is also demonstrated that inhibition of AChE reduces the incidence of streptozotocin-induced type 1 diabetes in mice through the inhibition of pathogenic T helper (Th)17 cells (George et al., 2016).

2.9 CONCLUSION

ACh is one of the major neurotransmitters of the nervous system. It is synthesized by enzyme-mediated processes in the cytoplasm of cholinergic neurons. The ACh is stored in membrane-bound vesicles in the nerve terminals of the cholinergic neurons. It is released in response to action potential by calcium-induced exocytosis in the synaptic cleft. The released ACh molecules produce a broad spectrum of biological effects by acting on cholinergic receptors. These receptors are nicotinic and muscarinic receptors, which mediate various actions of ACh. Alteration in cholinergic neurotransmission is responsible for several diseases such as myasthenia gravis, Parkinson's disease, Alzheimer's disease, epilepsy, schizophrenia, and Huntington's disease. Recently, the role of ACh in anti-inflammatory pathways of the body has also been reported. Furthermore, evidence is available suggesting the use of drugs that target cholinergic system for the management of these diseases.

REFERENCES

Abrams P, Andersson KE, Buccafusco JJ, Chapple C, de Groat WC, Fryer AD, et al. Muscarinic receptors: their distribution and function in body systems, and the implications for treating overactive bladder. Br J Pharmacol. 2006;148:565–78.

Aderem A, Underhill DM. Mechanisms of phagocytosis in macrophages. Annu Rev Immunol. 1999;17:593–623.

Aghamohammadi H, Mehrabi S, Beigi FMA. Prevention of bradycardia by atropine sulfate during urological laparoscopic surgery: a randomized controlled trial. Urol J. 2009;6:92–5.

Alagha K, Palot A, Sofalvi T, Pahus L, Gouitaa M, Tummino C, et al. Long-acting muscarinic receptor antagonists for the treatment of chronic airway diseases. Ther Adv Chronic Dis. 2014;5:85–98.

Bahr BA, Parsons SM. Demonstration of a receptor in *Torpedo* synaptic vesicles for the acetylcholine storage blocker L-trans-2-(4-phenyl[3,4-3H]-piperidino)cyclohexanol. Proc Natl Acad Sci U S A. 1986;83:2267–70.

Barar FSK. Essentials of Pharmacotherapeutics. New Delhi: S. Chand; 2000.

Berman JA, Talmage DA, Role LW. Cholinergic circuits and signaling in the pathophysiology of schizophrenia. Int Rev Neurobiol. 2007;78:193–223.

Bhandari PR. Pharmacology for Medical Undergraduates. 1st ed. New Delhi: Jaypee Brothers Medical Publishers; 2018.

Blennow K, Leon MJ de, Zetterberg H. Alzheimer's disease. Lancet. 2006;368:387–403.

Bonner TI, Buckley NJ, Young AC, Brann MR. Identification of a family of muscarinic acetylcholine receptor genes. Science.1987;237:527–32.

Broadley KJ, Kelly DR. Muscarinic receptor agonists and antagonists. Molecules. 2001;6:142–93.

Bruce G, Hersh LB. The phosphorylation of choline acetyltransferase. Neurochem Res. 1989;14:613–20.

Buels KS, Fryer AD. Muscarinic receptor antagonists: effects on pulmonary function. Handb Exp Pharmacol. 2012;204:317–41.

Collier B, Katz HS. Acetylcholine synthesis from recaptured choline by a sympathetic ganglion. J Physiol. 1974;238:639–55.

Cox MA, Bassi C, Saunders ME, Nechanitzky R, Morgado-Palacin I, Zheng C, et al. Beyond neurotransmission: acetylcholine in immunity and inflammation. J Intern Med. 2020;287:120–33.

Craig CR, Stitzel RE. Modern Pharmacology with Clinical Applications. 6th ed. New Delhi: Wolters Kluwer; 2004.

D'Souza GX, Waldvogel HJ. Targeting the cholinergic system to develop a novel therapy for Huntington's disease. J Huntingtons Dis. 2016;5:333–42.

Dale HH. The action of certain esters and ethers of choline and their relation to muscarine. J Pharmacol Exp Ther 1914;6:147–90.

Dani JA, Bertrand D. Nicotinic acetylcholine receptors and nicotinic cholinergic mechanisms of the central nervous system. Annu Rev Pharmacol Toxicol. 2007;47:699–729.

Davies P, Maloney AJ. Selective loss of central cholinergic neurons in Alzheimer's disease. Lancet. 1976;2:1403.

Davis KL, Powchik P. Tacrine. Lancet. 1995;345 :625–30.

Dixon RA, Kobilka BK, Strader DJ, Benovic JL, Dohlman HG, Frielle T, et al. Cloning of the gene and cDNA for mammalian β-adrenergic receptor and homology with rhodopsin. Nature. 1986;321:75–9.

Eisenman LM. Motion sickness may be caused by a neurohumoral action of acetylcholine. Med Hypotheses. 2009;73:790–3.

Ferreira-Vieira TH, Guimaraes IM, Silva FR, Ribeiro FM. Alzheimer's disease: targeting the cholinergic system. Curr Neuropharmacol. 2016;14:101–15.

Fetscher C, Fleichman M, Schmidt M, Krege S, Michel MC. M3 muscarinic receptors mediate contraction of human urinary bladder. Br J Pharmacol. 2002;136:641–4.

Fukuda K, Kubo T, Akiba I, Maeda A, Mishina M, Numa S. Molecular distinction between muscarinic acetylcholine receptor subtypes. Nature. 1987;327:623–5.

George JA, Bashir G, Qureshi MM, Mohamed YA, Azzi J, Al-Ramadi BK, et al. Cholinergic stimulation prevents the development of autoimmune diabetes: evidence for the modulation of Th17 effector cells via an IFNγ-dependent mechanism. Front Immunol. 2016;7:419.

Ghasemi M, Hadipour-Niktarash A. Pathologic role of neuronal nicotinic acetylcholine receptors in epileptic disorders: implication for pharmacological interventions. Rev Neurosci. 2015;26:199–223.

Gil DW, Krauss HA, Bogardus AM, WoldeMussie E. Muscarinic receptor subtypes in human iris-ciliary body measured by immunoprecipitation. Invest Ophthalmol Vis Sci. 1997;38:1434–42.

Goehler LE, Gaykema RPA, Hansen MK, Anderson K, Maier SF, Watkins LR. Vagal immune-to-brain communication: a visceral chemosensory pathway. Auton Neurosci. 2000;85:49–59.

Gupta D, Lacayo AA, Greene SM, Leahy JL, Jetton TL. β-Cell mass restoration by 7 nicotinic acetylcholine receptor activation. J Biol Chem. 2018;293:20295–306.

Gupta S, Galpalli N, Agrawal S, Srivastava S, Saxena R. Recent advances in pharmacotherapy of glaucoma. Indian J Pharmacol. 2008;40:197–208.

Hammer R, Berrie CP, Birdsall NJ, Burgen AS, Hulme EC. Pirenzepine distinguishes between different subclasses of muscarinic receptors. Nature. 1980;283:90–2.

Howland RD, Mycek MJ, Harvey RA, Champe PC. Lippincott's Illustrated Reviews: Pharmacology. Philadelphia, PA: Lippincott Williams & Wilkins; 2006.

Ishii M, Kurachi Y. Muscarinic acetylcholine receptors. Curr Pharm Des. 2006;12:3573–81.

Ji S, Tosaka T, Whitfield BH, Katchman AN, Kandil A, Knollmann BC, et al. Differential rate responses to nicotine in rat heart: evidence for two classes of nicotinic receptors. J Pharmacol Exp Ther. 2002;301:893–9.

Karrim N, Magula N, Saman Y. Antihistamines for motion sickness. Cochrane Database Syst Rev. 2017; 2017:CD012715.

Katzung BG, Trevor AJ. Basic and Clinical Pharmacology, 13th ed. San Francisco, CA: McGraw-Hill Medical; 2015.

Kawashima K, Fujii T, Moriwaki Y, Misawa H, Horiguchi K. Non-neuronal cholinergic system in regulation of immune function with a focus on α7 nAChRs. Int Immunopharmacol. 2015;29:127–34.

Kerr PM, Hillier K, Wallis RM, Garland CJ. Characterization of muscarinic receptors mediating contractions of circular and longitudinal muscle of human isolated colon. Br J Pharmacol. 1995;115:1518–24.

Kobayashi Y, Okuda T, Fujioka Y, Matsumura G, Nishimura Y, Haga T. Distribution of the high-affinity choline transporter in the human and macaque monkey spinal cord. Neurosci Lett. 2002;317:25–8.

Kuhar MJ, Murrin LC. Sodium-dependent, high affinity choline uptake. J Neurochem. 1978;30:15–21.

Leung AK, Hon KL. Motion sickness: an overview. Drugs Context. 2019;8:2019-9-4.

Lewartowski B, Mackiewicz U. The non-neuronal heart's acetylcholine in health and disease. J Physiol Pharmacol. 2015;66:773–8.

Liu D, Jin F, Shu G, Xu X, Qi J, Kang X, et al. Enhanced efficiency of mitochondria-targeted peptide SS-31 for acute kidney injury by pH-responsive and AKI-kidney targeted nanopolyplexes. Biomaterials. 2019;211:57–67.

Matsumoto Y, Miyazato M, Furuta A, Torimoto K, Hirao Y, Chancellor MB, et al. Differential roles of M2 and M3 muscarinic receptor subtypes in modulation of bladder afferent activity in rats. Urology. 2010;75:862–7.

Molina J, Rodriguez-Diaz R, Fachado A, Jacques-Silva MC, Berggren PO, Caicedo A. Control of insulin secretion by cholinergic signaling in the human pancreatic islet. Diabetes. 2014;63:2714–26.

Murdin L, Golding J, Bronstein A. Managing motion sickness. BMJ. 2011;343:d7430.

Noda M, Takahashi H, Tanabe T, Toyosato M, Furutani Y, Hirose T, et al. Primary structure of α-subunit precursor of *Torpedo californica* acetylcholine receptor deduced from cDNA sequence. Nature. 1982;299:793–7.

Olofsson PS, Steinberg BE, Sobbi R, Cox MA, Ahmed MN, Oswald M, et al. Blood pressure regulation by CD4+ lymphocytes expressing choline acetyltransferase. Nat Biotechnol. 2016;34:1066–71.

Parsons SM, Prior C, Marshall IG. Acetylcholine transport, storage, and release. Int Rev Neurobiol. 1993;35:279–390.

Pavlov VA, Tracey KJ. The cholinergic anti-inflammatory pathway. Brain Behav Immun. 2005;19:493–9.

Peralta EG, Winslow JW, Peterson GL, Smith DH, Ashkenazi A, Ramachandran J, et al. Primary structure and biochemical properties of an M2 muscarinic receptor. Science. 1987;236:600–5.

Potter PE, Hadjiconstantinou M, Meek JL, Neff NH. Measurement of acetylcholine turnover rate in brain: an adjunct to a simple HPLC method for choline and acetylcholine. J Neurochem. 1984;43:288–90.

Raftery MA, Hunkapiller MW, Strader CD, Hood LE. Acetylcholine receptor: complex of homologous subunits. Science. 1980;208:1454–7.

Rang HP, Dale MM, Ritter JM, Flower RJ. Rang and Dale's Pharmacology. 6th ed. London: Churchill Livingstone; 2007.

Reichrath S, Reichrath J, Moussa A, Meier C, Tschernig T. Targeting the non-neuronal cholinergic system in macrophages for the management of infectious diseases and cancer: challenge and promise. Cell Death Discov. 2016;2:16063.

Rocha-Resende C, Roy A, Resende R, Ladeira MS, Lara A, de Morais Gomes ER, et al. Non-neuronal cholinergic machinery present in cardiomyocytes offsets hypertrophic signals. J Mol Cell Cardiol. 2012;53:206–16.

Rodriguez-Diaz R, Dando R, Jacques-Silva MC, Fachado A, Molina J, Abdulreda MH, et al. Alpha cells secrete acetylcholine as a non-neuronal paracrine signal priming beta cell function in humans. Nat Med. 20111;17:888–92.

Rosas-Ballina M, Olofsson PS, Ochani M, Valdés-Ferrer SI, Levine YA, Reardon C, et al. Acetylcholine-synthesizing T cells relay neural signals in a vagus nerve circuit. Science. 2011;334:98–101.

Saw EL, Kakinuma Y, Fronius M, Katare R. The non-neuronal cholinergic system in the heart: a comprehensive review. J Mol Cell Cardiol. 2018;125:129–39.

Sihota R, Rajashekhar YL, Venkatesh P, Agarwal H. A prospective, long-term, randomized study of the efficacy and safety of the drug combination pilocarpine 1% with clonidine 0.06% or clonidine 0.125% versus timolol 0.25%. J Ocul Pharmacol Ther. 2002;18:499–506.

Smith R, Chung H, Rundquist S, Maat-Schieman MLC, Colgan L, Englund E, et al. Cholinergic neuronal defect without cell loss in Huntington's disease. Hum Mol Genet. 2006;15:3119–31.

Stanciu GD, Luca A, Rusu RN, Bild V, Chiriac SIB, Solcan C, et al. Alzheimer's disease pharmacotherapy in relation to cholinergic system involvement. Biomolecules. 2019;10:40.

Steer AE, Ozcan J, Emeto TI. The role of anticholinergic medication in the prevention of bradycardia during laparoscopic gynaecological surgery. Aust N Z J ObstetGynaecol. 2019;59:777–80.

Südhof TC. The synaptic vesicle cycle. Annu Rev Neurosci. 2004:27;509–47.

Südhof TC, Rizo J. Synaptic vesicle exocytosis. Cold Spring Harb Perspect Biol. 2011;3:a005637.

Terry AV Jr. Role of the central cholinergic system in the therapeutics of schizophrenia. Curr Neuropharmacol. 2008;6:286–92.

Tortora GJ, Derrickson BH. Principles of Anatomy and Physiology. 12th ed. Hoboken, NJ: Wiley; 2009.

Tracey KJ. The inflammatory reflex. Nature. 2002;420:853–9.

Tripathi KD. Essentials of Medical Pharmacology, 6th ed. New Delhi: Jaypee Brothers Medical Publishers; 2010.

Vickroy TW, Roeske WR, Gehlert DR, Wamsley JK, Yamamura HI. Quantitative light microscopic autoradiography of [3H] hemicholinium-3 binding sites in the rat central nervous system: a novel biochemical marker for mapping the distribution of cholinergic nerve terminals. Brain Res. 1985;329:368–73.

Wang P, Luthin GR, Ruggieri MR. Muscarinic acetylcholine receptor subtypes mediating urinary bladder contractility and coupling to GTP binding proteins. J Pharmacol Exp Ther. 1995;273:959–66.

Wang Y, Wang Y, Chen Z. The role of central cholinergic system in epilepsy. Zhejiang Da Xue Xue Bao Yi Xue Ban. 2017;46:15–21.

Watkins LR, Maier SF, Goehler LE. Cytokine-to-brain communication: a review & analysis of alternative mechanisms. Life Sci. 1995;57:1011–26.

Waugh A, Grant A. Ross and Wilson Anatomy and Physiology in Health and Illness. 10th ed. Edinburgh: Churchill Livingstone; 2006.

Xie G, Drachenberg C, Yamada M, Wess J, Raufman JP. Cholinergic agonist-induced pepsinogen secretion from murine gastric chief cells is mediated by M1 and M3 muscarinic receptors. Am J Physiol Gastrointest Liver Physiol. 2005;289:G521–9.

3 Dopamine Signaling

Aadil Javed
Department of Biotechnology, Graduate School of Natural and Applied Sciences, Ege University, Bornova, Izmir, Turkey

Fazlullah Khan
Faculty of Pharmacy, Capital University of Science & Technology, Islamabad 44000, Pakistan

Kamal Niaz
Department of Pharmacology and Toxicology, Faculty of Bio-Sciences, Cholistan University of Veterinary and Animal Sciences, Bahawalpur, Pakistan

CONTENTS

3.1 INTRODUCTION

Dopamine (3-hydroxytyramine) – a tyrosine metabolite – is a monoamine neurotransmitter, which is catecholaminergic in nature and is involved in the normal physiologic functions of the body (Carlsson 1987). Glutamate and gamma-aminobutyric acid (GABA) are known to facilitate the fast neurotransmission, which is modulated by dopamine in a slow manner for employing its overall effect on the circuitry of neurons. Four main dopaminergic pathways in brain – mesocortical, mesolimbic, nigrostriatal, and tuberoinfundibular pathways – contain cells with dopamine (Andén et al. 1964; Dahlstrom and Fuxe 1965). Different CNS functions are carried out by the cells (neurons) in these systems, including learning, memory, working, attention, voluntary movement, reward, feeding, affect, and sleep. The mechanisms of hormonal regulation, immune system, renal functions, cardiovascular functions, retinal processes, olfaction, and sympathetic systems are some of the important physiologic processes in which dopamine plays direct role (Carlsson 2001; Salamone and Correa 2012; Schultz 2007). As it is involved in vital physiologic functions, dysfunctions in dopamine are also correlated with human disorders. The absence of striatal innervations of dopaminergic origin is involved in one of the most recognized disorders [i.e., Parkinson's disease (PD)] (Segura-Aguilar et al. 2014). Dopamine is also involved in schizophrenia by providing a psychotomimetic effect, in which D2 dopamine receptor blockers are used as antipsychotics (Seeman et al. 1976). Tourette's syndrome and attention-deficit–hyperactivity disorder (ADHD) are also linked to the activity of dopamine (Gizer, Ficks, and Waldman 2009; Swanson et al. 2007). Drug abuse and addiction causing altered plasticity in mechanisms related to reward and its pathology are associated with dopamine regulation (Joutsa et al. 2012). Other major disorders linked to dopamine signaling include major depression, dyskinesia, and bipolar disorder (Money and Stanwood 2013).

DOI: 10.1201/9780429265198-4

Dopamine acts by initializing signaling via G protein-coupled receptors (GPCRs) known as dopamine receptors (D1–D5), when the presynaptic terminals release it. Dopamine activity is reported to be enhanced or blocked by agonists or antagonists, respectively, for these specific receptors. Different disorders are being cured using compounds that interfere with dopamine signaling at the ligand-binding level. A lot of research has been done to understand the mechanisms involved in the signaling of these receptors and how these receptors function in normal physiological conditions, by elucidating the structure, genetic elements, expression, and functions of these receptors at cellular and organ levels (Habibi 2016). There is a certain hierarchy involved in dopamine signaling between dopamine and other messengers, including serotonin and glutamate, leading to gene expression changes within the cells and causing phenotypic changes such as synaptic plasticity (long or short term; Ayano 2016). Intracellular signaling and machinery are involved in carrying out the functions that are intended by dopamine; therefore, clinical interventions can utilize the information of altered processes and even genetic errors to identify and target dopaminergic disorders (Schultz 2019). Clearly, there is a lot of literature on dopamine receptors that can be searched for in various databases. The aim of this chapter is to summarize the functions of dopamine and the dopamine receptor in a concise and brief manner to aid reader understanding. Here, we attempt to cover briefly the dopamine receptors, their classification, expression, and function; the general mechanisms of dopamine receptor signaling and the molecular mechanisms involved in its regulation ; and the pharmacology of dopamine signaling in human disorders.

3.2 DOPAMINE RECEPTORS: CLASSIFICATION, GENES, STRUCTURE, EXPRESSION, AND FUNCTIONS

Five GPCRs regulating the physiological attributes of dopamine are categorized further into two classes of receptors: the D1 and D2 dopamine receptors (Table 3.1). The classification of dopamine receptors is dependent on adenylyl cyclase (AC) modulation regulated by dopamine action (Stoof and Kebabian 1984; Vallone, Picetti, and Borrelli 2000). In early research, one group of dopamine receptors was shown to have positive coupling with AC; hence, the production of cyclic adenosine monophosphate (cAMP) is the criterion for the division of the D1 and D2 classes of dopamine receptor (Kebabian and Calne 1979; Spano, Govoni, and Trabucchi 1978). Genetic cloning methods have shown that dopamine can regulate multiple receptors; subtypes of these receptors are classified based on their biochemical and pharmacological attributes (e.g., D1 and D5 are D1 class receptors, while D2, D3, and D4 are D2 class receptors that contain a D5 dopamine receptor encoded by two pseudogenes that give rise to non-functional forms of the receptors) (Ayano 2016; Neve, Seamans, and Trantham-Davidson 2004). The subfamilies of these of classes of dopamine receptors contain a shared homology across genetic structures, especially in transmembrane domains (Beaulieu and Gainetdinov 2011). D1 class (D1 and D5) dopamine receptors regulate cAMP production by $G\alpha_{s/olf}$ activation, a subunit of heterotrimeric G proteins with predominant expression in olfactory system, by AC activity. D class receptors localize on dopamine-receptive cells in the striatum (spiny neurons) in a postsynaptic manner

TABLE 3.1
Dopamine Receptor Subtypes

	D1-Like		D2-Like		
	D1	D5	D2	D3	D4
Amino Acids	446 (human, rat)	477 (human) 475 (rat)	414/443 (human) 415/444 (rat)	400 (human) 446 (rat)	348 (human, rat)
Homology					
D1	100	82	44	44	42
D2	44	49	100	76	54
Localization	Caudate/putamen, nucleus, accumbens, olfactory tubercle, hypothalamus, thalamus, frontal cortex	Hippocampus, thalamus, lateral mammillary, nucleus, striatum, cerebral cortex	Caudate/putamen, nucleus, accumbens, olfactory tubercle, cerebral cortex	Nucleus, accumbens, olfactory tubercle, islands of Calleja, putamen, cerebral cortex	Frontal cortex, midbrain, hippocampus, hypothalamus, medulla, retina
Response	↑ Adenylyl cyclase	↑ Adenylyl cyclase	↓ Adenylyl cyclase	↓ Adenylyl cyclase	↓ Adenylyl cyclase
Introns in Gene	None	None	Yes	Yes	Yes
Sequence of Amino Acids	Putative third (short) carboxyl terminal tail (long)	Putative third (short) carboxyl terminal tail (long)	Putative third (long) Carboxyl terminal tail (short)	Putative third (long) carboxyl terminal tail (short)	Putative third (long) carboxyl terminal tail (short)
References	(Bhatia et al. 2020; Callier et al. 2003; Neve, 1997)	(Bhatia et al. 2020; Missale et al. 1998; Lavine et al. 2002; Neve, 1997)	(Bhatia et al. 2020; Aperia 2000; Carlsson 2001; Iversen and Iversen 2007; Witkovsky 2004; Neve, 1997)	(Bhatia et al. 2020; Sokoloff et al. 1992a; Pierre Sokoloff et al. 1992b; Neve, 1997)	(Bhatia et al. 2020; Oak, Oldenhof, and Van Tol 2000; Neve, 1997)

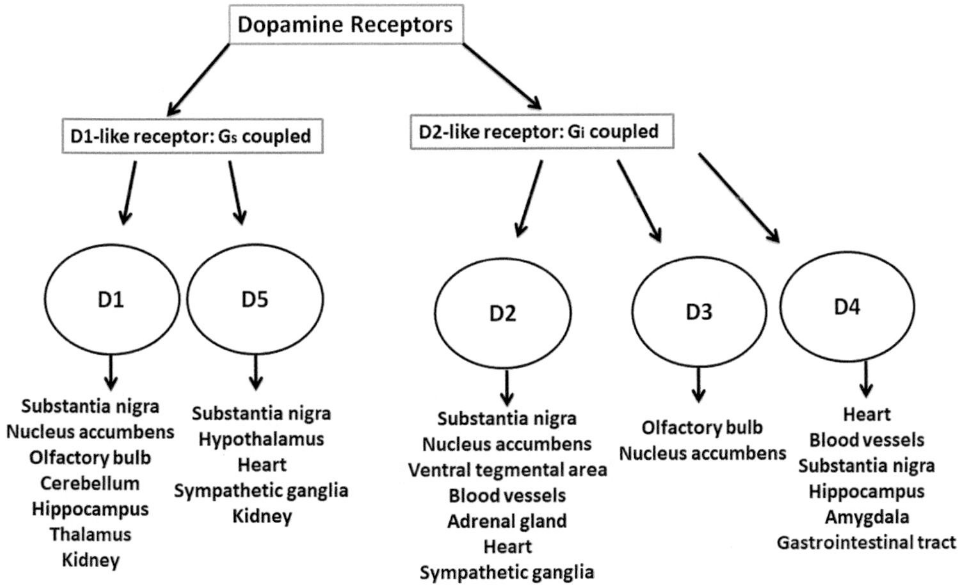

FIGURE 3.1 Subtype of dopamine receptors and their locations.

(Callier et al. 2003). However, D2, D3, and D4 dopamine receptors (D2 class receptors) are known to inhibit AC when they couple with the $G_{i/o}\alpha$ subunit of G proteins. Another difference between D1 and D2 class receptors is that the D2 and D3 subtypes of D2 class receptors are not only located on dopamine target cells in a postsynaptic manner, but also act in a presynaptic manner in dopaminergic neurons (Rankin et al. 2010; Rondou, Haegeman, and Van Craenenbroeck 2010), as shown in Figure 3.1. The genetic structure of both members of class 1 dopamine receptors is different as they do not contain introns inside a coding sequence; however, D2 class receptors have several, including six intronic regions in D2, five in D3, and three in the D4 dopamine receptor gene (Jackson and Westlind-Danielsson 1994; Gingrich and Caron 1993). The receptor splice variants are present in D2 class dopamine receptors. For example, the D2 dopamine receptor has two variants arising from alternative splicing between introns four and five, named D2-long (D2L) and D2-short (D2S; Monsma et al. 1989). The third intracellular loop, consisting of 29 amino acids, is the difference between these two isoforms, which have different pharmacological and signaling patterns. D2L has a postsynaptic function, while D2S possess presynaptic expression with autoreceptor functionality (De Mei et al. 2009; Usiello et al. 2000). Splice variants of D3 receptors are considered to be non-functional; however, one (D3nf) has been reported to reduce ligand binding to the receptor, leading to hetero-oligomerization of the receptor (Elmhurst et al.; Giros et al. 1991). D2 class D4 dopamine receptors possess polymorphic variants. The polymorphic variants contain repeat sequences in the third cytoplasmic loop with variation in numbers causing altered expression, leading to an altered state of functionality of the receptor (e.g., an affinity for antipsychotic drugs such as clozapine) (Schoots and Van Tol 2003; Van Tol et al. 1991; Wong and Van Tol 2003). The transmembrane domain of the D1 and D5 receptors shares 80% homology, and the D2 receptor shares 75% homology with the D3 receptor and 53% with the D4 receptor. All the dopamine receptors exhibit a similar number of amino acids in the N-terminal region; however, the C-terminal region for D2-class dopamine receptors are reported to be seven times shorter than the D1 class of dopamine receptors (Gingrich and Carron 1993; Vallone et al. 2000). Both the D1 and D2 class of dopamine receptors are activated by dopamine with varying degrees of affinity. Similarly, dopamine antagonists and agonists also show different actions toward different dopamine receptor subtypes (Civelli and Zhou 2001).

The expression of dopamine receptors has different patterns in different parts of the body (e.g., brain and the periphery). D1 expression levels are lower in certain regions of the brain (e.g., cerebellum, hippocampus, hypothalamus, and thalamic areas), while they are higher in others, including the frontal cortex, mesolimbic, nigrostriatal, and mesocortical regions (e.g., olfactory bulb, amygdala, and striatum) (Missale et al. 1998). Multiple brain regions such as the premotor cortex, cingulated cortex, hypothalamus, dentate gyrus, hippocampus, substantia nigra, and the prefrontal cortex region with pyramidal neurons show lower D5 dopamine receptor expression (Missale et al. 1998). The striatum, olfactory tubercle, and nucleus accumbens show the highest levels of D2 dopamine receptors. Other regions that show significant levels of D2 receptors include the substantia nigra, hypothalamus, septum, amygdala, hippocampus, ventral tegmental area, and cortical areas (Beaulieu, Espinoza, and Gainetdinov 2015; Dziedzicka-Wasylewska 2004; Rangel-Barajas, Coronel, and Florán 2015). Medium spiny neurons (MSNs) in the nucleus and accumbens are separated on the basis of differential expression of D1 and D2 dopamine receptors by employing the bacterial artificial chromosome (BAC) in transgenic mice containing green and red fluorescently tagged reporters (Shuen et al. 2008; Valjent et al. 2009). These cells can be distinguished based on the expression levels of the reporters and the projections sites; for example, the D1 receptor is selectively expressed in MSNs projecting to the substantia nigra pars reticulate and medial globus pallidus that forms striatonigral pathway. However, D2 receptors are selectively expressed in MSNs projecting toward the lateral globus pallidus forming the indirect striatopallidal pathway. There is another group of MSNs in dorsal striatum that expresses both D1 and D2 receptors but with relatively lower expression (range 5–15%; Valjent et al. 2009). In a separate study, pyramidal neurons in the prefrontal cortex have shown co-expression of D1 and D2 receptors in transgenic mice with the BAC (Zhang et al. 2010). The highest level of expression of D3 dopamine receptors is reported in limbic areas, including the islands of Calleja, the olfactory tubercle, and the shell of the nucleus accumbens; however; D3 receptors cover less territory in terms of expression and a limited expression, compared to D1 and D2 receptors (Sokoloff et al. 1992). The striatum, the ventral tegmental area, cortical areas, the septal area, the hippocampus, and the substantia nigra pars compacta show lower levels of D3 dopamine receptors (Sokoloff et al. 2006). D4 dopamine receptor expression has been recorded in the amygdala, hypothalamus, substantia nigra pars reticulate, thalamus, hippocampus, and frontal cortex; however, their overall expression in brain is lower than the other receptors (Oak, Oldenhof, and Van Tol 2000). D1, D2 and D2 dopamine receptors show observable expression in the retina, while the pituitary gland shows significant expression of D2 receptors. However, all subgroups of dopamine receptors are expressed in the periphery, to varying degrees, including the sympathetic ganglia, gastrointestinal tract, blood vessels, heart, adrenal glands, and kidney (Aperia 2000; Carlsson 2001; Iversen and Iversen 2007; Witkovsky 2004).

The functions of dopamine receptors involve various physiologic processes. Locomotor activity is regulated predominantly by the action of D1, D2, and D3 receptors (Sibley 1999), as shown in Table 3.2. D1 receptors have a stimulatory effect in postsynaptic neurons that almost exclusively express this receptor for locomotor activity. However, D2 and D3 are have both pre- and postsynaptic functions in locomotor activity, but in a more complicated manner than D1 receptors (Sibley 1999; Waters et al. 1993). A negative feedback mechanism regulating neuron firing rate, along with the synthesis and release of neurotransmitters due to changes in their levels across the extracellular space, are delivered by presynaptic autoreceptors (Wolf and Roth 1990). Diminished locomotor activity that is caused by reduced dopamine release is also sourced back to D2 class autoreceptor activity presynaptically (Waters et al. 1993; Wolf and Roth 1990). As mentioned earlier, two isoforms of the D2 receptor – D2S (presynaptic) and D2L (postsynaptic) – conform to the functions of D2 receptors according to their neuronal distributions (Lindgren et al. 2003). Neuronal firing, synthesis, and phasic release of dopamine can also be determined by the autoreceptor activity of D3 dopamine receptor, which may mimic the D2S variant of D2 in a presynaptic manner (Gainetdinov et al. 1996; Joseph et al. 2002; Zapata and Shippenberg 2002). Movement control by primary motor regions of the brain is not under the regulation of D4 and D5 receptors as there is minimal expression shown by these receptors in that part of the brain (Spooren et al. 2010), while activation of the postsynaptic dopamine receptors (D1 and D2) is required for locomotor activity (White et al. 1988).

Pharmacological and genetic interventions resulting in altered states of dopamine receptors often lead to the change in the body's response to addictive drugs or natural rewards; therefore, dopamine receptors are considered central to drug addiction research (Di Chiara and Bassareo 2007; Hyman, Malenka, and Nestler 2006; Koob and Volkow 2010). Working memory facilitated by the prefrontal cortex is associated with the activity of D1 dopamine receptors along with receptors that are considered important for learning behavior and mechanisms related to memory in general (Goldman-Rakic et al. 2004; Xu and Yao 2010). Cognitive functions facilitated by the hippocampus are associated with the activity of D3, D4, and, to some extent, D5 dopamine receptors (Rondou et al. 2010), as shown in Table 3.2. Antipsychotic drugs tend to block D2 dopamine receptors, highlighting the importance of D2 dopamine receptors in the clinical applicability of these receptors in controlling the psychotic reactions that play a role in the pathophysiology of bipolar disorder and schizophrenia (Roth, Sheffer, and Kroeze 2004). Brain dopamine receptor (D1, D2, and D3) activation is reported in the mechanisms of reinforcement and reward in behavioral biology. Furthermore, actions like affect, decision-making, sleep, motor learning, reproductive behaviors, impulse control, food intake, and attention have also been associated with different dopamine receptor subtypes (Missale et al. 1998). Dopamine receptors are also involved in functions that are not directly related to the CNS, including vision, hormonal regulation, and olfaction. D2 dopamine receptors that are localized in the pituitary mediate prolactin secretion. In the kidney, D1 dopamine receptors facilitate the secretion of renin. The D2 receptor in the adrenal gland is involved in the mechanism of aldosterone secretion. Other important functions of dopamine receptors include vasodilation, regulation of blood pressure, renal function, and motility in the gastrointestinal system (Cuevas et al. 2013; Pivonello et al. 2007).

3.3 GENERAL PRINCIPLES OF DOPAMINE RECEPTOR SIGNAL TRANSDUCTION AND REGULATION

The general mechanisms through which dopamine signal transduction and regulation are carried out include G protein-mediated signaling, the inactivation of G proteins, arrestins/GPCR kinases, and heterologous desensitization of the receptors. Since dopamine receptors are GPCRs, they share similarity with activating heterotrimeric G proteins for the induction of intracellular signaling (O'Hayre, Degese, and Gutkind 2014). G proteins are involved in the activation of dopamine receptors; however, G protein-independent mechanisms have also been shown to be involved in dopamine receptor signaling (Luttrell et al. 1999). Sixteen heterotrimeric G protein subtypes are further classified into

TABLE 3.2
Dopamine Receptor Classification, Localization, and Their Functions

Receptors	D1	D5	D2	D3	D4
Location	Striatum, nucleus accumbens. olfactory bulb, amygdala hippocampus, substantia nigra hypothalamus, frontal cortex	Cortex, substantia nigra, hypothalamus	Striatum, ventral tegmental area, olfactory bulb, cerebral cortex	Striatum, islands of Calleja, cortex	Frontal cortex, amygdala, hypothalamus, nucleus accumbens
Type	G_s-coupled	G_s-coupled	G_i-coupled	G_i-coupled	G_i-coupled
Mechanism	Increased intracellular cAMP by activated adenylate cyclase	Adenylate cyclase↑	Increased intracellular cAMP by activated adenylate cyclase	Adenylate cyclase↓	Adenylate cyclase↓
Function	Locomotion, learning and memory, attention, impulse control, sleep, regulation of renal function	Cognition, attention, decision-making, motor learning, renin secretion	Locomotion, learning and memory, attention, sleep, reproductive behavior	Locomotion, cognition, attention, impulse control, sleep, regulation of food intake	Cognition, impulse control, attention, sleep, reproductive behavior
Selective Agonist	SKF-38393, SKF-81297, fenoldopam (SKF-82526)	6-chloro-PB forskolin, (+)bromo-APB	Bromocriptine, pergolide, cabergoline, ropinirole	7-OH-DPAT, pramipexole, rotigotine,PD-128907	A-412997, ABT-670, PD-168077
Selective Antagonist	SCH-23390, SCH-39166, SKF-83566,	(+)SCH 23390	Haloperidol, raclopride, sulpiride, spiperone, risperidone	Nafadotride, GR-103691, GR-218231, SB-277011A, NGB-2904, PG-01037, ABT-127	A-381393, FAUC213, L-745870, L-750667
References	(Mishrate al. 2018; Enjalbert and Bockaert 1983; Trantham-Davidson et al. 2004)	(Mishrate al. 2018; Enjalbert and Bockaert 1983; Trantham-Davidson et al. 2004)	(Mishrate al. 2018; Bateup et al. 2008; Millan 2010)	(Mishrate al. 2018; Zintzaras et al. 2010; Millan 2010)	(Mishrate al. 2018)

four categories ($G\alpha_s$, $G\alpha_i$, $G\alpha_q$, and $G\alpha_{12}$) for the regulation of GPCRs. G proteins contain protein subunits (α, β, γ), with the α-subunit defining the overall characterization of the proteins in a functional manner (Gao and Wang 2006; Rajagopal, Rajagopal, and Lefkowitz 2010). In the absence of a ligand (agonist), the α-subunit remains bound with guanosine diphosphate (GDP), with its guanine nucleotide-binding site along with $\beta\gamma$-subunit for the formation of trimeric inactivated complex. When the ligand binds, GDP is released and α-subunit binds to guanosine triphosphate (GTP) causing the α-subunit to dissociate from $\beta\gamma$-subunit, leading to the activation of effector systems by the activity of α-subunit and $\beta\gamma$-subunit. AC is stimulated by $G\alpha_s$, while cAMP production is inhibited by $G\alpha_i$. The heterotrimeric, inactivated G protein complex is formed upon the hydrolysis of GTP, leading to binding of the GDP-bound α-subunit and its association with the $\beta\gamma$-subunit (McCudden et al. 2005; Ross and Wilkie 2000). The specific GTPase-activating proteins belonging to the family of G protein regulators also mediate GPCR signaling processes (Chasse and Dohlman 2003; Dohlman 2009). The proteins belonging to this family share a RH (Rubicon homology) domain, which speeds up the process of GTP hydrolysis as the proteins bind with GTP-bound G protein α-subunits (Dohlman 2009). Therefore, regulators of G protein signaling proteins act as inhibitors or negative modulators for dopamine receptor functions. The subsequent GTP hydrolysis by the action of regulators of G protein signaling (RGS) is carried out via binding with the α-subunits of $G_{i/o}$ and G_q (Berman and Gilman 1998).

The concentration or intensity of the signal alters the sensitivity of the receptors along with the dynamic regulation of GPCRs (Ferguson 2001). The prolonged duration of the signal from the agonist, or absence of the signal and sudden exposure can lead to desensitization and resensitization of the GPCRs, including dopamine receptors. The activity of GPCR kinases (GRKs) results in the phosphorylation of receptors that are activated for the purpose of recruiting other adaptor proteins (arrestins), which have multiple functions, including homologous desensitization of GPCRs (Lohse et al. 1990; Premont and Gainetdinov 2007). Followed by ligand activity leading to receptor activation, explicit amino acids of the receptor become phosphorylated by GRK activity within the intracellular loops or C-terminal region, resulting in the recruitment of arrestin proteins that bind the receptor and stop receptor activation. These proteins also bind to clathrin adaptor proteins (clathrin and adaptin) and leads to internalization of the receptor from the cellular membrane (Laporte et al. 2002). After clathrin-mediated endocytosis occurs, the receptors are either degraded inside the endosomes or recycled (resensitized) toward the cell membrane (Claing et al. 2002). D1 and D2 dopamine receptors are known to interact with calnexin (a chaperone protein in endoplasmic reticulum) for the regulation of newly synthesized receptors and folded for trafficking in the direction of cell surface (Free et al. 2007). There are seven GRKs, divided into three classes. The third class (GRK4-like), containing GRK4, GRK5, and GRK6 proteins, is involved in the regulation of GPCRs through their kinase activity (Krasel and Lohse 2007; Shukla et al. 2013). Arrestins act as scaffold-inducing adaptors for proteins involved in signaling cascades, including mitogen-activated protein kinases (MAPKs). The effect of arrestin/GRK signaling can either suppress or promote GPCR signaling (Shenoy and Lefkowitz 2003). Heterologous desensitization occurs when one GPCR activation leads to the desensitization of another GPCR located in the same cell. It may incorporate receptor feedback regulation through signaling components from second messenger-regulated kinases (e.g., in neurons; Hoelz et al. 2013). Protein kinase A (PKA), protein kinase C (PKC), and other kinases, including MAPKs are reported to regulate $G\alpha_{s/olf}$ receptors (Nobles, Benians, and Tinker 2005). PKC has been reported to phosphorylate, internalize, and desensitize D2 and D3 dopamine receptors (Cho et al. 2007; Namkung et al. 2009; Namkung and Sibley 2004). The role of ethanol in dopaminergic neurotransmission was revealed when D1 dopamine receptors was shown to be regulated in an ethanol-dependent manner by PKCγ and PKCδ, along with RanBP9 and RanBP10 (Ran binding proteins) (Rex et al. 2008, 2010).

3.4 DOPAMINE RECEPTOR SIGNALING

The dopamine signaling network is shown in Figure 3.2 in the form of a flowchart that describes the crosstalk between various key players involved in cellular signaling mechanisms.

3.4.1 cAMP, Protein Kinase A, DARPP-32, and Associated Proteins

Regulation of cAMP and PKA by G protein signaling are usually linked with dopamine receptor functionality. D1 and D5 dopamine receptors (D1 class receptors) cause the production of cAMP (a secondary messenger) by coupling to $G\alpha_{s/olf}$. However, D2, D3 and D4 dopamine receptors (D2 class receptors) facilitate cAMP production in a negative manner and reduce the activity of PKA as compared to D1 class dopamine receptors, which stimulate the kinase activity of PKA (Enjalbert and Bockaert 1983; Trantham-Davidson et al. 2004).

Dopamine receptor stimulation by a ligand affects the substrates of PKA, including ion channels, ionotropic glutamate receptors, and transcription factors such as cAMP response element-binding protein (CREB) (Greengard 2001). A PKA substrate named 'dopamine and cAMP-regulated phosphoprotein-32' (DARPP-32; a multifunctional protein) is expressed in MSNs, where they regulate dopamine signaling via PKA phosphorylation of threonine 34 (Svenningsson et al. 2004). The inhibitory function of DARPP-32 is stimulated by protein phosphatase 1 (PP1), which, in turn, is activated by protein kinase G, validated in MSNs that express D1 and D2 dopamine receptors (Bateup et al. 2008; Hemmings et al. 1984). Cyclin-dependent kinase 5 (CDK5) acts by phosphorylating threonine 75 of DARPP-32, which puts DARPP-32 into inhibition mode for PKA, as inhibition of PP1 is prevented after this phosphorylation (Bibb et al. 1999). Casein kinase 1 (CK1) acts by phosphorylating the DARPP-32

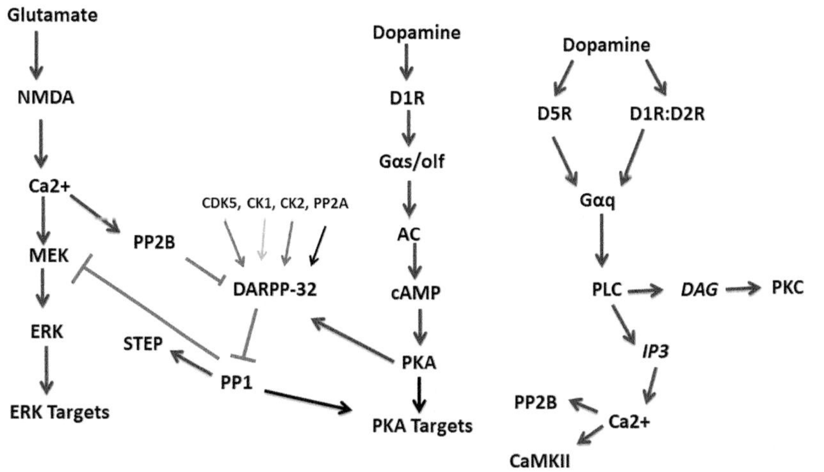

FIGURE 3.2 Dopamine signaling network. Dopamine exerts it activity by acting as a ligand for dopamine receptors, which, in turn, upon activation lead to complex signaling events involved in a cross-talk with other signaling cascades. The different layers of dopamine signaling include effectors like different G protein subtypes, G protein-coupled receptor kinases, β-arrestins, protein kinases, phosphatases, ion channels, and other receptors. The signaling cascade crosstalk with glutamate and serotonin (5-HT)/brain-derived neurotrophic factor (BDNF) is also highlighted in the flowchart of the dopamine signaling network. NMDA, *N*-methyl-D-aspartate; MEK, mitogen-activated protein kinase; PP2B, protein phosphatase-2B; CDK5, cyclin-dependent kinase 5; CK, creatine kinase; PP2A, protein serine/threonine phosphatase 2A; ERK, extracellular signal-regulated kinase; STEP, striatal-enriched protein tyrosine phosphatase; DARPP-32, dopamine and cAMP-regulated phosphoprotein 32; PP1, protein phosphatase 1; PKA, protein kinase A; cAMP, cyclic adenosine monophosphate; AC, adenylyl cyclase; D1R, dopamine 1 receptor; CaMKII, Ca^{2+}/calmodulin (CaM)-dependent protein kinase II; PLC, phospholipase C; DAG, diacylglycerol; PKC, protein, kinase C; IP3, inositol trisphosphate.

serine 137 position, reducing DARPP-2 phosphorylation at threonine 34 by PKA. Casein kinase 2 (CK2) phosphorylates DARPP-32 at serine 97/102, enhancing DARPP-2 phosphorylation of threonine 34 by the kinase action of PKA. PP2A dephosphorylates the serine 97/102 site on DARPP-32, leading to accumulation of the protein in the nucleus, causing inhibition of PP1 histone H3 dephosphorylation. These processes can lead to gene expression responses when D1 dopamine receptors are stimulated (Desdouits et al. 1995; Girault et al. 1989; Stipanovich et al. 2008).

PKA signaling is amplified by DARPP-32. The equilibrium of PKA and PP1 activity leads to phosphorylation states of different PKA substrates, including GABA receptors and ionotropic glutamate receptors (e.g., in the frontal cortex, dopamine receptors lead to synaptic plasticity due to these mechanisms) (Greengard, Allen, and Nairn 1999; Xu et al. 2009). Excessive dopamine receptor stimulation leading to increased CDK5 expression can be associated with the degeneration of MSNs, in which microtubule-associated tau proteins are also hyperphosphorylated, which, in turn, is associated with Alzheimer's disease (Cyr et al. 2003; Kim and Ryan 2010). Adverse iono-tropic effects are associated with the enhanced stimulation of ionotropic neurotransmitter receptors, especially glutamate receptors. When overstimulated. D1 dopamine receptors lead to enhanced stimulation of secondary receptors, such as metabotropic receptors, causing neurogenic phenotypes resembling disorders like Huntington's disease (Robinson, Lebel, and Cyr 2008). Cellular responses arise from dopamine receptor signaling dependent on coincidence detectors and play their role in synaptic plasticity by co-occurrences. MAPKs are known to act as integrators of signals that arise from dopamine, along with other neurotransmitter cascades (Bateup et al. 2008). Both D1 and D2 class dopamine receptors are reported in the regulation of MAPKs [e.g., extracellular signal-regulated kinases (ERK1 and ERK2)] (Sung et al. 2006; Wang et al. 2005). However, in striatum, D3 dopamine receptors are reported to inhibit ERK pathways (Zhang et al. 2004). Several signaling events converge in bringing *N*-methyl-D-aspartate (NMDA) receptors and D1 dopamine receptors to regulate ERK (MEK) phosphorylation. When the dopamine receptor is unstimulated, ERK activity is considered to be inhibited through striatal enriched tyrosine phosphatase (STEP), which, in turn, is dependent on PP1 for dephosphorylation to become active. Therefore, the inactivation of STEP by the inactivation of PP1 via D1 dopamine receptor/PKA/DARPP-32 can arise through dopamine stimulus. This integration of signals from NMDA and dopamine receptors informs us that ERK acts as an integrator for the transmission of signals from neurotransmitters, which has been especially shown in studies related to drug abuse (Valjent et al. 2000, 2005). ERK-mediated signaling has been reported to be involved in behavioral outcomes, leading to the phosphorylation of histone H3 and the activation of transcription factors such as c-Fos, Zif268, and CREB (Brami-Cherrier et al. 2005; Miller and Marshall 2005). Mammalian target of rapamycin (mTOR) complex 1 activation by the induction of dopamine D1 receptors has also been reported in parallel to the ERK signaling cascade for neurotransmitter signaling events leading to dyskinesia (Miller and Marshall 2005; Santini et al. 2009). Activation of exchange proteins (Epac1 and Epac2) via cAMP is reported in depression and synapse modeling by D1 dopamine receptor function (Woolfrey et al. 2009). Apart from PKA and cAMP-dependent activation of Epac, other cAMP-dependent proteins might be involved in the signaling of dopamine receptors (Podda et al. 2010).

3.4.2 Alternative G Protein Mechanisms

Phospholipase C (PLC) can be regulated by the coupling of dopamine receptors with $G\alpha_q$ for the production of inositol triphosphate (IP_3) along with diacylglycerol, which, in turn, activates PKC, resulting in enhanced movement of intracellular calcium (Lee et al. 2004; Sahu et al. 2009). It has been suggested that dopamine receptor activation may be involved in enhancing other GPCR signals via the PLC/IP_3 pathway (Dai et al. 2008). In a calcium-dependent signaling mechanism, dopamine D1 receptors contribute in modulating brain-derived neurotrophic factor (BDNF) production in a localized manner in the nucleus accumbens (Hasbi et al. 2009). A dopamine-dependent behavior study also revealed that the heterodimers of D1/D2 dopamine receptors could mediate calcium-dependent signaling in neurons in the prefrontal cortex (Perreault et al. 2010; Zhang et al. 2010). D2 class dopamine receptors that trigger the calcium release from intracellular stores through regulation of ion channels or calcium signaling are thought to be regulated by $G\beta\gamma$ subunits that are present in the heterotrimeric G proteins, activate PLC, and cause MSNs to have increased calcium in their cytoplasm (Hernádez-López et al. 2000). N-type calcium channels in striatal interneurons are regulated by the action of D2 dopamine receptor-regulated $G\beta\gamma$ subunits (Yan, Song, and Surmeier 1997). Potassium channels also contribute to the modulation of D2 dopamine receptor signaling, known as G protein-coupled inwardly rectifying potassium channels (GIRKs) (Lavine et al. 2002). Several in vivo functions of dopamine are regulated by the inhibition of neurons via activation of GIRKs led by D2 dopamine receptor activation and other GPCRs (Lüscher and Slesinger 2010).

3.4.3 Regulation of G Protein Activity

D2 receptor signaling is also regulated by RGSs, for example RGS9-2 in striatum neurons and heterologous cellular systems, which is implicated in Parkinson's disease (Martemyanov and Arshavsky 2009; Traynor et al. 2009). RGS9-2 in the striatum is associated with R7B9 (additional subunit) for its functionality, as depicted in models of locomotor sensitivity and coordination deficits (Anderson et al. 2007). It has been shown that the regulation of dopamine signaling in the striatum, especially due to the cooperative role of RGS7 and RGS9-2 (RGS proteins) and their activity, is stabilized by the R7BP subunit (Anderson et al. 2010). RGS2 and RGS4 are other members of RGS family, which are also reported in dopamine signaling in the regions of the brain where dopaminergic signals occur (Jules et al. 2015). A leucine zipper-containing protein known as prostate apoptosis response-4 (PAR-4) is another modulator protein for dopamine D2-receptor signaling (Glantz et al. 2010). The third cytoplasmic loop of the receptor contains a calmodulin-binding motif, which is associated with Par-2 to carry the dopamine D2-receptor signaling through cAMP (Sang et al. 2005).

3.4.4 Direct Interactions with Ion Channels and Associated Proteins

On apical dendrites of the prefrontal cortex, there is high expression and co-localization of D1 dopamine receptors and N-type calcium channels. The second intracellular loop of the receptors interacts with the C-terminal of the Cav2.2 subunit of the channels for regulating their activity by internalization of the channels (Kisilevsky et al. 2008; Kisilevsky and Zamponi 2008). Ionotropic glutamate and GABA receptors also interact with dopamine receptors. The C-terminal of D1 dopamine receptors interacts with NR1 and NR2A subunits of glutamatergic NMDA receptors. D2 receptors also interact with NR2B in striatal neurons, inhibiting the phosphorylation of the NMDA receptor by disrupting its association with CaMKII (Fiorentini et al. 2003; Nai et al. 2010). The NR2B subunit of the NMDA receptor has been shown to interact with D2 dopamine receptors in striatal neurons in a postsynaptic manner (Liu et al. 2006). Similarly, the C-terminal of D5 dopamine receptors is known to interact specifically with the $GABA_A$-$\gamma 2$ subunit in the hippocampus, reducing the $GABA_A$-mediated cell currents (Liu et al. 2000). D2 dopamine receptor interaction with NMDA receptors is associated with the regulation of working memory (Lisman 2017). NMDA receptor activation mediated by D1 receptor activation in response to dopamine affects synaptic plasticity by protein kinases (Trepanier, Jackson, and MacDonald 2012). D1 receptor expression on the cell surface and its trafficking are altered upon interaction with NMDA receptors in certain populations of neurons (Pei et al. 2004). Synaptic scaffolding protein (PSD-95) interacts with the dopamine D1 receptor in the C-terminal and leads to dynamin-dependent endocytosis to reduce its expression on the cell surface, which, in turn, leads to less interaction of the D1 receptor and NMDA (Zhang et al. 2007). Dissociation of the D1 receptor from the NMDA receptor by PSD-95 employs another level of regulation for dopamine receptor signaling as it interfere with the surface anchorage of the receptor by competing with NMDA for the C-terminal of the D1 dopamine receptor (Zhang et al. 2009). Dopamine receptor interaction with ion channels or with other associated proteins, and the action of associated proteins directly at the receptor, other downstream factors or interacting partners for the receptors add another layer of receptor signaling regulation.

3.4.5 β-Arrestins/G Protein-Coupled Receptor Kinases: From Dopamine Receptor Desensitization to Signaling

GRKs and β-arrestins phosphorylate dopamine receptors, desensitizing them (Gainetdinov et al. 2004). The effect on dopamine signaling by β-arrestins and GRKs is the result of the binding of one particular protein from either family of proteins or the action of a collection of different proteins from either, depending on their expression in different tissues (Watari, Nakaya, and Kurose 2014).

GRK2 kinase acts by phosphorylating D1, D2, and D3 dopamine receptors for the modulation of dopamine signaling. Fragile X mental retardation protein (FMRP) interacts with GRK2 to regulate D1 dopamine signaling (Wang, Kim, and Zhuo 2010). Neuronal calcium sensor-1 (NCS-1) interacts with GRK2 to desensitize D2 dopamine receptors, and is involved in bipolar disorder and schizophrenia (Kabbani et al. 2002). GRK3, another GRK, is involved in phosphorylation events and the subsequent modulation of D1, D2, and D3 dopamine receptors (Cho et al. 2006). GRK3 regulates the presynaptic control of dopamine release in a negative manner (Gurevich, Gainetdinov, and Gurevich 2016). Since GRK4 is not highly expressed in the brain, it has not been shown to be directly involved in the regulation of neuronal dopamine receptor signaling (Gurevich et al. 2016). Polymorphic variants of GRK4 are known to inactivate D1 dopamine receptors by phosphorylation in patients with hypertension caused by mechanisms related to receptor desensitization (Sanada et al. 2016). In the kidney, GRK4 regulates dopamine D3 receptors in human proximal tubule cells by phosphorylation (Van Villar et al. 2009). In PD animal models and cellular systems, GRK5 has been reported to be involved in the regulation of D1 and D2 dopamine receptors. GRK6 is highly expressed in the brain, most prominently in striatum and MSNs, has shown high affinity toward dopamine D2 receptors (Raehal et al. 2009). When introduced in animal models of PD, the physiological effects of prolonged dopamine treatment for dyskinesia appeared to be alleviated when GRK6 was administered (Rafiuddin Ahmed et al. 2015). Therefore, GRKs, while providing an additional layer for regulation of dopamine signaling, also act as therapeutic targets for disorders that arise from altered or abnormal dopamine signaling. Arrestin 1, which is involved in G protein-independent signaling of dopamine receptor desensitization, has 10–20-fold higher expression in the brain than β-arrestin-2 (Oakley et al. 2001). Arrestins also regulate the D1, D2, and D3 subgroups of dopamine receptor in heterologous cellular systems (Lan, Liu, et al. 2009; Lan, Teeter, et al. 2009). Arrestin 2 also regulates the desensitization of μ-opioid receptors and regulates dopamine receptor signaling in a positive manner by regulating the protein kinase B (Akt) and glycogen synthase kinase 3 (GSK3) pathway for D2 receptor signaling (Beaulieu, Gainetdinov, and Caron 2007; Bohn et al. 2004). Akt-mediated signaling responses that are controlled by

D2 receptors are also regulated by β-arrestin 2, while ERK signaling by D1 receptors is regulated by β-arrestin-2 (Urs, Daigle, and Caron 2011).

3.4.6 β-Arrestin-Mediated Signaling and the Regulation of Akt by Dopamine

β-arrestin 2 is considered to be an intermediate signaling molecule in cAMP-independent mechanisms for Akt/GSK3 regulation by dopamine (Beaulieu et al. 2006). PKB (or Akt, as it is also known) is activated upon recruitment to the cellular membrane by phosphatidylinositol phosphorylation events that include rictor–mTOR protein complexes, which is also associated with the activity of insulin, neurotrophins, insulin-like growth factor, BDNF, nerve growth factor, and neurotrophin 3, stimulating the receptor tyrosine kinase pathways (Chen et al. 2013; Manning and Toker 2017). Akt inhibits GSK3α and GSK3β by the action of different growth factors, signaling molecules, and hormones (Hermida, Dinesh Kumar, and Leslie 2017). β-arrestin 2 is involved in regulating Akt using dopamine receptors and this regulation involves protein complexes formed by Akt, β-arrestin 2, and PP2A for Akt dephosphorylation and inactivation, which is regulated by PP2A by the action of dopamine. Increased dopamine levels in the extracellular environment reduces Akt activity in the striatum and causes increased activity of GSK3α and GSK3β in mice lacking the dopamine transporter. Akt signaling is regulated by D2 dopamine receptors; cAMP is not involved in this regulation (Beaulieu et al. 2004; Gainetdinov 2008).

β-arrestin 2 and the actions of GPCR and their regulation on dopamine receptors are dependent on kinetics that are temporal in nature. ERK signaling mediated by β-arrestins is considered to have a delayed onset, compared to GPCR signaling (Heitzler et al. 2012). cAMP-dependent phosphorylation events for ERK, along with DARPP-32, are considered to take approximately 30 minutes upon dopaminergic drug treatment (Walaas et al. 2011), while the inhibition of Akt mediated by β-arrestin 2 takes 30–60 minutes (Beaulieu et al. 2005). Therefore, there are two phases of signaling events mediated by GPCRs: an early phase involving cAMP with the actions of G proteins causing speedy desensitization; and a late phase containing β-arrestin signaling of longer duration or late onset without involving desensitization of the receptors and slow synaptic transmission. Quick and tonic alterations in dopamine levels for dopamine neurotransmissions can be complemented with early and late phase of synaptic transmission involving these molecular mechanisms (Do et al. 2013).

Dopamine regulates ionotropic glutamate receptors in neurons which are heteromultimeric ion channels and regulate synaptic plasticity. GSK3 inhibits the expression of NMDA (NR2A/B subunits) in the hippocampus and cortical neurons through D2 dopamine receptors (Bradley et al. 2012). GSK3 is known to disturb AMPA (α-amino-3-hydroxy-5-methyl-4-isoxazolepropionic acid) receptor transport by kinesin cargo system at the synapse in the regulation of long-term depression. A subpopulation of MSNs containing D2 receptor stimulation regulated by the action of GSKs and other kinases can be responsible for glutamate response in physiological and pathological conditions or depolarization of neurons (Rangel-Barajas et al. 2015). In response to dopamine, GSK3 inhibits the excitatory glutamate receptors in MSNs that contain D2 dopamine receptors, leading to locomotion control (Yu et al. 2018). Dopamine activity that leads to the circadian rhythm is also linked to GSK3 activity via D2 dopamine receptor signaling, which increases the CLOCK-BMAL1 transcription factor complex activity in retinas that involves protein degradation pathways via ubiquitination (Li et al. 2019; Yujnovsky et al. 2006). Therefore, it is suggested that D2 receptor signaling mediated by GSK3 activity and its phosphorylation of downstream targets (including epigenetic responses) can regulate the behaviors resulting from changes in circadian regulation and synaptic plasticity (Borrelli et al. 2008). The complex mechanisms through which D2 dopamine receptors regulate their activity on mechanisms like circadian rhythm involve other signaling cascades like MAPKs (Dibner, Schibler, and Albrecht 2010).

3.5 PHARMACOLOGY OF DOPAMINE RECEPTORS

Irregularities in dopamine receptors are one of the major underlying causes of mental disorders. Schizophrenia is often linked to enhanced sensitivity of postsynaptic D2 dopamine receptors (Urs, Peterson, and Caron 2017). Dopamine-related psychostimulants provide sensitivity to patients with schizophrenia and their psychotic reactions are associated with the high affinity of dopamine and increased number of D2 dopamine receptors in the brain (Weinstein et al. 2018). Patients with depression showing anhedonia usually show decreased D2/D3 dopamine receptors and postsynaptic sensitivity in the brain regions involved in reward mechanisms (Willner 1997). Patients with bipolar disorder show higher D2 dopamine receptors, while D1 receptors are decreased in patients with major depression. Drug abusers and patients with Tourette's syndrome show decreased binding of the striatal D2 dopamine receptor (Frankle and Laruelle 2002; Nikolaus, Antke, and Müller 2009a; Volkow et al. 2009). Similarly, in PD, in the basal ganglia, dopamine D2 receptors are changed depending on disease stage, with the early stage showing higher expression and the later stage showing lower density of the receptor (Rinne et al. 1993). In progressive supranuclear palsy, D2 receptor binding is selectively reduced (Oyanagi et al. 2002). In Huntington's disease, D1 and D2 dopamine receptors have lowered expression, leading to postsynaptic degeneration in MSN GABA signals (Nikolaus, Antke, and Müller 2009b). D2, D3, and D4 dopamine receptor variants have been associated with schizophrenia, while D4 and D5 dopamine receptor genes are considered as candidate genes for ADHD (Klein et al. 2019). L-DOPA – an indirect dopamine receptor agonist and dopamine precursor – is considered a success in patients with PD (Poewe et al. 2017). In order to alleviate the symptoms of PD, efficient activators of dopamine receptors (chemical compounds) re used, including, but not limited to, pramipexole, apomorphine, pergolide, piribedil, bromocriptine, ropinirole, and rotigotine (Millan 2010). For the treatment of hyperprolactinemia, tumors of the pituitary, and similar conditions, cabergoline and bromocriptone are the D2 receptor agonists of choice (Huang et al. 2018). Restless leg syndrome is clinically treated with pramipexole and ropinirole (Zintzaras et al. 2010). Bipolar disorder and depression are clinically combated with rotigotine and pramipexole (Szmulewicz et al. 2017). Bromocriptine, another dopamine receptor agonist, is used in conditions that result from type 2 diabetes (Reusch and Manson 2017). Apomorphine, which is considered as a non-selective D1 and D2 receptor agonist, has been used in the treatment of erectile dysfunction (Simonsen, Comerma-Steffensen, and Andersson 2016). Hypertensive crises are clinically treated by a synthetic derivative of benzazepine called fenoldopam, the function of which is a modulatory agonist for D1 dopamine receptors located in the periphery (Alshami et al. 2018). Dopamine receptor D2 antagonism by the action of a receptor antagonist paved a way for different generations of compounds that act on psychotic disorders, including chlorpromazine for schizophrenia. However, antipsychotic drugs or compounds are usually associated with certain side effects, including dyskinesia, neuroleptic syndrome, or extrapyramidal syndrome (Désaméricq et al. 2014). Other antipsychotics include perazine, prochlorperazine, fluphenazine, thioridazine, haloperidol, triflupromazine, tiapride, perphenazine, droperidol, benperidol, clopenthixol, promazine, flupentixol, sultopride, trifluperidol, thiothixene, penfluridol, fluspirilene, pimozide, and trifluoperazine (Taipale et al. 2018). Atypical antipsychotics are also available. They are considered to be second-generation antipsychotics as they usually show limited side effects; examples include remoxipride, quetiapine, amisulpride, ziprasidone, sulpiride, risperidone, sertindole, aripiprazole, and olanzapine (Loy et al. 2017). Apart from the usual D2 dopamine receptors, these second-generation antipsychotics obtain their pharmacological actions from inhibiting 5-HT_{2A} receptors and other neurotransmitter receptors. Apart from the treatment of depression, episodes of mania that occur in bipolar disorder can also be treated with these antipsychotics (Singh et al. 2019). For general vomiting and nausea, D2 dopamine receptor antagonists are recommended as they also act as antiemetic drugs (Kovac 2000). D1 and D3 dopamine receptors are targeted by the compounds used against PD. The molecular mechanisms realized in response to the activation or inhibition of dopamine receptor signaling pathways can also be targeted to elucidate their importance in the pathology of human disorders (Moritz, Benjamin Free, and Sibley 2018). An extensive body of literature is available for a more comprehensive understanding of the targeting of mechanisms for dopamine signaling in therapeutics, and is outside of the limits and scope of this chapter (Enriquez-Barreto and Morales 2016; Klein et al. 2019; Leggio et al. 2016; Maiese 2016; Nishi and Shuto 2017; Serafino et al. 2017).

3.6 RECENT DEVELOPMENTS AND CHALLENGES

Motor neurons are activated upon dopamine's interaction with its receptor, leading to reward, memory, pleasure, motivation, and sexual behaviors (Meiser, Weindl, and Hiller 2013). Upon binding with different receptors, dopamine is able to transduce the signal inside the cell, which also includes mediators and signaling molecules like arrestins, GSK3, ERK, PKC, and ion channels. Hence, the overall effect of dopamine depends on a number of factors, including the genetic background, and the expression and functionality of these molecules, to effect a response to the cell and ultimately for the individual through the actions of neurons (Girault and Greengard 2004). The aim of this chapter was to summarize the hallmarks of dopamine signaling. By using nervous system-related models to study the features of dopamine signaling, a great deal of knowledge has been added to the literature. The scope of this chapter was limited and therefore agonist and antagonist actions on specific receptors and their effector molecules could not be discussed in further detail. Overall, dopamine signaling – a key mechanism for neurologic actions inside body – is involved in almost all neurologic disorders and pathologic conditions of the nervous system. However, there are still gaps in our knowledge as to how and to what extent various signaling molecules are interconnected; their interplay in complex pathological and physiological conditions remains to be fully understood. Various compounds that act on these signaling pathways have already cured a lot of people from certain nervous system-related ailments; however, more research is required for therapeutic interventions as new mechanisms in neurobiological research are being discovered with each passing year.

3.7 CONCLUSION

Signaling pathways that are dopaminergic in nature are involved in the physiologic responses of the central and peripheral nervous systems. The dopaminergic receptor signaling mechanism is more complex as it involves interactions with other cell signaling mechanisms that are not only involved in those mediators, but also in their regulation. GPCRs are a class of transmembrane receptors that include dopamine receptors, which act on dopamine stimulation. Various physiological processes in the body are under the regulation of different types of dopamine receptors, which keep the body healthy through different downstream mechanisms. Dysfunctions linked with neurodegenerative diseases and conditions arise from the unbalanced activities of these receptors. Dopamine receptors are known to facilitate the catecholaminergic functions of dopamine, and pharmacologic agents that target neurotransmission are clinically important for regulating psychiatric conditions such as schizophrenia, PD, Huntington's disease, and bipolar disorder, among others. A complete understanding of these complex and intricate interactions is still lacking. However, with the knowledge in hand, researchers worldwide have accumulated a vast body of literature that explains these mechanisms to the extent that we understand their actions in human disorders. Agonistic or antagonistic approaches against dopamine receptors are being employed in clinical settings, especially in psychotic and brain-related (CNS, MSNs) pathological conditions. Dopamine acts by initiating signaling via binding with metabotropic receptors in cells.

REFERENCES

Alshami, A., C. Romero, A. Avila, and J. Varon. 2018. "Management of Hypertensive Crises in the Elderly." *Journal of Geriatric Cardiology*.

Andén, N.E., A. Carlsson, A. Dahlström, K. Fuxe, N.Å. Hillarp, and K. Larsson. 1964. "Demonstration and Mapping out of Nigro-Neostriatal Dopamine Neurons." *Life Sciences*.

Anderson, G.R., Y. Cao, S. Davidson, H.V. Truong, M. Pravetoni, M.J. Thomas, et al. 2010. "R7BP Complexes with RGS9-2 and RGS7 in the Striatum Differentially Control Motor Learning and Locomotor Responses to Cocaine." *Neuropsychopharmacology*.

Anderson, G.R., A. Semenov, J.H. Song, and K.A. Martemyanov. 2007. "The Membrane Anchor R7BP Controls the Proteolytic Stability of the Striatal Specific RGS Protein, RGS9-2." *Journal of Biological Chemistry*.

Aperia, A.C. 2000. "Intrarenal Dopamine: A Key Signal in the Interactive Regulation of Sodium Metabolism." *Annual Review of Physiology*.

Ayano, G. 2016. "Dopamine: Receptors, Functions, Synthesis, Pathways, Locations and Mental Disorders: Review of Literatures." *Journal of Mental Disorders and Treatment*.

Bateup, H.S., P. Svenningsson, M. Kuroiwa, S. Gong, A. Nishi, N. Heintz, and P. Greengard. 2008. "Cell Type-Specific Regulation of DARPP-32 Phosphorylation by Psychostimulant and Antipsychotic Drugs." *Nature Neuroscience*.

Beaulieu, J.M. and R.R. Gainetdinov. 2011. "The Physiology, Signaling, and Pharmacology of Dopamine Receptors." *Pharmacological Reviews*.

Beaulieu, J.M., R.R. Gainetdinov, and M.G. Caron. 2007. "The Akt-GSK-3 Signaling Cascade in the Actions of Dopamine." *Trends in Pharmacological Sciences*.

Beaulieu, J.M., T.D. Sotnikova, W.D. Yao, L. Kockeritz, J.R. Woodgett, R.R. Gainetdinov, and M.G. Caron. 2004. "Lithium Antagonizes Dopamine-Dependent Behaviors Mediated by an AKT/Glycogen Synthase Kinase 3 Signaling Cascade." *Proceedings of the National Academy of Sciences of the United States of America*.

Beaulieu, J.M., T.D. Sotnikova, S. Marion, R.J. Lefkowitz, R.R. Gainetdinov, and M.G. Caron. 2005. "An Akt/β-Arrestin 2/PP2A Signaling Complex Mediates Dopaminergic Neurotransmission and Behavior." *Cell*.

Beaulieu, J.M., T.D. Sotnikova, R.R. Gainetdinov, and M.G. Caron. 2006. "Paradoxical Striatal Cellular Signaling Responses to Psychostimulants in Hyperactive Mice." *Journal of Biological Chemistry*.

Beaulieu, J.M., S. Espinoza, and R.R. Gainetdinov. 2015. "Dopamine Receptors – IUPHAR Review 13." *British Journal of Pharmacology*.

Benjamin Free, R., L.A. Hazelwood, D.M. Cabrera, H.N. Spalding, Y. Namkung, M.L. Rankin, and D.R. Sibley. 2007. "D1 and D2 Dopamine Receptor Expression is Regulated by Direct Interaction with the Chaperone Protein Calnexin." *Journal of Biological Chemistry*.

Berman, D.M. and A.G. Gilman. 1998. "Mammalian RGS Proteins: Barbarians at the Gate." *Journal of Biological Chemistry*.

Bibb, J.A., G.L. Snyder, A. Nishi, Z. Yan, L. Meijer, A.A. Flenberg, et al. 1999. "Phosphorylation of DARPP-32 by Cdk5 Modulates Dopamine Signalling in Neurons." *Nature*.

Bhatia, A., J.R. Lenchner, J.R. and A. Saadabadi. 2020. Biochemistry, Dopamine Receptors. StatPearls. Available at: www.ncbi.nlm.nih.gov/books/NBK538242/PMID: 30855830

Bohn, L. M., L.A. Dykstra, R.J. Lefkowitz, M.G. Caron, and L.S. Barak. 2004. "Relative Opioid Efficacy is Determined by the Complements of the G Protein-Coupled Receptor Desensitization Machinery." *Molecular Pharmacology*.

Borrelli, E., E.J. Nestler, C.D. Allis, and P. Sassone-Corsi. 2008. "Decoding the Epigenetic Language of Neuronal Plasticity." *Neuron*.

Bradley, C.A., S. Peineau, C. Taghibiglou, C.S. Nicolas, D.J. Whitcomb, Z.A. Bortolotto, et al. 2012. "A Pivotal Role of GSK-3 in Synaptic Plasticity." *Frontiers in Molecular Neuroscience*.

Brami-Cherrier, K., E. Valjent, D. Hervé, J. Darragh, J.C. Corvol, C. Pages, et al. 2005. "Parsing Molecular and Behavioral Effects of Cocaine in Mitogen- and Stress-Activated Protein Kinase-1-Deficient Mice." *Journal of Neuroscience*.

Callier, S., M. Snapyan, S. Le Crom, D. Prou, J.D. Vincent, and P. Vernier. 2003. "Evolution and Cell Biology of Dopamine Receptors in Vertebrates." *Biology of the Cell*.

Carlsson, A. 1987. "Perspectives on the Discovery of Central Monoaminergic Neurotransmission." *Annual Review of Neuroscience*.

Carlsson, A. 2001. "A Paradigm Shift in Brain Research." *Science*.

Chasse, S.A. and H.G. Dohlman. 2003. "RGS Proteins: G Protein-Coupled Receptors Meet Their Match." *Assay and Drug Development Technologies*.

Chen, C.H., V. Kiyan, A.A. Zhylkibayev, D. Kazyken, O. Bulgakova, K.E. Page, et al. 2013. "Autoregulation of the Mechanistic Target of Rapamycin (MTOR) Complex 2 Integrity is Controlled by an ATP-Dependent Mechanism." *Journal of Biological Chemistry*.

Di Chiara, G. and V. Bassareo. 2007. "Reward System and Addiction: What Dopamine Does and Doesn't Do." *Current Opinion in Pharmacology*.

Cho, D.I., S.R. Beom, H.H.M. Va Tol, M.G. Caron, and K.M. Kim. 2006. "Characterization of the Desensitization Properties of Five Dopamine Receptor Subtypes and Alternatively Spliced Variants of Dopamine D2 and D4 Receptors." *Biochemical and Biophysical Research Communications*.

Cho, E.Y., D.I. Cho, J.H. Park, H. Kurose, M.G. Caron, and K.M. Kim. 2007. "Roles of Protein Kinase C and Actin-Binding Protein 280 in the Regulation of Intracellular Trafficking of Dopamine D3 Receptor." *Molecular Endocrinology*.

Civelli, O. and Q.-Y. Zhou. 2001. "Dopamine Receptors: Molecular Pharmacology." In: *Encyclopedia of Life Sciences*. DOI: https://doi.org/10.1038/npg.els.0000097

Claing, A., S.A. Laporte, M.G. Caron, and R.J. Lefkowitz. 2002. "Endocytosis of G Protein-Coupled Receptors: Roles of G Protein-Coupled Receptor Kinases and β-Arrestin Proteins." *Progress in Neurobiology*.

Cuevas, S., A. Van Villar, P.A. Jose, and I. Armando. 2013. "Renal Dopamine Receptors, Oxidative Stress, and Hypertension." *International Journal of Molecular Sciences*.

Cyr, M., J.M. Beaulieu, A. Laakso, T.D. Sotnikova, W.D. Yao, L.M. Bohn, et al. 2003. "Sustained Elevation of Extracellular Dopamine Causes Motor Dysfunction and Selective Degeneration of Striatal GABAergic Neurons." *Proceedings of the National Academy of Sciences of the United States of America.*

Dahlstrom, A. and K. Fuxe. 1965. "Evidence for the Existence of Monoamine-Containing Neurons in the Central Nervous System." *Acta Physiologica Scandinavica.*

Dai, R., M.K. Ali, N. Lezcano, and C. Bergson. 2008. "A Crucial Role for CAMP and Protein Kinase A in D1 Dopamine Receptor Regulated Intracellular Calcium Transients." *NeuroSignals.*

De Mei, C., M. Ramos, C. Iitaka, and E. Borrelli. 2009. "Getting Specialized: Presynaptic and Postsynaptic Dopamine D2 Receptors." *Current Opinion in Pharmacology.*

Désaméricq, G., F. Schurhoff, A. Meary, A. Szöke, I. Macquin-Mavier, A.C. Bachoud-Lévi, and P. Maison. 2014. "Long-Term Neurocognitive Effects of Antipsychotics in Schizophrenia: A Network Meta-Analysis." *European Journal of Clinical Pharmacology.*

Desdouits, F., J.C. Siciliano, P. Greengard, and J.A. Girault. 1995. "Dopamine- and CAMP-Regulated Phosphoprotein DARPP-32: Phosphorylation of Ser-137 by Casein Kinase I Inhibits Dephosphorylation of Thr-34 by Calcineurin." *Proceedings of the National Academy of Sciences of the United States of America.*

Dibner, C., U. Schibler, and U. Albrecht. 2010. "The Mammalian Circadian Timing System: Organization and Coordination of Central and Peripheral Clocks." *Annual Review of Physiology.*

Do, J., J.I. Kim, J. Bakes, K. Lee, and B.K. Kaang. 2013. "Functional Roles of Neurotransmitters and Neuromodulators in the Dorsal Striatum." *Learning and Memory.*

Dohlman, H.G. 2009. "RGS Proteins the Early Days." *Progress in Molecular Biology and Translational Science.*

Dziedzicka-Wasylewska, M. 2004. "Brain Dopamine Receptors – Research Perspectives and Potential Sites of Regulation." *Polish Journal of Pharmacology.*

Elmhurst, J.L., Z. Xie, B.F. O'Dowd, and S.R. George. 2000. "The Splice Variant D3nf Reduces Ligand Binding to the D3 Dopamine Receptor: Evidence for Heterooligomerization." *Molecular Brain Research.*

Enjalbert, A. and J. Bockaert. 1983. "Pharmacological Characterization of the D2 Dopamine Receptor Negatively Coupled with Adenylate Cyclase in Rat Anterior Pituitary." *Molecular Pharmacology.*

Enriquez-Barreto, L. and M. Morales. 2016. "The PI3K Signaling Pathway as a Pharmacological Target in Autism Related Disorders and Schizophrenia." *Molecular and Cellular Therapies.*

Ferguson, S.S.G. 2001. "Evolving Concepts in G Protein-Coupled Receptor Endocytosis: The Role in Receptor Desensitization and Signaling." *Pharmacological Reviews.*

Fiorentini, C., F. Gardoni, P.F. Spano, M. Di Luca, and C. Missale. 2003. "Regulation of Dopamine D1 Receptor Trafficking and Desensitization by Oligomerization with Glutamate *N*-Methyl-D-Aspartate Receptors." *Journal of Biological Chemistry.*

Frankle, G.W. and M. Laruelle. 2002. "Neuroreceptor Imaging in Psychiatric Disorders." *Annals of Nuclear Medicine.*

Gainetdinov, R.R. 2008. "Dopamine Transporter Mutant Mice in Experimental Neuropharmacology." *Naunyn-Schmiedeberg's Archives of Pharmacology.*

Gainetdinov, R. R., R.T. Premont, L.M. Bohn, R.J. Lefkowitz, and M.G. Caron. 2004. "Desensitization of G Protein-Coupled Receptors and Neuronal Functions." *Annual Review of Neuroscience.*

Gainetdinov, R.R., T.D. Sotnikova, T.V. Grekhova, and K.S. Rayevsky. 1996. "In Vivo Evidence for Preferential Role of Dopamine D3 Receptor in the Presynaptic Regulation of Dopamine Release but not Synthesis." *European Journal of Pharmacology.*

Gao, Q.B. and Z.Z. Wang. 2006. "Classification of G-Protein Coupled Receptors at Four Levels." *Protein Engineering, Design and Selection.*

Gingrich, J.A. and Caron, M.G. 1993. "Recent Advances in the Molecular Biology of Dopamine Receptors." *Annual Review of Neuroscience.*

Girault, J.A. and P. Greengard. 2004. "The Neurobiology of Dopamine Signaling." *Archives of Neurology.*

Girault, J. A., H.C. Hemmings, K.R. Williams, A.C. Nairn, and P. Greengard. 1989. "Phosphorylation of DARPP-32, a Dopamine- and CAMP-Regulated Phosphoprotein, by Casein Kinase II." *Journal of Biological Chemistry.*

Giros, B., M.P. Martres, C. Pilon, P. Sokoloff, and J.C. Schwartz. 1991. "Shorter Variants of the D3 Dopamine Receptor Produced through Various Patterns of Alternative Splicing." *Biochemical and Biophysical Research Communications.*

Gizer, I.R., C. Ficks, and I.D. Waldman. 2009. "Candidate Gene Studies of ADHD: A Meta-Analytic Review." *Human Genetics.*

Glantz, L.A., J.H. Gilmore, D.H. Overstreet, K. Salimi, J.A. Lieberman, and L. Fredrik Jarskog. 2010. "Pro-Apoptotic Par-4 and Dopamine D2 Receptor in Temporal Cortex in Schizophrenia, Bipolar Disorder and Major Depression." *Schizophrenia Research.*

Goldman-Rakic, P.S., S.A. Castner, T.H. Svensson, L.J. Siever, and G.V. Williams. 2004. "Targeting the Dopamine D1 Receptor in Schizophrenia: Insights for Cognitive Dysfunction." *Psychopharmacology.*

Greengard, P. 2001. "The Neurobiology of Slow Synaptic Transmission." *Science.*

Greengard, P., P.B. Allen, and A.C. Nairn. 1999. "Beyond the Dopamine Receptor: The DARPP-32/Protein Phosphatase-1 Cascade." *Neuron.*

Gurevich, E.V., R.R. Gainetdinov, and V.V. Gurevich. 2016. "G Protein-Coupled Receptor Kinases as Regulators of Dopamine Receptor Functions." *Pharmacological Research.*

Habibi, M. 2016. "Dopamine Receptors." In: *The Curated Reference Collection in Neuroscience and Biobehavioral Psychology* pp. 326–329. DOI:10.1016/B978-0-12-809324-5.00551-4.

Hasbi, A., T. Fan, M. Alijaniaram, T. Nguyen, M.L. Perreault, B.F. O'Dowd, and S.R. George. 2009. "Calcium Signaling Cascade Links Dopamine D1-D2 Receptor Heteromer to Striatal BDNF Production and Neuronal Growth." *Proceedings of the National Academy of Sciences of the United States of America.*

Heitzler, D., G. Durand, N. Gallay, A. Rizk, S. Ahn, J. Kim, et al. 2012. "Competing G Protein-Coupled Receptor Kinases Balance G Protein and β-Arrestin Signaling." *Molecular Systems Biology.*

Hemmings, H.C., P. Greengard, H.Y.L. Tung, and P. Cohen. 1984. "DARPP-32, a Dopamine-Regulated Neuronal Phosphoprotein, is a Potent Inhibitor of Protein Phosphatase-1." *Nature.*

Hermida, M.A., J. Dinesh Kumar, and N.R. Leslie. 2017. "GSK3 and its Interactions with the PI3K/AKT/MTOR Signalling Network." *Advances in Biological Regulation.*

Hernádez-López, S., T. Tkatch, E. Perez-Garci, E. Galarraga, J. Bargas, H. Hamm, and D.J. Surmeier. 2000. "D2 Dopamine Receptors in Striatal Medium Spiny Neurons Reduce L-Type Ca^{2+} Currents and Excitability via a Novel PLCβ1-IP3-Calcineurin-Signaling Cascade." *Journal of Neuroscience*.

Hoelz, L.V.B., G.B.L. De Freitas, P.H.M. Torres, T.V.A. Fernandes, M.G. Albuquerque, J.F.M. Da Silva, et al. 2013. "G Protein-Coupled Receptors." *Revista Virtual de Quimica*.

Huang, H.Y., S.J. Lin, W.G. Zhao, and Z.B. Wu. 2018. "Cabergoline versus Bromocriptine for the Treatment of Giant Prolactinomas: A Quantitative and Systematic Review." *Metabolic Brain Disease*.

Hyman, S.E., R.C. Malenka, and E.J. Nestler. 2006. "Neural Mechanisms of Addiction: The Role of Reward-Related Learning and Memory." *Annual Review of Neuroscience*.

Iversen, S.D. and L.L. Iversen. 2007. "Dopamine: 50 Years in Perspective." *Trends in Neurosciences*.

Jackson, D.M. and A. Westlind-Danielsson. 1994. "Dopamine Receptors: Molecular Biology, Biochemistry and Behavioural Aspects." *Pharmacology and Therapeutics*.

Joseph, J. D., Y.M. Wang, P.R. Miles, E.A. Budygin, R. Picetti, R.R. Gainetdinov, et al. 2002. "Dopamine Autoreceptor Regulation of Release and Uptake in Mouse Brain Slices in the Absence of D3 Receptors." *Neuroscience*.

Joutsa, J., J. Johansson, S. Niemelä, A. Ollikainen, M.M. Hirvonen, P. Piepponen, et al. 2012. "Mesolimbic Dopamine Release is Linked to Symptom Severity in Pathological Gambling." *NeuroImage*.

Jules, J., S. Yang, W. Chen, and Y.P. Li. 2015. "Role of Regulators of G Protein Signaling Proteins in Bone Physiology and Pathophysiology." *Progress in Molecular Biology and Translational Science*.

Kabbani, N., L. Negyessy, R. Lin, P. Goldman-Rakic, and R. Levenson. 2002. "Interaction with Neuronal Calcium Sensor NCS-1 Mediates Desensitization of the D2 Dopamine Receptor." *Journal of Neuroscience*.

Kebabian, J.W. and D.B. Calne. 1979. "Multiple Receptors for Dopamine." *Nature*.

Kim, S.H. and T.A. Ryan. 2010. "CDK5 Serves as a Major Control Point in Neurotransmitter Release." *Neuron*.

Kisilevsky, A.E., S.J. Mulligan, C. Altier, M.C. Iftinca, D. Varela, C. Tai, et al. 2008. "D1 Receptors Physically Interact with N-Type Calcium Channels to Regulate Channel Distribution and Dendritic Calcium Entry." *Neuron*.

Kisilevsky, A.E. and G.W. Zamponi. 2008. "D2 Dopamine Receptors Interact Directly with N-Type Calcium Channels and Regulate Channel Surface Expression Levels." *Channels*.

Klein, M.O., D.S. Battagello, A.R. Cardoso, D.N. Hauser, J.C. Bittencourt, and R.G. Correa. 2019. "Dopamine: Functions, Signaling, and Association with Neurological Diseases." *Cellular and Molecular Neurobiology*.

Koob, G.F. and N.D. Volkow. 2010. "Neurocircuitry of Addiction." *Neuropsychopharmacology*.

Kovac, A.L. 2000. "Prevention and Treatment of Postoperative Nausea and Vomiting." *Drugs*.

Krasel, C. and M.J. Lohse. 2007. "G Protein Coupled Receptor Kinases." In: *xPharm: The Comprehensive Pharmacology Reference* 1–19.

Lan, H., Y. Liu, M.I. Bell, V.V. Gurevich, and K.A. Neve. 2009. "A Dopam/Ine D 2 Receptor Mutant Capable of G Protein-Mediated Signaling but Deficient in Arrestin Binding." *Molecular Pharmacology*.

Lan, H., M.M. Teeter, V.V. Gurevich, and K.A. Neve. 2009. "An Intracellular Loop 2 Amino Acid Residue Determines Differential Binding of Arrestin to the Dopamine D 2 and D 3 Receptors." *Molecular Pharmacology*.

Laporte, S.A., W.E. Miller, K.M. Kim, and M.G. Caron. 2002. "β-Arrestin/AP-2 Interaction in G Protein-Coupled Receptor Internalization. Identification of a β-Arrestin Binding Site B2-Adaptin." *Journal of Biological Chemistry*.

Lavine, N., N. Ethier, J.N. Oak, L. Pei, F. Liu, P. Trieu, et al. 2002. "G Protein-Coupled Receptors Form Stable Complexes with Inwardly Rectifying Potassium Channels and Adenylyl Cyclase." *Journal of Biological Chemistry*.

Lee, S.P., C.H. So, A.J. Rashid, G. Varghese, R. Cheng, A. José Lança, et al. 2004. "Dopamine D1 and D2 Receptor Co-Activation Generates a Novel Phospholipase C-Mediated Calcium Signal." *Journal of Biological Chemistry*.

Leggio, G.M., C. Bucolo, C.B.M. Platania, S. Salomone, and F. Drago. 2016. "Current Drug Treatments Targeting Dopamine D3 Receptor." *Pharmacology and Therapeutics*.

Li, M., H. Xu, G. Chen, S. Sun, Q. Wang, B. Liu, et al. 2019. "Impaired D2 Receptor-Dependent Dopaminergic Transmission in Prefrontal Cortex of Awake Mouse Model of Parkinson's Disease." *Brain*.

Lindgren, N., A. Usiello, M. Goiny, J. Haycock, E. Erbs, P. Greengard, et al. 2003. "Distinct Roles of Dopamine D2L and D2S Receptor Isoforms in the Regulation of Protein Phosphorylation at Presynaptic and Postsynaptic Sites." *Proceedings of the National Academy of Sciences of the United States of America*.

Lisman, J. 2017. "Glutamatergic Synapses are Structurally and Biochemically Complex Because of Multiple Plasticity Processes: Long-Term Potentiation, Long-Term Depression, Short-Term Potentiation and Scaling." *Philosophical Transactions of the Royal Society B: Biological Sciences*.

Liu, F., Q. Wan, Z.B. Pristupa, X.M. Yu, Y.T. Wang, and H.B. Niznik. 2000. "Direct Protein-Protein Coupling Enables Cross-Talk Between Dopamine D5 and γ-Aminobutyric Acid A Receptors." *Nature*.

Liu, X.Y., X.P. Chu, L.M. Mao, M. Wang, H.X. Lan, M.H. Li, et al. 2006. "Modulation of D2R–NR2B Interactions in Response to Cocaine." *Neuron*.

Lohse, M.J., J.L. Benovic, J. Codina, M.G. Caron, and R.J. Lefkowitz. 1990. "β-Arrestin: A Protein that Regulates β-Adrenergic Receptor Function." *Science*.

Loy, J.H., S.N. Merry, S.E. Hetrick, and K. Stasiak. 2017. "Atypical Antipsychotics for Disruptive Behaviour Disorders in Children and Youths." *Cochrane Database of Systematic Reviews*.

Lüscher, C. and P.A. Slesinger. 2010. "Emerging Roles for G Protein-Gated Inwardly Rectifying Potassium (GIRK) Channels in Health and Disease." *Nature Reviews Neuroscience*.

Luttrell, L.M., S.S.G. Ferguson, Y. Daaka, W.E. Miller, S. Maudsley, G.J. Della Rocca, et al. 1999. "β-Arrestin-Dependent Formation of B2 Adrenergic Receptor-Src Protein Kinase Complexes." *Science*.

McCudden, C.R., M.D. Hains, R.J. Kimple, D.P. Siderovski, and F.S. Willard. 2005. "G-Protein Signaling: Back to the Future." *Cellular and Molecular Life Sciences*.

Maiese, K. 2016. "Targeting Molecules to Medicine with MTOR, Autophagy and Neurodegenerative Disorders." *British Journal of Clinical Pharmacology*.

Manning, B.D. and A. Toker. 2017. "AKT/PKB Signaling: Navigating the Network." *Cell*.

Martemyanov, K.A. and V.Y. Arshavsky. 2009. "Biology and Functions of the RGS9 Isoforms." *Progress in Molecular Biology and Translational Science*.

Meiser, J., D. Weindl, and K. Hiller. 2013. "Complexity of Dopamine Metabolism." *Cell Communication and Signaling*.

Millan, M.J. 2010. "From the Cell to the Clinic: A Comparative Review of the Partial D2/D3 Receptor Agonist and A2-Adrenoceptor Antagonist, Piribedil, in the Treatment of Parkinson's Disease." *Pharmacology and Therapeutics*.

Miller, C.A. and J.F. Marshall. 2005. "Molecular Substrates for Retrieval and Reconsolidation of Cocaine-Associated Contextual Memory." *Neuron*.

Missale, C., S. Russel Nash, S.W. Robinson, M. Jaber, and M.G. Caron. 1998. "Dopamine Receptors: From Structure to Function." *Physiological Reviews*.

Mishra, A., S. Singh, and S. Shukla. 2018. "Physiological and Functional Basis of Dopamine Receptors and their Role in Neurogenesis: Possible Implication for Parkinson's Disease." *Journal of Experimental Neuroscience* 12, 1179069518779829. DOI: 10.1177/1179069518779829, PMID: 29899667; PMCID:PMC5985548.

Money, K.M. and G.D. Stanwood. 2013. "Developmental Origins of Brain Disorders: Roles for Dopamine." *Frontiers in Cellular Neuroscience*.

Monsma, F.J., L.D. McVittie, C.R. Gerfen, L.C. Mahan, and D.R. Sibley. 1989. "Multiple D2 Dopamine Receptors Produced by Alternative RNA Splicing." *Nature*.

Moritz, A.E., R. Benjamin Free, and D.R. Sibley. 2018. "Advances and Challenges in the Search for D2 and D3 Dopamine Receptor-Selective Compounds." *Cellular Signalling*.

Nai, Q., S. Li, S.H. Wang, J. Liu, F.J.S. Lee, P.W. Frankland, and F. Liu. 2010. "Uncoupling the D1-*N*-Methyl-D-Aspartate (NMDA) Receptor Complex Promotes NMDA-Dependent Long-Term Potentiation and Working Memory." *Biological Psychiatry*.

Namkung, Y., C. Dipace, J.A. Javitch, and D.R. Sibley. 2009. "G Protein-Coupled Receptor Kinase-Mediated Phosphorylation Regulates Post-Endocytic Trafficking of the D2 Dopamine Receptor." *Journal of Biological Chemistry*.

Namkung, Y. and D.R. Sibley. 2004. "Protein Kinase C Mediates Phosphorylation, Desensitization, and Trafficking of the D2 Dopamine Receptor." *Journal of Biological Chemistry*.

Neve, K.A., J.K. Seamans, and H. Trantham-Davidson. 2004. "Dopamine Receptor Signaling." *Journal of Receptors and Signal Transduction*.

Neve, K. and R.L. Neve. 1997. The Dopamine Receptors. Toronto, NJ: Humana Press J Receptors Signal Transduction, 24:3, 165-205, DOI: 10.1081/RRS-200029981.

Nikolaus, S., C. Antke, and H.W. Müller. 2009a. "In Vivo Imaging of Synaptic Function in the Central Nervous System: II. Mental and Affective Disorders." *Behavioural Brain Research*.

Nikolaus, S., C. Antke, and H.W. Müller. 2009b. "In Vivo Imaging of Synaptic Function in the Central Nervous System. I. Movement Disorders and Dementia." *Behavioural Brain Research*.

Nishi, A. and T. Shuto. 2017. "Potential for Targeting Dopamine/DARPP-32 Signaling in Neuropsychiatric and Neurodegenerative Disorders." *Expert Opinion on Therapeutic Targets*.

Nobles, M., A. Benians, and A. Tinker. 2005. "Heterotrimeric G Proteins Precouple with G Protein-Coupled Receptors in Living Cells." *Proceedings of the National Academy of Sciences of the United States of America*.

Oak, J.N., J. Oldenhof, and H.H.M. Van Tol. 2000. "The Dopamine D4 Receptor: One Decade of Research." *European Journal of Pharmacology*.

Oakley, R.H., S.A. Laporte, J.A. Holt, L.S. Barak, and M.G. Caron. 2001. "Molecular Determinants Underlying the Formation of Stable Intracellular G Protein-Coupled Receptor-β-Arrestin Complexes after Receptor Endocytosis." *Journal of Biological Chemistry*.

O'Hayre, M., M.S. Degese, and J.S. Gutkind. 2014. "Novel Insights into G Protein and G Protein-Coupled Receptor Signaling in Cancer." *Current Opinion in Cell Biology*.

Oyanagi, C., Y. Katsumi, T. Hanakawa, T. Hayashi, D.H.D. Thuy, K. Hashikawa, et al. 2002. "Comparison of Striatal Dopamine D2 Receptors in Parkinson's Disease and Progressive Supranuclear Palsy Patients Using [123I] Iodobenzofuran Single-Photon Emission Computed Tomography." *Journal of Neuroimaging*.

Pei, L., F.J.S. Lee, A. Moszczynska, B. Vukusic, and F. Liu. 2004. "Regulation of Dopamine D1 Receptor Function by Physical Interaction with the NMDA Receptors." *Journal of Neuroscience*.

Perreault, M.L., A. Hasbi, M. Alijaniaram, T. Fan, G. Varghese, P.J. Fletcher, et al. 2010. "The Dopamine D1-D2 Receptor Heteromer Localizes in Dynorphin/Enkephalin Neurons: Increased High Affinity State Following Amphetamine in Schizophrenia." *Journal of Biological Chemistry*.

Pivonello, R., D. Ferone, G. Lombardi, A. Colao, S.W.J. Lamberts, and L.J. Hofland. 2007. "Novel Insights in Dopamine Receptor Physiology." *European Journal of Endocrinology* (Supplement).

Podda, M.V., E. Riccardi, M. D'Ascenzo, G.B. Azzena, and C. Grassi. 2010. "Dopamine D1-like Receptor Activation Depolarizes Medium Spiny Neurons of the Mouse Nucleus Accumbens by Inhibiting Inwardly Rectifying K^+ Currents through a CAMP-Dependent Protein Kinase A-Independent Mechanism." *Neuroscience*.

Poewe, W., K. Seppi, C.M. Tanner, G.M. Halliday, P. Brundin, J.Volkmann, A.E. Schrag, and A.E. Lang. 2017. "Parkinson Disease." *Nature Reviews Disease Primers*.

Premont, R.T. and R.R. Gainetdinov. 2007. "Physiological Roles of G Protein-Coupled Receptor Kinases and Arrestins." *Annual Review of Physiology*.

Raehal, K.M., C.L. Schmid, I.O. Medvedev, R.R. Gainetdinov, R.T. Premont, and L.M. Bohn. 2009. "Morphine-Induced Physiological and Behavioral Responses in Mice Lacking G Protein-Coupled Receptor Kinase 6." *Drug and Alcohol Dependence*.

Rafiuddin Ahmed M., E. Bychkov, S. Kook, L. Zurkovsky, K.N. Dalby, and E.V. Gurevich. 2015. "Overexpression of GRK6 Rescues L-DOPA-Induced Signaling Abnormalities in the Dopamine-Depleted Striatum of Hemiparkinsonian Rats." *Experimental Neurology*.

Rajagopal, S., K. Rajagopal, and R.J. Lefkowitz. 2010. "Teaching Old Receptors New Tricks: Biasing Seven-Transmembrane Receptors." *Nature Reviews Drug Discovery*.

Rangel-Barajas, C., I. Coronel, and B. Florán. 2015. "Dopamine Receptors and Neurodegeneration." *Aging and Disease*.

Rankin, M.L., L.A. Hazelwood, R. Benjamin Free, Y. Namkung, E.B. Rex, R.A. Roof, and D.R. Sibley. 2010. "Molecular Pharmacology of the Dopamine Receptors." In: *Dopamine Handbook* 63–87, DOI: https://doi.org/10.1093/acprof:oso/9780195373035.003.0006.

Reusch, J.E.B. and J.A.E. Manson. 2017. "Management of Type 2 Diabetes in 2017 Getting to Goal." *JAMA*.

Rex, E.B., M.L. Rankin, M.A. Ariano, and D.R. Sibley. 2008. "Ethanol Regulation of D1 Dopamine Receptor Signaling is Mediated by Protein Kinase C in an Isozyme-Specific Manner." *Neuropsychopharmacology*.

Rex, E.B., M.L. Rankin, Y. Yang, Q. Lu, C.R. Gerfen, P.A. Jose, and D.R. Sibley. 2010. "Identification of RanBP 9/10 as Interacting Partners for Protein Kinase C (PKC) γ/δ and the D1 Dopamine Receptor: Regulation of PKC-Mediated Receptor Phosphorylation." *Molecular Pharmacology*.

Rinne, J.O., A. Laihinen, U.K. Rinne, K. Någren, J. Bergman, and U. Ruotsalainen. 1993. "PET Study on Striatal Dopamine D2 Receptor Changes During the Progression of Early Parkinson's Disease." *Movement Disorders*.

Robinson, P., M. Lebel, and M. Cyr. 2008. "Dopamine D1 Receptor-Mediated Aggregation of *N*-Terminal Fragments of Mutant Huntingtin and Cell Death in a Neuroblastoma Cell Line." *Neuroscience*.

Rondou, P., G. Haegeman, and K. Van Craenenbroeck. 2010. "The Dopamine D4 Receptor: Biochemical and Signalling Properties." *Cellular and Molecular Life Sciences: CMLS*.

Ross, E.M. and T.M. Wilkie. 2000. "GTPase-Activating Proteins for Heterotrimeric G Proteins: Regulators of G Protein Signaling (RGS) and RGS-Like Proteins." *Annual Review of Biochemistry*.

Roth, B.L., D.J. Sheffer, and W.K. Kroeze. 2004. "Magic Shotguns versus Magic Bullets: Selectively Non-Selective Drugs for Mood Disorders and Schizophrenia." *Nature Reviews Drug Discovery*.

Sahu, A., K.R. Tyeryar, H.O. Vongtau, D.R. Sibley, and A.S. Undieh. 2009. "D 5 Dopamine Receptors are Required for Dopaminergic Activation of Phospholipase C." *Molecular Pharmacology*.

Salamone, J.D. and M. Correa. 2012. "The Mysterious Motivational Functions of Mesolimbic Dopamine." *Neuron*.

Sanada, H., M. Yoneda, J. Yatabe, S.M. Williams, J. Bartlett, M.J. White, et al. 2016. "Common Variants of the G Protein-Coupled Receptor Type 4 are Associated with Human Essential Hypertension and Predict the Blood Pressure Response to Angiotensin Receptor Blockade." *Pharmacogenomics Journal*.

Sang, K.P., D.N. Minh, A. Fischer, M.P.S. Luke, B. El Affar, P.B. Dieffenbach, et al. 2005. "Par-4 Links Dopamine Signaling and Depression." *Cell*.

Santini, E., M. Heiman, P. Greengard, E. Valjent, and G. Fisone. 2009. "Inhibition of MTOR Signaling in Parkinson's Disease Prevents L-DOPA-Induced Dyskinesia." *Science Signaling*.

Schoots, O. and H.H.M. Van Tol. 2003. "The Human Dopamine D4 Receptor Repeat Sequences Modulate Expression." *Pharmacogenomics Journal*.

Schultz, W. 2007. "Multiple Dopamine Functions at Different Time Courses." *Annual Review of Neuroscience*.

Schultz, W. 2019. "Recent Advances in Understanding the Role of Phasic Dopamine Activity." *F1000Research*.

Seeman, P., T. Lee, M. Chau-Wong, and K. Wong. 1976. "Antipsychotic Drug Doses and Neuroleptic/Dopamine Receptors." *Nature*.

Segura-Aguilar, J., I. Paris, P. Muñoz, E. Ferrari, L. Zecca, and F.A. Zucca. 2014. "Protective and Toxic Roles of Dopamine in Parkinson's Disease." *Journal of Neurochemistry*.

Serafino, A., G. Sferrazza, A.C. Baldeschi, G. Nicotera, F. Andreola, E. Pittaluga, and P. Pierimarchi. 2017. "Developing Drugs that Target the Wnt Pathway: Recent Approaches in Cancer and Neurodegenerative Diseases." *Expert Opinion on Drug Discovery*.

Shenoy, S.K. and R.J. Lefkowitz. 2003. "Multifaceted Roles of β-Arrestins in the Regulation of Seven-Membrane-Spanning Receptor Trafficking and Signalling." *Biochemical Journal*.

Shuen, J.A., M. Chen, B. Gloss, and N. Calakos. 2008. "Drd1a-TdTomato BAC Transgenic Mice for Simultaneous Visualization of Medium Spiny Neurons in the Direct and Indirect Pathways of the Basal Ganglia." *Journal of Neuroscience*.

Shukla, A.K., A. Manglik, A.C. Kruse, K. Xiao, R.I. Reis, W.C. Tseng, et al. 2013. "Structure of Active β-Arrestin-1 Bound to a G-Protein-Coupled Receptor Phosphopeptide." *Nature*.

Sibley, D.R. 1999. "New Insights into Dopaminergic Receptor Function Using Antisense and Genetically Altered Animals." *Annual Review of Pharmacology and Toxicology*.

Simonsen, U., S. Comerma-Steffensen, and K.E. Andersson. 2016. "Modulation of Dopaminergic Pathways to Treat Erectile Dysfunction." *Basic and Clinical Pharmacology and Toxicology*.

Singh, R., Y. Bansal, B. Medhi, and A. Kuhad. 2019. "Antipsychotics-Induced Metabolic Alterations: Recounting the Mechanistic Insights, Therapeutic Targets and Pharmacological Alternatives." *European Journal of Pharmacology*.

Sokoloff, P., J. Diaz, B. Le Foll, O. Guillin, L. Leriche, E. Bezard, and C. Gross. 2006. "The Dopamine D3 Receptor: A Therapeutic Target for the Treatment of Neuropsychiatric Disorders." *CNS & Neurological Disorders Drug Targets*.

Sokoloff, P., B. Giros, M.P. Martres, M. Andrieux, R. Besancon, C. Pilon, et al. 1992a. "Localization and Function of the D3 Dopamine Receptor." in *Arzneimittel-Forschung/Drug Research*.

Sokoloff, P., M. Andrieux, R. Besançon, C. Pilon, M.P. Martres, B. Giros, and J. Charles Schwartz. 1992b. "Pharmacology of Human Dopamine D3 Receptor Expressed in a Mammalian Cell Line: Comparison with D2 Receptor." *European Journal of Pharmacology: Molecular Pharmacology*.

Spano, P.F., S. Govoni, and M. Trabucchi. 1978. "Studies on the Pharmacological Properties of Dopamine Receptors in Various Areas of the Central Nervous System." *Advances in Biochemical Psychopharmacology*.

Spooren, A., P. Rondou, K. Debowska, B. Lintermans, L. Vermeulen, B. Samyn, et al. 2010. "Resistance of the Dopamine D4 Receptor to Agonist-Induced Internalization and Degradation." *Cellular Signalling*.

Stipanovich, A., E. Valjent, M. Matamales, A. Nishi, J.H. Ahn, M. Maroteaux, et al. 2008. "A Phosphatase Cascade by Which Rewarding Stimuli Control Nucleosomal Response." *Nature*.

Stoof, J.C. and J.W. Kebabian. 1984. "Two Dopamine Receptors: Biochemistry, Physiology and Pharmacology." *Life Sciences*.

Sung, Y.K., C.C. Kyou, S.C. Min, H.K. Myoung, Y.K. Sa, Y.S. Na, et al. 2006. "The Dopamine D2 Receptor Regulates the Development of Dopaminergic Neurons via Extracellular Signal-Regulated Kinase and Nurr1 Activation." *Journal of Neuroscience*.

Svenningsson, P., A. Nishi, G. Fisone, J.-A. Girault, A.C. Nairn, and P. Greengard. 2004. "DARPP-32: An Integrator of Neurotransmission." *Annual Review of Pharmacology and Toxicology*.

Swanson, J.M., M. Kinsbourne, J. Nigg, B. Lanphear, G.A. Stefanatos, N. Volkow, et al. 2007. "Etiologic Subtypes of Attention-Deficit/Hyperactivity Disorder: Brain Imaging, Molecular Genetic and Environmental Factors and the Dopamine Hypothesis." *Neuropsychology Review*.

Szmulewicz, A.G., F. Angriman, C. Samamé, A. Ferraris, D. Vigo, and S.A. Strejilevich. 2017. "Dopaminergic Agents in the Treatment of Bipolar Depression: A Systematic Review and Meta-Analysis." *Acta Psychiatrica Scandinavica*.

Taipale, H., E.Mittendorfer-Rutz, K. Alexanderson, M. Majak, J. Mehtälä, F. Hoti, et al. 2018. "Antipsychotics and Mortality in a Nationwide Cohort of 29,823 Patients with Schizophrenia." *Schizophrenia Research*.

Van Tol, H.H.M., J.R. Bunzow, H.C. Guan, R.K. Sunahara, P. Seeman, H.B. Niznik, and O. Civelli. 1991. "Cloning of the Gene for a Human Dopamine D4 Receptor with High Affinity for the Antipsychotic Clozapine." *Nature*.

Trantham-Davidson, H., L.C. Neely, A. Lavin, and J.K. Seamans. 2004. "Mechanisms Underlying Differential D1 versus D2 Dopamine Receptor Regulation of Inhibition in Prefrontal Cortex." *Journal of Neuroscience*.

Traynor, J.R., D. Terzi, B.J. Caldarone, and V. Zachariou. 2009. "RGS9-2: Probing an Intracellular Modulator of Behavior as a Drug Target." *Trends in Pharmacological Sciences*.

Trepanier, C.H., M.F. Jackson, and J.F. MacDonald. 2012. "Regulation of NMDA Receptors by the Tyrosine Kinase Fyn." *FEBS Journal*.

Urs, N.M., T.L. Daigle, and M.G. Caron. 2011. "A Dopamine D1 Receptor-Dependent B-Arrestin Signaling Complex Potentially Regulates Morphine-Induced Psychomotor Activation but not Reward in Mice." *Neuropsychopharmacology*.

Urs, N.M., S.M. Peterson, and M.G. Caron. 2017. "New Concepts in Dopamine D2 Receptor Biased Signaling and Implications for Schizophrenia Therapy." *Biological Psychiatry*.

Usiello, A., J.H. Baik, F. Rougé-Pont, R. Picetti, A. Dierich, M. LeMeur, et al. 2000. "Distinct Functions of the Two Isoforms of Dopamine D2 Receptors." *Nature*.

Valjent, E., J.C. Corvol, C. Pages, M.J. Besson, R. Maldonado, and J. Caboche. 2000. "Involvement of the Extracellular Signal-Regulated Kinase Cascade for Cocaine-Rewarding Properties." *Journal of Neuroscience*.

Valjent, E., J. Bertran-Gonzalez, D. Hervé, G. Fisone, and J.A. Girault. 2009. "Looking BAC at Striatal Signaling: Cell-Specific Analysis in New Transgenic Mice." *Trends in Neurosciences*.

Valjent, E., V. Pascoli, P. Svenningsson, S. Paul, H. Enslen, J.C. Corvol, et al. 2005. "Regulation of a Protein Phosphatase Cascade Allows Convergent Dopamine and Glutamate Signals to Activate ERK in the Striatum." *Proceedings of the National Academy of Sciences of the United States of America*.

Vallone, D., R. Picetti, and E. Borrelli. 2000. "Structure and Function of Dopamine Receptors." *Neuroscience and Biobehavioral Reviews*.

Van Villar, A.M., J.E. Jones, I. Armando, C. Palmes-Saloma, P. Yu, A.M. Pascua, et al. 2009. "G Protein-Coupled Receptor Kinase 4 (GRK4) Regulates the Phosphorylation and Function of the Dopamine D3 Receptor." *Journal of Biological Chemistry*.

Volkow, N.D., J.S. Fowler, G.J. Wang, R. Baler, and F. Telang. 2009. "Imaging Dopamine's Role in Drug Abuse and Addiction." *Neuropharmacology*.

Walaas, S.I., H.C. Hemmings, P. Greengard, and A.C. Nairn. 2011. "Beyond the Dopamine Receptor: Regulation and Roles of Serine/Threonine Protein Phosphatases." *Frontiers in Neuroanatomy*.

Wang, C., D.C. Buck, R. Yang, T.A. Macey, and K.A. Neve. 2005. "Dopamine D2 Receptor Stimulation of Mitogen-Activated Protein Kinases Mediated by Cell Type-Dependent Transactivation of Receptor Tyrosine Kinases." *Journal of Neurochemistry*.

Wang, H., S.S. Kim, and M. Zhuo. 2010. "Roles of Fragile X Mental Retardation Protein in Dopaminergic Stimulation-Induced Synapse-Associated Protein Synthesis and Subsequent α-Amino-3-Hydroxyl-5-Methyl-4-Isoxazole-4-Propionate (AMPA) Receptor Internalization." *Journal of Biological Chemistry*.

Watari, K., M. Nakaya, and H. Kurose. 2014. "Multiple Functions of G Protein-Coupled Receptor Kinases." *Journal of Molecular Signaling*.

Waters, N., K. Svensson, S.R. Haadsma-Svensson, M.W. Smith, and A. Carlsson. 1993. "The Dopamine D3-Receptor: A Postsynaptic Receptor Inhibitory on Rat Locomotor Activity." *Journal of Neural Transmission*.

Weinstein, J. J., E. Van De Giessen, R. J. Rosengard, X. Xu, N. Ojeil, G. Brucato, et al. 2018. "PET Imaging of Dopamine-D2 Receptor Internalization in Schizophrenia." *Molecular Psychiatry*.

White, F.J., L.M. Bednarz, S.R. Wachtel, S. Hjorth, and R.J. Brooderson. 1988. "Is Stimulation of Both D1 and D2 Receptors Necessary for the Expression of Dopamine-Mediated Behaviors?" *Pharmacology, Biochemistry and Behavior*.

Willner, P.. 1997. "The Mesolimbic Dopamine System as a Target for Rapid Antidepressant Action." *International Clinical Psychopharmacology*.

Witkovsky, P. 2004. "Dopamine and Retinal Function." *Documenta Ophthalmologica*.

Wolf, M.E. and R.H. Roth. 1990. "Autoreceptor Regulation of Dopamine Synthesis." *Annals of the New York Academy of Sciences*.

Wong, A.H.C. and H.H.M. Van Tol. 2003. "The Dopamine D4 Receptors and Mechanisms of Antipsychotic Atypicality." *Progress in Neuro-Psychopharmacology and Biological Psychiatry*.

Woolfrey, K.M., D.P. Srivastava, H. Photowala, M. Yamashita, M.V. Barbolina, M.E. Cahill, et al. 2009. "EPAC2 Induces Synapse Remodeling and Depression and its Disease-Associated Forms Alter Spines." *Nature Neuroscience*.

Xu, T.X., T.D. Sotnikova, C. Liang, J. Zhang, J.U. Jung, R.D. Spealman, et al. 2009. "Hyperdopaminergic Tone Erodes Prefrontal Long-Term Potential via a D 2 Receptor-Operated Protein Phosphatase Gate." *Journal of Neuroscience*.

Xu, T.X. and W.D. Yao. 2010. "D1 and D2 Dopamine Receptors in Separate Circuits Cooperate to Drive Associative Long-Term Potentiation in the Prefrontal Cortex." *Proceedings of the National Academy of Sciences of the United States of America*.

Yan, Z., W.J. Song, and D.J. Surmeier. 1997. "D2 Dopamine Receptors Reduce *N*-Type Ca^{2+} Currents in Rat Neostriatal Cholinergic Interneurons Through a Membrane-Delimited, Protein-Kinase-C-Insensitive Pathway." *Journal of Neurophysiology* 77(2):1003-15. DOI: 10.1152/jn.1997.77.2.1003.

Yu, Z., W. Asaad, A. Nurmikko, and I. Ozden. 2018. "Optogenetics-Based Neuromodulation for the Treatment of Parkinson's Disease." In: 2017 International Conference on Intelligent Informatics and Biomedical Sciences (ICIIBMS).

Yujnovsky, I., J. Hirayama, M. Doi, E. Borrelli, and P. Sassone-Corsi. 2006. "Signaling Mediated by the Dopamine D2 Receptor Potentiates Circadian Regulation by CLOCK:BMAL1." *Proceedings of the National Academy of Sciences of the United States of America.*

Zapata, A. and T.S. Shippenberg. 2002. "D3 Receptor Ligands Modulate Extracellular Dopamine Clearance in the Nucleus Accumbens." *Journal of Neurochemistry.*

Zhang, J., A. Vinuela, M.H. Neely, P.J. Hallett, S.G.N. Grant, G.M. Miller, et al. 2007. "Inhibition of the Dopamine D1 Receptor Signaling by PSD-95." *Journal of Biological Chemistry.*

Zhang, J., T.X. Xu, P.J. Hallett, M. Watanabe, S.G.N. Grant, O. Isacson, and W.D. Yao. 2009. "PSD-95 Uncouples Dopamine-Glutamate Interaction in the D 1/PSD-95/NMDA Receptor Complex." *Journal of Neuroscience.*

Zhang, L., D. Lou, H. Jiao, D. Zhang, X. Wang, Y. Xia, et al. 2004. "Cocaine-Induced Intracellular Signaling and Gene Expression are Oppositely Regulated by the Dopamine D1 and D3 Receptors." *Journal of Neuroscience.*

Zhang, Z.W., M.W. Burke, N. Calakos, J.M. Beaulieu, and E. Vaucher. 2010. "Confocal Analysis of Cholinergic and Dopaminergic Inputs onto Pyramidal Cells in the Prefrontal Cortex of Rodents." *Frontiers in Neuroanatomy.*

Zintzaras, E., G.D. Kitsios, A.A. Papathanasiou, S. Konitsiotis, M. Miligkos, P. Rodopoulou, and G.M. Hadjigeorgiou. 2010. "Randomized Trials of Dopamine Agonists in Restless Legs Syndrome: A Systematic Review, Quality Assessment, and Meta-Analysis." *Clinical Therapeutics.*

Section II

Neurochemical Signaling and Pathologies

4 Neurochemical Signaling in Alzheimer's Disease

Mamta F. Singh
Sardar Bhagwan Singh University, Dehradun, Uttarakhand, India

Aman Upaganlawar
SNJB's Shriman Suresh Dada Jain College of Pharmacy, Chandwad, Nashik, India

Vinod Singh
Prabha Harjilal College of Pharmacy and Paraclinical Sciences, Jammu, J & K, India

Shradha Bisht
Amity Institute of Pharmacy, Amity University, Lucknow, India

Manoj K. Sarangi
Sardar Bhagwan Singh University, Dehradun, Uttarakhand, India

CONTENTS

4.1 INTRODUCTION

Dementia is a condition which is defined by gradual and age-related deterioration in at least two or more cognitive domains that include language, memory, personality, decision-making, and visuospatial functions and behavior that causes the loss of the ability of a person to execute basic routine activities of life. Among various forms of dementia, Alzheimer's disease (AD) is the most common and fatal escalating, degenerative disease, characterized by brain shrinkage and neuronal loss

DOI: 10.1201/9780429265198-6

specifically at the basal forebrain and hippocampus. German psychiatrists and neuropathologists discovered AD for the first time in 1906. Beta-amyloid (Aβ) plaque formation and the accumulation of neurofibrillary tangles (NFTs) are considered to be the prime pathologic indications of the disease (De-Paula et al., 2012). Enormous clinical and preclinical histopathologic studies on several models of AD have revealed the existence of NFTs and Aβ plaques in the basal forebrain, hippocampus, cerebral cortex, and frontal lobe (Armstrong, 2014). Definite genetic polymorphism and increasing age are familiar risk factors for AD. Short-term memory impairment is the first clinical sign, whereas the improvement of distant memory is preserved well into the pathway of the disease. With progression of the disease, other cognitive abilities are also damaged. Motor weakness, followed by muscular contractures, are features of advanced phases of AD. The prevalence of AD rises with age, at around 5% at the age of 65 years and >90% at 95 years of age. As per the World Alzheimer's Report 2015, 46.85 million people were estimated to be affected by AD, the number is likely to double by 2030 and by 2050 it will be three times the current number (Daulatzai, 2016; Ferri et al., 1984). Among all patients with dementia, AD comprises 60–80% of cases and leads to unavoidable death (Plassman et al., 2007).

Acetylcholine, a neurotransmitter synthesized in cholinergic neurons, plays a pivotal role in the process of learning and memory. Loss of cholinergic neurons in the hippocampal and cortical regions is associated with AD (Arendt et al., 2015). Apart from this, the functions of dopaminergic, serotonergic, adrenergic, and glutaminergic neurons are also altered in AD, and neuromolecules such as acetylcholine, dopamine, gamma-amino butyric acid (GABA), glutamate, and serotonin play a major role in the progression of AD (Danysz and Parsons, 2012). Aβ oligomers interact with neurotransmitter systems and, in turn, induce imbalance in neurotransmitter signaling, cell dysfunction, and various neurologic abnormalities leading to the death of neurons (Danysz and Parsons, 2012).

Several studies support the fact that aggregation of Aβ oligomers is closely related to synaptic damage, the formation of NFTs, and cholinergic deficit, which are the fundamentals of AD pathogenesis (Crews and Masliah, 2010). A complex interplay exists between the various proposed mechanisms and signaling pathways in the brain of patients with AD. Protein misfolding and aggregation, the generation of reactive oxygen/nitrogen free radicals, calcium dysregulation, mitochondrial dysfunction, impaired bioenergetics, neuroinflammatory processes, and disturbances of cellular and axonal transport are interlinked in a way that leads to neuronal dysfunction and death in AD. Abnormalities in various signaling pathways, such as glycogen synthase kinase-3 beta (GSK-3β), Fyn kinases, CDK5, mammalian target of rapamycin (mTOR), and 5′ adenosine monophosphate kinase (AMPK), may promote Aβ aggregation and tau phosphorylation, the activation of apoptotic pathways, and cytoskeletal abnormalities leading to synaptic failure (Kablar, 2019).

The aim of this chapter is to give an overview of the neurobiology and neurotransmission in AD, and to discuss various signaling pathways associated with the progression of AD, along with their recent developments and challenges.

4.2 GENETICS OF ALZHEIMER'S DISEASE

AD has a complex genetic basis. Genetic studies indicate that genetic mutations in amyloid precursor protein gene (*APP*), presenilin 1 (*PSEN1*), and presenilin 2 (*PSEN2*), and an apolipoprotein (*APOE*) E4/4 (*ApoE 4/4*) genotype increases with the formation and aggregation of Aβ, and are responsible for familial early-onset AD (EOAD). In patients with AD, approximately 30 missense mutations in *APP* are detected (Awada, 2015; Li et al., 2019). Mutations near the site cleaved by β-secretase may increases β-secretase activity, which shifts the processing of amyloid precursor protein (APP) to the amyloidogenic pathway and subsequently increases the production of Aβ. Mutations around the site cleaved by γ-secretase increase the ratio of Aβ42/40, leading to the formation of pathogenic Aβ42 (Suárez et al., 2014). γ-Secretase is a multimeric protein complex, that cleaves membrane proteins like APP and notch. It is made up of five subunits: nicastrin; presenilin 1 (PS-1); presenilin 2 (PS-2); presenilin enhancer 2 (PEN-2); and anterior pharynx defective 1 (Yonemura et al., 2011). Chromosomes 14 and 1 contain the genes for PS-1 and PS-2, respectively. *PSEN1* encodes PS-1 and mutation in this gene is considered to be the most important reason for the occurrence of EOAD (Kelleher and Shen, 2017). It has been postulated that mutations in *PSEN1* may enhance the cleavage of APP by γ-secretase, thereby increasing the formation of Aβ42 fragment.

APOE polymorphism is a major risk factor for late-onset AD (LOAD). The gene for apolipoprotein E (ApoE) is positioned on chromosome 19 and the gene has three isoforms, alleles *ApoE ε2*, *ApoE ε3*, and *ApoE ε4*. Of these isoforms, *ApoE ε2* and *ApoE ε3* are the most familiar and harmless, but the presence of *ApoE ε4* is considered to be a predisposing factor for LOAD. Individuals bearing homozygous *ApoE ε4* have a 10-fold increased chance for developing AD versus those with other *APOE* alleles. Post-mortem brain studies have revealed that Aβ has a greater tendency to link with the product of the *ApoE ε4* allele than the products of the other alleles. This causes more involvement of Aβ with the product of *ApoE ε4* rather *ApoE ε3*. The presence of *ApoE ε4* alleles is associated with impaired clearance of Aβ from the brain and tau protein homeostasis, leading to mitochondrial dysfunction and neuroinflammation in patients with AD (Shi et al., 2017).

4.3 NEUROBIOLOGY OF ALZHEIMER'S DISEASE

AD is marked by a deficiency in cholinergic neurons in the frontal cortex and hippocampus, and atrophy of the cerebral cortex, which is speculated to be the major cause of cognitive deficit and short-term memory loss that takes place in AD (Geulaet al., 2008). Two microscopic features, namely extracellular amyloid plaques, which consists of Aβ protein aggregates and intraneuronal phosphorylated tau (a microtubule associated protein) NFTs, are the hallmark of AD. These extra- and intracellular deposits are protein aggregates that are formed due to the misfolding of APP. The abundance of Aβ plaques and NFTs in the hippocampus and other associative areas of the cerebral cortex correlate to the clinical features of the deterioration of memory.

Accumulation of Aβ and tau hyperphosphorylation play a vital role in neuroinflammation, the formation of excessive reactive oxygen species (ROS), deregulation in mitochondrial calcium homeostasis, impairment in mitochondrial energy production, synaptic loss, and neurotransmitter deficiency in patients with AD, leading to memory loss and cognitive impairment (Kumar et al., 2018).

In patients with AD, the increased generation of ROS leads to a reduced antioxidant defense. The dysfunctioning of mitochondria and altered metal homeostasis that affects synaptic activity and neurotransmission leads to cognitive impairment (Huang et al., 2016, Tönnies and Trushina, 2017). However, further studies and research have revealed that no precise relationship exists between the abundance of amyloid plaques and the intensity of cognitive defects in patients with AD, indicating that some other factors might be responsible for the pathogenesis of the disease (Terry, 2006).

On the basis of animal and human data, various hypothesis for the pathogenesis of AD have been proposed. The cholinergic hypothesis of cognitive dysfunction is the oldest one, and was accepted by most scientists. Recent studies on microglial activation and inflammatory pathways in the brains of patients with AD have supported the inflammatory hypothesis (Kandimalla and Reddy, 2017; McGeer et al., 2016). Intracellular calcium overloading is also correlated with the production of Aβ and tau hyperphosphorylation, which induces neuronal cell death through necrosis and/or apoptosis (Kametani and Hasegawa, 2018). The most widely accepted pathologic mechanisms of AD are discussed in the following subsections.

4.3.1 The Beta Amyloid Hypothesis

The presence of Aβ aggregates are constant features of AD. The amyloid cascade hypothesis came into existence after the discovery of the fact that mutations in the gene are responsible for the expression of APP led to AD in families carrying the defective gene (Selkoe and Hardy, 2016). Excessive production of Aβ associated with disturbed APP metabolism was considered to be the critical event in sporadic and familial AD, with changes such as synaptic and neuronal loss and neurochemical dysfunction (Kar et al., 2006).

Under physiologic conditions, APP is cleaved into smaller peptide fragments of 40 or 42 residues. One of such protein is Aβ, which is formed by APP cleavage by a family of proteases called α-, β-, and γ-secretase enzyme protein complexes. Under physiologic conditions, α-secretase cleaves APP and releases extracellular domain-soluble APP (sAPPα), which remains in the extracellular space (Mucke and Palop, 2010). α-Secretase activity is exhibited by members of metalloprotease family (membrane bound) that mainly includes the ADAM (a disintegrin and metalloprotease) group of enzymes within the brain. sAPPα modulates excitability of neurons; enhances synaptic plasticity, learning and memory; and improves the neuronal resistance to oxidative and metabolic stress. However, APP cleavage by β- and γ-secretase at two different points causes formation of Aβ, which is pathogenic in nature. In 1999 a group of researchers identified a protease with β-secretase activity and named it BACE1 (β-site APP cleaving enzyme 1). Unlike the ADAM family of proteases, which are expressed uniformly throughout the body, BACE1 proteases are mainly found in neurons in the brain (Bi, 2010). It is shown in Figure 4.1 that, under pathologic conditions, APP is initially cleaved by β-secretase 1 into a 99-amino acid membrane-bound fraction (C99). It is then processed and subsequently cleaved by γ-secretase into either Aβ40 or Aβ42 peptide fragments (Mokhtar et al., 2013).

Some studies indicate that the production of Aβ40 or Aβ42 is a normal phenomenon in the brain, with Aβ40 mainly produced in the brain (Wang et al., 2006). Subsequently, these Aβ peptides are degraded by the ApoE2 and ApoE3 isoform. Neprilysin (NEP) and insulin-degrading enzyme (IDE) are further responsible for the complete degradation and clearance of Aβ from the brain through the blood–brain barrier (BBB). Mutations in *APP* affect the preferred cleavage point and *PSEN* mutation leads to an increase in the activity of γ-secretase favoring the formation of Aβ42. Mutation in ApoE4, a lipid transporter protein, also predisposes to AD by facilitating the aggregation of Aβ. A disproportionate increase in the generation and decrease in clearance of Aβ may be responsible for degeneration and neuronal death. In patients with AD, the abnormal clearance of Aβ is mainly responsible for its accumulation in the brain (Wang et al., 2006). Lipoprotein receptor related protein 1 (LRP-1) and P-glycoproteins are

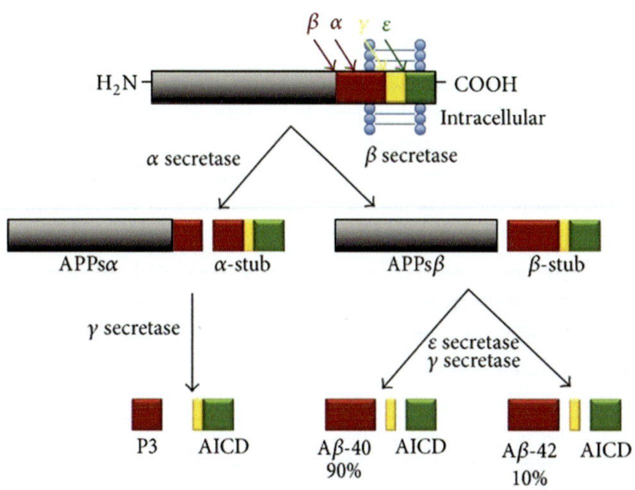

FIGURE 4.1 The processing of amyloid precursor protein (APP) through the beta-site APP cleaving enzyme 1 (BACE1), followed by presenilin 1 (PS-1). Sequential β- and γ-secretase cleavage of APP generates the synaptotoxic amyloid-β (Aβ) peptide species, Aβ1–40 and Aβ1–42 (adopted from Mokhtar et al., 2013). AICD, amyloid precursor protein intracellular domain.

responsible for the transport of Aβ peptide from the brain to the blood. In patients with AD, an increased level of Aβ leads to a decrease in the expression of LRP-1 receptors and is responsible for the aggregation of Aβ peptide in the brain. Advanced glycation end product receptors (RAGE) are responsible for the movement of Aβ peptide into various parts of brain. Decreased expression of LRP-1 receptors as a consequence of a high level of Aβ causes increased expression of RAGE. The interaction of RAGE with Aβ produces an inflammatory response in the endothelium, leading to endothelial cell apoptosis, suppression of long-term potentiation (LTP), and decreased cerebral blood flow (Origlia et al., 2008). It is known that both NEP and IDE are responsible for proteolytic cleavage and for the complete clearance of Aβ peptide from the brain. In patients with AD, decreased expression and activity of NEP and IDE results in the accumulation of Aβ peptide in the brain (Miners et al., 2009).

The formation of Aβ induces morphologic and metabolic changes in the pyramidal neurons of the neocortex and hippocampus, along with the changes in the prefrontal and cingulate cortices that are responsible for cognitive impairment. In AD, Aβ peptides may cause neural death via several mechanisms, including synaptic loss, alter energy metabolism, decreased neuronal plasticity, mitochondrial dysfunction and increased oxidative stress, disruptions in cellular calcium balance, and neuroinflammation, and presumably also increase hyperphosphorylation of tau and the formation of NFTs (Haas et al., 2012). Therefore, amyloid theory suggest that the formation, aggregation, and deposition of Aβ peptides (specifically Aβ42), is a primary event in the pathogenesis of AD and acts as a triggering factor in neurotoxicity and neurodegeneration.

4.3.2 The Tau Hypothesis

Another player in the biochemical pathway is tau protein. The presence of NFTs composed of hyperphosphorylated and aggregated tau proteins is the main pathologic hallmark of AD and other related tauopathies (Miao et al, 2019). Tau is a microtubule-associated protein that promotes the assembly of microtubules and maintains microtubule structure. In AD, the abnormal phosphorylation and intracellular deposition of tau occurs, which causes the formation of insoluble paired helical filaments (PHF). After the death of neurons, these filaments aggregate and form NFTs. Hyperphosphorylated tau interacts much less with microtubules, which increases the amount of the free form of tau protein, resulting in the increased formation and accumulation of NFTs and the malfunctioning of axonal transport (Kuret et al., 2005). Aβ plaques also enhance tau phosphorylation and, due to hyperphosphorylation, the microtubule-binding capacity of tau protein is lost resulting in cytoskeleton destabilization, which is responsible for neurodegeneration and neuronal death. In patients with AD, it has been found that the formation of NFTs starts in subcortical regions, transentorhinal areas, the entorhinal cortex, and in the hippocampus formation and neocortex. The presence of NFTs in the neocortex is correlated with the progression of AD (Braak and Del Tredici, 2011). It has recently been observed that tau hyperphosphorylation facilitates Aβ toxicity due to the activation of a Src kinase called Fyn (Nygaard, 2018). Various neurochemical studies have established a relationship between impaired insulin signaling and the level of tau protein hyperphosphorylation in AD. In diabetic individuals, impaired insulin signaling leads to GSK-3β activation, thereby resulting in increased tau phosphorylation followed by the subsequent generation of intracellular neurofibrillary tangles (Folch et al., 2018). In AD, an increased glycosylation of tau protein results in the formation of a tau protein that is more likely to be phosphorylated by GSK-3β, cyclic adenosine monophosphate (cAMP)-dependent protein kinase A (PKA), and CDK5. This tau protein is more sensitive to the downregulation of the dephosphorylation effect of protein phosphatase 2A (PP-2A). Decreased PP-2A activity in AD may be a factor responsible for the hyperphosphorylation of the tau protein (Magalingam et al., 2018). Recent studies have shown a crosslink between tau protein hyperphosphorylation and mitochondrial stress. Increased oxidative stress also contributes to the hyperphosphorylation of tau. As tau protein protects genomic DNA from oxidative stress-induced damage, the abnormalities in tau protein lead to the increased susceptibility of DNA in hippocampal neurons to oxidative damage in patients with AD (Melov et al., 2007).

4.3.3 The Mitochondrial Hypothesis and Oxidative Stress

Mitochondrial adenosine triphosphate (ATP) generation is crucial for cellular functioning, as well as for various signaling pathways and cellular activities. Due to energy requirement, neurons are at a high risk of injury and death as a result of mitochondrial dysfunctioning. The maintenance of metabolic homeostasis and the activation of stress reactions in the body also require appropriate functioning of mitochondria. Recent findings indicate that abnormalities in mitochondrial fusion, fission, and trafficking dynamics are responsible for mitochondrial dysfunction (Gao et al., 2017). Clinical and experimental data have revealed that mitochondrial dysfunction plays a vital role in the pathogenesis of AD. Mutations in genes that code for mitochondria, dynamic mitochondrial defects, an excessive formation of mitochondrial ROS, dysfunctions due to protein aggregation, and various environmental factors may change energy metabolism and are linked to AD (Moreira et al., 2010). The following evidence indicates that mitochondrial dysfunctions play a vital role in AD.

1. Decreased energy metabolism as a result of alterations in enzymes involved in oxidative phosphorylation.
2. Impaired calcium-buffering capacity and alterations in endoplasmic reticulum calcium channels develop a calcium imbalance leading to the activation of calmodulin-dependent kinases, calpain, and apoptosis.
3. Abnormal mitochondrial dynamics are observed due to the downregulation of *OPA1* and overexpression of dynamin-related protein 1 (DrP1). DrP1 is a GTPase of the dynamin family involved in fission of the

mitochondrial outer membrane. Approximately 30% of DrP1 is located in the cytosol. During apoptosis, the translocation of DrP1 takes place from the cytosol to the mitochondrial outer membrane at the point where constriction rings are formed. In AD, disturbed mitochondrial dynamics and synaptic degeneration occur due to the interaction of hyperphosphorylated tau and Aβ with DrP1, which causes mitochondrial fragmentation (Kandimalla and Reddy, 2017). This interaction increases with the progress of the disease, thereby causing the death of neurons.

4. Silent information regulator 1 (SIRT1), peroxisome proliferator-activated receptor-gamma coactivator 1-alpha (PGC-1α) and nuclear respiratory factor (NRF) regulates mitochondrial biogenesis. In patients with AD, a remarkable decrease in NRF1, NRF2, and PGC-1α levels occurs, indicating a role in AD (Lagouge et al., 2006).

Studies from transgenic mice and cell cultures indicates various mitochondrial defects such as decreased ATP synthesis, impairment in oxidative phosphorylation, increased production of ROS, and an impaired antioxidant enzyme system in patients with AD (Cabezas-Opazo et al., 2015). Experimental data from animal studies indicate an impaired distribution and localization of mitochondria due to overexpression and hyperphosphorylation of tau causing synaptic loss (Cai and Tammineni, 2017, Wang et al., 2015). Therefore, knowledge of the involvement of mitochondria in the pathogenesis of AD could help in the development of potential treatment strategies for it.

4.3.4 Neuroinflammation

Infections, diseases, trauma, and toxic substances initiate a series of inflammatory responses in the brain known as neuroinflammation. A wealth of knowledge indicates a relationship between AD and neuroinflammation. Microglia and astrocytes form a major part of the immune system of central nervous system (CNS), and it has been found that microglial cells are mainly involved in inflammatory events (Calsolaro and Edison, 2016). Studies of various serum and brain tissues from patients with AD indicate the presence of activated microglia, as well as inflammatory mediators like interleukin (IL)-6, IL-1β, tumor necrosis factor (TNF)-α, and cyclooxygenase 2 (COX-2). In a rat model of AD, it was observed that *COX2* expression is upregulated and associated with an increased level of inducible COX-2, followed by the production of various prostaglandins (PGs). PGs, by acting on PG receptors present on microglia, activate and stimulate the synthesis of proinflammatory cytokines (Caggiano and Kraig, 1999). Data from previous studies have shown that microglia, through the protein kinase C (PKC)-δ-dependent pathway, also stimulate the synthesis of neurotoxic proinflammatory mediators. The TNF-α level was also shown to be higher in the brains of patients with AD, compared to normal brains (Magalingam et al., 2018).

It is hypothesized that the presence of Aβ is primarily responsible for the activation of microglia. In the early stages of AD, microglia exert a protective effect through the phagocytosis of Aβ. However, sustained activation of microglia and prolonged activation of immune system exacerbates the symptoms of AD. This leads to Aβ accumulation and the activation of microglia, and increases the formation of inflammatory cytokines, TNF-α, and chemokines, which further activates primed microglia. An increased IL-1 level elevates the production of IL-6, which, in turn, activates CDK5, which is involved in tau phosphorylation. Shanya Jiang and Kiran Bhaskar postulated the role of complementary cytokine and chemokine systems that control synaptic functions and their dysfunction in AD (Jiang and Bhasker, 2017). They reported that Aβ aggregation and tau hyperphosphorylation causes neuroinflammation and synaptic, as well as neuronal, loss in patients with AD, followed by memory loss and cognitive impairment. Genetic studies have suggested that the neurotransmitter systems may be linked to changed immune pathways and synaptic dysfunction in AD. A reduced level of norepinephrine in the locus coeruleus is also considered to be a factor responsible for microglial activation and neuroinflammation (Jiang and Bhasker, 2017). It is hypothesized that synaptic pruning mediated by the immune system is also involved in the pathogenesis of AD. Microglia also have an important role in postnatal synaptic pruning in which proteins of the complement system play a pivotal role by controlling the extent of pruning (Stephan et al., 2012). Aβ oligomers cause the upregulation of complement protein C1q by up to 80-fold which, in turn, activates the complement cascade in the human AD brain. This effect leads to the cleavage and activation of C3 and other downstream complementary proteins, and the formation of the C5b–C9 membrane attack complex which stimulates synapse engulfment by microglia, microglial inflammatory signaling, and synaptic loss (Hong et al., 2016). Recent studies have revealed that mutations in triggering receptor expressed on myeloid cells 2 (TREM2) increases the likelihood of AD (Saber et al., 2017).

4.4 NEUROTRANSMISSION AND NEUROCHEMICAL ALTERATIONS IN ALZHEIMER'S DISEASE

Neurotransmitters are molecules synthesized in presynaptic neurons, stored in synaptic vesicles, and released in response to appropriate signals from neurons, initiating neurotransmission via the synaptic cleft by binding to appropriate postsynaptic receptors. The majority of neurotransmitter systems are involved in maintaining brain homeostasis (Ricci et al., 2009). Neurologic disorders are correlated with hypo- or hyperactivities of any of the neurotransmitter present in the brain. Hyper- or hypofunction of any of the neurotransmitters have been correlated with specific neurologic disorders. Evidence from the literature suggests that AD causes neurochemical changes in the levels of various neurotransmitters in the brain (Kaur, 2019). Specifically, the altered neurotransmission of cholinergic, GABAergic, serotonergic, adrenergic, and dopaminergic neurotransmitting systems play a major role in the progression of AD (Nava-Mesa et al., 2014). Molecules like norepinephrine, oxytocin, neuropeptide Y, enkephalins, melatonin, and glycine are not involved directly in the pathogenesis of AD, but may causes an increase in oxidative stress, which is a major factor in the pathogenesis of AD (Kandimalla and Reddy, 2017).

4.4.1 Cholinergic Transmission in Alzheimer's Disease

In the human brain, cholinergic synapses are found to be one of the most abundant synapses, mostly present in the thalamus, limbic system, striatum, and neocortex. Cholinergic transmission is probably essential for learning, memory, attention, and other brain functions. This system plays a vital role in age-associated cognitive decline, including AD. Acetylcholine (ACh) is a neurotransmitter in the cholinergic system that binds to either muscarinic or nicotinic receptors to facilitate cholinergic transmission in brain. In the CNS, muscarinic receptors are predominantly located in the cortex, hippocampus, and thalamus, whereas nicotinic receptors are mainly found in the cortex, striatum, lateral geniculate nucleus, superior colliculus, and cerebellum (Sarasamkan et al., 2016). The cholinergic hypothesis of AD indicates a gradual loss in cholinergic innervations in the limbic system and neocortex of patients with AD (Hampel et al., 2018). Although changes in various neurotransmitter systems have been observed in AD, damage to cholinergic neurons in the hippocampus and basal forebrain nuclei is the major characteristic feature associated with the severity of the disease. Deficits in the transcription and expression of muscarinic and nicotinic receptors, a deficient supply and support of cholinergic neurotrophic factors resulting in functional impairment of the nucleus basalis of Meynert (NBM), and medium spiny neurons are correlated with AD (Shen and Wu, 2015). The degeneration of cholinergic neurons in the NBM and its axons projecting to the cerebral cortex is the key factor determining the level of cholinergic loss in the brain. Depletion of cholinergic neurons augments Aβ deposition and tau protein hyperphosphorylation, which contributes to cognitive impairment (Ramos et al., 2013).

Cholinergic degeneration in the NBM and lesions also lead to changes in cortical nicotinic and muscarinic receptors, as well as the loss of nicotinic receptors (α7; Schroder et al., 1991), which is supported by evidence suggesting a role of nicotinic and muscarinic receptors in AD. By enhancing the firing rate of neurons, postsynaptic α7 nicotinic receptors lead to the formation of LTP in the hippocampus, an essential component of learning and memory. By activating either the phosphatidylinositol-3-kinase (PI3K)–Akt axis, by the downregulation of GSK-3β, or by activating antiapoptotic factor Bcl-2, nicotinic α7 receptors exert a protective effect against Aβ-induced neurotoxicity (Beaulieu, 2012). However, a significant decrease in the density of nicotinic receptors and their binding efficiency to ACh is observed with the progression of AD (Haga, 2013). Pertaining to cortical muscarinic receptors, it was found that a decrease in presynaptic M2 receptor activity takes place, with no changes in the number of postsynaptic M1 receptors in patients with AD (Mash et al., 1985). However, few studies have suggested that M1 receptors may be dysfunctional in the cerebral cortex (Jiang et al., 2014). Overactivation of GSK-3β is associated with various cellular events in AD, such as the formation of a high level of Aβ oligomers, tau hyperphosphorylation, and the formation of NFTs (Chu et al., 2017). Mounting evidence suggests that M1 receptor activation can downregulate the tau and amyloidogenic pathways. Nicotinic α7 receptors and M1 receptors coupled to PKC may also lead to the downregulation of GSK-3β-mediated tau hyperphosphorylation, thereby preventing the detrimental effects of tau in patients with AD (Espada et al., 2009). The destruction of afferent connections in cholinergic neurons may alter the BBB and thereby changes the transport and removal of Aβ from the brain. In patients with AD, cholinergic deafferentation may change the properties of the BBB and arterial, as well as lymphatic drainage of Aβ (Engelhardt et al., 2016). These findings show that there may exist a complex interaction between cholinergic degeneration, transmission, and the pathologic features of AD.

4.4.2 Glutaminergic Transmission in Alzheimer's Disease

Glutamate is the main excitatory neurotransmitter in the brain; it plays a crucial role in synaptic transmission, synaptic plasticity, neuronal growth and differentiation, learning, and memory. Glutamate binds to two glutamate receptors (i.e. ionotropic and metabotropic glutamate receptors) in the brain and facilitates transmission (Conn and Pin, 1997).

A significant wealth of knowledge indicates that, under pathologic conditions, glutamate – an excitatory neurotransmitter in the CNS – has been documented to play a vital role in changes in glutaminergic signaling linked to the disturbances in cholinergic transmission found in AD. This may be due to the fact that the glutaminergic and cholinergic systems interact significantly during neurotransmission (Lin et al., 2010). It has been found that glutaminergic abnormalities are initially observed at the entorhinal cortex followed by further defects associated with the neurotransmission in hippocampus, amygdala, frontal cortex, and parietal cortex (Mohandas et al., 2009). In the hippocampus, glutaminergic neurotransmission via the *N*-methyl-D-aspartate (NMDA) receptor is responsible for a calcium-mediated phenomenon called synaptic plasticity, and LTP, which is required for learning and memory consolidation. Mounting evidence correlates the overactivation of NMDA and the formation of amyloid plaques in AD (Rudy et al., 2015). The overactivation of NMDA receptors due to the deposition of Aβ and NFT formation leads to a sustained rise in the Ca^{2+} level in the cytoplasm. Increased cytoplasmic Ca^{2+} further causes mitochondrial dysfunction and excitotoxicity that is responsible for the onset of the neurodegenerative process and neuron death (Friedman, 2006). In AD, Aβ causes a sustained reduction in the activity of PKA and therefore inhibits the phosphorylation of cAMP response element binding protein (CREB; Teich et al., 2015). With the subsequent inhibition of CREB phosphorylation, the formation of prosurvival factors such as brain-derived neurotropic factor will also be decreased, making neurons more vulnerable to oxidative stress-induced death (Bathina and Das, 2015). Overactivation of NMDA receptors and increase in intraneural calcium is associated with the generation of long-lasting depression, mitochondrial calcium overload, activation of nitric oxide synthase, the formation of oxygen and nitrogen free radicals, apoptosis, and cell death (Fan et al., 2007).

Decreased glutamine synthetase activity inhibits glutamate transport presynaptically, as well as in glial cells, and an increased production of Aβ-mediated nitric oxide is observed in patients with AD. The inhibition of presynaptic and glial glutamate transport, reduced glutamine synthetase activity (converts glutamate to glutamine), the discrete depolarization of neurons, and stimulation of nitric oxide production by amyloid favor the involvement of glutaminergic transmission in AD (Mohandas et al., 2009).

4.4.3 Serotonergic Transmission in Alzheimer's Disease

Various neuropathologic studies have strongly proposed a link between functional and morphologic changes in the monoaminergic ascending system, mainly in serotonin and norepinephrine, and the pathophysiology of AD (Trillo et al., 2013). Serotonin or 5-hydroxytryptamine (5-HT), has been reported to play a pivotal role in the regulation of cognition, behavior, autonomic responses, sensory and emotional processes, and motor activity in the CNS (Geldenhuys and Van der Schyf, 2011). The serotonergic system mediates these effects through its interconnection with cholinergic, dopaminergic, glutaminergic, adrenergic, and GABAergic systems. Serotonin produces its effect by acting on specific receptors present in the septohippocampal nucleus basalis and magnocellularis–frontal cortex complexes involved in learning and memory retention (Buhot et al., 2000). Mounting evidence has indicated the role of the serotonergic system in the pathogenesis of AD. AD is linked to a reduced density of serotonergic neurons in median and dorsal raphe nuclei, with a major loss in the caudal part. Postmortem studies conducted on AD brain also show a constant reduction in the number of serotonergic neurons in the raphe nuclei (Lyness et al., 2003). Decreased 5-HT activity and the degeneration of serotonergic neurons are associated with the behavioral abnormalities seen in the progression of AD. The presence of Aβ plaques and NFTs in medial and dorsal raphe nuclei has been correlated with the progression of clinical symptoms in patients with AD. Decreased 5-HT levels have also been reported in the frontal and parietal cortex, temporal, amygdala, putamen, caudate nucleus, hippocampus, and raphe nucleus of patients with AD (Bowen et al., 2008). The midbrain raphe nuclei also shows enhanced NFT pathology, which may contribute to the death of serotonergic neurons. Mounting evidence has indicated a role of serotonergic receptors in the progression of AD. Among the seven families (5-HT_1–5-HT_7) of serotonergic receptors, 5-HT_6, 5-HT_3, 5-HT_1, and 5-HT_2 receptor subtypes has been involved in the pathogenesis of AD. The 5-HT_6 receptor regulates the activities of various other neurotransmitter molecules such as ACh and glutamate, which play a pivotal role in learning and memory process (Geldenhuys and Van der Schyf, 2008). 5-HT_3 receptor activation in the hippocampus is associated with a decrease in cholinergic transmission, which in AD acts synergistically in the progression of the disease. Additionally, 5-HT_4 receptor activation increases the release of ACh and also reduces Aβ toxicity (Passani and Blandina, 1998).

4.4.4 Dopaminergic Transmission in Alzheimer's Disease

The appearance of extrapyramidal symptoms in about 35–40% patients with AD supports the fact that dopamine (DA)-containing neurons also undergo degenerative changes in AD (Martorana and Koch, 2014). In patients with AD, the dopaminergic neurons of the nigrostriatal pathway show various pathologic changes such as Aβ plaques, NFTs, neuronal loss, and a decrease in DA content. All of these changes clearly suggest the involvement of DA in the cognitive and behavioral symptoms of AD. A decrease in glutamate release from hippocampi and frontal cortex further decreases DA release. The appearance of negative behavioral symptoms in the patients of AD and the elderly are considered to be due to impairment in dopaminergic transmission. Experimental data from transgenic AD mice indicate that dopaminergic pathology and the deposition of Aβ are related to each other, and have suggested their involvement in dopaminergic dysfunction (Perez et al., 2005). Mounting evidence for dopaminergic transmission has demonstrated a decreased expression of both D1- and D2-like receptors in the hippocampus and prefrontal cortex of patients with AD (Nyberg et al., 2016). Interestingly, the nucleus accumbens is also highly affected in patients with AD, and shows a decrease in D2-like DA receptor expression, and decreased expression of SDAT and the enzyme tyrosine hydroxylase (Joyce et al., 1997). On the basis of the available literature, it has been suggested that the formation of Aβ oligomers and the hyperphosphorylation of tau might be responsible for dysfunction in dopaminergic transmission in the early phases of AD. Various studies have correlated the overproduction of Aβ oligomers and their prolonged interaction with α7 nicotinic ACh receptors with dopaminergic dysfunction in the early stages of AD. Owing to prolonged exposure to Aβ oligomers, the physiologic release of glutamate and GABA are also impaired, followed by a further decrease in the release of DA in the hippocampus and neocortex, thus contributing to impaired memory, attention, and executive function (Posadas et al., 2013). In the later stages of the disease, prolonged nicotinic receptor dysfunctioning in substantia nigra is followed by extracellular regulated kinase–mitogen-activated protein kinase (MAPK) pathway inactivation. This inactivation favors hyperphosphorylation of tau protein, leading to neuronal degeneration and cell death (Wang and Sun, 2010). A few experimental studies have also suggested that, under pathologic conditions, Aβ oligomers interact with alpha-synuclein (α-syn) protein and induces its aggregation, leading to the formation of pentamers and hexamers with ring-like structures that then associate with membrane-forming pores and cause degeneration of dopaminergic neurons (Samuel et al., 2016).

4.4.5 Adrenergic Transmission and Alzheimer's Disease

Noradrenergic neurons are primarily situated in the locus coeruleus, and, via efferent projections, they spread adrenaline throughout the brain. Studies have reported changes in the noradrenergic systems of patients with AD (Ramos and Arnsten, 2007). The gradual degeneration of adrenergic neurons in subcortical nuclei leads to the deprivation of cortical and hippocampal neurons from their negative influence. Initially, in the early phase of AD, pathologic changes, such as the accumulation of NFTs on noradrenergic neurons in the locus coeruleus, are observed, which are followed by a profound loss of noradrenergic neurons in the later phase of the disease (Braak and Del Tredici, 2012). Degeneration of noradrenergic neurons in locus coeruleus activates the formation of Aβ plaques and NFTs, and increases the severity of AD.

In spite of the loss of locus coeruleus neurons and alterations in adrenergic receptor density in various brain areas, studies of alterations in the level of noradrenaline in the brain of patients with AD are still controversial (Gannon et al., 2015). A few studies have revealed that the level of noradrenaline remains constant or is even elevated in patients with AD (Sparks et al., 1988). Looking deeper into the matter, it has been found that these differences may exist due to the various stages of the disease and the ability of the brain to compensate for the loss of neurons. The expression of α and β adrenergic receptors in the brain is responsible for mediating functions related to cognition (Ramos and Arnsten, 2007). However, a significant decrease in β1 and an increase in β2 adrenergic receptor density is observed in the cortex; in contrast to this, an increase in both β1 and β2 receptors in the hippocampal region is reported in the brain of patients with AD (Kalaria et al., 1989). Reports from various studies have suggested a role of the α2A receptor subtype in the pathogenesis of AD and suggested that its stimulation enhances amyloidogenesis and Aβ-related neuropathologic changes (Chen et al., 2014).

Furthermore, noradrenergic projections in the locus coeruleus play a vital role in the uptake and clearance of Aβ by microglia cells, supporting the fact that these projections act as a connecting bridge between neurons and microglia to activate the host response to Aβ in AD (Kong et al., 2010). Under physiologic conditions, noradrenaline, by acting on its β receptor, suppresses the expression of various proinflammatory genes like *TNF*, human leukocyte antigen gene complex (HLA), *IL1B*, and inducible nitric oxide synthase (NOS2), and produces an anti-inflammatory effect. Noradrenaline also promotes the expression of some anti-inflammatory molecules like inhibitory IκB, nuclear factor kappa B (NF-κB), and heat shock protein 70 in microglia and astrocytes, and contribute to the anti-inflammatory effect (Marien et al. 2004). Evidence from various studies has suggested that locus coeruleus degeneration induces an inflammatory response that contributes to the progression of AD.

4.5 MOLECULAR SIGNALING MECHANISMS AND ALZHEIMER'S DISEASE

The molecular mechanisms of Aβ deposits, APP metabolites, and tau phosphorylation leading to synaptic damage are not clearly understood. Various possibilities under investigation include changes in signaling pathways associated with neurogenesis, synaptic plasticity, synaptic loss, and neuron death (Crews and Masliah, 2010). A large number of signaling pathways and molecules, including GSK-3β, Fyn kinases, CDK5, p21 activated kinase, and members of the MAPK family, are associated with the progression of AD. Metabolic pathways like Wnt signaling, mTOR, AMPK, PGC-1α, and SIRT1 are associated with AD and are known to regulate various pathologic events in AD. Abnormalities in signaling molecules are pathways that may promote Aβ aggregation and tau phosphorylation, the activation of apoptotic pathways, cytoskeletal abnormalities, and the activation of calcium calpain-dependent proteolysis, leading to synaptic failure and neurogenesis (Kablar, 2019).

4.5.1 Fyn Kinases and Alzheimer's Disease

Fyn kinase is a non-receptor tyrosine kinase that belongs to the Src family of kinases and is assumed to play a vital role in synaptic plasticity, a mechanism responsible for learning and memory (Yang, 2011). It is an important modulator of glutamate-induced LTP and synaptic transmission, and a regulator of the NMDA receptor in the brain. Overexpression of and alteration in the activity of Fyn contributes to Aβ-induced toxicity in AD (Chin, 2005). Studies conducted in vitro and in animal models indicate that Fyn kinase activation is responsible for the death of neurons, decreases the survival of APP/Ps1 transgenic mice, causes synaptic loss and the degeneration of 5-HT neurons, and is responsible for defects in spatial learning and memory (Larson et al., 2012). Glutamate, an excitatory neurotransmitter, is considered to be one of its major substrates. There is mounting evidence that Fyn becomes concentrated in postsynaptic density (PSD) protein in the brain. Interaction of Fyn with PSD-95 (a synaptic scaffolding protein), modulates the NR2B unit of the NMDA receptor. Fyn activation further causes phosphorylation of the NR2A and NR2B subunits of the NMDA glutamate receptor and increases the trafficking of NR2B and membrane stabilization, leading to an increase in synaptic expression (Trepanier et al., 2012). Various studies have reported that, in AD, Aβ oligomers bind to various cell surface receptors; overall, the receptors that have the highest affinity for Aβ oligomers are cellular prion proteins (PrPC). PrPC has less or no affinity for monomeric and fibrillary Aβ peptides. They are abundant in the cells of CNS and are highly expressed proteins present in the brain. It is also considered that Aβ oligomer binding to the PrPC activates Fyn and then triggers a downstream signaling cascade (Lacor et al., 2007). It has been reported that the association of Aβ oligomer to PrPC and further activation of intracellular Fyn kinase also requires activation of the metabotropic receptor of glutamate, mGluR5. Figure 4.2 shows that mGluR5 is required in neuronal cells for Aβ-mediated neuronal dysfunction and therefore Aβ oligomers trigger a signaling cascade from PrPC-mGluR5-Fyn kinase, which leads to dendritic loss and changes in synaptic plasticity and AD (Nygaard et al., 2014). In binding Aβ oligomers to the cell surface, PrPC causes a dose-dependent increase in the phosphorylation of the NMDA receptor, which is responsible for dendritic spine loss (Um et al., 2012). Figure 4.2 also shows that, besides mediating Aβ signaling, it also plays an important role in the phosphorylation of tau. Fyn activation leads to tyrosine phosphorylation of tau and this interaction exacerbates the progression of AD (Bhaskar et al., 2010; Nygaard et al., 2014). The association of tau with Fyn–PSD95–NMDA receptor is required for neurotoxicity. Numerous studies indicate that functional tau is required for locating Fyn to NMDA receptors and increasing its interaction with postsynaptic density protein PSD-95, which is critical for Aβ neurotoxicity (Barry et al., 2011). In the absence of tau, there is uncoupling of Fyn and NMDA receptors and the prevention of Aβ toxicity. A recent study has shown that p38 MAPK-mediated tau phosphorylation at T205 prevents coupling of the Fyn–PSD95–tau–NMDA receptor complex and improves Aβ toxicity in an animal model of AD (Nygaard, 2018). The association of Fyn with tau phosphorylates the tyrosine residue (Tyr18) of tau near the amino terminus; the presence of NFTs with phosphorylated Tyr18 in human AD brain supports this (Ittner, 2016).

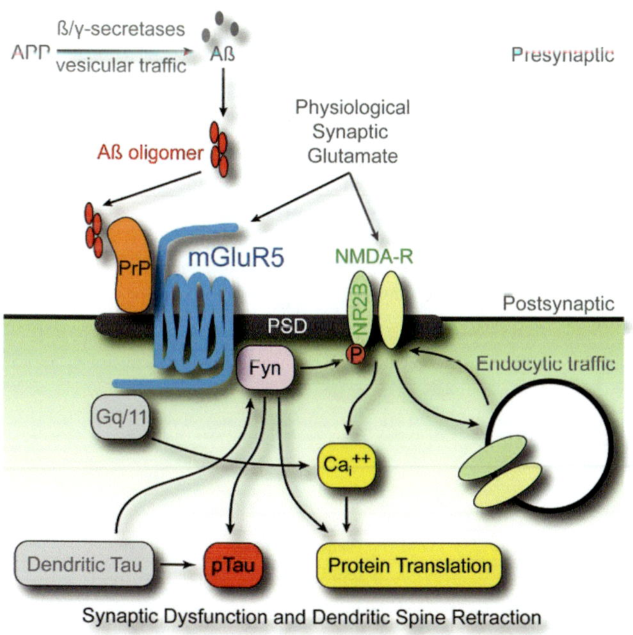

FIGURE 4.2 Diagrammatic illustration of role of Fyn in amyloid-beta (Aβ) oligomer (Aβo) toxicity. Association of Fyn with Aβo–metabotropic receptor of glutamate (GluR5) and cellular prion protein to initiate postsynaptic signaling. Aβo biding to cellular prion protein (PrPC) activates mGluR5-dependent signaling events. Tau also has an important role in confining Fyn and is a Fyn substrate. APP, amyloid precursor protein; NMDA-R, *N*-methyl-D-aspartate receptor; PSD, postsynaptic density (adopted from Nygaard and van Dyck, 2014).

4.5.2 Wnt Signaling and Alzheimer's Disease

Wnt is a cysteine-rich glycosylated protein named for the *Drosophila* protein 'wingless' and the mouse protein 'Int-1'. Wnt signaling contributes to various cellular processes and plays a vital role in cellular proliferation, adhesion, differentiation, apoptosis, and the survival of glial cell and neurons in the CNS (Nusse and Clevers, 2017). After binding to frizzled and co-receptors, Wnt LRP5/6 transmembrane receptors present on the cell surface activate three main signaling pathways:

1. Wnt/β catenin or the canonical Wnt pathway. This pathway regulates gene transcription through β-catenin. After the binding of Wnt ligands to their receptors, signalosomes are formed at the cell membrane due to recruitment of the scaffold proteins axin and disheveled, followed by the accumulation of β-catenin in the cytoplasm (Palomer et al., 2019). They migrate to the nucleus and, associated with T-cell factor/lymphoid enhancer factor, regulates Wnt target gene transcription (De, 2011).
2. The Wnt/Ca^{++} pathway is a non-canonical pathway that modulates intracellular Ca^{++} release and excitotoxicity. In the Wnt/Ca^{2+} signaling pathway, binding of Wnt to frizzled receptors causes the activation of phospholipase C, Ca^{2+} release from intracellular calcium stores, and further activation of calcium binding proteins PKC and CaMKII, which results in a change in transcription and actin remodeling (De, 2011).
3. Wnt/planar cell polarity (PCP)–Jun N-terminal kinase (JNK) pathway, or Wnt-cell polarity, or PCP. This pathway is modulated by JNK. In the Wnt/PCP–JNK pathway, association of Wnt ligand to frizzled transmembrane receptors is responsible for transcriptional changes and reorganization of the cytoskeleton. Wnt/PCP signaling activates RhoA (small GTPase) and Rac1, which further activate Rho kinase and JNK kinases, respectively (Seifert and Mlodzik, 2007).

Under physiologic conditions, the canonical and non-canonical Wnt/PCP signaling pathways are in balance to maintain synaptic functions. If canonical Wnt signaling is decreased, it leads to activation of the PCP pathway and results in the disassembly of synapses. Hence, Wnt signaling is considered to be a potential signaling pathway involved in the progression of AD, either directly or indirectly (McLeod and Salinas, 2018).

Numerous studies have shown that Wnt signaling protects the integrity of synapses from Aβ-induced toxicity (Marzo et al., 2016). Synaptic activities are directly affected by reduced Wnt signaling. Evidence from previous studies has shown that, in AD, Aβ oligomers cause the induction of Dickkoff-1 (Dkk1) and Dkk3, as well as Wnt antagonists in hippocampal neurons (Palomer et al., 2019). Dkk1 causes a reversible decrease in the number of presynaptic sites and synaptic proteins, leading to the disassembly of pre- and postsynaptic neurons (Boonen et al., 2009). As shown in Figure 4.3, by interacting with LRP5/6 Wnt co-receptors, Dkk1 inhibits canonical Wnt signaling and impairs the binding of Wnt ligand to frizzled and LRP5/6 receptors. The inhibition of canonical Wnt signaling causes enhanced activity of Gsk-3β and a decrease in the level of cytoplasmic β-catenin in the brains of patients with AD (Killick et al., 2014).

The ligand of non-canonical Wnt-5a improves the amplitude of excitatory postsynaptic currents and thereby prevents Aβ-induced synaptic damage. Wnt-5a also activates trafficking of NMDA and $GABA_A$ receptor in neurons, the growth of dendritic spines, and protects mitochondria from Aβ oligomers by activating the Wnt/Ca^{2+} pathways (Purro et al., 2012).

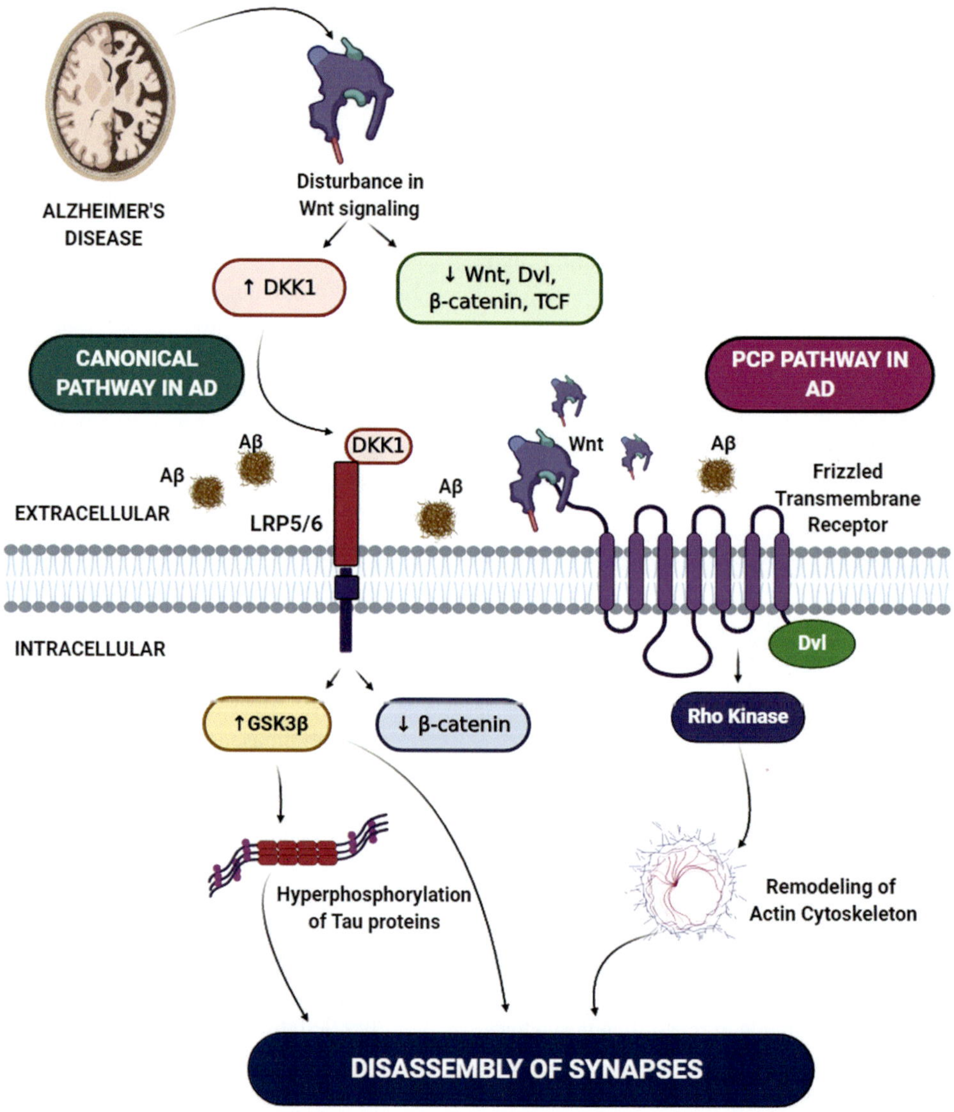

FIGURE 4.3 Deregulated Wnt signaling in Alzheimer's disease (AD). In AD, the components of Wnt signaling are deregulated. In AD, downregulation of β-catenin, Dvl, and T-cell factor (TCF), and upregulation of Dickkoff 1 (DKK1) occurs, leading to inhibition of the canonical pathway. Dkk1 further stimulates glycogen synthase kinase-3β and increases hyperphosphorylation of tau. Amyloid beta (Aβ) stimulates the Wnt planar cell polarity (PCP) signaling pathway and increases Rho kinase activity, and leads to synapse vulnerability (modified from Palomer et al., 2019). LRP, lipoprotein receptor-related protein 1; GSK, glycogen synthase kinase.

4.5.3 CDK5 and Alzheimer's Disease

CDK5 is one of the intracellular signaling pathways that has been discovered to play a major role in the progression of AD. Cyclin-dependent kinases are a family of kinases mainly involved in the regulation of the cell cycle, DNA replication, and the transition from one phase to another in the cell cycle. Contrary to other CDK family members, CDK5 has a vital role in the development of the mammalian CNS where it regulates synaptic plasticity and neurotransmission (Wilkaniec et al., 2018). In the course of embryogenesis, CDK5 is involved in axonal development, the migration of neurons, dendritic growth, cytoskeleton dynamics, neuronal survival, differentiation, and apoptosis. After birth, the activity of CDK5 is required for synaptic plasticity, learning, memory, and behavioral changes (Hernandez et al., 2016). Hypo- and hyperactivity of CDK5 is equally toxic to neurons and is responsible for various neurologic and neurodevelopmental disorders, including mental retardation, epilepsy, schizophrenia, AD, amyotrophic lateral sclerosis, and Parkinsonism (Meyer et al., 2014). Dysregulation of CDK5 signaling is linked to cytoskeletal changes, apoptosis, and an increase in calcium influx mediated by NMDA receptor activation. In AD, various neurotoxic signals cause an increase in the intracellular calcium level and activation of calpain, a calcium-dependent protease that further cleaves p35 and p39 (membrane-localized activator proteins) into p25 and p29, respectively. p25 has a very long half-life and prolongs the activation of CDK5. As p25 does not have membrane-anchoring signals, it causes constitutive activation and mislocation of the CDK5–p25 complex leading to phosphorylation and activation of various pathologic targets, and the formation of neurotoxic Aβ, NFTs, and neuron death (Liu et al., 2016).

The activity of CDK5 and formation of Aβ are closely related, and mutually affect each other. Pathologic activation and accumulation of CDK5 is associated with various pathologic events such as amyloid cascade, synaptic damage, neuronal dysfunction, hyperactivation of kinases, and neuronal loss. The phosphorylation of APP at Thr668 by CDK5 increases the formation of Aβ oligomers and also inhibits the binding of APP with the cytoplasmic adaptor protein Fe65, inhibiting the formation of Aβ oligomers (Liu et al., 2016). Deregulated CDK5 activity is also linked to the changes in the presenilin system, and a few studies have shown that CDK5, by phosphorylating PS-1 at thr354, promotes the level of presenilin and increases γ-secretase activity. CDK5 also enhances the amyloidogenic process of APP for the generation of Aβ by enhancing BACE1 transcription. In addition, CDK5, by deprivation of nerve growth factor, induces neuronal loss (Matrone et al., 2009). Therefore, Aβ may have a significant role in abnormal CDK5 activity. Aβ and CDK5 together form a positive feedback loop, which is accountable for various pathologic events of AD.

In AD, CDK5 activation is also linked to abnormalities in tau protein. The regulation of cytoskeletal integrity by tau is required either by CDK5/p35 or CDK5/p39 complexes. In AD, the activation of m-calpain cleaves p35 and p39 into p25 and p29 proteins. These proteins causes prolonged and overactivation of CDK5, leading to tau hyperphosphorylation for an extended period of time along with NFT formation, thereby causing disruption of cytoskeleton filaments and intracellular transport responsible for neurodegeneration (Peterson et al., 2010). As p25 lacks the myristoylation motif of p35, the CKD5/p25 complex loses the peripheral membrane distribution ability of CDK5/p35 and interacts with other substrates, which may indirectly increase hyperphosphorylation of tau.

Figure 4.4 shows that CDK5 also regulates the activity of a GSK-3β, which is directly associated with tau hyperphosphorylation (Castro-Alvarez et al., 2014). By phosphorylating Ser 235 and 404, it has been found that activated CDK5 increases the GSK-3β-mediated phosphorylation of tau. CDK5 also regulates the activity of GSK-3β through epidermal growth factor neuregulin and epidermal growth factor receptors (Li et al., 2003). In spite of evidence that links CDK5 and hyperphosphorylation of tau, the role of CDK5 in AD is controversial and needs more research to establish the mechanism completely.

4.5.4 PI3K/Akt/mTOR Signaling and Alzheimer's Disease

mTOR is a highly conserved serine/threonine kinase, that belongs to the PI3K/protein kinase B (Akt) family and is involved in the regulation of cell proliferation, growth, survival, and energy in various body tissues. Dysregulated mTOR signaling is associated with aging, and various diseases such as cancer, cardiovascular diseases, and neurodegenerative diseases (Orlova and Crino, 2010). During mTOR activation, the catalytic subunit of mTOR interacts with various proteins and leads to the formation of mTOR complex 1 (mTORC1) and mTOR complex 2 (mTORC2), the major intracellular signaling complexes (Wullschleger et al., 2006). The scaffolding proteins raptor and rictor are present in mTORC1 and mTORC2, respectively, and are responsible for connecting the complexes to various targets (Shimobayashi and Hall, 2014). mTORC1 is made up of mLST8 and raptor (that is sensitive to rapamycin), while mTOR2 complex is composed of mTOR, rictor (insensitive to rapamycin), mLST8, mLST1, and mSIN1.

Figure 4.5 indicates that mTORC1 is activated via the insulin receptor or receptor tyrosine kinase in response to a variety of extracellular signals such as insulin-like growth factor, insulin, and growth factors. Proteins like Ras, Raf, PIK3/Akt, Rheb/Rhes, and Epk1/2 activate mTORC1 and mediate cell growth, protein synthesis, glucose homeostasis, mitochondrial functions, and energy metabolism (Buerger, 2018; Yu and Cui, 2016,). mTORC2 is insensitive to rapamycin and is mainly associated with cytoskeleton reconstruction and cell survival (Uniyal et al., 2019).

Under physiological conditions, mTOR regulates neuronal recovery, synaptic plasticity, and memory retention by regulating protein synthesis in neurons. However, numerous

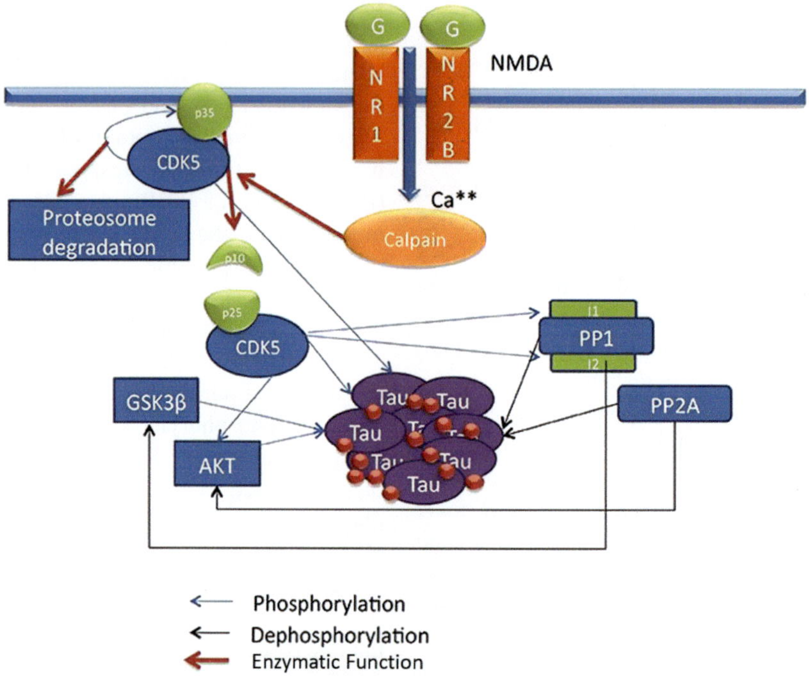

FIGURE 4.4 Scheme of the CDK5 signaling pathway involved in tau phosphorylation. This drawing shows the CDK5 signaling pathway, describing the direct and indirect phosphorylation of tau by CDK5 (adopted from Castro-Alvarez et al., 2014). NMDA, *N*-methyl-D-aspartate; PP1, protein phosphatase 1; PP2A, protein phosphatase 2A; GSK, glycogen synthase kinase.

studies correlate an association between mTOR signaling and the presence of Aβ plaques and tau phosphorylation, two hallmarks of AD pathogenesis (Di Domenico et al., 2018). It has been reported that mTOR signaling activation exacerbates the progression of AD by activating the formation of Aβ and NFTs in the AD brain (Perluigi et al., 2014). In AD, various pathologic changes, such as damage to dendrites, brain shrinkage, and the expansion of ventricles, are associated with hyperactivation of the PI3K/Akt/mTOR signaling pathway, as well as downstream effector molecules, such as 4E-BP1, p70S6K, and eukaryotic elongation factor 2. Though some studies link AD with reduced mTOR signaling and the hypo- or hyperactivation of the mTOR pathway in AD, this is still controversial (Griffin et al., 2005). However, a few studies conclude that, at a low level, Aβ increases mTOR signaling, while a significantly high level of Aβ decreases mTOR signaling (Oddo, 2012).

It was reported by Shahani et al. (2014) that a decreased level of the mTOR enhancer Rheb may be responsible for Aβ-induced dysregulation of mTOR signaling. Rheb regulates BACE1 degradation (through autophagy) and a reduced Rheb level is correlated with an increase in the activity of BACE1 and sustained production of Aβ. Hyperphosphorylation of tau is also correlated with hyperactivity of mTOR and its downstream targets in the brain of AD. Decreased degradation of phosphorylated tau due to inhibition of autophagy, an increase in tau mRNA translocation, and the regulation of tau phosphorylation by mTOR may contribute to the effect of hyperactive mTOR in AD (Di Domenico et al., 2018; Wang et al., 2009).

4.5.5 AMPK Signaling and Alzheimer's Disease

AMPK is a key energy sensor, important for maintaining cellular energy homeostasis. It is a regulator of energy metabolism and is involved in glucose uptake, fatty acid synthesis, and β-oxidation of fatty acids, as well as mitochondrial and glucose transporter 4 biogenesis (Wang et al., 2019). AMPK is expressed in various peripheral tissues and neuronal cells of the brain, and its activation inhibits cell anabolic activities and stimulates catabolic activities like autophagy. AMPK is a heterotrimeric complex made up of one catalytic α (α1 and α2) and two regulatory β (β1 and β2) and γ (γ1, γ2, and γ3) subunits (Cai et al., 2012). An increase in the adenosine monophosphate (AMP)/ATP ratio, a low energy level, hypoxia, and the formation of ROS are the main signals for its activation. When the ATP level is low, AMP binds to AMPKγ, inducing conformational changes and causing phosphorylation of AMPKα. An increase in AMP/ATP ratio and Ca^{+} -mediated activation of calmodulin-dependent protein kinase kinase-beta (CaMKKb) activates the liver kinase B1 complex (LKB1/STRAD/MO25), the primary kinase that phosphorylates the Thr-172 amino acid on the α-subunit (Viollet et al., 2009). Conversely, dephosphorylation of AMPK by protein phosphatase 2C (PP2C) converts it into an inactive form. In addition to energy homeostasis, AMPK controls cellular defense against stress through various downstream signaling pathways such as CREB, PGC-1α, mTOR, and SIRT1 (Cantó and Auwerx, 2010). AMPK inhibits mTORC1 and directly phosphorylates

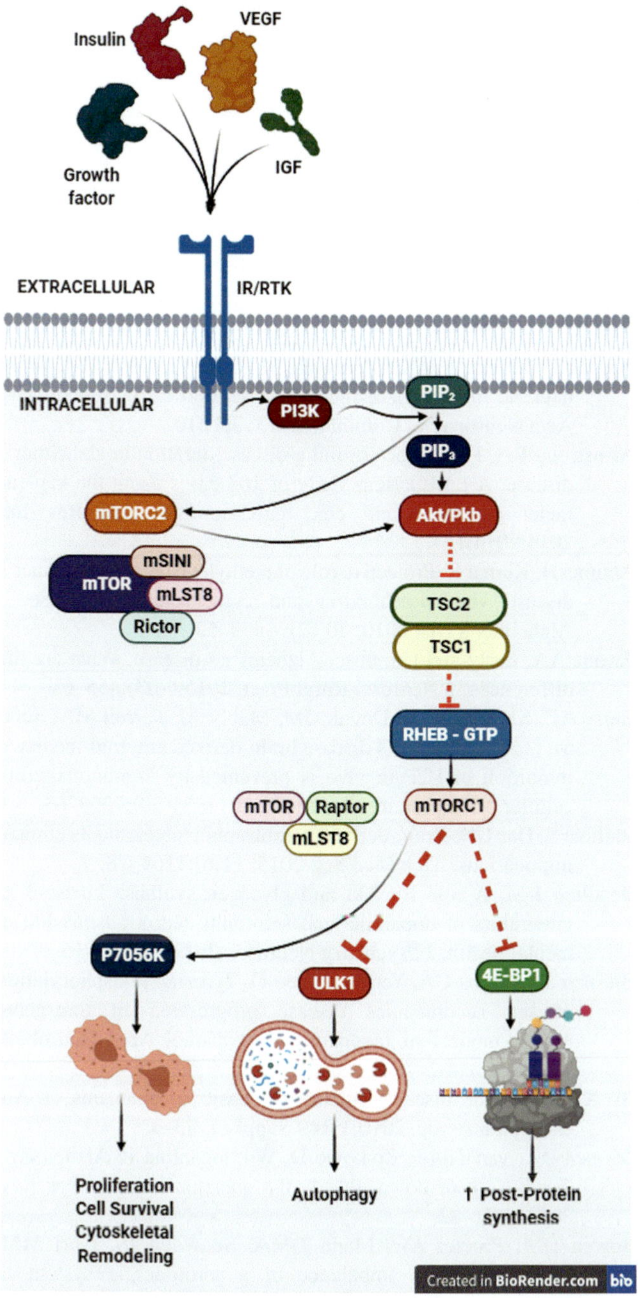

FIGURE 4.5 Diagrammatic representation of phosphatidylinositol-3-kinase (PI3K)/Akt/mammalian target of rapamycin (mTOR) signaling pathway. A more detailed description of the activation mechanism of mTOR in the regulation of protein synthesis and autophagy. VEGF, vascular endothelial growth factor; IGF, insulin-like growth factor; IR/RTK, insulin receptor/receptor tyrosine kinase; PIP, phosphatidylinositol 4,5-bisphosphate; TSC, tuberous sclerosis complex; RHEB, Ras homolog enriched in brain; GTP, guanosine triphosphate; mTORC1, mTOR complex 1; mTORC2, mTOR 2; ULK1, unc-51-like kinase 1; 4E-BP1, eukaryotic translation initiation factor 4E (eIF4E)-binding protein.

and activates Unc-51-like kinase 1 (ULK1), stimulating autophagy (Lombo et al., 2018).

Various studies have reported that abnormally high levels of AMPK in the brains of patients with AD and in animal models of AD indicate AMPK as an upstream driver for the disease (Vingtdeux et al., 2011). Hyperactivation of AMPK is associated with disturbance in synaptic plasticity, impaired learning, and memory AD progression (Vingtdeux et al., 2011). It is involved in Aβ degradation and phosphorylation of tau. Hyperphosphorylated tau is further self-assembled into PHFs and eventually forms NFTs. Besides this, Aβ peptides also activate AMPK for further enhancement of tau phosphorylation (Thornton et al., 2011). However few studies have also reported that it has an inhibitory effect on AMPK (Dong et al., 2016). The activation of AMPK is associated with the inhibition of BACE1 expression and therefore decreased generation of Aβ (Lu et al., 2010). A few studies have reported that AMPK activates SIRT1 and causes decreased acetylation of tau, facilitating its degradation and thereby inhibiting the accumulation of tau in the brain of patients with AD (Min et al., 2010). It also inhibits GSK3B and prevents the phosphorylation of tau. Therefore, the effect of AMPK on Aβ metabolism, the accumulation of NFTs, and AD pathogenesis is controversial and requires further studies to establish the concrete mechanism (Yang et al., 2019).

4.5.6 SIRT1, PGC-1α, AND ALZHEIMER'S DISEASE

Mitochondrial dysfunction and its correlation with the pathogenesis of AD is well known, and was discussed earlier in this chapter. It contributes to cognitive and memory impairment through various pathways (Atamna and Kumar, 2010). PGC-1α, SIRT1, and NRF regulate mitochondrial dynamics and biogenesis. PGC-1α is a transcriptional coactivator that regulates the expression of genes responsible for mitochondrial biogenesis and for the generation of ROS, and also regulates the gene expression of endogenous antioxidant enzymes such as glutathione peroxidase, superoxide dismutase, and catalase (Sweeney and Song, 2016).

PGC-1α regulates mitochondrial biogenesis by stimulating various transcription factors comprising mitochondrial transcription factor A (TFAM) and NRF-1 and NRF-2. NRF-1 and NRF-2 regulate the transcription of genes involved in electron transport, oxidative phosphorylation, mitochondrial DNA replication, and protein import and assembly (Mootha et al., 2004). The activity of PGC-1α is regulated by the histone deacetylase SIRT1. SIRT1 is a nicotinamide adenine dinucleotide (NAD^+)-dependent protein deacetylase that deacetylates various target proteins involved in cell metabolism, inflammation, and cell death (Zhou at al., 2018). It is essential for maintaining cell survival, cognition, synaptic plasticity, and glucose homeostasis. By deacetylating PGC-1α, SIRT1 enhances its activity and thereby promotes mitochondrial biogenesis and functions (Ren et al., 2019). PGC-1α/SIRT1 play an important role in the brain and protect the neurons from noxious stimuli. Mounting evidence indicates a remarkable decrease in NRF-1, NRF-2, PGC-1α, TEAM (triggering receptor expressed on myeloid cells), and SIRT1 levels in the hippocampal region of the brain of patients in AD, indicating a role for them in AD (Lagouge et al., 2006). Decreased activity of PGC-1α causes impairment in mitochondrial functions and triggers neuronal degeneration in AD (Weydt et al., 2006). Therefore, targeting PGC-1a and SIRT1 can improve mitochondrial function and cognition in AD.

4.6 FUTURE OPPORTUNITIES AND CHALLENGES

Despite intensive research to understand the pathogenesis of AD, an effective treatment approach to curing it is still missing (Frozza et al., 2018). The major focus of future research is to find effective treatment to reduce the global burden of the disease (Khachaturian et al., 2018). The key areas identified are provided below:

1. Understanding of the disease stage and the development of reliable early-detection biomarkers to improve the success rate of clinical trials.
2. Deeper understanding of cross-talk between various signaling pathways to develop novel and effective therapies.
3. Development of disease-modifying agents for preventing the disease in its early stage, and combination therapy by repurposing drugs and anti-inflammatory agents
4. Development of imaging biomarkers.

4.7 CONCLUSION

This chapter has summarized various pathologic mechanisms of neurodegeneration, which may lead to a series of events responsible for the manifestation of AD. An imbalance in the formation, aggregation, and clearance of Aβ from the brain and the subsequent formation of toxic oligomers of Aβ and NFTs are the major pathologic hallmarks of AD. These pathologic hallmarks are associated with synaptic damage, neurochemical alterations, and disturbances in cholinergic, glutaminergic, and monoaminergic neurotransmission, as well as disturbances in the signaling pathways of synaptic plasticity and neurogenesis. The chapter concludes that the neurobiological mechanisms and neurotransmission abnormalities discussed – along with disturbances in the activity of signaling proteins like Fyn kinase, CDK5, and GSK-3β, and signaling pathways, including the Wnt, mTOR, AMPK, PGC-1α, and SIRT1 pathways – have a significant role in the progression and modulation of various pathological events in AD. Exploring these pathways and their role in magnifying the pathologic mechanisms associated with AD may be helpful in finding a potential target for treatment. However, extensive investigations are needed to decide the exact contribution of these signaling molecules and pathways to AD pathogenesis.

ACKNOWLEDGEMENTS

The authors are immensely thankful to Ms. Lata Bisht, Ms. Swati Dobhal, and Mr. Piyush Verma for their help with drafting the figures.

REFERENCES

Arendt T, Brückner MK, Morawski M, Jäger C, Gertz H-J. Early neurone loss in Alzheimer's disease: cortical or subcortical? Acta Neuropathol Commun. 2015; 3(1):10.

Armstrong RA. Factors determining disease duration in alzheimer's disease: A postmortem study of 103 cases using the kaplan-meier estimator and cox regression. Biomed Res Int. 2014;2014:1–7.

Atamna H, Kumar R. Protective role of methylene blue in alzheimer's disease via mitochondria and cytochrome c oxidase. J Alzheimer's Dis. 2010; 20(s2): S439–52.

Awada AA. Early and late-onset Alzheimer's disease: What are the differences? J Neurosci Rural Pract. 2015;6(03):455–6.

Barry AE, Klyubin I, Mc Donald JM, Mably AJ, Farrell MA, Scott M, et al. Alzheimer's disease brain-derived amyloid-mediated inhibition of LTP in vivo is prevented by immunotargeting cellular prion protein. J Neurosci. 2011; 31(20):7259–63.

Bathina S, Das UN. Brain-derived neurotrophic factor and its clinical implications. Arch Med Sci. 2015; 11(6):1164–78.

Beaulieu J-M. A role for Akt and glycogen synthase kinase-3 as integrators of dopamine and serotonin neurotransmission in mental health. J Psychiatry Neurosci. 2012;37(1):7–16.

Bhaskar K, Hobbs GA, Yen S-H, Lee G. Tyrosine phosphorylation of tau accompanies disease progression in transgenic mouse models of tauopathy. Neuropathol Appl Neurobiol. 2010;36(6):462–77.

Bi X. Alzheimer disease: update on basic mechanisms. J Am Osteopath Assoc. 2010;110(9 Suppl 8): S3–9.

Boonen RA, van Tijn P, Zivkovic D. Wnt signaling in Alzheimer's disease: up or down, that is the question. Ageing Res Rev. 2009; 8:71–82.

Bowen DM, Procter AW, Mann DMA, Snowden JS, Esiri MM, Neary D, et al. Imbalance of a serotonergic system in frontotemporal dementia: implication for pharmacotherapy. Psychopharmacology (Berl). 2008; 196(4):603–10.

Braak H, Del Tredici K. Alzheimer's pathogenesis: is there neuron-to-neuron propagation? Acta Neuropathol. 2011;121(5):589–95.

Braak H., Del Tredici K. Where, when, and in what form does sporadic Alzheimer's disease begin? Curr. Opin. Neurol. 2012;25, 708–714.

Buerger C. Epidermal mTORC1 signaling contributes to the pathogenesis of psoriasis and could serve as a therapeutic target. Front Immunol. 2018; 9:1–7.

Buhot M-C, Martin S, Segu L. Role of serotonin in memory impairment. Ann Med. 2000; 32(3):210–21.

Cabezas-Opazo FA, Vergara-Pulgar K, Pérez MJ, Jara C, Osorio-Fuentealba C, Quintanilla RA. Mitochondrial dysfunction contributes to the pathogenesis of Alzheimer's disease. Oxid Med Cell Longev. 2015;2015:1–12.

Caggiano AO, Kraig RP. Prostaglandin E receptor subtypes in cultured rat microglia and their role in reducing lipopolysaccharide-induced interleukin-1 β production. J Neurochem. 1999;72(2):565–75.

Cai Q, Tammineni P. Mitochondrial aspects of synaptic dysfunction in Alzheimer's disease. J Alzheimer's Dis. 2017;57(4):1087–10.

Cai Z, Yan L-J, Li K, Quazi SH, Zhao B. Roles of AMP-activated protein kinase in Alzheimer's disease. Neuro Molecular Med. 2012; 14(1):1–14.

Calsolaro V, Edison P. Neuroinflammation in Alzheimer's disease: Current evidence and future directions. Alzheimer's Dement. 2016;12(6):719–32.

Cantó C, Auwerx J. AMP-activated protein kinase and its downstream transcriptional pathways. Cell Mol Life Sci. 2010; 67(20):3407–23.

Castro-Alvarez JF, Uribe-Arias SA, Mejãa-Raigosa D, Cardona-GÃ3mez GP. Cyclin-dependent kinase 5, a node protein in diminished tauopathy: A systems biology approach. Front Aging Neurosci. 2014; 6:1–13.

Chen Y, Peng Y, Che P, Gannon M, Liu Y, Li L, et al. α 2A adrenergic receptor promotes amyloidogenesis through disrupting APP-SorLA interaction. Proc Natl Acad Sci. 2014; 111(48):17296–301.

Chin J. Fyn kinase induces synaptic and cognitive impairments in a transgenic mouse model of Alzheimer's disease. J Neurosci. 2005;25(42):9694–70.

Chu J, Lauretti E, Praticò D. Caspase-3-dependent cleavage of Akt modulates tau phosphorylation via GSK3β kinase: Implications for Alzheimer's disease. Mol Psychiatry. 2017;22(7):1002–8.

Conn PJ, Pin J-P. Pharmacology and functions of metabotropic glutamate receptors. Annu Rev Pharmacol Toxicol. 1997; 37(1):205–37.

Crews L, Masliah E. Molecular mechanisms of neurodegeneration in Alzheimer's disease. Hum Mol Genet. 2010;19(R1):R12–20.

Danysz W, Parsons CG. Alzheimer's disease, β-amyloid, glutamate, NMDA receptors and memantine - searching for the connections. Br J Pharmacol. 2012; 167(2):324–52.

Daulatzai MA. Fundamental role of pan-inflammation and oxidative-nitrosative pathways in neuropathogenesis of Alzheimer's disease. Am J Neurodegener Dis. 2016; 5(2):102–30.

De A. Wnt/Ca2+ signaling pathway: A brief overview. Acta Biochim Biophys Sin. 2011; 43: 745–756.

De-Paula VJ, Radanovic M, Diniz BS, Forlenza OV. Alzheimer's disease. Subcell Biochem. 2012; 65: 329–52.

Di Domenico F, Tramutola A, Foppoli C, Head E, Perluigi M, Butterfield DA. mTOR in Down syndrome: Role in Aß and tau neuropathology and transition to Alzheimer disease-like dementia. Free Radic Biol Med. 2018; 114:94–101.

Dong W, Wang F, Guo W, Zheng X, Chen Y, Zhang W, et al. Aβ25–35 suppresses mitochondrial biogenesis in primary hippocampal neurons. Cell Mol Neurobiol. 2016; 36(1):83–91.

Engelhardt B, Carare RO, Bechmann I, Flügel A, Laman JD, Weller RO. Vascular, glial, and lymphatic immune gateways of the central nervous system. Acta Neuropathol. 2016;132(3):317–38.

Espada S, Rojo AI, Salinas M, Cuadrado A. The muscarinic M1 receptor activates Nrf2 through a signaling cascade that involves protein kinase C and inhibition of GSK-3beta: connecting neurotransmission with neuroprotection. J Neurochem. 2009;110(3):1107–19.

Fan M, Raymond L. N-Methyl-d-aspartate (NMDA) receptor function and excitotoxicity in Huntington's disease. Prog Neurobiol. 2007; 81; (5–6):272–93.

Ferri CP, Prince M, Brayne C, Brodaty H, Fratiglioni L, Ganguli M, et al. Global prevalence of dementia: a Delphi consensus study. Lancet. 2005; 17;366(9503):2112–7.

Folch J, Ettcheto M, Busquets O, Sánchez-López E, Castro-Torres R, Verdaguer E, et al. The implication of the brain insulin receptor in late onset Alzheimer's disease dementia. Pharmaceuticals. 2018;11(1):11.

Friedman LK. Calcium: A role for neuroproduction and sustained adaptation. Mol Interv. 2006; 6(6):315–29.

Frozza RL, Lourenco M V., De Felice FG. Challenges for Alzheimer's disease therapy: Insights from novel mechanisms beyond memory defects. Front Neurosci. 2018; 12:1–13.

Gannon M, Che P, Chen Y, Jiao K, Roberson ED, Wang Q. Noradrenergic dysfunction in Alzheimer's disease. Front Neurosci. 2015; 9:1–12.

Gao J, Wang L, Liu J, Xie F, Su B, Wang X. Abnormalities of mitochondrial dynamics in neurodegenerative diseases. Antioxidants. 2017;6(2):25.

Geldenhuys W, Van der Schyf C. Serotonin 5-HT6 receptor antagonists for the treatment of Alzheimer's disease. Curr Top Med Chem. 2008; 8(12):1035–48.

Geldenhuys WJ, Van der Schyf CJ. Role of serotonin in Alzheimer's disease. CNS Drugs. 2011; 25(9):765–81.

Geula C, Nagykery N, Nicholas A, Wu C-K. Cholinergic neuronal and axonal abnormalities are present early in aging and in Alzheimer disease. J Neuropathol Exp Neurol. 2008;67(4):309–18.

Griffin RJ, Moloney A, Kelliher M, Johnston JA, Ravid R, Dockery P, et al. Activation of Akt/PKB, increased phosphorylation of Akt substrates and loss and altered distribution of Akt and PTEN are features of Alzheimer's disease pathology. J Neuro chem. 2005; 93(1):105–17.

Haass C, Kaether C, Thinakaran G, Sisodia S. Trafficking and proteolytic processing of APP. Cold Spring Harb Perspect Med. 2012;2(5):a006270–a006270.

Haga T. Molecular properties of muscarinic acetylcholine receptors. Proc Japan Acad Ser B. 2013; 89(6):226–56.

Hampel H, Mesulam M-M, Cucllo AC, Farlow MR, Giacobini E, Grossberg GT, et al. The cholinergic system in the pathophysiology and treatment of Alzheimer's disease. Brain. 2018;141(7):1917–33.

Hernandez A, Tan C, Mettlach G, et al. Cdk5 modulates long-term synaptic plasticity and motor learning in dorsolateral striatum. Sci Rep. 2016; 6:29812.

Hong S, Beja-Glasser VF, Nfonoyim BM, Frouin A, Li S, Ramakrishnan S, Ramakrishnan, Merry KM, et al. Complement and microglia mediate early synapse loss in Alzheimer mouse models. Science. 2016 May 6; 352(6286):712–716.

Huang W-J, Zhang X, Chen W-W. Role of oxidative stress in Alzheimer's disease. Biomed Reports. 2016;4(5):519–22.

Ittner A, Chua SW, Bertz J, Volkerling A, van der Hoven J, Gladbach A, et al. Site-specific phosphorylation of tau inhibits amyloid-β toxicity in Alzheimer's mice. Science. 2016;354(6314):904–8.

Jiang S, Bhaskar K. Dynamics of the complement, cytokine, and chemokine systems in the regulation of synaptic function and dysfunction relevant to Alzheimer's disease. J Alzheimer's Dis. 2017;57(4):1123–35

Jiang S, Li Y, Zhang C, Zhao Y, Bu G, Xu H, et al. M1 muscarinic acetylcholine receptor in Alzheimer's disease. Neurosci Bull. 2014;30(2):295–307.

Joyce JN, Smutzer G, Whitty CJ, Myers A, Bannon MJ. Differential modification of dopamine transporter and tyrosine hydroxylase mRNAs in midbrain of subjects with Parkinson's, Alzheimer's with parkinsonism, and Alzheimer's disease. Mov Disord. 1997; 12(6):885–97.

Kablar NA. Signaling pathways perspective of neurodegenerative diseases: Parkinson's, Alzheimer's, and Huntington's diseases. Acta sci. microbial. 2019;2 (4): 16–21.

Kalaria RN, Andorn AC, Tabaton M, Whitehouse PJ, Harik SI, Unnerstall JR. Adrenergic receptors in aging and Alzheimer's disease: Increased? 2 -receptors in prefrontal cortex and hippocampus. J Neurochem. 1989; 53(6):1772–81.

Kametani F, Hasegawa M. Reconsideration of amyloid hypothesis and tau hypothesis in Alzheimer's disease. Front Neurosci. 2018;12:25.

Kandimalla R, Reddy PH. Therapeutics of neurotransmitters in Alzheimer's disease. J Alzheimer's Dis. 2017;57(4):1049–69.

Kar S, Li T, Hawkes C, Qureshi HY, Paudel HK. Cyclin-dependent protein kinase 5 primes microtubule-associated protein tau site-specifically for glycogen synthase kinase 3β †. Biochemistry. 2006;45(10):3134–45.

Kaur S, Das Gupta G, Singh S. Altered neurochemistry in Alzheimer's disease: Targeting neurotransmitter receptor mechanisms and therapeutic strategy. Neurophysiology. 2019; 51(4):293–309.

Kelleher RJ, Shen J. Presenilin-1 mutations and Alzheimer's disease. Proc Natl Acad Sci. 2017;114(4):629–31.

Khachaturian AS, Hayden KM, Mielke MM, Tang Y, Lutz MW, Gustafson DR, et al. Future prospects and challenges for Alzheimer's disease drug development in the era of the NIA-AA Research Framework. Alzheimer's Dement. 2018; 14(4):532–4.

Killick R, Ribe EM, Al-Shawi R, et al. Clusterin regulates β-amyloid toxicity via Dickkopf-1-driven induction of the wnt-PCP-JNK pathway. Mol Psychiatry. 2014; 19: 88–98.

Kimura N, Tokunaga C, Dalal S, Richardson C, Yoshino K, Hara K, et al. A possible linkage between AMP-activated protein kinase (AMPK) and mammalian target of rapamycin (mTOR) signalling pathway. Genes to Cells. 2003; 8(1):65–79.

Kong Y, Ruan L, Qian L, Liu X, Le Y. Norepinephrine promotes microglia to uptake and degrade amyloid peptide through upregulation of mouse formyl peptide receptor 2 and induction of insulin-degrading enzyme. J Neurosci. 2010; 30(35):11848–57.

Kumar K, Kumar A, Keegan RM, Deshmukh R. Recent advances in the neurobiology and neuropharmacology of Alzheimer's disease. Biomed Pharmacother. 2018; 98:297–307.

Kuret J, Congdon EE, Li G, Yin H, Yu X, ZQ. Evaluating triggers and enhancers of tau fibrillization. Microsc Res Tech. 2005;67(3–4):141–55.

Lacor PN, Buniel MC, Furlow PW, Sanz Clemente A, Velasco PT, Wood M, et al. Oligomer-induced aberrations in synapse composition, shape, and density provide a molecular basis for loss of connectivity in Alzheimer's disease. J Neurosci. 2007;27(4):796–807.

Lagouge M, Argmann C, Gerhart-Hines Z, Meziane H, Lerin C, Daussin F, et al. Resveratrol improves mitochondrial function and protects against metabolic disease by activating SIRT1 and PGC-1α. Cell. 2006; 127(6):1109–22.

Larson M, Sherman MA, Amar F, Nuvolone M, Schneider JA, Bennett DA, et al. The Complex PrPc-Fyn couples human oligomeric a with pathological tau changes in Alzheimer's disease. J Neurosci. 2012;32(47):16857–71.

Li BS, Ma W, Jaffe H, Zheng Y, Takahashi S, Zhang L, Kulkarni AB, Pant HC. Cyclin-dependent kinase-5 is involved in neuregulin-dependent activation of phosphatidylinositol 3-kinase and Akt activity mediating neuronal survival. J Biol Chem. 2003;278:35702–3570.

Li N-M, Liu K-F, Qiu Y-J, Zhang H-H, Nakanishi H, Qing H. Mutations of beta-amyloid precursor protein alter the consequence of Alzheimer's disease pathogenesis. Neural Regen Res. 2019;14(4):658–65.

Lin H, Vicini S, Hsu F-C, Doshi S, Takano H, Coulter DA, et al. Axonal 7 nicotinic ACh receptors modulate presynaptic NMDA receptor expression and structural plasticity of glutamatergic presynaptic boutons. Proc Natl Acad Sci. 2010; 107(38):16661–6.

Liu SL, Wang C, Jiang T, Tan L, Xing A, Yu JT. The role of Cdk5 in Alzheimer's disease. Mol Neurobiol. 2016; 53(7):4328–42.

Lombó GC, Schroder A, Reyes-Reyes EM and Franco, R. mTOR/AMPK signaling in the brain: Cell metabolism, proteostasis and survival. Current opinion in toxicology, 2018; 8:102–110.

Lu J, Wu D, Zheng Y, Hu B, Zhang Z, Shan Q, et al. Quercetin activates AMP-activated protein kinase by reducing PP2C expression protecting old mouse brain against high cholesterol-induced neurotoxicity. J Pathol. 2010; 222(2):199–212.

Lyness SA, Zarow C, Chui HC. Neuron loss in key cholinergic and aminergic nuclei in Alzheimer disease: a meta-analysis. Neurobiol Aging. 2003; 24(1):1–23.

McGeer PL, Rogers J, McGeer EG. Inflammation, antiinflammatory agents, and Alzheimer's disease: the last 22 years. J Alzheimer's Dis. 2016;54(3):853–7.

McLeod F and Salinas PC. Wnt proteins as modulators of synaptic plasticity. Curr Opin Neurobiol. 2018; 53: 90–95.

Magalingam KB, Radhakrishnan A, Ping NS, Haleagrahara N. Current concepts of neurodegenerative mechanisms in Alzheimer's disease. Biomed Res Int. 2018;2018:1–12.

Marien MR, Colpaert FC, Rosenquist AC. Noradrenergic mechanisms in neurodegenerative diseases: A theory. Brain Res Brain Res Rev. 2004; 45(1):38–78.

Martorana A, Koch G. Is dopamine involved in Alzheimer's disease? Front Aging Neurosci. 2014; 6:1–6.

Marzo A, Galli S, Lopes D, McLeod F, Podpolny M, Segovia-Roldan M et al. Reversal of synapse degeneration by restoring wnt signaling in the adult hippocampus. Curr Biol. 2016; 26:2551–2561.

Mash D, Flynn D, Potter L. Loss of M2 muscarine receptors in the cerebral cortex in Alzheimer's disease and experimental cholinergic denervation. Science. 1985;228(4703):1115–7.

Matrone C, Marolda R, Ciafrè S, Ciotti MT, Mercanti D and Calissano P. Tyrosine kinase nerve growth factor receptor switches from prosurvival to proapoptotic activity via Abeta-mediated phosphorylation. PNAS 2009; 106(27):11358–11363.

Melov S, Adlard PA, Morten K, Johnson F, Golden TR, Hinerfeld D, et al. Mitochondrial oxidative stress causes hyperphosphorylation of tau. PLoS One. 2007;2(6):e536.

Meyer DA, Torres-Altoro MI, Tan Z, Tozzi A, Di Filippo M, DiNapoli V, et al. Ischemic stroke injury is mediated by aberrant Cdk5. J Neurosci. 2014; 34: 8259–8267.

Miao J, Shi R, Li L, Chen F, Zhou Y, Tung YC, et al. Pathological tau from Alzheimer's brain induces site-specific hyperphosphorylation and SDS- and reducing agent-resistant aggregation of tau in vivo. Front Aging Neurosci. 2019;11(34):1–14.

Miners JS, Baig S, Tayler H, Kehoe PG, Love S. Neprilysin and insulin-degrading enzyme levels are increased in Alzheimer disease in relation to disease severity. J Neuropathol Exp Neurol. 2009;68(8):902–14.

Mohandas E, Rajmohan V, Raghunath B. Neurobiology of Alzheimer's disease. Indian J Psychiatry. 2009; 51(1):55–61.

Mokhtar SH, Bakhuraysah MM, Cram DS, Petratos S. The beta-amyloid protein of Alzheimer's disease: communication breakdown by modifying the neuronal cytoskeleton. Int J Alzheimers Dis. 2013; 2013:1–15.

Mootha VK, Handschin C, Arlow D, Xie X, St. Pierre J, Sihag S, et al. Err and Gabpa/b specify PGC-1 -dependent oxidative phosphorylation gene expression that is altered in diabetic muscle. Proc Natl Acad Sci. 2004; 101(17):6570–5.

Moreira PI, Carvalho C, Zhu X, Smith MA, Perry G. Mitochondrial dysfunction is a trigger of Alzheimer's disease pathophysiology. Biochim Biophys Acta - Mol Basis Dis. 2010;1802(1):2–10.

Mucke L, Palop JJ. Amyloid-β–induced neuronal dysfunction in Alzheimer's disease: from synapses toward neural networks. Nat Neurosci. 2010;13(7):812–8.

Nava-Mesa MO, Jiménez-Díaz L, Yajeya J, Navarro-Lopez JD. GABAergic neurotransmission and new strategies of neuromodulation to compensate synaptic dysfunction in early stages of Alzheimer's disease. Front Cell Neurosci. 2014; 8:1–19.

Nusse R and Clevers H. Wnt/b-catenin signaling, disease and emerging therapeutic modalities. Cell. 2017; 169:985–999.

Nyberg L, Karalija N, Salami A, Andersson M, Wåhlin A, Kaboovand N, et al. Dopamine D2 receptor availability is linked to hippocampal–caudate functional connectivity and episodic memory. Proc Natl Acad Sci. 2016; 113(28):7918–23.

Nygaard HB, van Dyck CH, Strittmatter SM. Fyn kinase inhibition as a novel therapy for Alzheimer's disease. Alzheimers Res Ther. 2014; 6(1):1–8.

Nygaard HB. Targeting fyn kinase in Alzheimer's disease. Biol Psychiatry. 2018;83(4):369–76.

Oddo S. The role of mTOR signaling in Alzheimer disease. Front Biosci. 2012; S4 (3):941–52.

Origlia N, Righi M, Capsoni S, Cattaneo A, Fang F, Stern DM, et al. Receptor for advanced glycation end product-dependent activation of p38 mitogen-activated protein kinase contributes to amyloid- -mediated cortical synaptic dysfunction. J Neurosci. 2008;28(13):3521–30.

Orlova KA, Crino PB. The tuberous sclerosis complex. Ann N Y Acad Sci. 2010; 1184(1):87–105.

Palomer E, Buechler J, Salinas PC. Wnt signaling deregulation in the aging and Alzheimer's brain. Front Cell Neurosci. 2019; 13:1–8.

Passani MB, Blandina P. Cognitive implications for H3 and 5-HT3 receptor modulation of cortical cholinergic function: A parallel story. Methods Find Exp Clin Pharmacol. 1998; 20(8):725.

Perez SE, Lazarov O, Koprich JB, Rodriguez-Menendez V, Lipton JW, Sisodia SS, et al. Nigrostriatal dysfunction in familial Alzheimer's disease-linked APPswe/PS1 E9 transgenic mice. J Neurosci. 2005; 25(44):10220–9.

Perluigi M, Di Domenico F, Butterfield DA. mTOR signalling in aging and neurodegeneration: At the crossroad between metabolism dysfunction and impairment of autophagy. Neurobiol Dis. 2015; 84:39–49.

Peterson DW, Ando DM, Taketa DA, Zhou H, Dahlquist FW, Lew J. No difference in kinetics of tau or histone phosphorylation by CDK5/p25 versus CDK5/p35 in vitro. Proc. Natl Acad Sci. 2010; 107:2884–2889.

Plassman BL, Langa KM, Fisher GG, Heeringa SG, Weir DR, Ofstedal MB, et al. Prevalence of dementia in the united states: The aging, demographics, and memory study. Neuroepidemiology. 2007;29(1–2):125–32.

Posadas I, Lopez-Hernandez B, Cena V. Nicotinic receptors in neurodegeneration. Curr Neuropharmacol. 2013; 11(3): 298–314.

Purro SA, Dickins EM, Salinas PC. The secreted Wnt antagonist Dickkopf-1 is required for amyloid β-mediated synaptic loss. J Neurosci. 2012; 32:3492–3498.

Ramos BP, Arnsten AFT. Adrenergic pharmacology and cognition: Focus on the prefrontal cortex. Pharmacol Ther. 2007; 113(3):523–36.

Ramos-Rodriguez JJ, Ortiz O, Jimenez-Palomares M, Kay KR, Berrocoso E, Murillo-Carretero MI, et al. Differential central pathology and cognitive impairment in pre-diabetic and diabetic mice. Psychoneuroendocrinology. 2013;38(11):2462–75.

Ren Z, He H, Zuo Z, Xu Z, Wei Z, Deng J. The role of different SIRT1-mediated signaling pathways in toxic injury. Cell Mol Biol Lett. 2019; 24(1):1–10.

Ricci G, Volpi L, Pasquali L, Petrozzi L, Siciliano G. Astrocyte–neuron interactions in neurological disorders. J Biol Phys. 2009; 35(4):317–36.

Rudy CC, Hunsberger HC, Weitzner DS, Reed MN. The role of the tripartite glutamatergic synapse in the pathophysiology of Alzheimer's disease. Aging Dis. 2015; 6(2):131–48.

Saber M, Cochran OK, Puntambekar SS, Lathia JD and Lamb BT. Triggering receptor expressed on myeloid cells 2 deficiency alters acute macrophage distribution and improves recovery after traumatic brain injury. J Neurotrauma. 2017 Jan 15; 34 (2):423–435.

Stephan A, Barres BA, Stevens B. The complement system: an unexpected role in synaptic pruning during development and disease Ann. Rev. Neurosci., 2012;35:369–389.

Sarasamkan J, Scheunemann M, Apaijai N, Palee S, Parichatikanond W, Arunrungvichian K, et al. Varying chirality across nicotinic acetylcholine receptor subtypes: Selective binding of quinuclidine triazole compounds. ACS Med Chem Lett. 2016; 7(10):890–5.

Schroder H, Giacobini E, Struble RG, Zilles K, Maelicke A. Nicotinic cholinoceptive neurons of the frontal cortex are reduced in Alzheimer's disease. Neurobiol Aging. 1991;12(3):259–62.

Seifert JRK and Mlodzik M. Frizzled/PCP signalling: A conserved mechanism regulating cell polarity and directed motility. Nat Rev Gent. 2007;8:126–138.

Selkoe DJ, Hardy J. The amyloid hypothesis of Alzheimer's disease at 25 years. EMBO Mol Med. 2016;8(6):595–608.

Shahani N, Pryor W, Swarnkar S, Kholodilov N, Thinakaran G, Burke RE, et al. RhebGTPase regulates β-secretase levels and amyloid β generation. J Biol Chem. 2014; 289(9):5799–808.

Shen J, Wu J. Nicotinic cholinergic mechanisms in Alzheimer's disease. Int Rev Neurobiol. 2015;124:275–92.

Shi Y, Yamada K, Liddelow SA, Smith ST, Zhao L, Luo W, et al. ApoE4 markedly exacerbates tau-mediated neurodegeneration in a mouse model of tauopathy. Nature. 2017;549(7673):523–7.

Shimobayashi M, Hall MN. Making new contacts: The mTOR network in metabolism and signalling crosstalk. Nat Rev Mol Cell Biol. 2014; 15(3):155–62.

Sparks DL, DeKosky ST, Markesbery WR. Alzheimer's disease: Aminergic-cholinergic alterations in hypothalamus. Arch Neurol. 1988; 45(9):994–9.

Stephan A, Barres BA, Stevens B. The complement system: an unexpected role in synaptic pruning during development and disease Ann. Rev. Neurosci., 2012;35:369–389.

Suárez-Calvet M, Belbin O, Pera M, Badiola N, Magrané J, Guardia-Laguarta C, et al. Autosomal-dominant Alzheimer's disease mutations at the same codon of amyloid precursor protein differentially alter Aβ production. J Neurochem. 2014;128(2):330–9.

Sweeney G, Song J. The association between PGC-1α and Alzheimer's disease. Anat Cell Biol. 2016;49(1):1–6.

Teich AF, Nicholls RE, Puzzo D, Fiorito J, Purgatorio R , Fa' M, et al. Synaptic therapy in Alzheimer's disease: A CREB-centric approach. Neurotherapeutics. 2015;12(1):29–41.

Terry RD. Alzheimer's disease and the aging brain. J Geriatr Psychiatry Neurol. 2006 Sep;19(3):125–8.

Thornton C, Bright NJ, Sastre M, Muckett PJ, Carling D. AMP-activated protein kinase (AMPK) is a tau kinase, activated in response to amyloid β-peptide exposure. Biochem J 2011; 434(3):503–12.

Tönnies E, Trushina E. Oxidative stress, synaptic dysfunction, and Alzheimer's disease. J Alzheimer's Dis. 2017;57(4):1105–2.

Trepanier CH, Jackson MF, MacDonald JF. Regulation of NMDA receptors by the tyrosine kinase Fyn. FEBS J. 2012;279(1):12–9.

Trillo L, Das D, Hsieh W, Medina B, Moghadam S, Lin B, et al. Ascending monoaminergic systems alterations in Alzheimer's disease. Translating basic science into clinical care. Neurosci Biobehav Rev. 2013; 37(8):1363–79.

Um JW, Nygaard HB, Heiss JK, Kostylev MA, Stagi M, Vortmeyer A, et al. Alzheimer amyloid-β oligomer bound to postsynaptic prion protein activates Fyn to impair neurons. Nat Neurosci. 2012;15(9):1227–35.

Uniyal JM, Wainwright J V., Mohan AL, Tobias ME, Murali R, Gandhi CD, et al. Diverse signaling mechanisms of mTOR complexes: mTORC1 and mTORC2 in forming a formidable relationship. Adv Biol Regul. 2019; 72:51–62.

Vingtdeux V, Davies P, Dickson DW, Marambaud P. AMPK is abnormally activated in tangle- and pre-tangle-bearing neurons in Alzheimer's disease and other tauopathies. Acta Neuropathol. 2011; 121(3):337–49.

Viollet B, Lantier L, Devin-Leclerc J, Hebrard S, Amouyal C, Mounier R, et al. Targeting the AMPK pathway for the treatment of Type 2 diabetes. Front Biosci (Landmark Ed). 2009; 14:3380–400.

Wang D-S, Dickson DW, Malter JS. β-Amyloid degradation and Alzheimer's disease. J Biomed Biotechnol. 2006;2006(3):1–12.

Wang JM, Sun C. Calcium and neurogenesis in Alzheimer's disease. Front Neurosci. 2010; 4:1–5.

Wang X, Zimmermann HR, Ma T. Therapeutic potential of AMP-activated protein kinase in Alzheimer's disease. J Alzheimer's Dis. 2019; 68(1):33–8.

Wang Y, Martinez-Vicente M, Krüger U, Kaushik S, Wong E, Mandelkow E-M, et al. Tau fragmentation, aggregation and clearance: the dual role of lysosomal processing. Hum Mol Genet. 2009; 18(21):4153–70.

Wang Z-X, Tan L, Yu J-T. Axonal transport defects in Alzheimer's disease. Mol Neurobiol. 2015;51(3):1309–21.

Weydt P, Pineda VV., Torrence AE, Libby RT, Satterfield TF, Lazarowski ER, et al. Thermoregulatory and metabolic defects in Huntington's disease transgenic mice implicate PGC-1α in Huntington's disease neurodegeneration. Cell Metab. 2006; 4(5):349–62.

Wilkaniec A, Gąssowska-Dobrowolska M, Strawski M, Adamczyk A, Czapski GA. Inhibition of cyclin-dependent kinase 5 affects early neuroinflammatory signalling in murine model of amyloid beta toxicity. J Neuroinflammation. 2018; 15:1–18.

Wullschleger S, Loewith R, Hall MN. TOR signaling in growth and metabolism. Cell. 2006; 124(3):471–84.

Yang K, Belrose J, Trepanier CH, Lei G, Jackson MF, MacDonald JF. Fyn, a potential target for Alzheimer's disease. J Alzheimer's Dis. 2011; 27(2):243–52.

Yang W, Tian Z-K, Yang H-X, Feng Z-J, Sun J-M, Jiang H, et al. Fisetin improves lead-induced neuroinflammation, apoptosis and synaptic dysfunction in mice associated with the AMPK/SIRT1 and autophagy pathway. Food Chem Toxicol. 2019; 134:110824.

Yonemura Y, Futai E, Yagishita S, Suo S, Tomita T, Iwatsubo T, et al. Comparison of presenilin 1 and presenilin 2 γ-secretase activities using a yeast reconstitution system. J Biol Chem. 2011;286(52):44569–75.

Yu JSL, Cui W. Proliferation, survival and metabolism: the role of PI3K/AKT/mTOR signalling in pluripotency and cell fate determination. Development. 2016; 143(17):3050–60.

Zhou Y, Wang S, Li Y, Yu S, Zhao Y. SIRT1/PGC-1α signaling promotes mitochondrial functional recovery and reduces apoptosis after intracerebral hemorrhage in rats. Front Mol Neurosci. 2018; 10:1–14.

5 Alzheimer's Disease

Pathogenesis and Therapeutics Advancements Targeting Potential Neurotransmitters and Neuronal Peptides

Md. Sayeed Akhtar and Moteb Khobrani
Department of Clinical Pharmacy, College of Pharmacy, King Khalid University, Abha, Saudi Arabia

Faheem Hyder Pottoo
Department of Pharmacology, College of Clinical Pharmacy, Imam Abdulrahman Bin Faisal University, Dammam, Saudi Arabia

Jawad ur Rahman
Department of Microbiology, College of Medicine, Imam Abdulrahman Bin Faisal University, Dammam, Saudi Arabia

CONTENTS

DOI: 10.1201/9780429265198-7

5.1 INTRODUCTION

In the early twentieth century, a German psychiatrist, Alois Alzheimer, explained the pathologic findings of an elderly woman as psychiatric disturbances, memory loss, and sluggish speech (Zvěřová, 2019). The World Health Organization has stated that almost 50 million people has been affected by the Alzheimer's disease (AD), and this number is anticipated to raise threefold over the next 30 years (Garre-Olmo, 2018; Nichols et al., 2019). The percentage of mortality due to AD is around 70% higher than stroke and prostate cancer in the elderly (Alzheimer's Association, 2016). It is also the sixth major cause of death in the U.S.A., and 5.7 million people are estimated to be living with AD. The annual pharmacoeconomic loss in the U.S.A. alone has been estimated to be \$290 billion. However, the worldwide estimated expenditure on AD and related dementia is reported to be around \$605 billion (Norins, 2019). The human brain is a highly complex structure involved in various activities and thus it is difficult to resolve any brain disorder. An understanding of the complexity is of utmost important in fixing errors in such a machine (Buffalo et al., 2019). In particular, neuronal damage is seen more in the hippocampus, amygdala, thalamus, and cortical area, which are greatly involved in the control of learning and memory processes in the human brain (Franklin and Grossberg, 2017). AD has been considered to be a major form of dementia, as well as learning debility (Sharma et al., 2020). In this context, the involvement of neurotransmitters in the development of AD is also under investigation.

Neurotransmitters are endogenous biochemicals that transmit neuronal signals between the neuronal synapses and neuromuscular junctions (NMJs). Acetylcholine (ACh), dopamine, gamma-amino butyric acid (GABA), serotonin, *N*-methyl-D-aspartate (NMDA), glutamate, norepinephrine, and endorphins are among those neurotransmitters that play a vital role in learning, cognition, behavioral activities, and muscular movements (Purves et al., 2001). These neurotransmitters are released from synaptic vesicles by exocytosis and bind with the respective neurotransmitter receptors for transmission of neuronal signals (Lodish et al., 2000b). Two types of neurotransmitter receptors are involved in neuronal transmission, and are called metabotropic (muscarinic ACh receptor, $GABA_B$ receptor, serotonin receptor) and ionotropic neurotransmitter receptors (nicotinic ACh receptor, $GABA_A$ receptor) (Lodish et al., 2000a). The former act through secondary messengers and the latter via ligand-gated channels. New target molecules like ACh esterase (AChE), dopamine, glutamate, GABA, noradrenaline, beta secretase 1 (BACE-1), glycogen synthase kinase 3β (GSK-3β), monoamine oxidases (MAOs), metal ions in the brain, NMDA receptor, 5-hydroxytryptamine (5-HT/serotonin) receptors, histamine-3 (H-3) receptor, cannabinoid receptor, and phosphodiesterases have been considered as a major therapeutic opportunity for AD (Agis-Torres et al., 2014; Bortolato et al., 2008; Companys-Alemany et al., 2020; Grill and Cummings, 2010). Among these targets, AChE inhibitors (AChEIs), in particular, are the key targets and have currently achieved preference over other receptor-based target strategies.

On the basis of all this evidence, the U.S. Food and Drug Administration (FDA) approved the three key AChEIs – galantamine, rivastigmine, and donepezil – and a NMDA antagonist – memantine – for clinical use in AD (Koola et al., 2018; Parsons et al., 2013). Meanwhile, the earlier-approved AChEI tacrine lost its approval due to its hepatotoxicity. Keeping these facts in mind, we discuss the basics of different neuronal transmission related to AD in this review. We also provide an update on the present status of receptor-based, and related, therapeutic approaches in neurotransmission modulation (Figure 5.1).

5.2 HUMAN BRAIN

The human brain is the central and most vital organ of our central nervous system, shielded by the skull bones. It is made up of billions of neurons and glial cells that remain interconnected as a complex network (Buffalo et al., 2019). Being the command and control system, it directs and coordinates our body system to perform tasks related to our daily life activities. A human brain has three main parts: the cerebrum, the cerebellum, and the brain stem (NIH Curriculum Supplement Series, 2007; Johns, 2014). The cerebrum is the largest part of the brain and is divided into two hemispheres: the left and the right. Each hemisphere is further divided into four lobes: frontal, temporal, parietal, and occipital. Each lobe is further subdivided into many areas based on the specific functions and tasks they perform. Generally, the cerebrum is involved in controlling and interpreting sensory signals of touch, vision, hearing, movement, reasoning, emotions, learning, and speech (Abhang et al., 2016). The cerebellum is located just below the cerebrum. It regulates muscle movement, maintaining body posture and its equilibrium. The brain stem connects the cerebrum and cerebellum to the peripheral nervous system (spinal cord), and consists of the midbrain, pons, and medulla oblongata. It functions as a relay and controls autonomous brain functions like breathing,

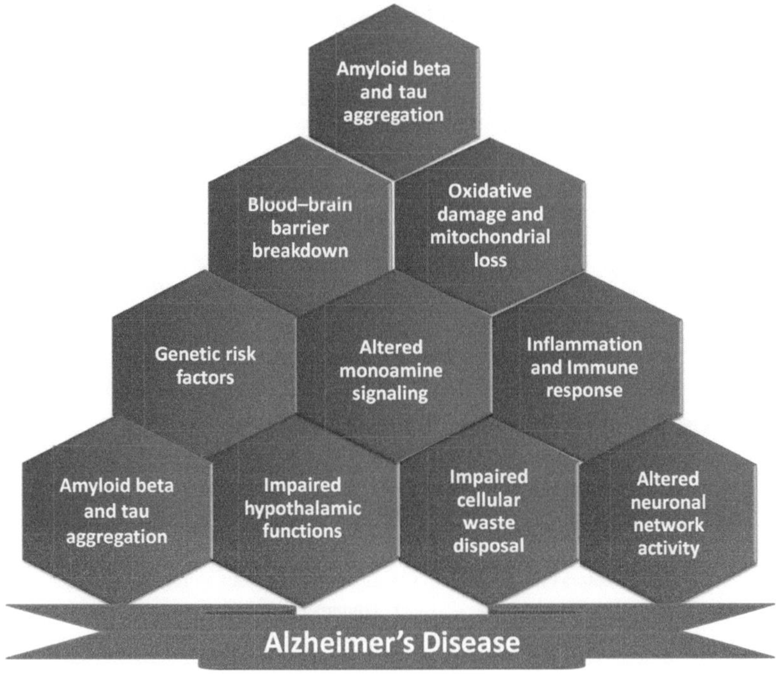

FIGURE 5.1 Pathogenesis of Alzheimer's disease.

swallowing, heart rate, blood pressure, digestion, sneezing, coughing, vomiting, swallowing, wake and sleep cycles, and so on (Suchan and Daum, 2001). The structural development of the human nervous system begins on embryonic day 21, from the ectoderm layer of the embryo and it is continuous from childhood through adulthood. The in utero conditions are critical for brain development; any perturbation in the in utero environment or postnatal nutrition will significantly affect brain development in the infant (Elshazzly and Caban, 2019; Silbereis et al., 2016; Stiles and Jernigan, 2010). Brain development is a highly organized process in which the pattern of events is crucial in terms of sequence and timing. These events include neurogenesis, neuronal migration, synaptogenesis, gliogenesis, and neuronal wiring. Each of these events – either sequential or overlapping – has to occur in a proper temporal window. Any deviation from this pattern has the potential to cause aberrant brain development that significantly compromises brain function (Jiang and Nardelli, 2016).

5.3 ALZHEIMER'S DISEASE

AD creates traumatic and stressful issues for both patients and their caregivers. From a clinical perspective, AD is described by progressive, irreversible dementia because of concurrent neurodegeneration and cognitive decline (Mir et al., 2021; Pottoo, Sharma, et al., 2020). The brain pathology of AD includes neurodegeneration, extracellular accrual of amyloid β (Aβ) plaques, and the intracellular development of axonal tangles of hyperphosphorylated tau protein (Ibrahim et al., 2020; Mir et al., 2020; Bennett et al., 2004). In general, AD includes typical cognitive issues like short-term memory deficit, aphasia, praxis, and executive dysfunction (Zvěřová, 2019). A major risk factor is age, but the presence of apolipoprotein E4 (ApoE4) alleles, a family history of AD, low occupational and educational skills, metabolic disorder and brain trauma are also significantly involved in the initiation and progression of AD (Livingston et al., 2017). Taken together, the quality of life of patients with AD, as well as their families, worsens and even affects their socioeconomic status; therefore, in AD, the utmost priority must be given to treatment (Barbe et al., 2018; Kahle-Wrobleski et al., 2017).

5.4 GENETICS OF ALZHEIMER'S DISEASE

The genetics of AD will be explained before a discussion of the different pathologic hypotheses and other molecular alterations related to AD (Nelson et al., 2015). Based on age, AD is classified into early-onset AD (EOAD; in patients aged <60 years) and late-onset AD (LOAD; in those aged ≥60 years). In 58–79% of patients with LOAD and in >90% of patients with EOAD, genetic inheritability plays a significant role in the initiation and development of AD. Thus, genetic decoding may help in its early prediction, as well as with better recognition of the etiologic complexity related to AD. More than 50 loci related to different pathological changes are involved in AD (Sims et al., 2020). Overall, 5% of cases comprise EOAD; of these, 10–15% carry autosomal dominant inheritance. Around 95% of AD cases are LOAD and around 79% of cases exhibit polygenic inheritance. All cases of EOAD have a family history of a dementia-related disorder. LOAD cases are more complex and regulated by polygenic inheritance. Greater advancement in microarray and genome sequencing has revealed the complete genetic structure of AD.

This further helps in identifying different novel drug target for therapeutic use (Bagyinszky et al., 2020).

5.5 NEUROCHEMICAL INVOLVEMENT IN ALZHEIMER'S DISEASE

Neurochemicals like ACh, dopamine, glutamate, GABA, serotonin, noradrenaline, and endocannabinoids have been shown to regulate the behavior, learning, and cognition in humans (Table 5.1).

5.5.1 Acetylcholine in Alzheimer's Disease

ACh is secreted after activation of the cholinergic neuronal system and has been reported to be involved in memory (Contestabile, 2011). A decreased level of ACh in the brain concurrent to the degradation of cholinergic neurons in the basal forebrain is the major reason for the loss of cognitive function in AD (Rossor et al., 1982). Many studies have indicated a strong correlation between reduced ACh concurrent to ACh receptors and synthesis, release and deposition of Aβ,

TABLE 5.1
Various Neurochemical and Associated Functions in Alzheimer's Disease

Neurotransmitters and Neuropeptides	Basic Mechanism	Major Functions	References
Acetylcholine cholinergic/parasympathetic neurotransmitters	Binding to cholinergic receptors (M_{1-5}) Mediated via IP3 synthesis and cAMP inhibition	• Synaptic survival • Learning and memory regulation • Glutamatergic neuronal regulation • Dopaminergic regulation	Contestablie, 2011; Rossor et al., 1982; Majdi et al., 2020; Foster et al., 2014; Seeger et al., 2004; Lombardo and Maskos, 2015; Hoover, 2017; Ferreira-Vieira et al., 2016
Dopamine (inhibitory/excitatory dopaminergic neurotransmitter)	Binding to D1 and D2 receptor of substantia nigra pars compacta and ventral tegmental area	• Maintain cognition and learning • Control motor activity and extrapyramidal symptoms	Nobili et al., 2017; Martorana and Koch, 2014; Rangel-Barajas et al., 2015
Glutamate (excitatory neurotransmitter)	Binding to NMDARs Increased cytosolic free $[Ca^{2+}]$i Modulation of excitatory glutamatergic neurotransmitter	• Neuronal survival and plasticity • Excitotoxicity and neuronal death (NMDAR hyperactivity)	Hawkins, 2013; Valtcheva and Venanee, 2019; Luo et al., 2011; Rudy et al., 2015
GABA (inhibitory GABAergic neurotransmitter)	Binding to $GABA_A$ and $GABA_B$ receptors in brain	• Maintains neurotransmission • Aβ-related neuronal death • Impaired neurogenesis	Li et al., 2016; Xu et al., 2020; Tang et al., 2019; Kwakowsky et al., 2018
Serotonin (5-HT/monoamine neurotransmitter)	Binding to serotonin receptor ($5\text{-}HT_{1-5}$)	• Loss of cognition via decreased cAMP activity and Ca^{2+} influx • Decreased Aβ accumulation via increased secretase (α/β/γ) activity • Decreased inflammatory activity and modulation of learning and memory via activation of adenylyl cyclase and cAMP	Polter and Li, 2010; Ledo et al., 2013, 2016; Andrews et al., 2018; Ramírez, 2013; Marner et al., 2011
Noradrenaline (adrenergic/ sympathetic neurotransmitter)	Binding to α_2/β_2 Adrenergic receptor activation	• Aβ production • Neuronal degeneration	Chen et al., 2014; Toneff et al., 2013; Samuels and Szabadi, 2008
Histamine (histaminergic neurotransmitter)	Binding to H-3R	• Maintenance of cognitive function via decreased adenylyl cyclase activity, decreased activation of high-voltage Ca^{2+} channel and increased MAPK activity	Brown et al., 2001; Tiligada et al., 2011; Notcovich et al., 2010; Saraiva et al., 2019; Naddafi and Mirshafiey, 2013; Zlomuzica et al., 2016
Adenosine (purine nucleoside base)	A_2A receptor	• Aβ-related neurotoxicity • Modulation of cholinergic and GABAergic activity	Haas and Selbach, 2000; Prediger and Takahashi, 2005; Fitzgerald et al., 2004; Stockwell et al., 2017
Cannabinoids	Binding to CB-1 and CB-2 receptors	• Aβ synthesis and its accumulation • Neuronal survival due to decreased inflammation, mitochondrial damage, and oxidative stress	Eubanks et al., 2006; Savonenko et al., 2015; Galve-Roperh et al., 2008; Aso and Ferrer, 2016

IP3, inositol triphosphate; cAMP, cyclic adenosine monophosphate; NMDAR, N-methyl-d-aspartate receptor; GABA, γ-aminobutyric acid; Aβ, amyloid beta; 5-HT, 5-hydroxytryptamine; H-3R, histamine-3 receptor; MAPK, mitogen-activated protein kinase; A2A, adenosine receptor; CB, cannabinoid.

as well as the formation of neurofibrillary tangles (NFTs) in AD (Majdi et al., 2020).

The expression and activation of M1, M3, and M5 receptors causes augmentation of Ca^{2+} entry and inositol trisphosphate generation, and M2 and M4 receptors inhibit formation of cyclic adenosine monophosphate (cAMP). Activation of M1 receptors affects synaptic plasticity, amyloidogenesis, the modulation of glutamatergic neurons, and neuronal differentiation in learning and cognition (Foster et al., 2014). M2 receptors possess an autoinhibitory character for the GABA-mediated signaling system and have been shown to exhibit inhibitory modulation on dopaminergic neurons (Seeger et al., 2004). M3 receptors indirectly activate the accumulation of Aβ plaques. However, diminished expression of M3 receptors reduces learning and cognition (Ahmed et al., 2017; Rossner et al., 1998). M4 receptor degeneration causes behavioral and locomotor disturbances, as well as memory deficits (Foster et al., 2014). The involvement of M4 receptors is also implicated in the development of Parkinson's disease, psychosis, and schizophrenia (Foster et al., 2016; Verma et al., 2018). Reduced blood flow in neurons due to M5 receptor degeneration may cause cerebral ischemia, which may indirectly initiate the development of AD (Lebois et al., 2018). Nicotinic receptors consist of five pentameric subunits that remain present mainly at the NMJ, as well as the brain, including neurons and synapses of hippocampus. Each subunit has 16 small loops formed by α (1–7), α (9, 10), β (1–4), γ, δ, and ε components (Zoli et al., 2015). These nicotinic receptor subtypes possess high Ca^{2+} permeability, glutamatergic transmission activity, and the potential for modulation of neuronal plasticity (Lombardo and Maskos, 2015). Moreover, it also regulates the release of ACh, inflammatory cytokines, apoptotic processes, Aβ synthesis, and Aβ deposition (Dineley et al., 2015; Ferreira-Vieira et al., 2016; Hoover, 2017) (Figure 5.2).

5.5.2 Dopamine in Alzheimer's Disease

By releasing dopamine, the dopaminergic neuron has been shown to be involved in cognition and synaptic plasticity (Nobili et al., 2017). Severe injury of the dopaminergic neurons in the ventral tegmental area (VTA), together with striatum, develop memory loss and extrapyramidal symptoms (EPSs) (Martorana and Koch, 2014). Dopamine and related drugs act in the hippocampus zone of the brain and regulate activities related to memory and cognition (Rangel-Barajas et al., 2015). The loss of dopaminergic neurons in the substantia nigra pars compacta, as well as the VTA, produces EPSs in the later stages of AD (Alberico et al., 2015; Luo and Huang, 2016).

5.5.3 Glutamate in Alzheimer's Disease

The role of glutamate, an excitatory neurotransmitter, in AD is widely reported by various studies. Glutamate helps in neuronal protection, synaptic plasticity, and the coordination of memory (Hawkins, 2013; Valtcheva and Venance, 2019). NMDA is a type of ionotropic receptor and acts through Ca^{2+} influx. The activity of NMDA receptors depends on their distribution in presynaptic, synaptic, and extrasynaptic regions in neuronal wiring. The NMDA receptors (synaptic) are activated by the release of glutamate and are reported to control neuronal survival via antioxidant and antiapoptotic actions (Rudy et al., 2015). Meanwhile, the extrasynaptic NMDA-dependent activation of the 2B subunits of NMDA receptors also prevents apoptosis, as well as neuronal death (Luo et al., 2011).

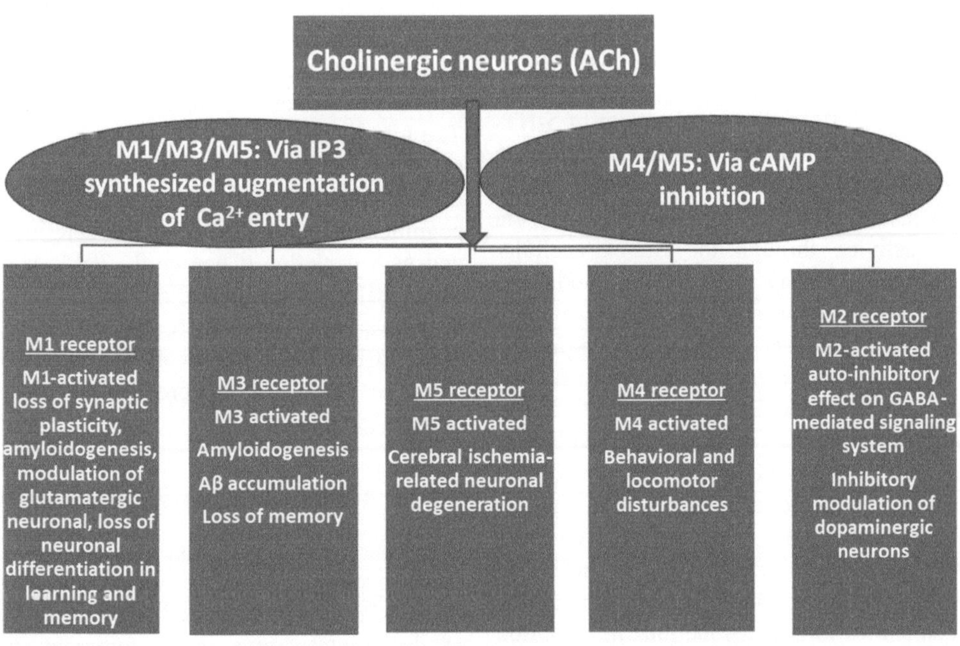

FIGURE 5.2 Role of muscarinic receptors in Alzheimer's disease. ACh, acetylcholine; IP3, inositol triphosphate; cAMP, cyclic adenosine monophosphate; GABA, γ-aminobutyric acid.

5.5.4 γ-Aminobutyric Acid in Alzheimer's Disease

GABA is an inhibitory neurotransmitter that is distributed widely in the human brain. The GABA receptor has been shown to control neurotransmission and Aβ-related excitotoxicity and neuronal cell death. An interrelation between NMDA activation and increased GABA-mediated neurotransmission has been also observed (Lanctôt et al., 2004; Li et al., 2016; Xu et al., 2020). However, an imbalance between GABAergic and glutamatergic neuronal transmission has been shown to impair neurogenesis and routine neuronal functioning(Sun et al., 2009; Tang, 2019; Xu et al., 2020). Taken together, in the GABA signaling system, GABA levels, GABA neurotransmission, and the appearance of GABA receptor-mediated functioning is reported to be damaged in AD (Kwakowsky et al., 2018).

5.5.5 Serotonin and Monoamine Signaling in Alzheimer's Disease

Serotonergic neurons remain present on raphe nuclei and secrete serotonin, which controls mood swings, emotions, the sleep cycle, and human personality in relation to their behaviors (Fakhoury, 2016). Serotoninergic neuronal cells interconnect with the hippocampus area with the raphe nuclei and limbic arrangements (Charnay and Leger, 2010). 5-HT_{1A} activation inhibits Ca^{2+} influx by inhibiting its channel opening via reduced formation of cAMP (Polter and Li, 2010). In a 5-HT_{1A} receptor knockout experimental model, loss of cognition was observed in AD, which was in contrast to memory improvement after the administration of the 5-HT_{1A} mimetic 8-hydroxy-2-(di-n-propylamino) tetralin (Bonaventure et al., 2002). The study revealed an interrelation between 5-HT_{2A} activation and functioning of the NMDA receptor via raised Ca^{2+} levels. Downregulation of 5-HT_{2A} receptor expression and an increased level of Aβ in hippocampus area causes impaired cognition (Marner et al., 2011; Zhang et al., 2013). The involvement of 5-HT_4 receptors in the regulation of learning and memory, and are present in the hippocampus area. This indicates that 5-HT_4 receptors take part in the development of AD probably via adenylate cyclase activity, through the Src/extracellular regulated kinase (ERK) pathway, and exert a modifying action on the expression of cAMP response element-binding protein (CREB), protein kinase B (Akt), and brain-derived neurotrophic factor (BDNF) in neuronal cells (Hagena and Manahan-Vaughan, 2017; Pascual-Brazo et al., 2012). 5-HT_4 receptor-activated effects have been reported to be involved in reduced Aβ accumulation by enhancing the activity of α- and γ-secretase, and in controlling the release of ACh, dopamine, and GABA. Furthermore, the 5-HT_6 receptor agonist has been reported to cause neuronal cell death mainly in the dorsal hippocampus, entorhinal cortex, and striatum area of the brain, and has been shown to impair functions in cognition and memory via Fyn protein kinase, which stimulates ERK-1/2 and enhances tau phosphorylation in AD (Andrews et al., 2018; Ramírez, 2013). 5-HT_7 receptors remain in the area of the hypothalamus, thalamus, hippocampus, cerebral cortex, and raphe nuclei; triggering of these receptors leads to neuronal excitation via the increased release of cAMP (Stiedl et al., 2015). Moreover, their activation has been shown to be responsible for neuronal activation mediated by activating Rho kinase and mitogen-activating protein (MAP) (Zareifopoulos and Papatheodoropoulos, 2016). Various evidence supports the fact that microglia activation and brain tissue inflammation cause mood disorder and depression, which lead to frequent AD related comorbidities. In correlation, studies have shown that altered memory and mood in AD is also due to Aβ oligomers (AβOs) inducing serotonin-related depression and memory deficits in various experimental models of AD (Ledo et al., 2013, 2016). Fascinatingly, patients with AD have been shown to have reduced tryptophan levels and increased quinolinic acid, as well as indolamine-2,3-dioxygenase (IDO) in plasma, which activates microglia (Bonda et al., 2010; Gulaj et al., 2010). Collectively, this evidence points to serotonin, tryptophan, quinolinic acid, and IDO as having a strong correlation in the development of AD. In addition, a dopaminergic alteration, including reduced levels of dopamine, as well as dopamine receptors, is well evidenced in patients with AD and different experimental models (Krashia et al., 2019; Morgese and Trabace, 2019; Pan et al., 2019).

5.5.6 Noradrenaline in Alzheimer's Disease

It is evident that noradrenergic neuronal cells in the brain (locus coeruleus) are mostly responsible to control fear-, fight-, and flight-related functions in humans (Samuels et al., 2008). This noradrenergic neuron directs the release of various neurotransmitters, namely epinephrine (adrenaline), norepinephrine (noradrenaline), and dopamine, which are either degraded by catechol-O-methyltransferase (COMT) and MAO, or are reabsorbed inside the synapses through the Na^+/K^+ pump (Zareifopoulos and Papatheodoropoulos, 2016). The loss of noradrenergic neurons in Korsakoff's syndrome (memory deficit-related disease) point to the importance of the adrenergic/noradrenergic system in memory and cognition, which may be further deteriorated by dopamine and ACh depletion (Chen et al., 2014; Mayes, 2001; Toneff et al., 2013). β_2 adrenergic receptors regulate Aβ production by activating γ-secretase activity in transgenic mice (Chen et al., 2014). Thus, the involvement of noradrenaline in AD needs to be further explored in various experimental and clinical settings.

5.5.7 Histamine and Alzheimer's Disease

Histamine-releasing neurons mainly remain present in the tuberomammillary nucleus, which is located in the posterior part of the hypothalamus of the brain, especially in the area of the frontal cortex, basal forebrain, hippocampus, and amygdala (Blandina et al., 2012). The enzyme L-histidine decarboxylase converts histidine to histamine, which acts on four different subtypes of histamine receptors: H1R, H2R, H3R, and H4R (Haas and Panula, 2003; Passani et al., 2014). However, among these, H3R is reported to modulate the production and secretion of other neurotransmitters like ACh, GABA, dopamine, and serotonin. H1R is a G protein-coupled receptor (GPCR) and possesses a neuronal excitatory effect via phospholipase C activation, which leads to the release of inositol triphosphate (IP3) and Ca^{2+}, along with the formation of diacylglycerol (Brown et al., 2001; Notcovich et al., 2010; Saraiva et al., 2019). H2R is also a type of GPCR, like H1R, and causes adenylyl cyclase activation, which causes the formation of cAMP, protein kinase A (PKA), and MAP kinase (MAPK; Brown et al., 2001; Tiligada et al., 2011). In contrast, H3R stimulation blocks adenylyl cyclase, which activates MAPK and inhibits high-voltage activated Ca^{2+} channels, leading to interference in the exocytosis process and is much less effective in the direct treatment of AD (Brown et al., 2001; Celanire et al., 2005; Kubo et al., 2015; Łażewska and Kieć-Kononowicz, 2014). Importantly, blood histamine levels have been observed to be higher in EOAD than in LOAD, signaling its early benefits in AD (Alvarez et al., 1996; Fernández-Novoa et al., 1994; Naddafi and Mirshafiey, 2013). In patients with AD, the level of histamine-releasing factor was reported to remain diminished in an age-matched control study (Kim et al., 2001). In addition, the study reported reduced H1R binding capacity in patients with AD, which has a significant association with the severity of their cognitive symptoms. In an in vitro PC12 cell model of AD, histamine reversed the $A\beta_{42}$-induced neurotoxicity by activating H2R and/or H3R (Fu et al., 2007). Interestingly, in many another neuronal diseases, such as Tourette's syndrome, addiction, Parkinson's disease, and Huntington's disease, the neuronal histaminergic system has been reported to play a major role in disease initiation and progression (Panula and Nuutinen, 2013; Shan et al., 2012). Many pieces of evidence have indicated the involvement of the neuronal histaminergic system in the loss of cognition and memory associated with AD (Naddafi and Mirshafiey, 2013; Zlomuzica et al., 2016). However, in a clinical trial, histamine congeners have shown only trifling effects on cognition and memory associated with AD (Grove et al., 2014; Nathan et al., 2013). A more systematic approach in the investigation of novel histamine-related drugs is still a major area of concern and can be explored further.

5.5.8 Adenosine and Alzheimer's Disease

Adenosine is a ribonucleoside-containing purine ring that remains present mainly in neuronal cells, especially in the hippocampus region of the brain, and has been reported to modulate the diseases related to neuronal degeneration. These endogenous nucleotides control both neuronal excitability and synaptic transmission via GPCRs of adenosine receptor subtypes, A1, A2A, A2B, and A3 (Abbracchio and Cattabeni, 1999). Adenosine not only regulates neuronal activity, but also maintains the sleep cycle, memory, cognition, and neuronal survival (Haas and Selbach, 2000; Prediger and Takahashi, 2005). Pharmacological studies have been conducted encompassing the involvement of adenosine A1 and A2A receptors in AD (Arendash et al., 2006; Rahman, 2009; Rebola et al., 2003). Adenosine has also been reported to increase the release of ACh by modulating these receptors and thus may act as a key target for the development of new molecules (Fitzgerald et al., 2004). Caffeine and an A2A receptor blocker have been shown to activate the cholinergic pathway, and reduce Aβ-related neurotoxicity and might add symptomatic and cognitive enhancement value in AD (Batalha et al., 2016; Hsu et al., 2010). Caffeine is a nonselective antagonist with great potency; it has similar efficacy in neuroprotection as selective A2A receptor antagonists (Chen et al., 2001). Moreover, endogenous adenosine involvement in reducing seizure in epilepsy is well studied (Weltha et al., 2019). In contrast, overexpression of A2AR has been reported in AD brain and this receptor obstructs memory, synaptic plasticity, Aβ synthesis, and the synthesis of NFTs in different experimental models of AD (Calon et al., 2004; Stockwell et al., 2017). Either A2AR blockade or its genetic inactivation protects against neuronal damage and delays the memory loss in AD. A molecular-level understanding, especially of the differential expression of receptors, is of the utmost importance for A2A receptor-related neuroprotection and it will help in the design of more clinical trials. However, adenosine-based therapies need more exploration with regard to the role of adenosine in patients with AD.

5.5.9 Cannabinoids in Alzheimer's Disease

Endogenous cannabinoids (CB) like arachidonoylethanolamine, 2-arachidonoylglycerol, and others (2-arachidonylglyceryl ether, virodhamine, and *N*-arachidonoyl dopamine) are degradation by-products of phospholipids (Bedse et al., 2015). Various evidence suggests its involvement in many functions, like cognition, memory, emotions, and pain sensitivity mediated by CB-1 and CB-2 receptors. CB-2 activation causes Aβ synthesis and accumulation (Galve-Roperh et al., 2008; Savonenko et al., 2015). Moreover, exogenous CB has been reported to increase neuronal survival in AD (Galve-Roperh et al., 2008). The administration of a CB receptor agonist reduces Aβ aggregation and facilitates its clearance (Aso and Ferrer, 2016; Eubanks et al., 2006). Tetrahydrocannabinol (THC) has been shown to inhibit AChE and increase cholinergic activity in AD (Eubanks et al., 2006). Also, CB signaling works against neuronal inflammation and degeneration, mitochondrial damage, and oxidative stress, as well as cognitive decline, either via directly or through GABAergic activity (Aso and Ferrer, 2016).

5.6 PATHOLOGICAL ALTERATIONS IN ALZHEIMER'S DISEASE

Over the past decades, the understanding of cellular and molecular facets of AD have been largely expanded (Uddin et al., 2020a). However, pathological advances in preventing the progression of AD effectively are still lacking. Evidence

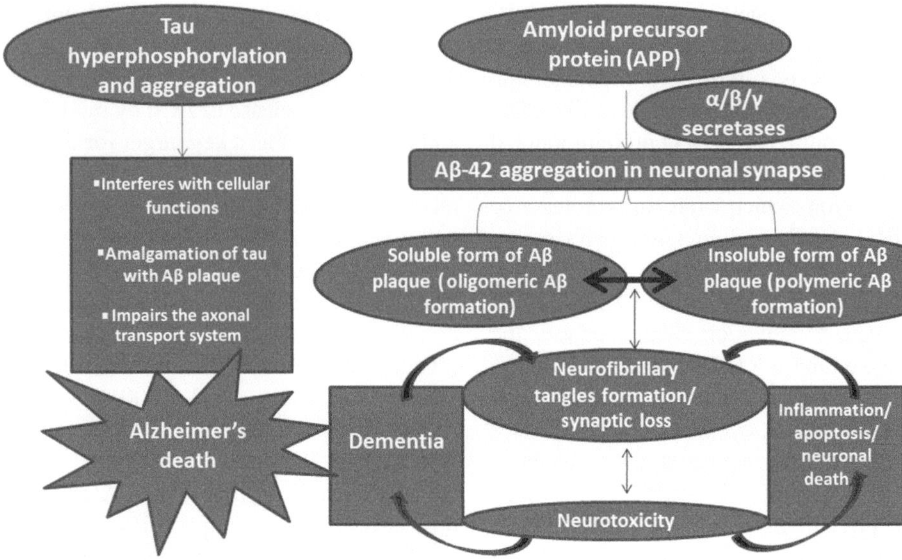

FIGURE 5.3 Amyloid beta peptide and tau in Alzheimer's disease.

for the understanding of the initiation and prevention of AD and its related conditions is limited. Emerging data indicate that healthy physiological communication between neurons, astrocytes, and microglia, as well as vascular cells, is a prerequisite for proper brain functioning (Hansson and Gouras, 2016). Detection of early changes helps in protecting neural circuits from damage. However, further damage may be caused if changes are not treated properly (Figure 5.3; Table 5.2).

5.6.1 Amyloid β and Alzheimer's Disease

As evidenced, AD includes Aβ plaque deposition, the formation of axonal tangles of tau protein, and neurodegeneration, leading to destabilizing microtubules and compromising axonal transport (Bennett et al., 2004; Mamun et al., 2020). Aβ peptides are produced after breaking off from their larger precursor, amyloid precursor protein (APP), by the action of β- and γ-secretases. APP, expressed mainly in neuronal synapses, regulates the formation and repair of synapses, neuronal transportation, and the export of iron. APP is made up of three components, a single membrane-spanning domain, a large extracellular glycosylated N-terminus, and a shorter cytoplasmic C-terminus (van der Kant and Goldstein, 2015). Aβ rapidly starts to tangle as fibrils and begins to deposit around amyloid plaques. This continuous accumulation further causes dementia via neurotoxicity and amyloidosis in AD. Aβ is actively secreted, less degraded, and locally accumulated in high concentrations after release from neighboring synapses (Coronel et al., 2018). This overaccumulation of Aβ initiates further damage to the neuronal synapses and aggravates the cyclic deposition of Aβ and related local damage. Meanwhile, microglia attracted to the Aβ plaque start to phagocytize the dead or damaged synapses. This short-term benefit initiates devastating effects on Aβ clearance and focal damage. Interestingly, the overall functions of the brain in AD are least affected due to the low plaque load. Thus, the Aβ hypothesis postulates that brain functioning in AD can remain unaffected if the axonal part of the neuron will not be damaged even after synaptic damage (Kung, 2012).

5.6.2 Tau and Alzheimer's Disease

Tau is a major phosphoprotein associated with microtubules in normal neurons; six tau isoforms are detected in the human brain. These isoforms are produced by the splicing of pre-mRNA, become localized on chromosome 17, and remain coded by only one gene. In tauopathy – a related neurodegenerative disease – tau protein becomes hyperphosphorylated and aggregated into filament bundles. Similarly, both helical and straight NFTs begin to form between neurons, especially axons, in AD. The adult human brain has unique twisted fibrillary tangle 80 nm periodicity having both helical and straight NFTs (Greenberg and Davies, 1990). In AD, neurofibrillary degeneration is indicated by the deposition of insoluble hyperphosphorylated tau protein tangles in the neuronal body. These twisted tau aggregates interfere with cellular functions and the axonal transport system, leading to the development of AD. Altogether, it is well evidenced that pathological modifications in tau reflect the downstream phenomenon of Aβ deposition. It is also reasonable that both tau and Aβ work in parallel in developing neurotoxicity and initiating the pathogenic events of AD (Bennett et al., 2004; Spires-Jones and Hyman, 2014).

5.6.3 Inflammation and Alzheimer's Disease

The raised levels of various inflammatory markers, especially interleukin-6, tumor necrosis factor (TNF)-α, and other various cytokines in the cerebral part of the brain, suggest a pivotal role in both the initiation and the progression of

TABLE 5.2
Pathological Changes Due to Alzheimer's Disease (AD) in Different Cells in the Brain

Cell type	Pathological Alteration in Cellular Function Related to AD	References
Neurons	• Intracellular tau aggregation • Aβ and tau accrual at neuronal synapses • Altered synaptic transmission and its plasticity along with neuronal loss	Hansson and Gouras, 2016; Uddin et al., 2020a; Ibrahim et al., 2020, Silbereis et al., 2016; Spires-Jones and Hyman, 2014
Microglial cells	• Cellular multiplication • Impairment of tau/Aβ clearance • Release of inflammatory mediators • Synaptic loss by phagocytic process • Reduced clearance of debris • Altered surveillance system in brain	Ibrahim et al., 2020; Perry and Holmes, 2014; Salter and Stevens, 2017
Astrocytes	• Cellular multiplication, impairment of tau/Aβ clearance • Release of inflammatory mediators • Synaptic loss by phagocytic process • Altered homeostasis at synapses • Degradation of BBB due to astrocytes damage • Altered synaptic modulation	De Strooper et al., 2016; Steele and Robinson, 2012; Liddelow et al., 2017
Oligodendrocytes	• Cellular damage and altered neuronal plasticity • Loss of myelin sheath due to decreased production of myelin	Da Mesquita et al., 2016; Cai and Xiao, 2016; Quintela-López et al., 2019
Endothelial cells, perivascular macrophages, smooth muscles	• Degradation of BBB due to cellular loss reduced clearance of toxins due to damage of macrophages and reduced blood flow due to alteration in smooth muscles	Cai et al., 2018; Zenaro et al., 2017

Aβ, amyloid beta; BBB, blood–brain barrier.

various neurological disorders, including AD (Androsova et al., 2013; Kinney et al., 2018; Nigar et al., 2016). The active microglial cells initiate proinflammatory signaling activity, cause synaptic damage, and reduce the clearance of Aβ. This further promotes the aggregation of Aβ, tau phosphorylation, and the formation of NFTs (Salter and Stevens, 2017; Mamun, et al., 2020). Furthermore, these inflammatory mediators, notably TNF-α, which is produced from AβOs, induce the unfolded protein response and defective proteostasis. Recent evidence suggests that AβOs activate eukaryotic initiation factor 2 (eIF2)α phosphorylation that is center point of cognition loss related to AD (Lourenco et al., 2015; Oliveira and Lourenco, 2016). Altogether, the deposition of misfolded protein, overactivation of brain immune responses, and altered proteostasis lead to deformed synapses and loss of learning, as well as memory, in AD.

5.6.4 Impaired Glucose Metabolism and Alzheimer's Disease

Disturbed peripheral metabolism, hyperglycemia, and insulin resistance are the fundamental features of type 2 diabetes mellitus (T2DM) and has been reported as a major cause of the development of AD (Crane et al., 2013). However, persistent hyperglycemia poses a greater risk to the development of AD-related pathological changes and there is an evident correlation between AD and T2DM (Janson et al., 2004; Mukherjee and Soto, 2017). In AD brain tissue, reduced plasma insulin, decreased insulin receptor expression/insulin sensitivity, as well as a reduction in cerebral glucose metabolism, has been well reported (Arnold et al., 2018; Ferreira et al., 2018; Lyra e Silva et al., 2019; Talbot, 2014). Thus, the excessive generation of proinflammatory cytokines, brain inflammation, defective neuronal insulin signaling, and peripheral metabolic deregulation play a major role in the progression of AD.

5.7 ADVANCES IN THE PHARMACOLOGIC APPROACH IN ALZHEIMER'S DISEASE

Considering the correlation of allied risk and issues, one-third of AD cases have been observed to be modifiable. The incidence of AD could be decreased by interventions like imparting health education and promoting physical activity, smoking cessation, the management of risk factors related to metabolic diseases, cognitive stimulation, and social engagement (Gronek et al., 2019; Xiu et al., 2019). A vascular dementia project (The Systolic Hypertension in Europe trial) has shown decreased stroke cases related to a 20% reduction in systolic blood pressure (SBP). This study also indicated that the occurrence of dementia was reduced by 50% after the reduction in SBP. Similarly, the Framingham Heart Study showed a reduced incidence of dementia concurrent with improved cardiovascular health over a period. The PROSPER trial also reported hyperlipidemia as a controversial risk factor in developing AD. A randomized controlled trial (Action to Control Cardiovascular Risk in Diabetes Trial with Memory in Diabetes study) exhibited nonsignificant differences in cognition and persistent glycemic control (glycosylated hemoglobin <6%). However, greater brain volume was observed in diabetics with controlled blood glucose (Heart Protection Study Collaborative Group, 2002; Hsu and Marshall, 2017; Satizabal et al., 2016). Interventions like lifestyle modifications, a Mediterranean diet, exercise, cognitive training, and vitamin supplementation may reduce the risk of AD by around 50% (Morris et al., 2015). Future clinical trials are needed to determine the effects of lifestyle modifications and possible multidimensional interventions in the prevention of AD. Based on the different evidence and possible targets, major pharmacological interventions came into existence. Evidence has suggested that these targets to related signaling pathways have a major impact on both initiation and progression of AD (Uddin et al., 2020b, Uddin et al., 2021). Based on the above-discussed targets and other evidence, various molecules have been developed for the treatment of dementia-related disorders (Table 5.3). These can be delivered by exploiting bio-inspired, bioengineered, and biomimetic drug delivery carriers. However, successful product development, and subsequent prompt regulatory approvals, can also be achieved by utilizing the principles of quality by design (Yoo et al., 2011; Javed et al., 2019).

5.7.1 Therapies Targeted at the Acetylcholine Receptor and Acetylcholinesterase

AChEIs are reversible inhibitors of both butyrylcholinesterase and AChE and prevent the degradation of ACh, indirectly increase the ACh level in the synaptic cleft, and thus preserve neuronal transmission. Guidelines recommend three AChEIs – donepezil, rivastigmine, and galantamine – for use in the early/middle stages of AD (Andrieu et al., 2015; McGleenon et al., 1999). Many synthetic chemical compounds and flavonoids like rutin, myricetin, quercetin, kaempferol, brucine xanomeline, and macluraxanthone have been tested

TABLE 5.3
Advances in the Targeted Therapeutic Management of Alzheimer's Disease

Therapeutic Strategies	Mechanism of Drug Targets	Target Agents and Effects	References
Aβ-42 accumulation and fibrin tangle formation	Inhibition of Aβ accumulation, Reduction in Aβ formation, Neutralizing Aβ-mediated neurotoxicity	Aβ peptide fibrils, β-secretase inhibitors, α-secretase inhibitors, γ-secretase inhibitors, anti-inflammatory agents	Kennedy et al., 2016; Hu et al. 2018; Netzer et al., 2003; Gillman et al., 2010; Folch et al., 2016; Bachurin et al., 2017; Abushouk et al., 2017
Tau-based therapeutic approach	Tau stabilizations and aggregation inhibitors, tau post-translational modification, anti-tau immunotherapy	Tau anti-aggregation and interruption of tau oligomerization, Glycogen synthase kinase 3 beta inhibitor, cyclin dependent kinase 5 inhibitors, prevent uptake and progression of abnormal tau formation	Hung and Fu, 2017; Baddeley et al., 2015; Wischik et al., 2015; Bakota and Brandt et al., 2016; Sontag and Sontag, 2014
Neuronal transmission modulation	Aβ-peptide induced modulation of neuronal transmission	Acetyl cholinesterase inhibitors, NMDA antagonist, modulation of GABA noradrenaline, dopamine, and serotonin	Liu et al., 2019; Pottoo et al., 2020; Wu and Sun, 2015; Fernandes et al., 2015; Feinstein et al., 2016; Hattori et al., 2019; Andrieu et al., 2015; Grove et al., 2014.
ApoE-targeted approach	Interruption of binding between ApoE and Aβ by peptide fragments	Restricts further accumulation of brain Aβ	Hanson et al., 2015; Liu et al., 2014; Williams et al., 2020
NTs and NT receptor-based approach	Targeting p75, tropomyosin receptor kinase A, tropomyosin receptor kinase B	Restrict degenerating signaling system of target p75	Allen et al., 2011; Diniz and Teixeira, 2011; Chao et al., 2006
Oxidative stress, neuronal inflammation, and neuroprotection	Reduction in free radicals (ROS), anti-inflammatory agents, regulation of metal ions	Antioxidants, MAO-B inhibitors, metal chelating agents	von Bernhardi et al., 2015; Salloway et al., 2008; Malinski et al., 2007; Khan et al., 2019; Ayaz et al., 2019
Symptomatic cognition enhancers	Blockade of AChE and NMDA receptor, α-7 nicotinic acid agonist, 5-HT_6 antagonist, H-3 receptor antagonist	Cognitive enhancement by modulating different neurotransmitters	Joe and Ringman, 2019

Aβ, amyloid beta; NMDA, *N*-methyl-D-aspartate; GABA, γ-aminobutyric acid; ApoE, apolipoprotein E; NT, neurotrophin; ROS, reactive oxygen species; MAO, monoamine oxidase; AChE, acetylcholinesterase; NMDAR, NMDA receptor; 5-HT, 5-hydroxytryptamine; H-3, histamine-3.

for their efficacy and safety in protecting against neuronal damage in AD. Drugs like 6-chloropyridonezepil, AF102B, AF267B, AF150, AF292B, 77-LH-28-1, VU0357017, and VU364572 target the M1 receptor and have been shown to more specifically protect cholinergic transmission by increasing neuronal viability (Kaur et al., 2019; Kumar et al., 2016). Cognitive enhancement by means of symptomatic therapies has focused on brain functioning modulation related to both the cholinergic and glutamatergic systems. Thus, both AChEIs and NMDA receptor antagonists together have been employed in preventing the development of AD. Evidence suggests that ACHEIs moderately improve the cognitive and global functional status of patients with mild-to-moderate. However, efficacy wanes with long-term treatment due to its own side effects, such as weight loss and syncope. Memantine blocks overexcited NMDA receptors to prevent glutamate release, thereby inhibiting neurotoxicity. It also inhibits and reverses tau hyperphosphorylation with only mild adverse effects. Combination therapy of memantine with ACHEIs has been approved for the treatment of patients with moderate-to-severe AD (Buckley and Salpeter, 2015; Patel and Grossberg, 2011). Furthermore, the alpha-7 nicotinic acid agonist encenicline, the serotonin 5-HT_6 antagonist idalopirdine, and histamine H3 receptor antagonists have shown positive results on primary outcome measures. However, phase III trials suggest no benefit to cognitive function with treatment, and thus they need to be explored efficiently (Grove et al., 2014; Thomsen et al., 2010; Wilkinson et al., 2014).

5.7.2 Therapies Targeted at the Serotonin Receptor and Monoamine Oxidase Inhibitors

Evidence suggests that treatment with selective serotonin reuptake inhibitors (SSRIs) increases the expression of 5-HT_{1B} receptors mRNA in transgenic mice and are reliable antidepressants in the treatment of AD symptoms (Tajeddinn et al., 2015). In a preclinical study, TSB-2 (5-HT_{2A} selective agonist) administration showed improvement in fear-associated memory (Tajeddinn et al., 2016). Similarly, the administration of SSRIs like fluoxetine and paroxetine correct the impaired learning and memory caused by SB221284 (5-$HT_{2B/2C}$ receptor antagonist; Gibbs and Hertz, 2014). Additionally, an MAO-B inhibitor (rasagiline) is in a phase III clinical trial for treating AD-related brain function (Hattori et al., 2019). More preclinical and clinical studies are required to explore the therapies targeted at serotonin receptors and MAO inhibitors.

5.7.3 Therapies Targeted at the Dopamine Receptor

Various evidence indicates that decreased dopamine release, the degeneration of dopamine receptors, reduced dopamine transporter expression, or diminished tyrosine hydroxylase activity is responsible for the development of AD (Nobili et al., 2017). The protection related to dopamine in AD can be achieved by facilitating dopaminergic protuberance of the hippocampal area (Nobili et al., 2017). The upregulation of MAO-A activity has been reported in patients with AD (Lam et al., 2017). In different animal studies, MAO-B inhibitors like isocarboxazid, phenelzine, selegiline, and tranylcypromine, and COMT inhibitors like entacapone, opicapone, stalevo, tasmar, and tolcapone prevented the synthesis of Aβ and reversed the loss of learning ability, memory, and behavioral changes (Olgiati, 2012; Yeung et al., 2019). Numerous drugs targeting dopamine activity modulation like atomoxetine, venlafaxine, desipramine, and sertraline have been reported to be effective in neuroprotection and loss of memory related to AD (Feinstein et al., 2016). Similarly, newer molecules need to be studied for their benefits in AD.

5.7.4 Therapies Targeted at the Glutamate Receptor

There are eight subtypes of glutamate receptors and they have been shown to be involved in the formation of AβO. Similar to MTEP (3-[(2-methyl-1,3-thiazol-4-yl) ethynyl] pyridine), CTEP 2-(chloro-4-((2, 5-dimethyl-1-(4-(trifluoromethoxy) phenyl)-1H-imidazol-4-yl) ethynyl) pyridine) has been reported to have a negative allosteric modulating property on metabotropic glutamate receptor 5 (mGluR5) and exhibits an antagonistic action in Aβ synthesis and accumulation in AD (Jaeschke et al., 2015; Ossowska et al., 2005). Memantine increases the excitatory neurotransmission and has already been approved clinically. Many more, like huperzine A, neramexane (an uncompetitive NMDA antagonist), MK-801 (an NMDA-blocking agent), ifenprodil (a GluN2B antagonist), and RO 25-6981 (a GluN2B antagonist), have been reported to modulate neuronal transmission in AD (Fernandes et al., 2015; Hu et al., 2009; Zhang, 2012). However, contradictory to the above drugs and their related mechanisms, $GABA_A$ agonists like etazolate, muscimol, propofol, MRK-016, CGS9896, Ro-4938581, and Ro-4882224, as well as $GABA_B$ antagonists like SGS742 and CGP55845, have been shown to be effective in reverting learning and cognition (Lasarge et al., 2009; Marcade et al., 2008; Pilipenko et al., 2015; Shao et al., 2014).

5.7.5 Therapies Targeted at γ-Aminobutyric Acid

GABA is a major inhibitory neurotransmitter involved in neurodegenerative disorders (Pottoo, Tabassum, et al., 2020; Wu and Sun, 2015). It is well evidenced that general anesthetics like isoflurane, pentobarbital, propofol, and chloral hydrate bind to the $GABA_A$ receptor and have recently been reported to contribute to the initiation, progression, and aggravation of AD via increased tau phosphorylation. One study has indicated a strong correlation between the GABAergic signaling system and Aβ toxicity, tau hyperphosphorylation, as well as an ApoE4 effect in the development of AD (Whittington et al., 2013).

Etazolate, a $GABA_A$ receptor modulator, has been shown to exhibit a neuronal protective effect in a phase II clinical trial (Vellas et al., 2011). This causes $GABA_A$ receptor-mediated activation of α-secretase in breaking APP into sAPPα, which exhibits both neurotrophic and neuroprotective effects against Aβ-mediated loss of cognition (Drott et al., 2010; Marcade et al., 2008). Other $GABA_A$ receptor agonists, such as muscimol and propofol, have been shown to inhibit $Aβ_{25-35}$-induced neuronal apoptotic death (Shao et al., 2014). In this context, bicuculine (a $GABA_A$ receptor antagonist) has been observed to entirely invert the neuronal benefits of muscimol by inhibiting glutamate release, the generation of reactive oxygen species, and caspase-3 activity (Lee et al., 2005). Chronic activation of the $GABA_A$ receptor by propofol showed neuroprotection by decreased expression of BACE-1, and decreases both Aβ generation and accelerating Aβ degradation, along with preventing neuronal apoptosis (Shao et al., 2014). In various studies, the $GABA_A$ α5 receptor has also been observed to be connected to cognition and memory (Collinson et al., 2002; Crestani et al., 2002; Gabriella and Giovanna, 2010). Based on $GABA_A$ α5 receptor subunit selectivity, MRK-016, α5IA, and α5IA-II have been reported to have an encouraging effect on memory disorder in animal and clinical studies (Atack, 2010; Atack et al., 2009; Dawson et al., 2006; Guerrini et al., 2013). Compounds like Ro-4938581 and Ro-4882224 have been shown to significantly reverse scopolamine-induced working memory impairment (Knust et al., 2009). As $GABA_B$ receptor antagonists, molecules like CGP55845 improve memory and learning activity via increased expression of mRNA, nerve growth factor, and BNDF in rats and humans (Froestl et al., 2004; Getova and Bowery, 2001; Helm et al., 2005).

5.7.6 Therapies Targeted at Noradrenaline-Related Neurotransmission

The noradrenergic system releases norepinephrine (noradrenaline) in the brain and is reported to modulate memory-related activities. DSP-4 (*N*-(2-chloroethyl)-*N*-ethyl-2-bromobenzylamine hydrochloride) is a selective neurotoxin for norepinephrine-releasing neurons, and causes loss of cognition after central norepinephrine depletion (Rajabi et al., 2012). This loss of cognition may become further aggravated after depletion of dopamine and ACh neurotransmitters. In an animal study, the administration of norepinephrine has been observed to ameliorate both spatial and recognition memory (Chalermpalanupap et al., 2013; Liu et al., 2019; Murchison et al., 2004). This noradrenaline receptor-targeted drug discovery needs to be explored.

5.7.7 Therapies Targeted at Cannabinoid Receptors

The pathologic involvement of CB receptors has been a major advancement in drug development for AD (Cristino et al., 2020). The blockade of CB receptors has been observed to revert amnesia due to Aβ fragments (Mazzola et al., 2003). Moreover, in vitro studies have shown that endogenous CBs like dronabinol, anandamide, and noladin ether increases neuronal survival in Aβ-treated cells (Aso et al., 2013). A synthetic cannabinoid, HU-210 has been also shown to decrease excitotoxicity after binding to NMDA receptors in AD (Ferrer, 2014). In another study, different CBs were studied in combination with THC and more-than-synergistic potential in neuroprotection was seen. Together, this evidence significantly extends the benefits of nonpsychoactive CBs and can be lead molecules in neurodegenerative diseases.

5.7.8 Therapies Targeted at Amyloid β Synthesis and Clearance

As evidenced from the pathogenesis of AD, modulation of β- and γ-secretase can be used to inhibit Aβ formation. Inhibition of BACE-1 has also shown a promising effect in the inhibition of Aβ formation in animal models (Kennedy et al., 2016). Unfortunately, BACE-1 inhibitors failed in phase II/III clinical trials due to lack of efficacy and undesirable side effects. However, a recent study exhibited BACE-1 inhibition, reversing deposited amyloid and improving cognitive functions (Hu et al., 2018). Selective γ-secretase inhibitors include notch-sparing γ-secretase inhibitors and γ-secretase modulators. Gleevec – the ABL kinase inhibitor – was found to reduce Aβ production but spare notch cleavage by γ-secretase in primary neuronal cells and animals (Netzer et al., 2003). Avagacestat was reported to inhibit APP processing more preferably than notch cleavage (Gillman et al., 2010). A recently developed notch-sparing γ-secretase inhibitor, pinitol, was derived from a natural product and was reported to have insulin sensitization properties. It is currently in a phase II trial for the treatment of AD (Folch et al., 2016). Nonsteroidal anti-inflammatory drugs (NSAIDs) were the first γ-secretase modulators shown to shift Aβ production from the agreeable form ($Aβ_{42}$) to a more soluble form ($Aβ_{38}$; Weggen et al., 2001). Immunotherapy via active and passive vaccines against Aβ have been utilized as a therapeutic approach in AD (Bachurin et al., 2017). Active immunotherapy targeted at Aβ failed until now due to autoimmune response; a small portion of the cohort developed severe meningoencephalitis with the AN-1792 vaccine targeting the full-length $Aβ_{1-42}$ peptide (Lemere, 2009). The subsequent development of vaccines was tailored to target specific Aβ epitopes such as ACC-001, which targets an N-terminal $Aβ_{1-7}$ peptide fragment (Pasquier et al., 2016). Currently, second-generation active immunotherapy vaccines have been progressed into clinical trials. The passive anti-Aβ immunization may be a more promising strategy for treating AD. Bapineuzumab is an antibody targeted at the N-terminus of Aβ peptide acting on soluble Aβ. However, no beneficial effect was found with both bapineuzumab and solanazumab in phase III clinical studies of patients with mild-to-moderate AD (Abushouk et al., 2017). Crenezumab, gantenerumab, and aducanumab, targeting both soluble and aggregated Aβ species (both oligomeric and fibrillar Aβ), are currently in phase III trials in patients with prodromal, mild, and early AD, respectively.

5.7.9 Therapies Targeted at Tau Stabilizations, Aggregation, and Post-Translational Modifications

Most tau-stabilizing agents (e.g. paclitaxel and epothilone D) manifest undesirable toxic side effects (Hung and Fu, 2017). Several tau aggregation inhibitors tested have failed in clinical trials due to either unwanted side effects or a lack of efficacy, for example Rember™ in phase II (Baddeley et al., 2015; Wischik et al., 2015) and TRx0237 (LMTM) in three phase III trials of patients with mild-to-moderate AD and behavioral variant frontotemporal dementia (Bakota and Brandt, 2016). Recently, abeotaxane was evaluated in patients with a range of AD stages, and their outcomes are to be reported (Hattori et al., 2019). There are three ways to aim for post-translational tau modifications to develop a target-based treatment: inhibition of kinases responsible for tau hyperphosphorylation (GSK-3β and CKD5); induction of tau dephosphorylation enzymes (protein phosphate 2A); and tau acetylation and cis-transformation modulation (Pei et al., 2008; Sontag and Sontag, 2014). Unfortunately, no tau kinase inhibitors, excluding GSK-3 inhibitors, and tau dephosphorylation inducers or modulators have passed clinical trials because of safety concerns and efficacy profiles, and need to be investigated further.

5.7.10 Therapies Targeted at Anti-Tau Immunotherapy

The rationale behind the application of immunization against pathologic tau variants is to interrupt the uptake and propagation processes of abnormal tau. Using a high-affinity antibody against phosphorylated tau is an approach that should not disturb the function of physical tau from active immunization of phosphorylated tau (Yanamandra et al., 2013). Several studies using active immunization against phospho-tau peptides modulating tau pathology showed positive results in tauopathies in mouse models of AD (Chukwu et al., 2018; Rajamohamedsait et al., 2017). Tau-targeted passive immunization is an exciting development in AD because it offers the possibility of halting the spread of tau pathology through the extracellular space. Currently, ABBV-8E12, an antibody against aggregated extracellular Tau, is in two phase II trials targeting patients with progressive supranuclear palsy and patients with mild cognitive impairment with positive amyloid scans (West et al., 2017). In addition, different tau antibodies against tau pathology are under investigation in different phases of clinical trials, but they needed to be explored with caution.

5.7.11 Therapies Targeted at Apolipoprotein E

The ApoE genotype is one of the strongest genetic risk factors for developing sporadic AD. It should be noted that in clinical trials the ApoE4 genotype is an important determinant of therapeutic responses in AD (Hanson et al., 2015). Different therapeutic strategies targeted at ApoE have been tested in vitro and in vivo in preclinical animal models. Blocking the interaction between ApoE and Aβ with a peptide fragment of Aβ reduced brain amyloid accumulation, ameliorated memory deficits, and reduced brain-insoluble tau levels in transgenic mouse models of AD (Kuszczyk et al., 2013; Liu et al., 2014). Many newer therapeutic agents like CS-6253, CN-105, ApoE antibody, bexarotene, probucol, AGB 101, rosiglitazone, and epigallocatechin gallate, along with diet and physical activity interventions, has been reported to ameliorate ApoE functions (Williams et al., 2020).

5.7.12 Therapies Targeted at Neurotrophins

Due to the pleiotropy of neurotrophins (NTs), a receptor-targeted treatment approach in AD has been looked at for a decade and may emerge as a pivotal target in the treatment of AD. Due to inappropriate pharmacokinetic properties of NTs such as the short half-life, decreased permeability to the blood–brain barrier, reduced bioavailability, and restricted diffusion in brain tissues, NTs and their receptor-based therapeutic approach are matters of concern (Chao et al., 2006; Allen et al., 2011). Recently, the idea of gene transfer technology has been applied for obtaining new chemical entities for AD therapy based on NTs. These strategies mainly target p75, tropomyosin receptor kinase A, and tropomyosin receptor kinase B receptors. Importantly, targeting p75 as a therapeutic approach needs the discovery of selective molecules of small size, to restrict the degenerative signaling system of target p75. Thus, even in the absence of NTs, it may surely promote brain cell survival. LMA11A-31A is a small molecule that penetrates the brain cells and work against Aβ and tau-mediated toxicity to put off synaptic dysfunction, neurotic degeneration, microglia activation, and altered cognition (Allen et al., 2011; Diniz and Teixeira, 2011). Therefore, NTs and receptor-based therapies possess synergistic intracellular signaling potential and the possibility of fighting against multidimensional pathological alterations in AD. However, these strategies need a critical approach to developing and using the best molecules in clinical practice.

5.7.13 Therapies Targeted at Oxidative Stress, Inflammation, and Neuroprotection

Various evidence support the involvement of neuroinflammation, as well as microglia, in the pathogenesis of AD. A model of microglial priming states that the activation of microglia by proinflammatory mediators followed phenotypic amplification of astrocytes, aggravating neuronal damage (Perry and Holmes, 2014; Sudduth et al., 2013; Tuppo and Arias, 2005). However, regarding neuroinflammation, different ways have been explored to modulate the functioning of immune cell-like reducing cytokine expression, restricting the release and interrupting the binding of cytokines to their receptors (von Bernhardi et al., 2015). Moreover, this is well evidenced from in vitro and in vivo animal studies, which showed decreased injury due to oxidative stress mediated by the inhibition of cyclooxygenase-2, and inducible nitric oxide synthase ameliorated the symptoms of AD (Malinski, 2007). Low doses of NSAIDs like aspirin have been reported to protect against AD (Broe et al., 2000). In addition, it is evident that synaptic dysfunctions due to simultaneous triggers like toxic Aβ aggregates, age-related degeneration, neuroinflammation, and the unknown trigger hypothesis is a hallmark in the development of AD. The search for antioxidants as a neuroprotective therapy has focused on targeting Aβ/tau in most clinical trials and have been developed to find out the neurodegenerative mechanism (Salloway et al., 2008). Phytoconstituents like quercetin, resveratrol, *Ginkgo biloba* extract, and many more have been reported as neuroprotective agents in AD therapy and need proper clinical investigations (Khan et al., 2019; Gomes et al., 2018; Singh et al., 2019; Ayaz et al., 2019).

5.8 RECENT DEVELOPMENTS AND CHALLENGES

Future studies to gain knowledge about the underlying pathology and progression of AD may direct us towards new drug discoveries. The involvement of different neurotransmitters and other targets has been well established by several studies and were needed for complete clarity for the development of targeted treatment if AD. However, several studies have indicated that not only Aβ, but also tau-modifying agents and many other target-based entities, require treatment in AD. Concurrently, the development and use of diagnostic biomarkers and various imaging modalities are required to identify the chance of disease initiation and any other pathologic abnormalities. A tau positron emission tomography imaging technique has emerged to aid knowledge about the preclinical stages of AD. It is now well understood that too much neurodegeneration may lead to irreversible damage of neural networks. Implementation of an effective diagnostic and therapeutic algorithm is required for the use of selective biomarkers, as well as to foresee the risk in the population for getting dementia and its progression to AD. Currently, the exploration of AD management has shifted toward multitargeted drug application for earlier interception in AD.

5.9 CONCLUSION

From the above discussion, the role of various neurotransmitters along with the other targets and their loopholes became an area of study. In this review, various neurotransmitters, their association with pathologic alterations, and a related receptor-based pharmacologic approach have been discussed. The limited availability of approved drugs like AChEIs and NMDA antagonists urges exploration of other targets and opens up a new pharmacologic approach to the treatment of AD. Meanwhile, along with neurotransmitters and neuropeptides like dopamine, glutamate, GABA, noradrenaline, and serotonin, other receptor-based approaches like H-3, CB, and adenosine have been observed to impart neuroprotective effects in AD. Consequently, the different facets related to neurodegeneration and its treatment need to be explored; only then can a multi-approach therapy come into existence. Also, the exploration of immunity and gene-based therapeutics may play a significant role in overcoming AD-related neurological alterations.

REFERENCES

Abbracchio MP and Cattabeni F (1999) Brain adenosine receptors as targets for therapeutic intervention in neurodegenerative diseases. Annals of the New York Academy of Sciences 890: 79–92. DOI: 10.1111/j.1749-6632.1999.tb07983.x.

Abhang PA, Gawali BW and Mehrotra SC (2016) Introduction to emotion, electroencephalography, and speech processing. In: Abhang PA, Gawali BW, and Mehrotra SC (eds) Introduction to EEG- and Speech-Based Emotion Recognition. Academic Press, pp. 1–17. DOI: 10.1016/B978-0-12-804490-2.00001-4.

Abushouk AI, Elmaraezy A, Aglan A, et al. (2017) Bapineuzumab for mild to moderate Alzheimer's disease: a meta-analysis of randomized controlled trials. BMC Neurology 17: 66. DOI: 10.1186/s12883-017-0850-1.

Agis-Torres A, Sölhuber M, Fernandez M, et al. (2014) Multi-target-directed ligands and other therapeutic strategies in the search of a real solution for Alzheimer's disease. Current Neuropharmacology 12: 2–36. DOI: 10.2174/1570159x113116660047.

Ahmed T, Zahid S, Mahboob A, et al. (2017) Cholinergic system and post-translational modifications: an insight on the role in Alzheimer's disease. Current Neuropharmacology 15: 480–494. DOI: 10.2174/1570159X14666160325121145.

Alberico SL, Cassell MD and Narayanan NS (2015) The vulnerable ventral tegmental area in Parkinson's disease. Basal Ganglia 5: 51–55. DOI: 10.1016/j.baga.2015.06.001.

Allen SJ, Watson JJ and Dawbarn D (2011) The neurotrophins and their role in Alzheimer's disease. Current Neuropharmacology 9: 559–573. DOI: 10.2174/157015911798376190.

Alvarez XA, Franco A, Fernández-Novoa L, et al. (1996) Blood levels of histamine, IL-1β, and TNF α in patients with mild to moderate Alzheimer disease. Molecular and Chemical Neuropathology 29: 237–252. DOI: 10.1007/BF02815005.

Alzheimer's Association (2016) 2016 Alzheimer's disease facts and figures. Alzheimers & Dementia 12: 459–509.

Andrews M, Tousi B and Sabbagh MN (2018) 5HT6 antagonists in the treatment of Alzheimer's dementia: current progress. Neurology and Therapy 7: 51–58. DOI: 10.1007/s40120-018-0095-y.

Andrieu S, Coley N, Lovestone S, et al. (2015) Prevention of sporadic Alzheimer's disease: lessons learned from clinical trials and future directions. The Lancet Neurology 14: 926–944. DOI: 10.1016/S1474-4422(15)00153-2.

Androsova LV, Mikhaĭlova NM, Zozulia SA, et al. (2013) [Inflammatory markers in Alzheimer's disease and vascular dementia]. Zhurnal Nevrologii I Psikhiatrii Imeni S.S. Korsakova 113: 49–53.

Arendash GW, Schleif W, Rezai-Zadeh K, et al. (2006) Caffeine protects Alzheimer's mice against cognitive impairment and reduces brain β-amyloid production. Neuroscience 142: 941–952. DOI: 10.1016/j.neuroscience.2006.07.021.

Arnold SE, Arvanitakis Z, Macauley-Rambach SL, et al. (2018) Brain insulin resistance in type 2 diabetes and Alzheimer disease: concepts and conundrums. Nature Reviews Neurology 14: 168–181. DOI: 10.1038/nrneurol.2017.185.

Aso E and Ferrer I (2016) CB2 Cannabinoid receptor as potential target against Alzheimer's disease. Frontiers in Neuroscience 10: 243. DOI: 10.3389/fnins.2016.00243.

Aso E, Juvés S, Maldonado R, et al. (2013) CB 2 cannabinoid receptor agonist ameliorates Alzheimer-like phenotype in AβPP/PS1 mice. Journal of Alzheimer's Disease 35: 847–858. DOI: 10.3233/JAD-130137.

Atack JR (2010) Preclinical and clinical pharmacology of the GABAA receptor α5 subtype-selective inverse agonist α5IA. Pharmacology & Therapeutics 125: 11–26. DOI: 10.1016/j.pharmthera.2009.09.001.

Atack JR, Maubach KA, Wafford KA, et al. (2009) In vitro and in vivo properties of 3-tert-butyl-7-(5-methylisoxazol-3-yl)-2-(1-methyl-1h-1,2,4-triazol-5-ylmethoxy)-pyrazolo[1,5-d]-[1,2,4]triazine (MRK-016), a GABAA receptor α5 subtype-selective inverse agonist. Journal of Pharmacology and Experimental Therapeutics 331: 470–484. DOI: 10.1124/jpet.109.157636.

Ayaz M, Sadiq A, Junaid M, et al. (2019) Flavonoids as prospective neuroprotectants and their therapeutic propensity in aging associated neurological disorders. Frontiers in Aging Neuroscience 11: 155. DOI:10.3389/fnagi.2019.00155

Bachurin SO, Bovina EV and Ustyugov AA (2017) Drugs in clinical trials for Alzheimer's disease: the major trends. Medicinal Research Reviews 37: 1186–1225. DOI: 10.1002/med.21434.

Baddeley TC, McCaffrey J, Storey JMD, et al. (2015) Complex disposition of methylthioninium redox forms determines efficacy in tau aggregation inhibitor therapy for Alzheimer's disease. Journal of Pharmacology and Experimental Therapeutics 352: 110–118. DOI: 10.1124/jpet.114.219352.

Bagyinszky E, Giau VV and An SA (2020) Transcriptomics in Alzheimer's disease: aspects and challenges. International Journal of Molecular Sciences 21: 3517. DOI: 10.3390/ijms21103517.

Bakota L and Brandt R (2016) Tau biology and tau-directed therapies for Alzheimer's disease. Drugs 76: 301–313. DOI: 10.1007/s40265-015-0529-0.

Barbe C, Jolly D, Morrone I, et al. (2018) Factors associated with quality of life in patients with Alzheimer's disease. BMC Geriatrics 18: 159. DOI: 10.1186/s12877-018-0855-7.

Batalha VL, Ferreira DG, Coelho JE, et al. (2016) The caffeine-binding adenosine A 2A receptor induces age-like HPA-axis dysfunction by targeting glucocorticoid receptor function. Scientific Reports 6: 1.

Bedse G, Romano A, Lavecchia AM, et al. (2015) The role of endocannabinoid signaling in the molecular mechanisms of neurodegeneration in Alzheimer's disease. Journal of Alzheimer's disease 43: 1115–1136. DOI: 10.3233/JAD-141635.

Bennett DA, Schneider JA, Wilson RS, et al. (2004) Neurofibrillary tangles mediate the association of amyloid load with clinical Alzheimer disease and level of cognitive function. Archives of Neurology 61: 378–384. DOI: 10.1001/archneur.61.3.378.

Blandina P, Provensi G, Munari L, et al. (2012) Histamine neurons in the tuberomammillary nucleus: a whole center or distinct subpopulations? Frontiers in Systems Neuroscience 6. DOI: 10.3389/fnsys.2012.00033.

Bonaventure P, Nepomuceno D, Kwok A, et al. (2002) Reconsideration of 5-hydroxytryptamine (5-HT)(7) receptor distribution using [(3)H]5-carboxamidotryptamine and [(3)H]8-hydroxy-2-(di-n-propylamino)tetraline: analysis in brain of 5-HT(1A) knockout and 5-HT(1A/1B) double-knockout mice. The Journal of Pharmacology and Experimental Therapeutics 302: 240–248. DOI: 10.1124/jpet.302.1.240.

Bonda DJ, Mailankot M, Stone JG, et al. (2010) Indoleamine 2,3-dioxygenase and 3OH-kynurenine modifications are found in the neuropathology of Alzheimer disease. Redox Report 15: 161–168. DOI: 10.1179/174329210X12650506623645.

Bortolato M, Chen K and Shih JC (2008) Monoamine oxidase inactivation: from pathophysiology to therapeutics. Advanced Drug Delivery Reviews 60: 1527–1533. DOI: 10.1016/j.addr.2008.06.002.

Broe GA, Grayson DA, Creasey HM, et al. (2000) Anti-inflammatory drugs protect against Alzheimer disease at low doses. Archives of Neurology 57:1586–1591. DOI:10.1001/archneur.57.11.1586.

Brown RE, Stevens DR and Haas HL (2001) The physiology of brain histamine. Progress in Neurobiology 63: 637–672. DOI: 10.1016/S0301-0082(00)00039-3.

Buckley JS and Salpeter SR (2015) A risk–benefit assessment of dementia medications: systematic review of the evidence. Drugs & Aging 32: 453–467. DOI: 10.1007/s40266-015-0266-9.

Buffalo EA, Movshon JA and Wurtz RH (2019) From basic brain research to treating human brain disorders. Proceedings of the National Academy of Sciences of the United States of America 116: 26167–26172. DOI: 10.1073/pnas.1919895116.

Cai Z, Qiao PF, Wan CQ, et al. (2018) Role of blood-brain barrier in Alzheimer's disease. Journal of Alzheimers Disease 63: 1223–1234. doi: 10.3233/JAD-180098. PMID: 29782323.

Cai Z and Xiao M. (2016) Oligodendrocytes and Alzheimer's disease. International Journal of Neuroscience 126: 97–104. doi: 10.3109/00207454.2015.1025778.

Calon F, Dridi M, Hornykiewicz O, et al. (2004) Increased adenosine A2A receptors in the brain of Parkinson's disease patients with dyskinesias. Brain 127: 1075–1084. DOI: 10.1093/brain/awh128.

Celanire S, Wijtmans M, Talaga P, et al. (2005) Keynote review: Histamine H3 receptor antagonists reach out for the clinic. Drug Discovery Today 10: 1613–1627. DOI: 10.1016/S1359-6446(05)03625-1.

Chalermpalanupap T, Kinkead B, Hu WT, et al. (2013) Targeting norepinephrine in mild cognitive impairment and Alzheimer's disease. Alzheimer's Research & Therapy 5: 21. DOI: 10.1186/alzrt175.

Chao MV, Rajagopal R and Lee FS. (2006) Neurotrophin signalling in health and disease. Clinical Science 110: 167–173. doi: 10.1042/CS20050163.

Charnay Y and Leger L (2010) Brain serotonergic circuitries. Dialogues in Clinical Neuroscience 12: 471–487.

Chen JF, Xu K, Petzer JP, et al. (2001) Neuroprotection by caffeine and A(2A) adenosine receptor inactivation in a model of Parkinson's disease. The Journal of Neuroscience 21: RC143.

Chen Y, Peng Y, Che P, et al. (2014) α2A adrenergic receptor promotes amyloidogenesis through disrupting APP-SorLA interaction. Proceedings of the National Academy of Sciences of the United States of America 111: 17296–17301. DOI: 10.1073/pnas.1409513111.

Chukwu JE, Pedersen JT, Pedersen LØ, et al. (2018) Tau antibody structure reveals a molecular switch defining a pathological conformation of the tau protein. Scientific Reports 8: 6209. DOI: 10.1038/s41598-018-24276-4.

Collinson N, Kuenzi FM, Jarolimek W, et al. (2002) Enhanced learning and memory and altered GABAergic synaptic transmission in mice lacking the α5 subunit of the GABAA receptor. The Journal of Neuroscience 22: 5572–5580. DOI: 10.1523/JNEUROSCI.22-13-05572.2002.

Companys-Alemany J, Turcu AL, Bellver-Sanchis A, et al. (2020) A novel NMDA receptor antagonist protects against cognitive decline presented by senescent mice. Pharmaceutics 12: 284. DOI: 10.3390/pharmaceutics12030284.

Contestabile A (2011) The history of the cholinergic hypothesis. Behavioural Brain Research 221: 334–340. DOI: 10.1016/j.bbr.2009.12.044.

Coronel R, Bernabeu-Zornoza A, Palmer C, et al. (2018) Role of amyloid precursor protein (APP) and its derivatives in the biology and cell fate specification of neural stem cells. Molecular Neurobiology 55: 7107–7117. DOI: 10.1007/s12035-018-0914-2.

Crane PK, Walker R, Hubbard RA, et al. (2013) Glucose levels and risk of dementia. New England Journal of Medicine 369: 540–548. DOI: 10.1056/NEJMoa1215740.

Crestani F, Keist R, Fritschy JM, et al. (2002) Trace fear conditioning involves hippocampal α5 GABAA receptors. Proceedings of the National Academy of Sciences of the United States of America 99: 8980–8985. DOI: 10.1073/pnas.142288699.

Cristino L, Bisogno T and Di Marzo V. (2020) Cannabinoids and the expanded endocannabinoid system in neurological disorders. Nature Reviews Neurology 16: 9–29. DOI: 10.1038/s41582-019-0284-z.

Da Mesquita S, Ferreira AC, Sousa JC, et al. (2016) Insights on the pathophysiology of Alzheimer's disease: the crosstalk between amyloid pathology, neuroinflammation and the peripheral immune system. Neuroscience and Biobehavioral Reviews 68: 547–562.

Dawson GR, Maubach KA, Collinson N, et al. (2006) An inverse agonist selective for α5 subunit-containing GABAA receptors enhances cognition. Journal of Pharmacology and Experimental Therapeutics 316: 1335–1345. DOI: 10.1124/jpet.105.092320.

De Strooper B and Karran E (2016) The cellular phase of Alzheimer's disease. Cell 164: 603–615. https://doi.org/10.1016/j.cell.2015.12.056

Dineley KT, Pandya AA and Yakel JL (2015) Nicotinic ACh receptors as therapeutic targets in CNS disorders. Trends in Pharmacological Sciences 36: 96–108. DOI: 10.1016/j.tips.2014.12.002.

Diniz BS and Teixeira AL (2011) Brain-derived neurotrophic factor and Alzheimer's disease: physiopathology and beyond. Neuromolecular Medicine 13: 217–222. DOI: 10.1007/s12017-011-8154-x.

Drott J, Desire L, Drouin D, et al. (2010) Etazolate improves performance in a foraging and homing task in aged rats. European Journal of Pharmacology 634: 95–100. DOI: 10.1016/j.ejphar.2010.02.036.

Elshazzly M and Caban O (2019) Embryology, central nervous system. In: StatPearls. Treasure Island, FL: StatPearls Publishing. Available at: www.ncbi.nlm.nih.gov/books/NBK526024/ (accessed 19 October 2019).

Eubanks LM, Rogers CJ, Beuscher AE, et al. (2006) A molecular link between the active component of marijuana and Alzheimer's disease pathology. Molecular Pharmaceutics 3: 773–777. DOI: 10.1021/mp060066m.

Fakhoury M (2016) Revisiting the serotonin hypothesis: implications for major depressive disorders. Molecular Neurobiology 53: 2778–2786. DOI: 10.1007/s12035-015-9152-z.

Feinstein DL, Kalinin S and Braun D (2016) Causes, consequences, and cures for neuroinflammation mediated via the locus coeruleus: noradrenergic signaling system. Journal of Neurochemistry 139(S2): 154–178. DOI: 10.1111/jnc.13447.

Fernandes A, Wojcik T, Baireddy P, et al. (2015) Inhibition of in vivo [(3)H]MK-801 binding by NMDA receptor open channel blockers and GluN2B antagonists in rats and mice. European Journal of Pharmacology 766: 1–8. DOI: 10.1016/j.ejphar.2015.08.044.

Fernández-Novoa L, Alvarez XA, Franco-Maside A, et al. (1994) CDP-choline-induced blood histamine changes in Alzheimer's disease. Methods and Findings in Experimental and Clinical Pharmacology 16: 279–284.

Ferreira LSS, Fernandes CS, Vieira MNN, et al. (2018) Insulin resistance in Alzheimer's disease. Frontiers in Neuroscience 12: 830. DOI: 10.3389/fnins.2018.00830.

Ferreira-Vieira TH, Guimaraes IM, Silva FR, et al. (2016) Alzheimer's disease: targeting the cholinergic system. Current Neuropharmacology 14: 101–115. DOI: 10.2174/1570159X13666150716165726.

Ferrer I (2014) Cannabinoids for treatment of Alzheimer's disease: moving toward the clinic. Frontiers in Pharmacology 5: 37. DOI: 10.3389/fphar.2014.00037.

Fitzgerald RS, Shirahata M, Wang H-YJ, et al. (2004) The impact of adenosine on the release of acetylcholine, dopamine, and norepinephrine from the cat carotid body. Neuroscience Letters 367: 304–308. DOI: 10.1016/j.neulet.2004.06.019.

Folch J, Petrov D, Ettcheto M, et al. (2016) Current research therapeutic strategies for Alzheimer's disease treatment. Neural Plasticity 2016: 8501693. DOI: 10.1155/2016/8501693.

Foster DJ, Choi DL, Conn PJ, et al. (2014) Activation of M_1 and M_4 muscarinic receptors as potential treatments for Alzheimer's disease and schizophrenia. Neuropsychiatric Disease and Treatment 10: 183–191. DOI: 10.2147/NDT.S55104.

Foster DJ, Wilson JM, Remke DH, et al. (2016) Antipsychotic-like effects of M4 positive allosteric modulators are mediated by CB2 receptor-dependent inhibition of dopamine release. Neuron 91:1244–1252. DOI: 10.1016/j.neuron.2016.08.017.

Franklin DJ and Grossberg S (2017) A neural model of normal and abnormal learning and memory consolidation: adaptively timed conditioning, hippocampus, amnesia, neurotrophins, and consciousness. Cognitive, Affective, & Behavioral Neuroscience 17: 24–76. DOI: 10.3758/s13415-016-0463-y.

Froestl W, Gallagher M, Jenkins H, et al. (2004) SGS742: the first GABAB receptor antagonist in clinical trials. Biochemical Pharmacology 68: 1479–1487. DOI: 10.1016/j.bcp.2004.07.030.

Fu Q-L, Dai H-B, Shen Y, et al. (2007) [Reversing effect of histamine on neurotoxicity induced by beta-amyloid1-42]. Journal of Zhejiang University Medical Sciences 36: 146–149.

Gabriella G and Giovanna C (2010) γ-Aminobutyric acid type A (GABAA) receptor subtype inverse agonists as therapeutic agents in cognition. In: Conn PM (ed.) Methods in Enzymology. Constitutive Activity in Receptors and Other Proteins, Part B. Academic Press, pp. 197–211. DOI: 10.1016/B978-0-12-381296-4.00011-7.

Galve-Roperh I, Aguado T and Guzman JP and M (2008) Mechanisms of control of neuron survival by the endocannabinoid system. Current Pharmaceutical Design 14: 2279–2288.

Garre-Olmo J (2018) [Epidemiology of Alzheimer's disease and other dementias]. Revista De Neurologia 66: 377–386.

Getova D and Bowery N (2001) Effects of high-affinity GABAB receptor antagonists on active and passive avoidance responding in rodents with gamma-hydroxybutyrolactone-induced absence syndrome. Psychopharmacology 157: 89–95. DOI: 10.1007/s002130100766.

Gibbs ME and Hertz L (2014) Serotonin mediation of early memory formation via 5-HT2B receptor-induced glycogenolysis in the day-old chick. Frontiers in Pharmacology 5: 54. DOI: 10.3389/fphar.2014.00054.

Gillman KW, Starrett JE, Parker MF, et al. (2010) Discovery and evaluation of BMS-708163, a potent, selective and orally bioavailable γ-secretase inhibitor. ACS Medicinal Chemistry Letters 1: 120–124. DOI: 10.1021/ml1000239.

Gomes BAQ, Silva JPB, Romeiro CFR, et al. (2018) Neuroprotective mechanisms of resveratrol in Alzheimer's disease: role of SIRT1. Oxidative Medicine and Cellular Longevity 2018: 8152373. DOI:10.1155/2018/8152373

Greenberg SG and Davies P (1990) A preparation of Alzheimer paired helical filaments that displays distinct tau proteins by polyacrylamide gel electrophoresis. Proceedings of the National Academy of Sciences of the United States of America 87: 5827–5831. DOI: 10.1073/pnas.87.15.5827.

Grill JD and Cummings JL (2010) Novel targets for Alzheimer's disease treatment. Expert Review of Neurotherapeutics 10: 711–728. DOI: 10.1586/ern.10.29.

Gronek P, Balko S, Gronek J, et al. (2019) Physical activity and Alzheimer's disease: a narrative review. Aging and Disease 10: 1282–1292. DOI: 10.14336/AD.2019.0226.

Grove RA, Harrington CM, Mahler A, et al. (2014) A randomized, double-blind, placebo-controlled, 16-week study of the H3 receptor antagonist, GSK239512 as a monotherapy in subjects with mild-to-moderate Alzheimer's disease. Current Alzheimer Research 11: 47–58.

Guerrini G, Ciciani G, Costanzo A, et al. (2013) Synthesis of novel cognition enhancers with pyrazolo[5,1-c][1,2,4]benzotriazine core acting at γ-aminobutyric acid type A (GABAA) receptor. Bioorganic & Medicinal Chemistry 21: 2186–2198. DOI: 10.1016/j.bmc.2013.02.027.

Gulaj E, Pawlak K, Bien B, et al. (2010) Kynurenine and its metabolites in Alzheimer's disease patients. Advances in Medical Sciences 55: 204–211. DOI: 10.2478/v10039-010-0023-6.

Haas H and Panula P (2003) The role of histamine and the tuberomammillary nucleus in the nervous system. Nature Reviews Neuroscience 4: 121–130. DOI: 10.1038/nrn1034.

Haas HL and Selbach O (2000) Functions of neuronal adenosine receptors. Naunyn-Schmiedeberg's Archives of Pharmacology 362: 375–381. DOI: 10.1007/s002100000314.

Hagena H and Manahan-Vaughan D (2017) The serotonergic 5-HT4 receptor: a unique modulator of hippocampal synaptic information processing and cognition. Neurobiology of Learning and Memory 138: 145–153. DOI: 10.1016/j.nlm.2016.06.014.

Hanson AJ, Craft S and Banks WA (2015) The APOE genotype: modification of therapeutic responses in Alzheimer's disease. Current Pharmaceutical Design 21: 114–120. DOI: 10.2174/1381612820666141020164222.

Hansson O and Gouras G (2016) Brain activity and Alzheimer's disease: a complex relationship. Brain 139: 2109–2110. DOI: 10.1093/brain/aww146.

Hattori N, Takeda A, Takeda S, et al. (2019) Long-term, open-label, phase 3 study of rasagiline in Japanese patients with early Parkinson's disease. Journal of Neural Transmission 126: 299–308. DOI: 10.1007/s00702-018-1964-3.

Hawkins RD (2013) Possible contributions of a novel form of synaptic plasticity in aplysia to reward, memory, and their dysfunctions in mammalian brain. Learning & Memory 20: 580–591. DOI: 10.1101/lm.031237.113.

Heart Protection Study Collaborative Group (2002) MRC/BHF Heart Protection Study of cholesterol lowering with simvastatin in 20,536 high-risk individuals: a randomised placebo-controlled trial. Lancet 360: 7–22. DOI: 10.1016/S0140-6736(02)09327-3.

Helm KA, Haberman RP, Dean SL, et al. (2005) GABAB receptor antagonist SGS742 improves spatial memory and reduces protein binding to the cAMP response element (CRE) in the hippocampus. Neuropharmacology 48: 956–964. DOI: 10.1016/j.neuropharm.2005.01.019.

Hoover DB (2017) Cholinergic modulation of the immune system presents new approaches for treating inflammation. Pharmacology & Therapeutics 179: 1–16. DOI: 10.1016/j.pharmthera.2017.05.002.

Hsu CW, Wang CS and Chiu TH (2010) Caffeine and a selective adenosine A2A receptor antagonist induce sensitization and cross-sensitization behavior associated with increased striatal dopamine in mice. Journal of Biomedical Science 17: 4. DOI: 10.1186/1423-0127-17-4.

Hsu D and Marshall GA (2017) Primary and secondary prevention trials in Alzheimer disease: looking back, moving forward. Current Alzheimer Research 14: 426–440. DOI: 10.2174/1567205013666160930112125.

Hu NW, Klyubin I, Anwyl R, et al. (2009) GluN2B subunit-containing NMDA receptor antagonists prevent Aβ-mediated synaptic plasticity disruption in vivo. Proceedings of the National Academy of Sciences of the United States of America 106: 20504–20509. DOI: 10.1073/pnas.0908083106.

Hu X, Das B, Hou H, et al. (2018) BACE1 deletion in the adult mouse reverses preformed amyloid deposition and improves cognitive functions. The Journal of Experimental Medicine 215: 927–940. DOI: 10.1084/jem.20171831.

Hung SY and Fu W-M (2017) Drug candidates in clinical trials for Alzheimer's disease. Journal of Biomedical Science 24: 47. DOI: 10.1186/s12929-017-0355-7.

Ibrahim AM, Pottoo FH, Dahiya ES, et al. (2020) Neuron-glia interactions: molecular basis of Alzheimer's disease and applications of neuroproteomics. The European Journal of Neuroscience 52: 2931–2943. DOI: 10.1111/ejn.14838.

Jaeschke G, Kolczewski S, Spooren W, et al. (2015) Metabotropic glutamate receptor 5 negative allosteric modulators: discovery of 2-chloro-4-[1-(4-fluorophenyl)-2,5-dimethyl-1H-imidazol-4-ylethynyl]pyridine (basimglurant, RO4917523), a promising novel medicine for psychiatric diseases. Journal of Medicinal Chemistry 58: 1358–1371. DOI: 10.1021/jm501642c.

Janson J, Laedtke T, Parisi JE, et al. (2004) Increased risk of type 2 diabetes in Alzheimer disease. Diabetes 53: 474–481. DOI: 10.2337/diabetes.53.2.474.

Javed MN, Alam MS, Waziri A, et al. (2019) QbD Applications for the development of nanopharmaceutical products. In: Beg S and Hasnain MS (eds) Pharmaceutical Quality by Design. Academic Press, pp. 229–253. DOI: 10.1016/B978-0-12-815799-2.00013-7.

Jiang X and Nardelli J (2016) Cellular and molecular introduction to brain development. Neurobiology of Disease 92: 3–17. DOI: 10.1016/j.nbd.2015.07.007.

Joe E, Ringman JM (2019). Cognitive symptoms of Alzheimer's disease: clinical management and prevention. BMJ 367: l6216. doi: 10.1136/bmj.l6217.

Johns P (2014) Overview of the nervous system. In: Johns P (ed.) Clinical Neuroscience. Churchill Livingstone, pp. 1–17. DOI: 10.1016/B978-0-443-10321-6.00001-1.

Kahle-Wrobleski K, Ye W, Henley D, et al. (2017) Assessing quality of life in Alzheimer's disease: implications for clinical trials. Alzheimer's & Dementia Diagnosis, Assessment & Disease Monitoring 6: 82–90. DOI: 10.1016/j.dadm.2016.11.004.

Kaur S, DasGupta G and Singh S. (2019). Altered neurochemistry in Alzheimer's disease: targeting neurotransmitter receptor mechanisms and therapeutic strategy. Neurophysiology 51: 293–309. DOI:10.1007/s11062-019-09823-7.

Kennedy ME, Stamford AW, Chen X, et al. (2016) The BACE1 inhibitor verubecestat (MK-8931) reduces CNS β-amyloid in animal models and in Alzheimer's disease patients. Science Translational Medicine 8: 363ra150. DOI: 10.1126/scitranslmed.aad9704.

Khan H, Ullah H, Aschner M, et al. (2019) Neuroprotective effects of quercetin in Alzheimer's disease. Biomolecules 10: 59. DOI:10.3390/biom10010059

Kim SH, Cairns N, Fountoulakisc M, et al. (2001) Decreased brain histamine-releasing factor protein in patients with Down syndrome and Alzheimer's disease. Neuroscience Letters 300: 41–44. DOI: 10.1016/S0304-3940(01)01545-2.

Kinney JW, Bemiller SM, Murtishaw AS, et al. (2018) Inflammation as a central mechanism in Alzheimer's disease. Alzheimer's & Dementia 4: 575–590. DOI: 10.1016/j.trci.2018.06.014.

Knust H, Achermann G, Ballard T, et al. (2009) The discovery and unique pharmacological profile of RO4938581 and RO4882224 as potent and selective GABAA α5 inverse agonists for the treatment of cognitive dysfunction. Bioorganic & Medicinal Chemistry Letters 19: 5940–5944. DOI: 10.1016/j.bmcl.2009.08.053.

Koola MM, Nikiforuk A, Pillai A, et al. (2018) Galantamine-memantine combination superior to donepezil-memantine combination in Alzheimer's disease: critical dissection with an emphasis on kynurenic acid and mismatch negativity. Journal of Geriatric Care and Research 5: 57–67.

Krashia P, Nobili A and D'Amelio M (2019) Unifying hypothesis of dopamine neuron loss in neurodegenerative diseases: focusing on Alzheimer's disease. Frontiers in Molecular Neuroscience 12: 123. DOI: 10.3389/fnmol.2019.00123.

Kubo M, Kishi T, Matsunaga S, et al. (2015) Histamine H3 receptor antagonists for Alzheimer's disease: a systematic review and meta-analysis of randomized placebo-controlled trials. Journal of Alzheimer's Disease 48: 667–671. DOI: 10.3233/JAD-150393.

Kumar Ashwini, Nisha CM, Silakari C, et al. (2016) Current and novel therapeutic molecules and targets in Alzheimer's disease. Journal of the Formosan Medical Association 115: 3–10. DOI: 10.1016/j.jfma.2015.04.001.

Kung HF (2012) The β-amyloid hypothesis in Alzheimer's disease: seeing is believing. ACS Medicinal Chemistry Letters 3: 265–267. DOI: 10.1021/ml300058m.

Kuszczyk MA, Sanchez S, Pankiewicz J, et al. (2013) Blocking the interaction between apolipoprotein E and Aβ reduces intraneuronal accumulation of Aβ and inhibits synaptic degeneration. The American Journal of Pathology 182: 1750–1768. DOI: 10.1016/j.ajpath.2013.01.034.

Kwakowsky A, Calvo-Flores Guzmán B, Govindpani K, et al. (2018) Gamma-aminobutyric acid A receptors in Alzheimer's disease: highly localized remodeling of a complex and diverse signaling pathway. Neural Regeneration Research 13: 1362–1363. doi: 10.4103/1673-5374.235240.

Lam CS, Li JJ, Tipoe GL, et al. (2017) Monoamine oxidase A upregulated by chronic intermittent hypoxia activates indoleamine 2,3-dioxygenase and neurodegeneration. PLoS One 12: 177940. DOI: 10.1371/journal.pone.0177940.

Lanctôt KL, Herrmann N, Mazzotta P, et al. (2004) GABAergic function in Alzheimer's disease: evidence for dysfunction and potential as a therapeutic target for the treatment of behavioural and psychological symptoms of dementia: The Canadian Journal of Psychiatry 49: 439–453. DOI: 10.1177/070674370404900705.

Lasarge CL, Bañuelos C, Mayse JD, et al. (2009) Blockade of GABA(B) receptors completely reverses age-related learning impairment. Neuroscience 164: 941–947. DOI: 10.1016/j.neuroscience.2009.08.055.

Łażewska D and Kieć-Kononowicz K (2014) New developments around histamine H3 receptor antagonists/inverse agonists: a patent review (2010 – present). Expert Opinion on Therapeutic Patents 24: 89–111. DOI: 10.1517/13543776.2014.848197.

Lebois EP, Thorn C, Edgerton JR, et al. (2018) Muscarinic receptor subtype distribution in the central nervous system and relevance to aging and Alzheimer's disease. Neuropharmacology 136: 362–373. DOI: 10.1016/j.neuropharm.2017.11.018.

Ledo JH, Azevedo EP, Beckman D, et al. (2016) Cross talk between brain innate immunity and serotonin signaling underlies depressive-like behavior induced by Alzheimer's amyloid-β oligomers in mice. Journal of Neuroscience 36: 12106–12116. DOI: 10.1523/JNEUROSCI.1269-16.2016.

Ledo JH, Azevedo EP, Clarke JR, et al. (2013) Amyloid-β oligomers link depressive-like behavior and cognitive deficits in mice. Molecular Psychiatry 18: 1053–1054. DOI: 10.1038/mp.2012.168.

Lee BY, Ban JY and Seong YH (2005) Chronic stimulation of GABAA receptor with muscimol reduces amyloid β protein (25–35)-induced neurotoxicity in cultured rat cortical cells. Neuroscience Research 52: 347–356. DOI: 10.1016/j.neures.2005.04.008.

Lemere CA (2009) Developing novel immunogens for a safe and effective Alzheimer's disease vaccine. Progress in Brain Research 175: 83–93. DOI: 10.1016/S0079-6123(09)17506-4.

Li Y, Sun H, Chen Z, et al. (2016) Implications of GABAergic neurotransmission in Alzheimer's disease. Frontiers in Aging Neuroscience 8: 31. DOI: 10.3389/fnagi.2016.00031.

Liddelow SA, Guttenplan KA, Clarke LE, et al. (2017) Neurotoxic reactive astrocytes are induced by activated microglia. Nature 541: 481–487.

Liu S, Breitbart A, Sun Y, et al. (2014) Blocking the apolipoprotein E/amyloid β interaction in triple transgenic mice ameliorates

Alzheimer's disease related amyloid β and tau pathology. Journal of Neurochemistry 128: 577–591. DOI: 10.1111/jnc.12484.

Liu XH, Zhu R-T, Hao B, et al. (2019) Norepinephrine induces PTSD-like memory impairments via regulation of the β-adrenoceptor-cAMP/PKA and CaMK II/PKC systems in the basolateral amygdala. Frontiers in Behavioral Neuroscience 13: 43. DOI: 10.3389/fnbeh.2019.00043.

Livingston G, Sommerlad A, Orgeta V, et al. (2017) Dementia prevention, intervention, and care. The Lancet 390: 2673–2734. DOI: 10.1016/S0140-6736(17)31363-6.

Lodish H, Berk A, Zipursky SL, et al. (2000a) Neurotransmitter receptors. In: Darnell JE, Lodish H, Zipursky L, et al. (eds). Molecular Cell Biology. 4th edition. W. H. Freeman.

Lodish H, Berk A, Zipursky SL, et al. (2000b) Neurotransmitters, synapses, and impulse transmission. In: Darnell JE, Lodish H, Zipursky L, et al. (eds). Molecular Cell Biology. 4th edition. W.H. Freeman.

Lombardo S and Maskos U (2015) Role of the nicotinic acetylcholine receptor in Alzheimer's disease pathology and treatment. Neuropharmacology 96: 255–262. DOI: 10.1016/j.neuropharm.2014.11.018.

Lourenco MV, Ferreira ST and De Felice FG (2015) Neuronal stress signaling and eIF2α phosphorylation as molecular links between Alzheimer's disease and diabetes. Progress in Neurobiology 129: 37–57. DOI: 10.1016/j.pneurobio.2015.03.003.

Luo SX and Huang EJ (2016) Dopaminergic neurons and brain reward pathways: from neurogenesis to circuit assembly. The American Journal of Pathology 186: 478–488. DOI: 10.1016/j.ajpath.2015.09.023.

Luo T, Wu W-H and Chen B-S (2011) NMDA receptor signaling: death or survival? Frontiers in Biology 6: 468–476. DOI: 10.1007/s11515-011-1187-6.

Lyra e Silva N de M, Gonçalves RA, Boehnke SE, et al. (2019) Understanding the link between insulin resistance and Alzheimer's disease: insights from animal models. Experimental Neurology 316: 1–11. DOI: 10.1016/j.expneurol.2019.03.016.

McGleenon BM, Dynan KB and Passmore AP (1999) Acetylcholinesterase inhibitors in Alzheimer's disease. British Journal of Clinical Pharmacology 48: 471–480. DOI: 10.1046/j.1365-2125.1999.00026.x.

Majdi A, Sadigh-Eteghad S, Aghsan SR, et al. (2020) Amyloid-β, tau, and the cholinergic system in Alzheimer's disease: seeking direction in a tangle of clues. Reviews in the Neurosciences 31: 391–413. DOI: 10.1515/revneuro-2019-0089.

Malinski T (2007) Nitric oxide and nitroxidative stress in Alzheimer's disease. Journal of Alzheimer's Disease 11: 207–218.

Mamun AA, Uddin MdS, Mathew B, et al. (2020) Toxic tau: structural origins of tau aggregation in Alzheimer's disease. Neural Regeneration Research 15: 1417–1420. DOI: 10.4103/1673-5374.274329.

Marcade M, Bourdin J, Loiseau N, et al. (2008) Etazolate, a neuroprotective drug linking GABAA receptor pharmacology to amyloid precursor protein processing. Journal of Neurochemistry 106: 392–404. DOI: 10.1111/j.1471-4159.2008.05396.x.

Marner L, Knudsen GM, Madsen K, et al. (2011) The reduction of baseline serotonin 2A receptors in mild cognitive impairment is stable at two-year follow-up. Journal of Alzheimer's Disease 23: 453–459. DOI: 10.3233/JAD-2010-100903.

Martorana A and Koch G (2014) Is dopamine involved in Alzheimer's disease? Frontiers in Aging Neuroscience 6: 252. DOI: 10.3389/fnagi.2014.00252.

Mayes A (2001) Korsakoff's syndrome. In: Smelser NJ and Baltes PB (eds) International Encyclopedia of the Social & Behavioral Sciences. Oxford: Pergamon, pp. 8162–8166. DOI: 10.1016/B0-08-043076-7/03573-7.

Mazzola C, Micale V and Drago F. (2003) Amnesia induced by beta-amyloid fragments is counteracted by cannabinoid CB1 receptor blockade. European Journal of Pharmacology 477: 219–225. DOI: 10.1016/j.ejphar.2003.08.026. PMID: 14522360.

Mir RH, Sawhney G, Pottoo FH, et al. (2020) Role of environmental pollutants in Alzheimer's disease: a review. Environmental Science and Pollution Research International 27: 44724–44742. DOI: 10.1007/s11356-020-09964-x.

Mir RH, Shah AJ, Mohi-Ud-Din R, et al. (2021) Natural anti-inflammatory compounds as drug candidates in Alzheimer's disease. Current Medicinal Chemistry 28: 4799–4825. DOI: 10.2174/0929867327666200730213215.

Morgese MG and Trabace L (2019) Monoaminergic system modulation in depression and Alzheimer's disease: a new standpoint? Frontiers in Pharmacology 10: 483. DOI: 10.3389/fphar.2019.00483.

Morris MC, Tangney CC, Wang Y, et al. (2015) MIND diet associated with reduced incidence of Alzheimer's disease. Alzheimer's & Dementia 11: 1007–1014. DOI: 10.1016/j.jalz.2014.11.009.

Mukherjee A and Soto C (2017) Prion-like protein aggregates and type 2 diabetes. Cold Spring Harbor Perspectives in Medicine 7: a024315. DOI: 10.1101/cshperspect.a024315.

Murchison CF, Zhang XY, Zhang WP, et al. (2004) A distinct role for norepinephrine in memory retrieval. Cell 117: 131–143. DOI: 10.1016/S0092-8674(04)00259-4.

Naddafi F and Mirshafiey A (2013) The neglected role of histamine in Alzheimer's disease. American Journal of Alzheimer's Disease & Other Dementias 28: 327–336. DOI: 10.1177/1533317513488925.

Nathan PJ, Boardley R, Scott N, et al. (2013) The safety, tolerability, pharmacokinetics and cognitive effects of GSK239512, a selective histamine H_3 receptor antagonist in patients with mild to moderate Alzheimer's disease: a preliminary investigation. Current Alzheimer Research 10: 240–251. DOI: 10.2174/1567205011310030003.

Nelson MR, Tipney H, Painter JL, et al. (2015) The support of human genetic evidence for approved drug indications. Nature Genetics 47: 856–860. DOI: 10.1038/ng.3314.

Netzer WJ, Dou F, Cai D, et al. (2003) Gleevec inhibits beta-amyloid production but not Notch cleavage. Proceedings of the National Academy of Sciences of the United States of America 100: 12444–12449. DOI: 10.1073/pnas.1534745100.

Nichols E, Szoeke CEI, Vollset SE, et al. (2019) Global, regional, and national burden of Alzheimer's disease and other dementias, 1990–2016: a systematic analysis for the Global Burden of Disease Study 2016. The Lancet Neurology 18: 88–106. DOI: 10.1016/S1474-4422(18)30403-4.

Nigar S, Pottoo FH, Tabassum N, et al. (2016) Molecular insights into the role of inflammation and oxidative stress in epilepsy. Journal of Advances in Medical and Pharmaceutical Sciences 10: 1–9.

NIH Curriculum Supplement Series (2007) Information about the brain. National Institutes of Health (US). Available at: www.ncbi.nlm.nih.gov/books/NBK20367/ (accessed 25 September 2019).

Nobili A, Latagliata EC, Viscomi MT, et al. (2017) Dopamine neuronal loss contributes to memory and reward dysfunction in a model of Alzheimer's disease. Nature Communications 8: 14727. DOI: 10.1038/ncomms14727.

Norins LC (2019) Predicted economic damage from a quick, simple Alzheimer's disease cure. Medical Hypotheses 133: 109398. DOI: 10.1016/j.mehy.2019.109398.

Notcovich C, Diez F, Tubio MR, et al. (2010) Histamine acting on H1 receptor promotes inhibition of proliferation via PLC, RAC, and JNK-dependent pathways. Experimental Cell Research 316: 401–411. DOI: 10.1016/j.yexcr.2009.11.002.

Olgiati A and Serretti P (2012) Catechol-O-methyltransferase and Alzheimer's disease: a review of biological and genetic findings. CNS & Neurological Disorders Drug Targets 11: 299–305.

Oliveira MM and Lourenco MV (2016) Integrated stress response: connecting ApoE4 to memory impairment in Alzheimer's disease. The Journal of Neuroscience 36: 1053–1055. DOI: 10.1523/JNEUROSCI.4110-15.2016.

Ossowska K, Konieczny J, Wolfarth S, et al. (2005) MTEP, a new selective antagonist of the metabotropic glutamate receptor subtype 5 (mGluR5), produces antiparkinsonian-like effects in rats. Neuropharmacology 49: 447–455. DOI: 10.1016/j.neuropharm.2005.04.002.

Pan X, Kaminga AC, Wen SW, et al. (2019) Dopamine and dopamine receptors in Alzheimer's disease: a systematic review and network meta-analysis. Frontiers in Aging Neuroscience 11: 175. DOI: 10.3389/fnagi.2019.00175.

Panula P and Nuutinen S (2013) The histaminergic network in the brain: basic organization and role in disease. Nature Reviews Neuroscience: 472–487. DOI: 10.1038/nrn3526.

Parsons CG, Danysz W, Dekundy A, et al. (2013) Memantine and cholinesterase inhibitors: complementary mechanisms in the treatment of Alzheimer's disease. Neurotoxicity Research 24: 358–369. DOI: 10.1007/s12640-013-9398-z.

Pascual-Brazo J, Castro E, Díaz Á, et al. (2012) Modulation of neuroplasticity pathways and antidepressant-like behavioural responses following the short-term (3 and 7 days) administration of the 5-HT4 receptor agonist RS67333. International Journal of Neuropsychopharmacology 15: 631–643. DOI: 10.1017/S1461145711000782.

Pasquier F, Sadowsky C, Holstein A, et al. (2016) Two phase 2 multiple ascending-dose studies of vanutide cridificar (ACC-001) and QS-21 adjuvant in mild-to-moderate Alzheimer's disease. Journal of Alzheimer's Disease 51: 1131–1143. DOI: 10.3233/JAD-150376.

Passani MB, Panula P and Lin J-S (2014) Histamine in the brain. Frontiers in Systems Neuroscience 8: 64. DOI: 10.3389/fnsys.2014.00064.

Patel L and Grossberg GT (2011) Combination therapy for Alzheimer's disease. Drugs & Aging 28: 539–546. DOI: 10.2165/11591860-000000000-00000.

Pei J-J, Björkdahl C, Zhang H, et al. (2008) p70 S6 kinase and tau in Alzheimer's disease. Journal of Alzheimer's Disease 14: 385–392.

Perry VH and Holmes C (2014) Microglial priming in neurodegenerative disease. Nature Reviews Neurology 10: 217–224. DOI: 10.1038/nrneurol.2014.38.

Pilipenko V, Pupure J, Rumaks J, et al. (2015) GABAA agonist muscimol ameliorates learning/memory deficits in streptozocin-induced Alzheimer's disease non-transgenic rat model. SpringerPlus 4: P36. DOI: 10.1186/2193-1801-4-S1-P36.

Polter AM and Li X (2010) 5-HT1A receptor-regulated signal transduction pathways in brain. Cellular Signalling 22: 1406–1412. DOI: 10.1016/j.cellsig.2010.03.019.

Pottoo FH, Sharma S, Javed MN, et al. (2020) Lipid-based nanoformulations in the treatment of neurological disorders. Drug Metabolism Reviews 52: 185–204. DOI: 10.1080/03602532.2020.1726942.

Pottoo FH, Tabassum N, Javed MN, et al. (2020) Raloxifene potentiates the effect of fluoxetine against maximal electroshock induced seizures in mice. European Journal of Pharmaceutical Sciences 146: 105261. DOI: 10.1016/j.ejps.2020.105261.

Prediger RDS and Takahashi RN (2005) Modulation of short-term social memory in rats by adenosine A1 and A2A receptors. Neuroscience Letters 376: 160–165. DOI: 10.1016/j.neulet.2004.11.049.

Purves D, Augustine GJ, Fitzpatrick D, et al. (2001) Peptide neurotransmitters. neuroscience. 2nd edition. Sinauer Associates. Available at: www.ncbi.nlm.nih.gov/books/NBK10873/ (accessed 23 September 2020).

Quintela-López T, Ortiz-Sanz C, Serrano-Regal MP. et al. (2019) Aβ oligomers promote oligodendrocyte differentiation and maturation via integrin β1 and Fyn kinase signaling. Cell Death & Disease 10: 445.

Rahman A (2009) The role of adenosine in Alzheimer's disease. Current Neuropharmacology 7: 207–216. DOI: 10.2174/157015909789152119.

Rajabi S, Shamsizadeh A, Amini H, et al. (2012) Effect of DSP-4 induced central noradrenergic depletion on tactile learning in rat. Neurological Research 34: 80–84. DOI: 10.1179/1743132811Y.0000000050.

Rajamohamedsait H, Rasool S, Rajamohamedsait W, et al. (2017) Prophylactic active tau immunization leads to sustained reduction in both tau and amyloid-β pathologies in 3xTg mice. Scientific Reports 7: 17034. DOI: 10.1038/s41598-017-17313-1.

Ramírez MJ (2013) 5-HT6 receptors and Alzheimer's disease. Alzheimer's Research & Therapy 5: 15. DOI: 10.1186/alzrt169.

Rangel-Barajas C, Coronel I, Florán B (2015) Dopamine receptors and neurodegeneration. Aging and Disease 6: 349–368. DOI: 10.14336/AD.2015.0330.

Rebola N, Sebastião AM, de Mendonça A, et al. (2003) Enhanced adenosine A2A receptor facilitation of synaptic transmission in the hippocampus of aged rats. Journal of Neurophysiology 90: 1295–1303. DOI: 10.1152/jn.00896.2002.

Rossner S, Ueberham U, Schliebs R, et al. (1998) The regulation of amyloid precursor protein metabolism by cholinergic mechanisms and neurotrophin receptor signaling. Progress in Neurobiology 56: 541–569. DOI: 10.1016/S0301-0082(98)00044-6.

Rossor MN, Garrett NJ, Johnson AL, et al. (1982) A post-mortem study of the cholinergic and GABA systems in senile dementia. Brain 105: 313–330. DOI: 10.1093/brain/105.2.313.

Rudy CC, Hunsberger HC, Weitzner DS, et al. (2015) The role of the tripartite glutamatergic synapse in the pathophysiology of Alzheimer's disease. Aging and Disease 6: 131–148. DOI: 10.14336/AD.2014.0423.

Salloway, S., Mintzer, J., Weiner, M. F., & Cummings, J. L. (2008). Disease-modifying therapies in Alzheimer's disease. Alzheimer's & dementia: the journal of the Alzheimer's

Association, 4(2), 65–79. https://doi.org/10.1016/j.jalz.2007.10.001.

Salter MW and Stevens B (2017) Microglia emerge as central players in brain disease. Nature Medicine 23: 1018–1027. DOI: 10.1038/nm.4397.

Samuels ER, Szabadi E. (2008) Functional neuroanatomy of the noradrenergic locus coeruleus: its roles in the regulation of arousal and autonomic function part I: principles of functional organisation. Current Neuropharmacology 6:235–253. DOI:10.2174/157015908785777229.

Saraiva C, Barata-Antunes S, Santos T, et al. (2019) Histamine modulates hippocampal inflammation and neurogenesis in adult mice. Scientific Reports 9: 8384. DOI: 10.1038/s41598-019-44816-w.

Satizabal CL, Beiser AS, Chouraki V, et al. (2016) Incidence of dementia over three decades in the Framingham Heart Study. New England Journal of Medicine 374: 523–532. DOI: 10.1056/NEJMoa1504327.

Savonenko AV, Melnikova T, Wang Y, et al. (2015) Cannabinoid CB2 receptors in a mouse model of Aβ amyloidosis: immunohistochemical analysis and suitability as a PET biomarker of neuroinflammation. PLoS One 10: 129618. DOI: 10.1371/journal.pone.0129618.

Seeger T, Fedorova I, Zheng F, et al. (2004) M2 muscarinic acetylcholine receptor knock-out mice show deficits in behavioral flexibility, working memory, and hippocampal plasticity. The Journal of Neuroscience 24: 10117–10127. DOI: 10.1523/JNEUROSCI.3581-04.2004.

Shan L, Hofman MA, van Wamelen DJ, et al. (2012) Diurnal fluctuation in histidine decarboxylase expression, the rate limiting enzyme for histamine production, and its disorder in neurodegenerative diseases. Sleep 35: 713–715. DOI: 10.5665/sleep.1838.

Shao H, Zhang Y, Dong Y, et al. (2014) Chronic treatment with anesthetic propofol improves cognitive function and attenuates caspase activation in both aged and Alzheimer's disease transgenic mice. Journal of Alzheimer's Disease 41: 499–513. DOI: 10.3233/JAD-132792.

Sharma P, Sharma A, Fayaz F, et al. (2020) Biological signatures of Alzheimer's disease. Current Topics in Medicinal Chemistry 20: 770–781. DOI: 10.2174/1568026620666200228095553.

Silbereis JC, Pochareddy S, Zhu Y, et al. (2016) The cellular and molecular landscapes of the developing human central nervous system. Neuron 89: 248–268. DOI: 10.1016/j.neuron.2015.12.008.

Sims R, Hill M and Williams J (2020) The multiplex model of the genetics of Alzheimer's disease. Nature Neuroscience 23: 311–322. DOI: 10.1038/s41593-020-0599-5.

Singh SK, Srivastav S, Castellani RJ, et al. (2019) Neuroprotective and antioxidant effect of *Ginkgo biloba* extract against AD and other neurological disorders. Neurotherapeutics 16: 666–674. DOI:10.1007/s13311-019-00767-8.

Sontag J-M and Sontag E (2014) Protein phosphatase 2A dysfunction in Alzheimer's disease. Frontiers in Molecular Neuroscience 7: 16. DOI: 10.3389/fnmol.2014.00016.

Spires-Jones TL and Hyman BT (2014) The intersection of amyloid beta and tau at synapses in Alzheimer's disease. Neuron 82: 756–771. DOI: 10.1016/j.neuron.2014.05.004.

Steele ML, Robinson SR (2012) Reactive astrocytes give neurons less support: implications for Alzheimer's disease. Neurobiology of Aging 33: e421.

Stiedl O, Pappa E, Konradsson-Geuken Å, et al. (2015) The role of the serotonin receptor subtypes 5-HT1A and 5-HT7 and its interaction in emotional learning and memory. Frontiers in Pharmacology Frontiers 6: 162. DOI: 10.3389/fphar.2015.00162.

Stiles J and Jernigan TL (2010) The basics of brain development. Neuropsychology Review 20: 327–348. DOI: 10.1007/s11065-010-9148-4.

Stockwell J, Jakova E and Cayabyab FS (2017) Adenosine A1 and A2A receptors in the brain: current research and their role in neurodegeneration. Molecules 22: 676. DOI: 10.3390/molecules22040676.

Suchan B and Daum I (2001) Cerebellum: cognitive functions. In: Smelser NJ and Baltes PB (eds) International Encyclopedia of the Social & Behavioral Sciences. Oxford: Pergamon, pp. 1631–1634. DOI: 10.1016/B0-08-043076-7/03457-4.

Sudduth TL, Schmitt FA, Nelson PT, et al. (2013) Neuroinflammatory phenotype in early Alzheimer's disease. Neurobiology of Aging 34: 1051–1059. DOI: 10.1016/j.neurobiolaging.2012.09.012.

Sun B, Halabisky B, Zhou Y, et al. (2009) Imbalance between GABAergic and glutamatergic transmission impairs adult neurogenesis in an animal model of Alzheimer's disease. Cell Stem Cell 5: 624–633. DOI: 10.1016/j.stem.2009.10.003.

Tajeddinn W, Persson T, Calvo-Garrido J, et al. (2016) Pharmacological modulations of the serotonergic system in a cell-model of familial Alzheimer's disease. Journal of Alzheimer's Disease 53: 349–361. DOI: 10.3233/JAD-160046.

Tajeddinn W, Persson T, Maioli S, et al. (2015) 5-HT1B and other related serotonergic proteins are altered in APPswe mutation. Neuroscience Letters 594: 137–143. DOI: 10.1016/j.neulet.2015.03.064.

Talbot K (2014) Brain insulin resistance in Alzheimer's disease and its potential treatment with GLP-1 analogs. Neurodegenerative Disease Management 4: 31–40. DOI: 10.2217/nmt.13.73.

Tang BL (2019) Amyloid precursor protein (APP) and GABAergic neurotransmission. Cells 8: 550. DOI: 10.3390/cells8060550.

Thomsen MS, Hansen HH, Timmerman DB, et al. (2010) Cognitive improvement by activation of alpha7 nicotinic acetylcholine receptors: from animal models to human pathophysiology. Current Pharmaceutical Design 16: 323–343. DOI: 10.2174/138161210790170094.

Tiligada E, Kyriakidis K, Chazot PL, et al. (2011) Histamine pharmacology and new CNS drug targets. CNS Neuroscience & Therapeutics 17: 620–628. DOI: 10.1111/j.1755-5949.2010.00212.x.

Toneff T, Funkelstein L, Mosier C, et al. (2013) Beta-amyloid peptides undergo regulated co-secretion with neuropeptide and catecholamine neurotransmitters. Peptides 46: 126–135. DOI: 10.1016/j.peptides.2013.04.020.

Tuppo EE and Arias HR (2005) The role of inflammation in Alzheimer's disease. The International Journal of Biochemistry & Cell Biology 37: 289–305. DOI: 10.1016/j.biocel.2004.07.009.

Uddin MS, Al Mamun A, Kabir MdT, et al. (2021) Multi-target drug candidates for multifactorial Alzheimer's disease: AChE and NMDAR as molecular targets. Molecular Neurobiology 58: 28–303. DOI: 10.1007/s12035-020-02116-9.

Uddin MS, Al Mamun A, Rahman MA, et al. (2020) Emerging proof of protein misfolding and interaction in multifactorial

Alzheimer's disease. Current Topics in Medicinal Chemistry 20:2380–2390. DOI: 10.2174/1568026620666200601161703.
Uddin MS, Kabir MT, Mamun AA, et al. (2020) Pharmacological approaches to mitigate neuroinflammation in Alzheimer's disease. International Immunopharmacology 84: 106479. DOI: 10.1016/j.intimp.2020.106479.
Valtcheva S and Venance L (2019) Control of long-term plasticity by glutamate transporters. Frontiers in Synaptic Neuroscience 11: 10. DOI: 10.3389/fnsyn.2019.00010.
van der Kant R and Goldstein LSB (2015) Cellular functions of the amyloid precursor protein from development to dementia. Developmental Cell 32: 502–515. DOI: 10.1016/j.devcel.2015.01.022.
Vellas B, Sol O, Snyder PJ, et al. (2011) EHT0202 in Alzheimers disease: a 3-month, randomized, placebo-controlled, double-blind study. Current Alzheimer's Research 8: 203–212.
Verma S, Kumar Ashwini, Tripathi T, et al. (2018) Muscarinic and nicotinic acetylcholine receptor agonists: current scenario in Alzheimer's disease therapy. The Journal of Pharmacy and Pharmacology 70: 985–993. DOI: 10.1111/jphp.12919.
von Bernhardi R, Cornejo F, Parada GE, et al. (2015) Role of TGFβ signaling in the pathogenesis of Alzheimer's disease. Frontiers in Cellular Neuroscience 9: 426. DOI: 10.3389/fncel.2015.00426.
Weggen S, Eriksen JL, Das P, et al. (2001) A subset of NSAIDs lower amyloidogenic Abeta42 independently of cyclooxygenase activity. Nature 414: 212–216. DOI: 10.1038/35102591.
Weltha L, Reemmer J and Boison D (2019) The role of adenosine in epilepsy. Brain Research Bulletin 151: 46–54. DOI: 10.1016/j.brainresbull.2018.11.008.
West T, Hu Y, Verghese PB, et al. (2017) Preclinical and clinical development of ABBV-8E12, a humanized anti-tau antibody, for treatment of Alzheimer's disease and other tauopathies. The Journal of Prevention of Alzheimer's Disease 4: 236–241. DOI: 10.14283/jpad.2017.36.
Whittington RA, Bretteville A, Dickler MF, et al. (2013) Anesthesia and tau pathology. Progress in Neuro-Psychopharmacology and Biological Psychiatry 47: 147–155. DOI: 10.1016/j.pnpbp.2013.03.004.
Wilkinson D, Windfeld K and Colding-Jørgensen E (2014) Safety and efficacy of idalopirdine, a 5-HT6 receptor antagonist, in patients with moderate Alzheimer's disease (LADDER): a randomised, double-blind, placebo-controlled phase 2 trial. The Lancet Neurology 13: 1092–1099. DOI: 10.1016/S1474-4422(14)70198-X.
Williams, T., Borchelt, D. R., & Chakrabarty, P. (2020). Therapeutic approaches targeting Apolipoprotein E function in Alzheimer's disease. Molecular neurodegeneration, 15(1), 8. https://doi.org/10.1186/s13024-020-0358-9
Wischik CM, Staff RT, Wischik DJ, et al. (2015) Tau aggregation inhibitor therapy: an exploratory phase 2 study in mild or moderate Alzheimer's disease. Journal of Alzheimer's Disease 44: 705–720. DOI: 10.3233/JAD-142874.
Wu C and Sun D (2015) GABA receptors in brain development, function, and injury. Metabolic Brain Disease 30: 367–379. DOI: 10.1007/s11011-014-9560-1.
Xiu S, Liao Q, Sun L, et al. (2019) Risk factors for cognitive impairment in older people with diabetes: a community-based study. Therapeutic Advances in Endocrinology and Metabolism 10. DOI: 10.1177/2042018819836640.
Xu Y, Zhao M, Han Y, et al. (2020) GABAergic inhibitory interneuron deficits in Alzheimer's disease: implications for treatment. Frontiers in Neuroscience 14: 660. DOI: 10.3389/fnins.2020.00660.
Yanamandra K, Kfoury N, Jiang H, et al. (2013) Anti-tau antibodies that block tau aggregate seeding in vitro markedly decrease pathology and improve cognition in vivo. Neuron 80: 402–414. DOI: 10.1016/j.neuron.2013.07.046.
Yeung AWK, Georgieva MG, Atanasov AG, et al. (2019) Monoamine oxidases (MAOs) as privileged molecular targets in neuroscience: research literature analysis. Frontiers in Molecular Neuroscience 12: 143. DOI: 10.3389/fnmol.2019.00143.
Yoo JW, Irvine DJ, Discher DE et al. (2011) Bio-inspired, bioengineered and biomimetic drug delivery carriers. Nature Reviews Drug Discovery 10:521–35. DOI:10.1038/nrd3499. PMID: 21720407.
Zareifopoulos N and Papatheodoropoulos C (2016) Effects of 5-HT-7 receptor ligands on memory and cognition. Neurobiology of Learning and Memory 136: 204–209. DOI: 10.1016/j.nlm.2016.10.011.
Zenaro E, Piacentino G, & Constantin G. (2017). The blood–brain barrier in Alzheimer's disease. Neurobiology of Disease 107: 41–56. DOI:10.1016/j.nbd.2016.07.007
Zhang G, Ásgeirsdóttir HN, Cohen SJ, et al. (2013) Stimulation of serotonin 2A receptors facilitates consolidation and extinction of fear memory in C57BL/6J mice. Neuropharmacology 64: 403–413. DOI: 10.1016/j.neuropharm.2012.06.007.
Zhang H (2012) New insights into huperzine A for the treatment of Alzheimer's disease. Acta Pharmacologica Sinica 33: 1170–1175. DOI: 10.1038/aps.2012.128.
Zlomuzica A, Dere D, Binder S, et al. (2016) Neuronal histamine and cognitive symptoms in Alzheimer's disease. Neuropharmacology 106: 135–145. DOI: 10.1016/j.neuropharm.2015.05.007.
Zoli M, Pistillo F and Gotti C (2015) Diversity of native nicotinic receptor subtypes in mammalian brain. Neuropharmacology 96: 302–311. DOI: 10.1016/j.neuropharm.2014.11.003.
Zvěřová M (2019) Clinical aspects of Alzheimer's disease. Clinical Biochemistry 72: 3–6. DOI: 10.1016/j.clinbiochem.2019.04.015.

6 Cholinergic Neurotransmission System

Signaling Pathways and Their Role in Parkinson's Disease

Anushree Anushree, Md Zeeshan Ali, and Jawaid Ahsan
Department of Biotechnology, Central University of South Bihar, Panchanpur, Gaya, Bihar, India

CONTENTS

6.1 INTRODUCTION

The human brain is the controlling center and is composed of highly specialized components. These components form a complex structure that performs specific functions, making humans unique in their capabilities (Herculano-Houzel, 2012). Neuroscientists and researchers are continuously performing the daunting task of deciphering the complexity of brain. The process of releasing neurotransmitters, their diffusion, and binding to receptors is collectively referred to as transmission (Miller and Yeh, 2017).

Cholinergic neurotransmission is a form of neurotransmission that is most important in modulating the autonomous nervous systems in mammals, including humans, and has been extensively studied (Woolf, 1991; McKinney, 2005). The cholinergic system locations in the brain include the basal forebrain, cranial motor nuclei, striatum, spinal motor neurons, mesopontine tegmental nuclei, pedunculopontine tegmental, laterodorsal tegmental, and parabigeminal nuclei (Oda, 2000). In the central nervous system (CNS), the brainstem cholinergic network is composed of the pedunculopontine tegmental nucleus (PPN) and the laterodorsal tegmental nucleus, which projects the neuron to the thalamus, cerebellum, and brainstem (Yarnall et al., 2011; Ren et al., 2011; Mesulam et al., 1983). Basal forebrain cholinergic neurons are projected to the prefrontal cortex, striatum, and thalamus (Mesulam, 1995; Zaborszky, 2002; Zaborszky et al., 2008).

DOI: 10.1201/9780429265198-8

TABLE 6.1
Various Neurological Disorders, Their Cholinergic Effects, and Cholinergic Pharmacological Agents

Neurological Disease	Cholinergic Effect	Possible Pharmacological Treatment
Parkinson's disease	Reduced microglia activation and neuroinflammation Altered cholinergic striatal tone Degeneration of PPN	Anticholinesterase, antimuscarinic drugs, antinicotinic drugs
Alzheimer's disease	Amyloid plaques, neurofibrillary tangles Neuroinflammation	Anticholinesterase
Myasthenia gravis	Cholinesterase functional impairment	Anticholinesterase
Dementia	Cholinesterase functional impairment	Anticholinesterase
Bradycardia	Nicotinic and muscarinic receptor impairment	Antimuscarinic drugs, antinicotinic drugs

PPN, pedunculopontine tegmental nucleus.

The complexity of the cholinergic system derives from its ability to regulate diverse interpretative functions such as memory, control of the heartbeat, and regulation of rapid eye movement (REM) sleep cycles, blood pressure, movement, motor control, learning, and cognitive functions (Chaudhuri et al., 2006; Langston, 2006; Chaudhuri and Schapira, 2009). In 1906, Dr. Aloysius Alzheimer concluded that the cholinergic neurotransmission network is significant in both normal and abnormal cognitive functioning (Alzheimer, 1906). In response to a nerve impulse in the axon terminal, a neurotransmitter is released into the synaptic cleft (Kandel and Siegelbaum, 2000). It is distinct due to the actions of acetylcholine (ACh) and therefore ACh is referred to as neuromodulatory (Picciotto et al., 2012). Postsynaptic electrical responses are evoked by neurotransmitters binding to their receptors.

According to Dale (1914), the interaction of ACh with families of ionotropic and metabotropic receptors modulates functioning of cholinergic signaling. These two families of receptors are the nicotinic and muscarinic receptors, based on the type of ligand. Desensitization of these receptors due to the impairment of cholinergic neurotransmission network results in neurologic disorders like Alzheimer's disease (AD) and Parkinson's disease (PD; Table 6.1). Deterioration in cholinergic signaling impairs memory, while the escalation of cholinergic transmission improves cognitive function, specifically attentional performances (Furey et al., 2008; Herrero et al., 2008; Sarter, 2015). PD is a neurodegenerative disorders that poses a serious concern due to an increasing number of patients worldwide. It is the second most common neurodegenerative disease, with an average of 14 per 10,000 population in developed countries aged 65 years and older having the disease (Hirtz et al., 2007). PD was discovered in 1817 by James Parkinson (Parkinson, 1969). To date, no cure has been found. The main causative agent for this disease is the loss of dopaminergic nigral neurons.

This chapter focuses on cholinergic neurotransmission, along with deficits in the cholinergic system and signaling that lead to PD, which includes cognitive control disorder, age-related learning impairments, as well as memory loss, along with their signaling pathways and recent developments in this field.

6.2 ACETYLCHOLINE: A CHOLINERGIC NEUROTRANSMITTER

ACh was first recognized in 1914 by Henry Hallett Dale for its action on heart tissue. ACh was initially named 'Vagusstoff' by Otto Loewi as it was released from the vagus nerve and slowed the heart rate. Loewi recognized it as a neurotransmitter and it has become an important research chemical. In 1936, the Nobel Prize in Medicine was awarded to Dale and Loewi for discovering ACh. ACh is also referred to as a chemical messenger that functions in the autonomic nervous system (ANS) by imparting by fast neurotransmission in the sympathetic and parasympathetic nervous systems; in the peripheral nervous system (PNS) at neuromuscular junctions (NMJs), where ACh acts as a neurotransmitter between the motor nerve and skeletal muscle; and, lastly, in the CNS in carrying out diverse functions. It is involved in inhibitory functions. It is not directly engaged in synaptic transmission between specific neurons; rather it acts as a neuromodulator throughout the nervous system. It plays significant role in the normal functioning of the brain and body.

6.2.1 Acetylcholine Biosynthesis and Storage

ACh is composed of an acetate and choline molecule, which also acts as the precursor for its synthesis. Choline is found in vegetables, fruits, whole grains, fish, eggs, and so on. Also, in humans it is synthesized endogenously in the liver but not in a sufficient amount to fulfill the body's need, so ACh has to be supplemented by the diet. Choline is also obtained through the breakdown of the phosphatidylcholine reaction, catalyzed by phospholipase D. Free choline molecule circulating in the blood can cross the blood–brain barrier easily and are further taken up by cholinergic receptor for signal transmission. Those receptors have a very high affinity for ACh; their K_m value ranges between 1 and 5 μM (Lendvai, 2008). Cholinergic neurotransmission requires a highly synchronized and regulated process of ACh synthesis, its storage, release, breakdown, and reuptake into nerve terminals (English and Jones, 2012).

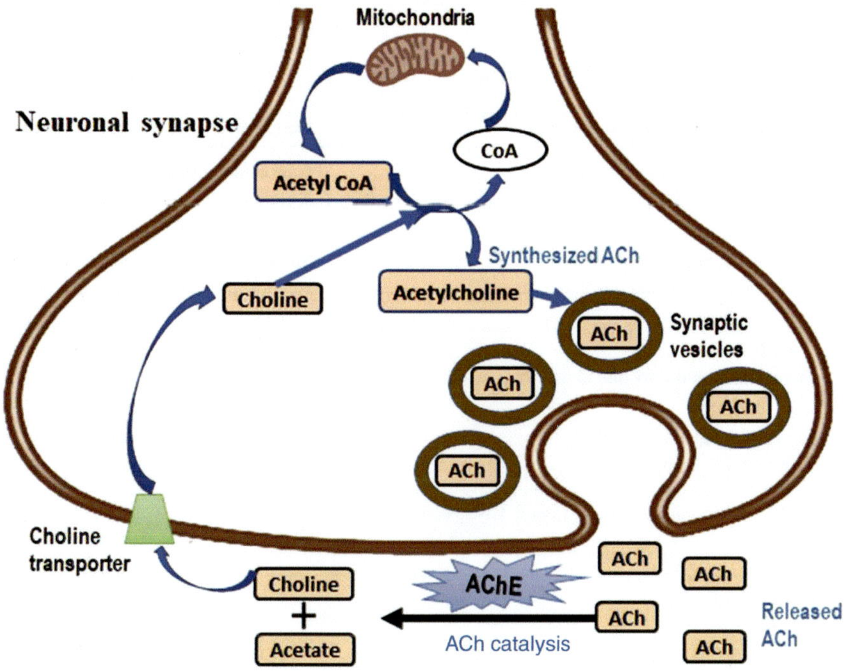

FIGURE 6.1 Acetylcholine (ACh) biosynthesis and catalysis (adapted from Kummer et al., 2008). AChE, acetylcholinesterase.

ACh is biosynthesized by the interaction between precursors, choline, and acetyl coenzyme A (acetyl CoA). This single-step reaction is catalyzed by choline acetyltransferase (ChAT) at nerve terminals (Figure 6.1). Ninety percent of ChAT at nerve terminals is in soluble form, while 20 percent is membrane bound (Pahud G et al., 1998; Oda, 1999). The high concentration of cyclic adenosine monophosphate (cAMP) is directly proportional to increased ChAT activity (Dolezal et al., 2001). The precursor acetyl-CoA is synthesized from glucose metabolism, while choline is taken up by neurons through a high-affinity sodium ion choline transporter from the plasma. ChAT is synthesized in the cell body of cholinergic neurons and is transported to nerve endings through the axon. ChAT and ACh are highly concentrated at the axon terminal. ChAT is present only in cholinergic neurons; hence, it even acts as a marker (Amenta and Tayebati, 2008).

ACh is mainly stored at the synaptic vesicles, which are about 100 μm in size, present in the presynaptic cholinergic neuron axon terminal (English et al., 2012). These vesicles have vesicular ACh transporter (VAChT), which transports ACh outside and protons inside. A small amount of ACh is present in the cytosol. Voltage-sensitive Ca^{2+} channels generate an action potential, which further leads to depolarization of neurons. The stored ACh is then released into the synaptic cleft. The released ACh then binds to the cholinergic receptors to mediate the signal.

6.2.2 Acetylcholine Catabolism

Released ACh can either be transported back into the presynaptic neuron from the synaptic cleft through the ACh transporter, or, within the extracellular space, ACh is degraded into small molecules with the help of acetylcholinesterase (AChE) (Kandel et al., 2000; Hecker and Konradi, 2003). ACh is hydrolyzed into choline and acetic acid, and the choline is taken back for the synthesis of ACh again. AChE has a very high catalysis rate.

Synaptogenesis required for glutamatergic signaling is regulated by the concentration of AChE during development (Dong et al., 2004). Anticholinesterases inhibit AChE activity, leading to the accumulation of ACh within the extracellular space and further delays the binding of ACh to the receptor by 1–30 ms. Anticholinesterases are even used in the treatment of AD.

6.3 CHOLINERGIC OR ACETYLCHOLINE RECEPTORS

There are two types of cholinergic receptors – nicotinic receptor (nAChRs) and muscarinic receptors (mAChRs) (Picciotto et al., 2000; Wess, 2003; Nys et al., 2013; Barrantes, 2015). These receptors mediate the postsynaptic effects of ACh neurotransmission based on nicotine and muscarine, which mimic the effects of ACh at the receptor ligand binding site. These receptors play an important role in synaptic plasticity, as reviewed by Giocomo and Hasselmo (2007) and McKay et al. (2007), by mediating their intracellular effects downstream of these receptors (Berg and Conroy, 2002; Lanzafame et al., 2003 Cancela, 2001). Cholinergic transmission causes long-term potentiation (LTP) due to the simultaneous stimulation of both postsynaptic activity and Ca^{2+} signaling (Navarrete et al., 2013). Table 6.2 details their properties, as well as how the two receptors differ from each other.

The two receptors are discussed in brief below.

TABLE 6.2
Characteristics of Nicotinic and Muscarinic Receptors

	Nicotinic acetylcholine receptor	Muscarinic acetylcholine receptor
Receptor type	Ligand-gated ion channel	G protein-coupled receptor
Interaction with agonist – acetylcholine (ACh)	Two molecules of ACh bind to the receptor for activation	A single molecule of ACh binds to the receptor for activation
Location	NMJ, ANS, and small area of CNS	Smooth muscles, discrete regions of CNS, and myocardial muscles
Receptor response	Sodium ion/potassium ion channel conduction increases	Heterotrimeric G proteins are activated due to receptor ligand binding
Transmission time	1 ms	400 ms
Neuronal response	Depolarization occurs due to flow of i ons through ion channel	Secondary messengers like cAMP, IP3, DAG, etc., are released
Stimulation	Excitatory signal to target cell	Both excitatory and inhibitory signal to target cell
Antagonist	Curare (tubocurarine)	Atropine

NMJ, neuromuscular junction; ANS, autonomic nervous system; CNS, central nervous system; cAMP, cyclic adenosine monophosphate; IP3, inositol triphosphate; DAG, diacylglycerol.

6.3.1 Nicotinic Acetylcholine Receptors

nAChRs are non-selective, pentameric, ligand-gated, cation channel neurotransmitter receptors (Changeux et al., 1998; Picciotto et al., 2001). These receptors are mainly found at postsynaptic terminals of the NMJ and autonomic ganglia for neurotransmission, and at presynaptic terminals throughout the CNS, cell bodies, and axons to release neurotransmitter (Kawai et al., 2007; Del Mar Arroyo-Jimenez et al., 1999; McGehee et al., 1995; Vidal and Changeux, 1993; Hill et al., 1993). Nicotine is a natural alkaloid and acts as a stimulant. It is known to stimulate CNS by binding to the nAChR and mediating the same effects. Albuquerque et al. (2009) reported that nAChRs are also found in non-neuronal cells, such as in the peripheral immune system, epidermal keratin cells, endothelial cells, and cells of the digestive and respiratory systems. These receptors also act as a neuromodulator in the brain, as reported by Picciotto (2003).

The neuronal nicotinic receptor is composed of two types of polypeptide subunit types – alpha (α) and beta (β) – and these subunits form either homo or heteromer receptors. The neuronal nAChR gene constitutes nine α (α2 - α10) and three β (β2-β4) subunits, which are arranged in such a way to form pentameric functional receptors (Picciotto et al., 2000; Role and Berg, 1996). nAChRs composed of α7 and the α4β2 subtypes are mainly found in the CNS, while nAChRs composed of α3β4 subtypes are found in PNS. The molecular mass of each subunit in the pentameric receptor ranges between 40,000 and 60,000 Da (Tiwari et al., 2013). They are ~8 nm in diameter and ~16 nm in length as per the side view observed under a cryoelectron microscope (Corringer et al., 2008). The five subunits appear to be arranged like a rosette when viewed from the top.

Methyllycaconitine and α-bungarotoxin are selective antagonists of nAChRs as they block channels and hence neurotransmission through them. Choline is the selective agonist that favors neurotransmission through these receptors.

6.3.1.1 Mechanism of nAChR Activation and Neurotransmission

ACh binding with the nAChR activates ACh, which further assists in opening the channel. The receptor is highly permeable to sodium ions (Na^+), potassium ions (K^+), and calcium ions (Ca^{2+}) in the CNS in comparison to nAChRs at the NMJ (Vernino et al., 1992). According to Seguela et al. (1993), the average permeability ratio of Ca^{2+}/Na^+ is > 1. In other words, Ca^{2+} influx is comparatively higher than Na^+ influx, which further leads to the pumping out of Ca^{2+} and an increase in Na^+ influx. This results in the opposite direction of transport (i.e., instead of release of the transmitter, uptake occurs). The consequence of ligand receptor binding is an increase in the Na^+ level within the target cell, which further increases the action of voltage-sensitive Ca^{2+} channels in different tissues. This is followed by depolarization and generation of the action potential. This is a type of rapid transmission. Desensitization depends on the type of cell. In neurons, desensitization is complete within a few seconds in nicotine-induced neurotransmission. After desensitization, 10–30 seconds of recovery time is required by nAChRs. Neurotransmission through presynaptic nAChRs is dominated by Ca^{2+} influx (McGehee et al., 1995; Gray et al., 1996; Maggi et al., 2003). Lonotropic (nAChR) neurotransmission mechanism and pathway is depicted in Figure 6.2.

6.3.2 Muscarinic Acetylcholine Receptors

mAChRs are monomeric G protein-coupled receptors (GPCRs). GPCRs are seven transmembrane spanning domains which transmit their effects by G coupled proteins. Muscarine, which is a by-product of the mushroom *Amanita muscarina*, binds to the receptor it is named after. There are five different types of mAChRs, M1–M5 (Table 6.3). They are numbered on the basis of sequence of their discovery. Based

TABLE 6.3
Types of Muscarinic Receptors and Their Properties

Receptor	Coupled G Protein	Location in the Brain	Mechanism of Action
Muscarinic M1	$G_{q/11}$	Striatum and cerebral cortex of CNS	PLC and PKC enzymes activation due to increase in Ca^{2+} and signal stimulation
Muscarinic M2	G_i	Striatum, cerebral cortex of CNS, myocardial muscles	Adenylyl cyclase decreases and increase in Na^+/K^+ ion channel conduction
Muscarinic M3	$G_{q/11}$	Smooth muscles, sweat glands	PLC and PKC activation due to increase in Ca^{2+} and signal stimulation
Muscarinic M4	G_i	Striatum CNS	Adenylyl cyclase decreases and increase in Na^+/K^+ ion channel conduction
Muscarinic M5	$G_{q/11}$	Striatum CNS	PLC and PKC activation due to increase in Ca^{2+} and signal stimulation

CNS, central nervous system; PLC, phospholipase C; PKC, protein kinase C.

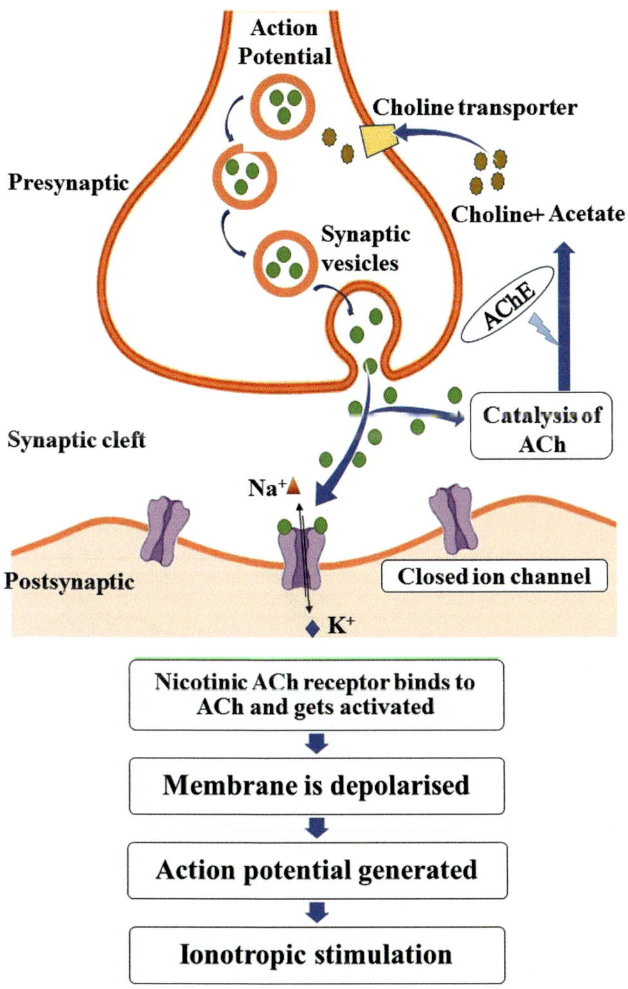

FIGURE 6.2 Ionotropic neurotransmission mechanism and pathway (adapted from Resende and Adhikari, 2009). AChE, acetylcholinesterase; ACh, acetylcholine.

on the their pharmacology and signaling properties, these receptors are divided into two groups. M2 and M4 belong to one group, while M1, M3, and M5 belong to the other (Wess, 2003). A single neuron can express all five types of mAChR (Hamilton et al., 1997). The action potential frequency adapted in the CNS is regulated by mAChRs (Nicoll et al., 1990).

M1, M3, and M5 receptors are coupled to the G_q/G_{11} protein, which transmits the signal by activating the enzyme phospholipase C (PLC) (Guo and Schofield, 2003; Jaakola et al., 2005). The secondary messengers released after phosphorylation of receptors in their signaling pathway are inositol triphosphate (IP3) and diacylglycerol (DAG; Haga and Haga, 1990). These receptors mediate the mobilization of the calcium ion within neurons through the G_q–PLC pathway, which further affects Ca^{2+}-regulated events. M2 and M4 receptors are involved in modulating specific ion channels and inhibit the activity of adenylyl cyclase (Douglas et al., 2002; Raiteri et al., 1984). They are coupled to the G_i/G_0 G protein and the secondary messenger involved in their signaling pathway is cAMP. These receptors are predominantly presynaptic in the CNS. M1 and M5 receptors participate in the stimulation of cortical neurons postsynaptically (Douglas et al., 2002) and stimulate release of dopamine (DA) from striatal synapses (Zhang et al., 2002). The functions of all five subtypes of muscarinic receptors are briefly discussed below.

M1 receptors are involved in seizure activity and it is reported that they are also responsible for increasing DA release, which plays an important role in the early stage of PD (Wess, 2003). M1 receptors affect learning and memory-related cognitive function (Anagnostaras et al., 2003).

M2 receptors inhibit the release of ACh at presynaptic synapses. They are responsible for inducing akinesia and central tremor. These receptors are located in brain regions like cerebellum, olfactory bulb, diencephalon, medulla pons, and basal forebrain (Décossas et al., 2005; Rouse et al., 2000; Bernard et al., 1998; Mrzljak et al., 1993).

M3 receptors play an important role in regulating daily food consumption, as well as appetite (Yamada et al., 2001a). They mediate smooth muscle contraction. These receptors are found in wider areas of the brain (Hersch et al., 1994).

M4 receptors modulate the release of DA with their presence at presynaptic heteroreceptors (Hersch et al., 1994; Bernard et al., 1999). The cholinergic signaling pathway is inhibited by them in a similarly way to M2 receptor inhibition of ACh. They also function in limiting locomotor activity and analgesia (Lendvai, 2008).

M5 receptors are the only receptors to be found in substantia nigra of dopaminergic neurons. The loss of nigrostriatal dopaminergic neurons leads to PD. Hence, this receptor can pose as a potential drug/ligand target in disease therapy. These receptors participate in cerebral blood vessel dilation (Yamada et al., 2001b), which might help in research of the development of AD.

A single neuron may possess more than one type of mAChR. Levey et al. (1995) reported that all five types of mAChR are present in hippocampal pyramidal neurons. mAChRs are single polypeptide chains composed of 460–590 amino acids, with an amino terminal outside and a carboxyl terminal inside the neuron. The five subtypes of mAChRs have highly conserved membrane-spanning regions; their shared sequence similarity is about 90 percent (Aronstam and Patil, 2009).

These receptors are not only mainly involved in parasympathetic neurotransmission, but also in a few sympathetic neurotransmissions like sweating and piloerection. mAChRs and nAChRs differ from each other in function and structure. Muscarinic receptors present in the brain surpass nicotinic receptors by a factor of 10–100. These receptors carry out either excitatory or inhibitory neurotransmission.

6.3.2.1 Mechanism of mAChR Activation and Neurotransmission

The muscarinic receptor signaling pathway depends on the subtype, the location, and their nature. Transducer G-coupled proteins, secondary messengers, and the presence of a specific protein kinase substrate, depending on their location in the tissue, also play major roles in neurotransmission. Cross-talk between transduction pathways also occurs Hornigold et al., 2003, which creates a problem with identifying the signaling event, whether the signal is proximal or indirect. This happens either due to a lack of specificity, or the presence of indirect and secondary effects. Muscarinic neurotransmission mechanism is shown in Figure 6.3.

The signaling pathways of the odd-numbered mAChR subtypes (M1, M3, and M5) involve pertussis toxin-insensitive G-coupled proteins. The signaling cascade involves PLC-β as an effector molecule. The secondary messenger DAG activates protein kinase C, while IP3 is involved in the release of Ca^{2+} ions by stimulating specific receptors on the endoplasmic reticulum, which is an intracellular calcium ion store (Gulledge and Stuart, 2005). These receptors also stimulate neurotransmission through the mitogen-activated protein kinase pathway and PLC-β (Wess et al., 2007).

The even-numbered mAChR subtypes (M2 and M4) transmit signals by activating the G_i-coupled receptor. Further, the signaling cascade involves the inhibition of adenylyl cyclase and release of the secondary messenger cAMP. These receptors activate specific ion channels (K^+ channel) on the plasma membrane, which leads to hyperpolarization, inhibiting the signaling pathway. When these receptors are located at presynaptic cholinergic neurons, they are considered to be autoreceptors, but when they are located at axon terminals containing neurotransmitter, they are considered heteroreceptors. Cross-talk between M1 pathway and these receptors occur as the M1-induced increase in the intracellular free Ca^{2+} level also leads to hyperpolarization by activation of the K^+ current.

Four major effects of mAChRs are involved in depolarization and hyperpolarization, either directly or indirectly, and can be summarized as follows (Aronstam and Patil, 2009):

(1) M1, M3, and M5 receptors modulate the conduction of calcium-dependent potassium and cations.
(2) M1 and M3 receptors inhibit voltage- and time-dependent potassium conduction.
(3) M2 and M4 receptors stimulate potassium ion conduction.
(4) M2 and M4 receptors inhibit calcium ion conduction either directly through G_o or indirectly through cyclic-guanosine monophosphate reduction.

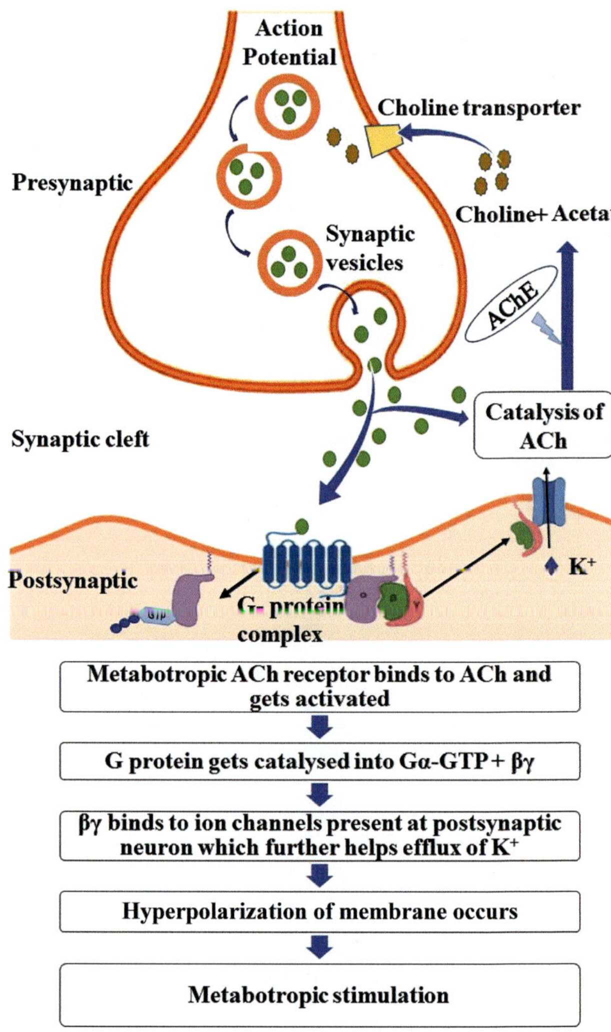

FIGURE 6.3 Metabotropic (muscarinic) neurotransmission mechanism and pathway (adapted from Resende and Adhikari, 2009). AChE, acetylcholinesterase; ACh, acetylcholine; GTP, guanosine triphosphate.

6.4 MOLECULAR AND CELLULAR MECHANISMS OF THE CHOLINERGIC SIGNALING PATHWAY

Cholinergic neurotransmission at neuronal synapses is initiated by binding of the ligand ACh. ACh as a neurotransmitter is found in neuronal synapses of the CNS and the ANS. Cholinergic signaling is found in both uni- and multicellular organisms (Kawashima et al., 2007; Wessler and Kirkpatrick, 2008). Cholinergic molecular and cellular signalling pathway is shown in Figure 6.4. Therefore, it is possible to hypothesise that ACh can even work in cells other than neurons.

The role of ACh as a neurotransmitter at the neuronal synapses of the ANS and CNS has been an intriguing topic of study. Researchers and neuroscientists have revealed that the neurotransmission and signaling pathways at these locations are much more complicated than initially documented. ACh is synthesized at the axon terminal and stored in synaptic vesicles. Further, it is released into a synapse in the form of a packaged vesicle, referred to as quanta. The smallest stimulation between neurons is generated by a quantum, also known as miniature end plate potential.

Depending upon the stimulation, an action potential is either generated or repressed. These stimulations generated at the synapses are referred to as inhibitory postsynaptic potentials (IPSPs) or excitatory postsynaptic potential (EPSPs).

Muscarinic receptors transmit signals in the presence of ACh, its receptor, and a G protein, to which guanosine diphosphate (GDP) is bound, forming a ternary complex. ACh binds to the receptor with high affinity. After the ligand binds to the receptor, GDP is replaced by guanosine triphosphate (GTP), bound to the coupled G protein. This leads to changes in the conformation of the receptor, which results in less receptor affinity for ACh. After this, the complex dissociates into its various components. Dissociation of the complex with lower affinity is more rapid from the binding sites than the complex with higher affinity.

The cholinergic neurotransmission rate varies with respect to time [i.e., a few milliseconds (extremely rapid) to several minutes (extremely slow)]. Rapid transmission requires physiochemical reactions to be fast, which can only be achieved when the receptor has a low affinity for the ligand (Katz et al., 1989): as time increases, sensitivity decreases (Ballestero et al., 2011). Neurotransmission at NMJs and nerve-electro plaque junctions (NEJs) is 10–50 times faster per impulse than at cholinergic neuronal synapses. The various time-dependent neurotransmissions researched using optogenetic approaches (Bennett et al., 2012; Hay et al., 2016; Hefft et al., 1999; Turrini et al., 2001; Takács et al., 2013) are discussed in the following subsections.

6.4.1 Ultra-Fast Cholinergic Neurotransmission

Ultrafast cholinergic neurotransmission is favored at the fastest cholinergic synapses (Dunant and Gisiger, 2017), include NMJs and NEJs, which are discussed in detail below.

The NMJ is also known as myoneural junction. It is a synaptic joint between the terminal end of a motor neuron and a muscle fiber (smooth or cardiac). The motor neuron transmits an action potential to the muscle fiber and signals the muscles for contraction. They form giant synapses in vertebrates as the contact area between motor neuron and a muscle fiber extends over several hundreds of square micrometers (μm^2). According to Couteaux et al. (1970), NMJs are structured into active zones and a double row of synaptic vesicles are present, which further helps in their characterization. As

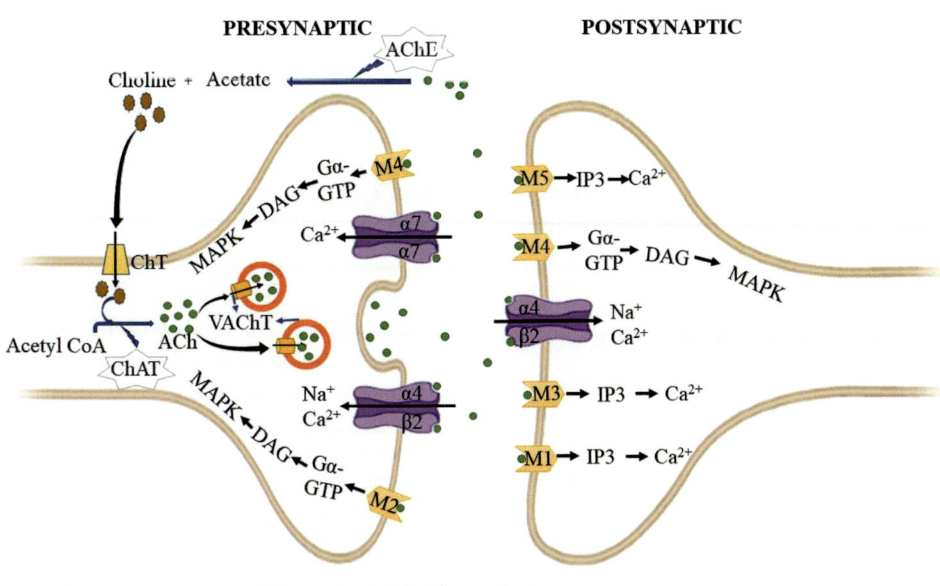

FIGURE 6.4 Cholinergic molecular and cellular signaling pathway (adapted from Jones et al., 2012). AChE, acetylcholinesterase; ChT, choline acetyl transfarase; MAPK, mitogen-activated protein kinase; DAG, diacylglycerol; GTP, guanosine triphosphate; ChAT, choline acetyltransferase; ACh, acetylcholine; VAChT, vesicular ACh transporter.

described by Kuno et al. (1971), one quantum is released per active zone. In the NMJ, the transmission of a single impulse takes approximately 2–4 ms (Dunant and Gisiger, 2017). For sustained muscle contraction the frequency of transmission should be up to 100 Hz, which requires an interval between consecutive impulses in the order of 10 ms. The rapid transmission of the action potential of individual impulses is required for ultra-fast transmission, which relies on low-affinity reactions.

A specific and asymmetric form of AChE found in vertebrate NMJs is A12-AChE, which is located at the synaptic cleft (Massoulié et al., 2008; Massoulie et al., 1993; Campanari et al., 2016). Its tail is composed of collagen Q, which has domains for anchoring AChE to the basal lamina and confers an important role in neurotransmission – specifically ultra-fast cholinergic neurotransmission. AChE is located in the synaptic cleft, which lies between the motor nerve ending and the receptor.

Muscles contain another form of AChE, which is tetrameric globular in nature and localized at perijunctional regions anchored to a proline-rich membrane (Gisiger and Stephens, 1988; Bernard et al., 2011); it is referred to as G4-AChE. The number of G4-AChE molecules in muscles is directly proportional to muscle activity; the higher the activity, the more G4-AChE molecules are present (Gisiger et al., 1994). G4-AChE functions by inhibiting the desensitization of nicotinic receptors and regulating the concentration of ACh at the junction (Gisiger and Stephens, 1988; Riad et al.,1998; Jasmin and Gisiger, 1990).

NEJs are found in certain fish species like *Torpedo marmorata* (Gautron, 1970; Witzemann and Boustead, 1982). These fish species have electric organs, which are modified forms of NMJs. Cholinergic neurotransmission in *Torpedo* NEJs is similar to NMJs containing collagen-tailed A12-AChE at the synaptic cleft in high concentrations (Gisiger and Stephens, 1988). NEJs are regarded as the purest form of ultra-fast cholinergic transmission and transmit signals through ultra-fast ACh hydrolysis. The duration of a single action potential generated by an average sized *Torpedo* is 2–4 ms. At 15°C, these fish species deliver repetitive electric currents at frequencies of 100–200 Hz for either attack or defense purposes (Dunant and Gisiger, 2017).

6.4.2 Rapid Cholinergic Neurotransmission

Rapid cholinergic neurotransmission is found in the PNS and CNS. An initial rapid nicotinic transmission of 10–50 ms is followed by various muscarinic or nicotinic effects. In high concentrations of ACh, the receptors are desensitized (Katz et al., 1957). In other words, receptors will not open in response to ACh. At synapses, AChE is not concentrated and hinders ultrafast transmission. Therefore, there is a delay in response regulated by the globular form of AChE (i.e., G4-AChE). G4-AChE may be membrane bound or secreted within the neuron.

Rapid cholinergic neurotransmission at the NMJ and NEJ has the following major properties (Dunant et al., 2000):

(a) An action potential generated from several afferent nerve fibers are received by a single cholinergic neuron.
(b) The initial nicotinic EPSP is 10–50 times slower than the neuron terminal end plate potential of both NMJs and NEJs.
(c) Depending on the type of tissue (neuronal, etc.), the initial EPSP is followed by either the muscarinic or nicotinic signal transmission pathway.
(d) The A12-AChE molecular form of AChE is not concentrated at the synaptic cleft; instead, the globular G4-AChE form of AChE is present in the endoplasmic reticulum of neurons of both the NMJ and NEJ.

6.4.3 Slow Cholinergic Processes in the Nervous Systems

Transmission in slow cholinergic processes in the nervous system is not so brief, and is also not a controlled process. In slow neurotransmission, the ACh neurotransmitter acts through 'metabotropic' mAChRs. mAChRs are coupled to a G protein and have a relatively high binding affinity for their ligand, ACh.

Response mechanisms mediated by mAChRs through G proteins involve the following.

6.4.3.1 Adenylate Cyclase Inhibition

M1 mAChRs activate an alpha inhibitory G protein subunit and so adenylyl cyclase remains switched off. Low levels of mAChR activate stimulatory subunit, which, in turn, activate adenylyl cyclase, controlling the quantity of cAMP in the cell. A low level of cAMP leads to reduced activation of cAMP-dependent protein kinase and therefore reduced heart rate and contraction stimulation.

6.4.3.2 Activation of Phospholipase C

The enzyme PLC-β is activated by an interaction between Gα-GTP binding proteins and the muscarinic receptor. PLC hydrolyzes phosphatidyl inositol phosphate to yield two types of secondary messengers: IP3 and DAG. DAG helps to activate the enzyme protein kinase C (PKC), after which PKC phosphorylates target proteins and further influences cellular response. IP3 binds with IP3 receptors present within the smooth endoplasmic reticulum, which leads to increased Ca^{2+} release from the intracellular storage site. This increase in downstream calcium ions has various cellular effects, depending on the type of cell or tissue.

6.4.3.3 Activation of K^+ Channels

When a muscarinic cholinergic receptor is activated, Gα-GTP binding protein interacts with K^+ ion channels to increase K^+ ion conductance, after which the resting membrane potential is increased in myocardial and other cell membranes, leading to inhibition.

6.5 DEFICIT OF CHOLINERGIC NEUROTRANSMISSION IN PARKINSON'S DISEASE

PD is the second most common neurodegenerative disorder worldwide. It is estimated that more than 10 million people are suffering from this neurological disorder at present. The chance of PD occurring increases with age. Only 4 percent of people are estimated to be diagnosed with PD before the age of 50 years, but the disease increases with age, particularly around 80 years (Bower et al., 2000; Pringsheim et al., 2014; Ascherio and Schwarzschild, 2016; Hirsch et al., 2016). Men are 1.5 times more prone to having PD than women worldwide (Zhang et al., 2005; Tan et al., 2007; Tian et al., 2011).

The essential neuropathologic characteristics of PD that can be diagnosed include loss of dopaminergic neurons of the substantia nigra of the pars compacta, associated with the presence of Lewy bodies in the remaining nigral neurons (Gibb and Lees, 1988; Fearnley and Lees, 1991; Young and Inouye, 2007; Dickson et al., 2009; Berg et al., 2014). The clinical syndrome of PD is referred to as 'Parkinsonism', which includes slow movements, tremors, stiff muscle, bradykinesia, gait difficulties, and postural instability (Reich and Savitt, 2019). The prognosis of PD involves major impacts on motor and cognitive functions. PD symptoms are classified in two categories. First, motor symptoms, which include bradykinesia, tremor, rigidity, and imbalance. Second, nonmotor symptoms, which include depression, cognitive difficulties, and disturbed REM sleep behavior (Chaudhuri et al., 2006; Langston, 2006; Chaudhuri and Schapira, 2009)

Neuroinflammation provokes a reduction in the number of dopaminergic neurons and hence plays an important role in the development of PD (Simone and Tan, 2011). Research and various clinical observations have made it obvious that a neurotransmitter like ACh, along with dopaminergic neurons, play a crucial role in the regulation of PD.

The cholinergic neurotransmitter ACh is reported to be responsible for controlling mobility (Figure 6.5). People suffering from PD display a lower level of ACh in than people without PD of the same age. Hence, the ACh neurotransmitter became a prominent aspect of research in PD due to its severe impact on cognitive impairment during the early stage of PD. ACh mainly affects the attention mechanism and executive functions of the frontal lobe. One to two percent of striatal cells are cholinergic interneurons, which help in regulating glutamatergic, γ-aminobutyric acid (GABAergic), and dopaminergic pathways. The effects of cholinergic system degeneration in PD are cortical cholinergic denervation due to basal forebrain degeneration and also thalamic cholinergic denervation due to brainstem degeneration.

6.5.1 Cholinergic Deficit and Impairment in Parkinson's Disease

Cholinergic deficits and impairments are responsible for affecting the specific function of the human brain and cause PD. Functions that are affected are discussed below in detail with their symptoms and causes; the available therapeutic approaches are also discussed.

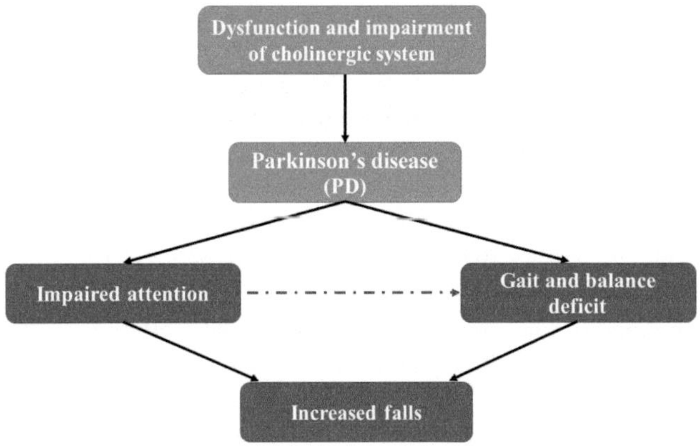

FIGURE 6.5 Cholinergic deficit causing Parkinson's disease leads to impaired attention and postural imbalance. These two symptoms are directly related to an increase in falls (adapted from Morris et al., 2019).

6.5.1.1 Motor Function

Motor functions are impaired in patients with PD, who exhibit various motor symptoms like bradykinesia, tremor, postural instability, and rigidity (Fahn et al., 2008; Soto et al., 2003; Spillantini et al., 1997; Cooper et al., 2006). Signals sent from dopaminergic neurons of the substantia nigra are received by the dorsal striatum. This dorsal striatum is responsible for regulating motor functions. A reduction in DA concentration causes various daily life problems in patients with PD by impairing their motor and cognitive skills.

A post- and presynaptic bidirectional interaction exists between the two neurotransmitters (i.e., ACh and DA), which are responsible for cognitive responses and also in selecting either reward or aversion motor responses (Threlfell and Cragg, 2011; Threlfell et al., 2012). This interaction is possible due to the presence of cholinergic interneurons, which constitute about 1 percent of total striatal neurons (Gonzales et al., 2015). Although striatal neurons are few, their network is wide and covers a major portion of striatal territory (Zhou et al., 2002). The DA concentration increases within the striatum due to the activation of dopaminergic neurons. A high level of DA affects cholinergic striatal neuron activity. nAChRs are reported to be responsible for uncontrolled DA release (Grace et al., 2007; Lobb et al., 2010).

ChAT cholinergic interneurons control the GABAergic network present in striata, which plays a major role in movement and attention (Faust et al., 2015). The latest research has revealed that fast GABAergic and cholinergic interneurons operate through nAChRs synaptic inputs (Faust et al., 2015). Anticholinergics, either in monotherapy or in combination with levodopa, are used to treat motor symptoms in PD (Fox et al., 2011).

6.5.1.2 Gait Impairment

Gait impairment is an important feature of PD, along with difficulties in balancing and falling, which further damages quality of life and increases the mortality rate (Boonstra et al., 2008; Perez-Lloret et al., 2014). The PPN is involved in coordinating the motor system (MacLaren et al., 2014). It was reported in a study that, in PD, a deficit function of cholinergic PPN leads to deficits in attention, gait, and balance, further enhancing the risk of falling (Grabli et al., 2012).

Donepezil, which inhibits AChE activity, is used in patients with PD to alter the cholinergic pathway, which results in a decrease in falls (Chung et al., 2010). Drugs that improve cholinergic tone may help in reducing falls effectively, but their efficacy on other gait impairments are yet to be studied.

6.5.1.3 Levodopa-Induced Dyskinesias

Levodopa (L-DOPA), also known as l-3,4-dihydroxyphenylalanine, is used for treating patients with PD. Long-term use of this dopaminergic drug by patients with PD lead to side effects which cause dyskinesia, complicating motor symptoms, and have an adverse impact on quality of life (Del Sorbo et al., 2008). Denervation of striatal niagra neurons increases ACh release (Ding et al., 2006) and facilitates LTP (Bonsi et al., 2008). Animal models of PD are reported to display dyskinetic behavior associated with LTP (Picconi et al., 2003). nAChRs play a prominent role in inducing levodopa-induced dyskinesia (LID). Several experiments have been conducted to understand the involvement of specific nicotinic receptors in LID (Quik et al., 2013; Quik et al., 2014; Bordia et al., 2015). Alpha7 nAChRs inhibit the occurrence of LID. The involvement of muscarinic receptors is yet to studied. There are very few data on LID, and therefore there is scope for more studies.

6.5.1.4 Cognitive Impairment and Mood

Severe cholinergic and dopaminergic damage is responsible for causing cognitive impairment. nAChRs, specifically alpha7 nAChRs in the hippocampus region, are affected in cognitive impairment (Hasselmo et al., 1996; Palma et al., 2013; Wallace and Porter, 2011; Gotti et al., 2006). The most abundant type of nAChR is the $\alpha4\beta2$ in the CNS; its huge loss leads to the cognitive impairment in patients with AD (Flynn and Mash, 1986; Burghaus et al., 2000). The various symptoms of PD are the result of the abnormal functioning of different oligomeric forms of the brain's nAChRs (Barrantes, 1998; Aubert et al., 1992; Banerjee et al., 2000; Vallés et al., 2014; Nikiforuk et al., 2015; Sadigh-Eteghad et al., 2015). In PD, cholinergic system deficit, along with abnormal basal forebrain volume and cholinergic denervation specifically in the cortex and thalamus, is responsible for impaired cognition function.

Therapeutic improvement in cognitive impairment, along with symptomatic improvement, requires nAChRs – specifically alpha 7 and its ligands – to be studied further and several compounds that enhance memory, as well cognitive function, can be researched (Francis et al., 1999; Babic et al., 1999; Barage and Sonawane, 2015; Kar et al., 2004). Cholinergic neurotransmission-enhancing drugs are effective in improving cognitive function in patients with PD. However, the drugs that block muscarinic receptors are report to cause dementia and chronic cognitive impairment (Yarnall et al., 2011).

6.5.1.5 Lewy Body Dementia and Delirium

Short-latency afferent inhibition is used to measure inhibitory cholinergic activity in PD brains (Manganelli et al., 2009). Lewy body dementia (LBD) is reported to involve visual hallucinations associated with dementia in PD (Chaudhuri et al., 2009). Delirium is specified by a deficit in attention and cognitive impairment, specifically in PD (Vardy et al., 2015) due to an abnormal cholinergic system. The available knowledge on LBD and delirium is insufficient. More research and study are required for this symptom of PD.

6.5.1.6 REM Sleep Behavior Disorder

The most common nonmotor symptom in PD is sleep disturbance (Chaudhuri et al., 2009). Normal PPN of the cholinergic system is responsible for REM sleep (Van Dort et al., 2015). REM sleep behavior disorder (RBD) and reduced REM sleep occurs due to cholinergic neuron dysfunction, specifically in the basal forebrain and brain stem (Suzuki et al., 2015). RBD is considered to be a premotor symptom of PD (Postuma et al., 2006). Cholinesterase inhibitors might have a promising impact in treating this disorder in PD, but further evidence is required to support it.

6.5.1.7 Olfactory Dysfunction

Approximately 90 percent of the population with PD is affected by abnormal olfaction (Doty, 2012). One of the premotor symptoms when dopaminergic degeneration is not yet evident is olfactory impairment. Abnormal cholinergic neurotransmission along with nondopaminergic neurotransmitter system dysfunction in PD is responsible for the olfactory loss. Abnormal cholinergic system pathology is similar to alpha-synuclein pathology. ChAT density in the olfactory bulb inner layer is reported to be very low in PD (Mundiñano et al., 2013).

6.6 RECENT DEVELOPMENTS AND CHALLENGES

ACh has become an important area of study due to its possible involvement in different pathologies and its ubiquitous nature. Evidence suggests it has diverse effects in brain functioning, such as memory processes, neuroplasticity, cognitive function, and so on (Perez-Lioret et al., 2014). These finding could play a major role in preventing age-related memory loss, cognitive dysfunction, and finding the relevant therapeutic approach. Considering the emerging tools helps with a better understanding of neurologic diseases, and finding effective treatments has become possible (Kondabolu et al., 2016; Stauffer et al., 2016; Zare-Shahabadi et al., 2016). The identification of new pharmacologic approaches to regulating cholinergic system components could revolutionize the present scenario of various neurological disorders. In future, this area of research needs to be worked on extensively to acknowledge the increasing number of people suffering from neurologic deficits. The role of various molecular forms of AChE in ultra-fast, rapid, or slow cholinergic neurotransmission helps in understanding the cellular and molecular basis of cholinergic transmission (Dunant and Gisiger, 2017). This area is still open for further exploration, including membrane fusion and neurotransmitter release (Peters et al., 2001; Dunant et al., 2000).

Various motor and nonmotor symptoms of PD severely affect the quality of life of patients (Morris et al., 2019). Future work on understanding the influence of the cholinergic system on PD neuroscience with the aim of improving the quality of life of patients is required to find a therapeutic approach.

6.7 CONCLUSION

This chapter reviewed the biochemistry of cholinergic neurotransmission, pathways, and impairment of the cholinergic system in detail. The contribution of mAChRs and nAChRs to cholinergic neurotransmission and the different types present in the brain, performing specific functions, were discussed. The abnormal structure and function of receptors, cholinergic neurotransmitters, ACh hydrolytic enzyme, and neurotransmission pathways are majorly involved in causing mental and neurologic disorders. These areas of cholinergic neurotransmission provide future research perspectives for drug development and treatment. PD is the second most common neurodegenerative disease in which the cholinergic neurotransmission system is severely affected. ACh involvement in abnormal motor functions and also in different pathologies make it an important area of research. Various motor and nonmotor clinical symptoms of PD, such as motor symptoms, gait dysfunction, cognitive deterioration, dementia, sleep abnormalities, and altered olfactory function, were discussed along with their symptoms, cause, and available therapeutic approaches.

REFERENCES

Albuquerque EX, Pereira EFR, Alkondon M, Rogers SW. Mammalian nicotinic acetylcholine receptors: From structure to function. Physiol Rev. 2009;89:73–120.

Alzheimer A. Über einen eigenartigen schweren Erkrankungsproze β der Hirnrincle. *Neurol Central*. 1906; 25: 1134 (in German).

Amenta F, Tayebati S. Pathways of acetylcholine synthesis, transport and release as targets for treatment of adult-onset cognitive dysfunction. Curr Med Chem. 2008;15:488–98.

Anagnostaras SG, Murphy GG, Hamilton SE, Mitchell SL, Rahnama NP, Nathanson NM, et al. Selective cognitive dysfunction in acetylcholine M1 muscarinic receptor mutant mice. Nat Neurosci. 2003;6:51–8.

Aronstam RS, Patil P. Muscarinic receptors: autonomic neurons. Encycl Neurosci. 2009;1141–9.

Ascherio A, Schwarzschild MA. The epidemiology of Parkinson's disease: risk factors and prevention. Lancet Neurol. 2016;15:1257–72.

Aubert I, Araujo DM, Cécyre D, Robitaille Y, Gauthier S, Quirion R. Comparative alterations of nicotinic and muscarinic binding sites in Alzheimer's and Parkinson's diseases. J Neurochem. 1992;58:529–41.

Babic T. The cholinergic hypothesis of Alzheimer's disease: a review of progress. J Neurol Neurosurg Psychiatry. 1999;67(4):558.

Ballestero, J., Zorrilla de San Martín, J., Goutman, J., Elgoyhen, A. B., Fuchs, P. A., & Katz, E. Short-term synaptic plasticity regulates the level of olivocochlear inhibition to auditory hair cells. J Neurosci. 2011 31(41), 14763–74.

Banerjee C, Nyengaard JR, Wevers A, De Vos RAI, Steur ENHJ, Lindstrom J, et al. Cellular expression of α7 nicotinic acetylcholine receptor protein in the temporal cortex in Alzheimer's and Parkinson's disease-a stereological approach. Neurobiol Dis. 2000;7:666–72.

Barage SH, Sonawane KD. Amyloid cascade hypothesis: pathogenesis and therapeutic strategies in Alzheimer's. Neuropeptides. 2015;52:1–18.

Barrantes FJ. The Nicotinic Acetylcholine Receptor: Current Views and Future Trends. Springer, 1998, pp. 175–213.

Barrantes FJ. Phylogenetic conservation of protein-lipid motifs in pentameric ligand-gated ion channels. Biochim Biophys Acta. 2015; 1848:1796–1805.

Bennett C, Arroyo S, Berns D, Hestrin S. Mechanisms generating dual-component nicotinic EPSCs in cortical interneurons. J Neurosci. 2012;32:17287–96.

Berg D, Postuma RB, Bloem B, Chan P, Dubois B, Gasser T, et al. Time to redefine PD? Introductory statement of the MDS Task Force on the definition of Parkinson's disease. Mov Disord. 2014;29:454–62.

Berg DK, Conroy WG. Nicotinic α7 receptors: Synaptic options and downstream signaling in neurons. J Neurobiol. 2002;53:512–23.

Bernard V, Girard E, Hrabovska A, et al. Distinct localization of collagen Q and PRiMA forms of acetylcholinesterase at the neuromuscular junction. Mol Cell Neurosci. 2011;46(1):272–81.

Bernard V, Laribi O, Levey AI, Bloch B. Subcellular redistribution of m2 muscarinic acetylcholine receptors in striatal interneurons in vivo after acute cholinergic stimulation. J Neurosci. 1998;18:10207–18.

Bernard V, Levey AI, Bloch B. Regulation of the subcellular distribution of m4 muscarinic acetylcholine receptors in striatal neurons in vivo by the cholinergic environment: Evidence for regulation of cell surface receptors by endogenous and exogenous stimulation. J Neurosci. 1999;19:10237–49.

Bonsi P, Martella G, Cuomo D, Platania P, Sciamanna G, Bernardi G, et al. Loss of muscarinic autoreceptor function impairs long-term depression but not long-term potentiation in the striatum. J Neurosci. 2008;28:6258–63.

Boonstra TA, Van Der Kooij H, Munneke M, Bloem BR. Gait disorders and balance disturbances in Parkinson's disease: Clinical update and pathophysiology. Curr Opin Neurol. 2008;21:461–71.

Bordia T, McGregor M, McIntosh JM, Drenan RM, Quik M. Evidence for a role for α6(*) nAChRs in l-dopa-induced dyskinesias using Parkinsonian α6(*) nAChR gain-of-function mice. Neuroscience. 2015;295:187–97.

Bower JH, Maraganore DM, McDonnell SK, Rocca WA. Influence of strict, intermediate, and broad diagnostic criteria on the age- and sex-specific incidence of Parkinson's disease. Mov Disord. 2000;15(5):819–25.

Burghaus L, Schütz U, Krempel U, De Vos RAI, Jansen Steur ENH, Wevers A, et al. Quantitative assessment of nicotinic acetylcholine receptor proteins in the cerebral cortex of Alzheimer's patients. Mol Brain Res. 2000;76:385–8.

Campanari ML, García-Ayllón MS, Ciura S, Sáez-Valero J, Kabashi E. Neuromuscular junction impairment in amyotrophic lateral sclerosis: Reassessing the role of acetylcholinesterase. Front Mol Neurosci. 2016;9:1–8.

Cancela JM. Specific Ca^{2+} signaling evoked by cholecystokinin and acetylcholine: The roles of NAADP, cADPR, and IP3. Annu Rev Physiol. 2001;63:99–117.

Changeux JP, Bertrand D, Corringer PJ, Dehaene S, Edelstein S, Léna C, et al. Brain nicotinic receptors: Structure and regulation, role in learning and reinforcement. Brain Res Rev. 1998;26:198–216.

Chaudhuri KR, Healy DG, Schapira AH V. Non-motor symptoms of Parkinson's disease: diagnosis and management. Lancet Neurol. 2006;5:235–45.

Chaudhuri KR, Schapira AH. Non-motor symptoms of Parkinson's disease: dopaminergic pathophysiology and treatment. Lancet Neurol. 2009;8:464–74.

Chung KA, Lobb BM, Nutt JG, Horak FB. Effects of a central cholinesterase inhibitor on reducing falls in Parkinson's disease. Neurology. 2010;75:1263–9.

Cooper AA, Gitler AD, Cashikar A, Haynes CM, Kathryn J, Bhullar B, et al. Alpha-synuclein blocks ER-Golgi traffic and Rab1 rescues neuron loss in Parkinson's models. Science. 2006;313:324–8.

Corringer, Pierre-Jean and Jean Pierre Changeux. "Nicotinic acetylcholine receptors." Scholarpedia 3 (2008): 3468.

Dale HH. The action of certain esters and ethers of choline and their relation to muscarine. J Pharmacol Exp Ther. 1914;6.

Décossas M, Doudnikoff É, Bloch B, Bernard V. Aging and subcellular localization of m2 muscarinic autoreceptor in basalocortical neurons in vivo. Neurobiol Aging. 2005;26:1061–72.

Del Mar Arroyo-Jiménez M, Bourgeois JP, Marubio LM, Le Sourd AM, Ottersen OP, Rinvik E, et al. Ultrastructural localization of the α4-subunit of the neuronal acetylcholine nicotinic receptor in the rat substantia nigra. J Neurosci. 1999;19:6475–87.

Del Sorbo F, Albanese A. Levodopa-induced dyskinesias and their management. J Neurol. 2008;255(SUPPL. 4):32–41.

Dickson DW, Braak H, Duda JE, Duyckaerts C, Gasser T, Halliday GM, et al. Neuropathological assessment of Parkinson's disease: refining the diagnostic criteria. Lancet Neurol. 2009;8:1150–7.

Ding J, Guzman JN, Tkatch T, Chen S, Goldberg JA, Ebert PJ, et al. RGS4-dependent attenuation of M4 autoreceptor function in striatal cholinergic interneurons following dopamine depletion. Nat Neurosci. 2006;9:832–42.

Dolezal V, Castell X, Tomasi M, Diebler MF. Stimuli that induce a cholinergic neuronal phenotype of NG108-15 cells upregulate ChAT and VAChT mRNAs but fail to increase VAChT protein. Brain Res Bull. 2001;54:363–73.

Dong H, Xiang YY, Farchi N, Ju W, Wu Y, Chen L, et al. Excessive expression of acetylcholinesterase impairs glutamatergic synaptogenesis in hippocampal neurons. J Neurosci. 2004;24:8950–60.

Doty RL. Olfactory dysfunction in Parkinson's disease. Nat Rev Neurol. 2012;8:329–39.

Douglas CL, Baghdoyan HA, Lydic R. Postsynaptic muscarinic m1 receptors activate prefrontal cortical EEG of C57BL/6J mouse. J Neurophysiol. 2002;88:3003–9.

Dunant Y, Gisiger V. Ultrafast and slow cholinergic transmission. Different involvement of acetylcholinesterase molecular forms. Molecules. 2017;22:1300.

Dunant Y, Israël M. Neurotransmitter release at rapid synapses. Biochimie. 2000;82(4):289–302. doi:10.1016/s0300-9084(00)00194-2.

English BA, Jones CK. Cholinergic Neurotransmission. 3rd ed. Primer on the Autonomic Nervous System. Elsevier, 2012, pp. 71–74. Available from: http://dx.doi.org/10.1016/B978-0-12-386525-0.00014-7

Fahn S. The history of dopamine and levodopa in the treatment of Parkinson's disease. Mov Disord. 2008;23(Suppl. 3):S497–508.

Faust TW, Assous M, Shah F, Tepper JM, Koós T. Novel fast adapting interneurons mediate cholinergic-induced fast GABAA inhibitory postsynaptic currents in striatal spiny neurons. Eur J Neurosci. 2015;42(2):1764–74.

Fearnley JM, Lees AJ. Ageing and Parkinson's disease: Substantia nigra regional selectivity. Brain. 1991;114:2283–301.

Flynn DD, Mash DC. Characterization of l-[3H]nicotine binding in human cerebral cortex: Comparison between Alzheimer's and the normal. J Neurochem. 1986;47:1948–54.

Fox SH, Katzenschlager R, Lim SY, Ravina B, Seppi K, Coelho M, et al. The movement disorder society evidence-based medicine review update: Treatments for the motor symptoms of Parkinson's disease. Mov Disord. 2011;26(SUPPL. 3):2–41.

Francis PT, Palmer AM, Snape M, Wilcock GK. The cholinergic hypothesis of Alzheimer's: A review of progress. J Neurol Neurosurg Psychiatry. 1999;66:137–47.

Furey ML, Pietrini P, Haxby JV, Drevets WC. Selective effects of cholinergic modulation on task performance during selective attention. Neuropsychopharmacology. 2008;33(4):913–923. doi:10.1038/sj.npp.1301461.

Gautron J. Localisation des cholinestérases au niveau de la jonction nerf-électroplaque de l'organe électrique de la Torpille marbrée. C R Acad Sci Paris. 1970;271:714–17 (in French).

Gibb WRG, Lees AJ. A comparison of clinical and pathological features of young- and old-onset Parkinson's disease. Neurology. 1988;38:1402–6.

Giocomo LM, Hasselmo ME. Neuromodulation by glutamate and acetylcholine can change circuit dynamics by regulating the relative influence of afferent input and excitatory feedback. Mol Neurobiol. 2007;36:184–200.

Gisiger V, Stephens HR. Localization of the pool of G4 acetylcholinesterase characterizing fast muscles and its alteration in murine muscular dystrophy. J Neurosci Res. 1988;19:62–78.

Gisiger V, Bélisle M, Gardiner PF. Acetylcholinesterase adaptation to voluntary wheel running is proportional to the volume of activity in fast, but not slow, rat hindlimb muscles. Eur J Neurosci. 1994;6:673–80.

Gonzales KK, Smith Y. Cholinergic interneurons in the dorsal and ventral striatum: anatomical and functional considerations in normal and diseased conditions. Ann N Y Acad Sci. 2015;1349(1):1–45.

Gotti C, Riganti L, Vailati S, Clementi F. Brain neuronal nicotinic receptors as new targets for drug discovery. Curr Pharm Des. 2006;12:407–28.

Grabli D, Karachi C, Welter M, Lau B, Hirsch EC, Vidailhet M, et al. Normal and pathological gait: What we learn from Parkinson's disease. J Neurol Neurosurg Psychiatry. 2013;83:979–85.

Grace AA, Floresco SB, Goto Y, Lodge DJ. Regulation of firing of dopaminergic neurons and control of goal-directed behaviors. Trends Neurosci. 2007;30:220–7.

Gray R, Rajan AS, Radcliffe KA, Yakehiro M, Dani JA. Hippocampal synaptic transmission enhanced by low concentrations of nicotine. Nature. 1996;383(6602):713–716.

Guo J, Schofield GG. Activation of muscarinic m5 receptors inhibits recombinant KCNQ2/KCNQ3 K^+ channels expressed in HEK293T cells. Eur J Pharmacol. 2003;462:25–32.

Gulledge AT, Stuart GJ. Cholinergic inhibition of neocortical pyramidal neurons. J Neurosci. 2005;25(44):10308–20.

Haga K, Haga T. Dual regulation by G proteins of agonist-dependent phosphorylation of muscarinic acetylcholine receptors. FEBS Lett. 1990;268(1):43–47.

Hamilton SE, Loose MD, Qi M, Levey AI, Hille B, Mcknight GS, et al. Disruption of the m1 receptor gene ablates muscarinic receptor-dependent M current regulation and seizure activity in mice. Proc Natl Acad Sci U S A. 1997;94:13311–16.

Hasselmo ME, Wyble BP, Wallenstein GV. Encoding and retrieval of episodic memories: Role of cholinergic and GABAergic modulation in the hippocampus. Hippocampus. 1996;6:693–708.

Hay YA, Lambolez B, Tricoire L. Nicotinic transmission onto layer 6 cortical neurons relies on synaptic activation of non-α7 receptors. Cereb Cortex. 2016;26:2549–62.

Hecker S, Konradi C. Synaptic function and biochemical neuroanatomy. In: Martin A, Scahill L, Charney DS, et al., editors. Pediatric Psychopharmacology. New York: Oxford University Press, 2003, pp. 20–32.

Hefft S, Hulo S, Bertrand D, Muller D. Synaptic transmission at nicotinic acetylcholine receptors in rat hippocampal organotypic cultures and slices. J Physiol. 1999;515:769–76.

Herculano-Houzel S. The remarkable, yet not extraordinary, human brain as a scaled-up primate brain and its associated cost. Proc Natl Acad Sci U S A. 2012;109(SUPPL.1):10661–8.

Herrero JL, Roberts MJ, Delicato LS, Gieselmann MA, Dayan P, Thiele A. Acetylcholine contributes through muscarinic receptors to attentional modulation in V1. Nature. 2008;454:1110–14.

Hill JA, Zoli M, Bourgeois JP, Changeux JP. Immunocytochemical localization of a neuronal nicotinic receptor: The β2-subunit. J Neurosci. 1993;13:1551–68.

Hirsch L, Jette N, Frolkis A, Steeves T, Pringsheim T. The incidence of Parkinson's disease: A systematic review and meta-analysis. Neuroepidemiology. 2016;46:292–300.

Hirtz D, Thurman DJ, Gwinn-Hardy K, Mohamed M, Chaudhuri AR, Zalutsky R. How common are the "common" neurologic disorders? Neurology. 2007;68:326–37.

Hornigold DC, Mistry R, Raymond PD, Blank JL, Challiss RAJ. Evidence for cross-talk between M 2 and M 3 muscarinic acetylcholine receptors in the regulation of second messenger and extracellular signal-regulated kinase signalling pathways in Chinese hamster ovary cells. Br J Pharmacol. 2003;138:1340–50.

Jaakola VP, Vainio M, Sen S, Rehn M, Heimo H, Scheinin M, et al. Intracellularly truncated human α2B-adrenoceptors: Stable and functional GPCRs for structural studies. J Recept Signal Transduct. 2005;25:99–124.

Jasmin BJ, Gisiger V. Regulation by exercise of the pool of G4 acetylcholinesterase characterizing fast muscles: Opposite effect of running training in antagonist muscles. J Neurosci. 1990;10:1444–54.

Jones CK, Byun N, Bubser M. Muscarinic and nicotinic acetylcholine receptor agonists and allosteric modulators for the treatment of schizophrenia. Neuropsychopharmacology. 2012;37:16–42.

Kandel ER, Siegelbaum SA. Overview of synaptic transmission. In: Kandel ER, Schwartz JH, Jessell TM, editors. Principles of Neural Science. 4th ed. New York: McGraw Hill, 2000, pp. 175–86.

Kar S, Slowikowski SPM, Westaway D, Mount HTJ. Interactions between β-amyloid and central cholinergic neurons: Implications for Alzheimer's. J Psychiatry Neurosci. 2004;29:427–41.

Katz B, Thesleff S. On the factors which determine the amplitude of the miniature end-plate potential. *J Physiol.* 1957;137(2):267–78

Kawai H, Lazar R, Metherate R. Nicotinic control of axon excitability regulates thalamocortical transmission. Nat Neurosci. 2007;10:1168–75.

Kawashima K, Misawa H, Moriwaki Y, Fujii YX, Fujii T, Horiuchi Y, et al. Ubiquitous expression of acetylcholine and its biological functions in life forms without nervous systems. Life Sci. 2007;80:2206–9.

Katz, B. Looking back at the neuromuscular junction. In Neuromuscular junction; Selling, LC, Libelius, R, Thesleff, S, eds. Amsterdam: Elsevier Science, 1989, pp. 3–9.

Kondabolu K, Roberts EA, Bucklin M, McCarthy MM, Kopell N, Han X. Striatal cholinergic interneurons generate beta and gamma oscillations in the corticostriatal circuit and produce motor deficits. Proc Natl Acad Sci U S A. 2016;113:3159–68.

Kummer W, Lips KS, Pfeil U. The epithelial cholinergic system of the airways. Histochem Cell Biol. 2008;130:219–34.

Kuno M, Turkanis SA, Weakly JN. Correlation between nerve terminal size and transmitter release at the neuromuscular junction of the frog. J Physiol. 1971;213(3):545–56.

Langston JW. The Parkinson's complex: Parkinsonism is just the tip of the Iceberg. Ann Neurol. 2006;59:591–6.

Lanzafame AA, Christopoulos A, Mitchelson F. Cellular signaling mechanisms for muscarinic acetylcholine receptors. Recept Channels. 2003;9:241–60.

Lendvai B. 5 Cholinergic Transmission. Handbook of Neurochemistry and Molecular Neurobiology, 2008;113–27. 2008.

Levey AI, Edmunds SM, Koliatsos V, Wiley RG, Heilman CJ. Expression of m1–m4 muscarinic acetylcholine receptor proteins in rat hippocampus and regulation by cholinergic innervation. J Neurosci. 1995;15:4077–92.

Lobb CJ, Wilson CJ, Paladini CA. A dynamic role for GABA receptors on the firing pattern of midbrain dopaminergic neurons. J Neurophysiol. 2010;104:403–13.

Massoulié J, Pezzementi L, Bon S, Krejci E, Vallette FM. Molecular and cellular biology of cholinesterases. Prog Neurobiol. 1993;41(1):31–91.

McGehee DS, Heath MJS, Gelber S, Devay P, Role LW. Nicotine enhancement of fast excitatory synaptic transmission in CNS by presynaptic receptors. 1995;269:1692–6.

McKay BE, Placzek AN, Dani JA. Regulation of synaptic transmission and plasticity by neuronal nicotinic acetylcholine receptors. Biochem Pharmacol. 2007;74:1120–33.

McKinney M. Brain cholinergic vulnerability: Relevance to behavior and disease. Biochem Pharmacol. 2005;70:1115–24.

Maclaren DAA, Santini JA, Russell AL, Markovic T, Clark SD. Deficits in motor performance after pedunculopontine lesions in rats - impairment depends on demands of task. Eur J Neurosci. 2014;40:3224–36.

Maggi L, Le Magueresse C, Changeux JP, Cherubini E. Nicotine activates immature "silent" connections in the developing hippocampus. Proc Natl Acad Sci U S A. 2003;100:2059–64.

Manganelli F, Vitale C, Santangelo G, Pisciotta C, Iodice R, Cozzolino A, et al. Functional involvement of central cholinergic circuits and visual hallucinations in Parkinson's disease. Brain. 2009;132:2350–5.

Massoulié J, Perrier N, Noureddine H, Liang D, Bon S. Old and new questions about cholinesterases. Chem Biol Interact. 2008;175:30–44.

Mesulam MM, Mufson EJ, Levey AI, Wainer BH. Cholinergic innervation of cortex by the basal forebrain: Cytochemistry and cortical connections of the septal area, diagonal band nuclei, nucleus basalis (Substantia innominata), and hypothalamus in the rhesus monkey. J Comp Neurol. 1983;214:170–97.

Mesulam MM. Structure and function of cholinergic pathways in the cerebral cortex, limbic system, basal ganglia, and thalamus of the human brain. In: Bloom FE, Kupfer DJ, eds. Psychopharmacology: The Fourth Generation of Progress. New York: Raven Press, 1995.

Miller SL, Yeh HH. Neurotransmitters and neurotransmission in the developing and adult nervous system. In: Conn PM, ed. Conn's Translational Neuroscience. New York: Elsevier, 2017, pp. 49–84.

Morris R, Martini DN, Madhyastha T, Kelly VE, Grabowski TJ, Nutt J, et al. Overview of the cholinergic contribution to gait, balance and falls in Parkinson's disease. Park Relat Disord. 2019;63:20–30.

Mrzljak L, Levey AI, Goldman-Rakic PS. Association of m1 and m2 muscarinic receptor proteins with asymmetric synapses in the primate cerebral cortex: Morphological evidence for cholinergic modulation of excitatory neurotransmission. Proc Natl Acad Sci U S A. 1993;90:5194–8.

Mundiñano IC, Hernandez M, Dicaudo C, Ordoñez C, Marcilla I, Tuñon MT, et al. Reduced cholinergic olfactory centrifugal inputs in patients with neurodegenerative disorders and MPTP-treated monkeys. Acta Neuropathol. 2013;126:411–25.

Navarrete M, Perea G, Maglio L, Pastor J, García De Sola R, Araque A. Astrocyte calcium signal and gliotransmission in human brain tissue. Cereb Cortex. 2013;23:1240–6.

Nicoll RA, Malenka RC, Kauer JA. Functional comparison of neurotransmitter receptor subtypes in mammalian central nervous system. Physiol Rev. 1990;70:513–65.

Nikiforuk A, Kos T, Potasiewicz A, Popik P. Positive allosteric modulation of alpha 7 nicotinic acetylcholine receptors enhances recognition memory and cognitive flexibility in rats. Eur Neuropsychopharmacol. 2015;25(8):1300–13.

Nys M, Kesters D, Ulens C. Structural insights into Cys-loop receptor function and ligand recognition. Biochem Pharmacol. 2013;86:1042–53.

Oda Y. Choline acetyltransferase: The structure, distribution and pathologic changes in the central nervous system. Pathol Int. 1999;49:921–37.

Pahud G, Salem N, van de Goor J, Medilanski J, Pellegrinelli N, Eder-Colli L. Study of subcellular localization of membrane-bound choline acetyltransferase in *Drosophila* central nervous system and its association with membranes. Eur J Neurosci. 1998;10:1644–53.

Palma E, Conti L, Roseti C, Limatola C. Novel approaches to study the involvement of α7-nAChR in human diseases. Curr Drug Targets. 2013;13:579–86.

Parkinson J. An essay on the Shaking Palsy. Arch Neurol. 1969;20:441–5.

Perez-Lloret S, Negre-Pages L, Damier P, Delval A, Derkinderen P, Destée A, et al. Prevalence, determinants, and effect on quality of life of freezing of gait in Parkinson's disease. JAMA Neurol. 2014;71:884–90.

Peters C, Bayer MJ, Bühler S, Andersen JS, Mann M, Mayer A. Trans-complex formation by proteolipid channels in the terminal phase of membrane fusion. Nature. 2001;409:581–8.

Picciotto MR. Nicotine as a modulator of behavior: Beyond the inverted U. Trends Pharmacol Sci. 2003;24:493–9.

Picciotto MR, Caldarone BJ, Brunzell DH, Zachariou V, Stevens TR, King SL. Neuronal nicotinic acetylcholine receptor subunit knockout mice: Physiological and behavioral phenotypes and possible clinical implications. Pharmacol Ther. 2001;92:89–108.

Picciotto MR, Caldarone BJ, King SL, Zachariou V. Nicotinic receptors in the brain: Links between molecular biology and behavior. Neuropsychopharmacology. 2000;22:451–65.

Picciotto MR, Higley MJ, Mineur YS. Acetylcholine as a neuromodulator. Neuron. 2012;76:116–29.

Picconi B, Centonze D, Håkansson K, Bernardi G, Greengard P, Fisone G, et al. Loss of bidirectional striatal synaptic plasticity in L-DOPA-induced dyskinesia. Nat Neurosci. 2003;6:501–6.

Postuma RB, Lang AE, Massicotte-Marquez J, Montplaisir J. Potential early markers of Parkinson's disease in idiopathic REM sleep behavior disorder. Neurology. 2006;66:845–51.

Pringsheim T, Jette N, Frolkis A, Steeves TDL. The prevalence of Parkinson's disease: A systematic review and meta-analysis. Mov Disord. 2014;29:1583–90.

Quik M, Campos C, Grady SR. Multiple CNS nicotinic receptors mediate L-dopa-induced dyskinesias: studies with parkinsonian nicotinic receptor knockout mice. Biochem Pharmacol. 2013;86(8):1153–62.

Quik M, Zhang D, Perez XA, Bordia T. Role for the nicotinic cholinergic system in movement disorders; therapeutic implications. Pharmacol Ther. 2014;144:50–9.

Riad M, Tong XK, el Mestikawy S, Hamon M, Hamel E, Descarries L. Endothelial expression of the 5-hydroxytryptamine1B antimigraine drug receptor in rat and human brain microvessels. *Neuroscience*. 1998;86(4):1031–1035.

Raiteri M, Leardi R, Marchi M. Heterogeneity of presynaptic muscarinic receptors regulating neurotransmitter release in the rat brain. J Pharmacol Exp Ther. 1984;228:209–214.

Reich SG, Savitt JM. Parkinson's disease. Med Clin North Am. 2019;103:337 50.

Ren J, Qin C, Hu F, Tan J, Qiu L, Zhao S, et al. Habenula "cholinergic" neurons corelease glutamate and acetylcholine and activate postsynaptic neurons via distinct transmission modes. Neuron. 2011;69:445–52.

Resende RR, Adhikari A. Cholinergic receptor pathways involved in apoptosis, cell proliferation and neuronal differentiation. Cell Commun Signal. 2009;7:1–20.

Role LW, Berg DK. Nicotinic receptors in the development and modulation of CNS synapses. Neuron. 1996;16(6):1077–85.

Rouse ST, Edmunds SM, Yi H, Gilmor ML, Levey AI. Localization of M2 muscarinic acetylcholine receptor protein in cholinergic and non-cholinergic terminals in rat hippocampus. Neurosci Lett. 2000;284:182–6.

Sadigh-Eteghad S, Talebi M, Mahmoudi J, Babri S, Shanehbandi D. Selective activation of α7 nicotinic acetylcholine receptor by PHA-543613 improves Aβ25-35-mediated cognitive deficits in mice. Neuroscience. 2015;298:81–93.

Sarter M. Behavioral-cognitive targets for cholinergic enhancement. Curr Opin Behav Sci. 2015;4:22–6.

Seguela P, Wadiche J, Dineley-Miller K, Dani JA, Patrick JW. Molecular cloning, functional properties, and distribution of rat brain α7: A nicotinic cation channel highly permeable to calcium. J Neurosci. 1993;13:596–604.

Simone MJ, Tan ZS. The role of inflammation in the pathogenesis of delirium and dementia in older adults: A review. CNS Neurosci Ther. 2011;17:506–13.

Soto C. Unfolding the role of protein misfolding in neurodegenerative diseases. Nat Rev Neurosci. 2003;4:49–60.

Spillantini MG, Schmidt ML, Lee VMY, Trojanowski JQ, Jakes R, Goedert M. α-synuclein in Lewy bodies. Nature. 1997;388:839–40.

Stauffer WR, Lak A, Yang A, Borel M, Paulsen O, Boyden ES, et al. Dopamine neuron-specific optogenetic stimulation in rhesus macaques. Cell. 2016;166:1564–1571.

Stelzhammer V, Ozcan S, Gottschalk MG, Steeb H, Hodes GE, Guest PC, et al. Central and peripheral changes underlying susceptibility and resistance to social defeat stress – A proteomic profiling study. Diagn Neuropsychiatry. 2015;1:1–7.

Suzuki K, Miyamoto M, Miyamoto T, Hirata K. Parkinson's disease and sleep/wake disturbances. Curr Neurol Neurosci Rep. 2015;15: 8.

Takács VT, Freund TF, Nyiri G. Neuroligin 2 is expressed in synapses established by cholinergic cells in the mouse brain. PLoS One. 2013;8:e72450.

Tan LCS, Venketasubramanian N, Jamora RDG, Heng D. Incidence of Parkinson's disease in Singapore. Park Relat Disord. 2007;13:40–3.

Threlfell S, Cragg SJ. Dopamine signaling in dorsal versus ventral striatum: The dynamic role of cholinergic interneurons. Front Syst Neurosci. 2011;5:1–10.

Threlfell S, Lalic T, Platt NJ, Jennings KA, Deisseroth K, Cragg SJ. Striatal dopamine release is triggered by synchronized activity in cholinergic interneurons. Neuron. 2012;75:58–64.

Tian YY, Tang CJ, Wu J, Zhou JS. Parkinson's disease in China. Neurol Sci. 2011;32:23–30.

Tiwari P, Dwivedi S, Singh MP, Mishra R, Chandy A. Basic and modern concepts on cholinergic receptor: A review. Asian Pacific J Trop Dis. 2013;3:413–20.

Turrini P, Casu MA, Wong TP, De Koninck Y, Ribeiro-da-Silva A, Cuello AC. Cholinergic nerve terminals establish classical synapses in the rat cerebral cortex: Synaptic pattern and age-related atrophy. Neuroscience. 2001;105:277–85.

Vallés AS, Borroni MV, Barrantes FJ. Targeting brain α7 nicotinic acetylcholine receptors in Alzheimer's: Rationale and current status. CNS Drugs. 2014;28:975–87.

Van Dort CJ, Zachs DP, Kenny JD, Zheng S, Goldblum RR, Gelwan NA, et al. Optogenetic activation of cholinergic neurons in the PPT or LDT induces REM sleep. Proc Natl Acad Sci U S A. 2015;112:584–9.

Vardy ERLC, Teodorczuk A, Yarnall AJ. Review of delirium in patients with Parkinson's disease. J Neurol. 2015;262:2401–10.

Vernino S, Amador M, Luetje CW, Patrick J, Dani JA. Calcium modulation and high calcium permeability of neuronal nicotinic acetylcholine receptors. Neuron. 1992;8:127–34.

Vidal C, Changeux JP. Nicotinic and muscarinic modulations of excitatory synaptic transmission in the rat prefrontal cortex in vitro. Neuroscience. 1993;56(1):23–32.

Wallace TL, Porter RHP. Targeting the nicotinic alpha7 acetylcholine receptor to enhance cognition in disease. Biochem Pharmacol. 2011;82:891–903.

Wess J. Novel insights into muscarinic acetylcholine receptor function using gene targeting technology. Trends Pharmacol Sci. 2003;24:414–20.

Wess J, Eglen RM, Gautam D. Muscarinic acetylcholine receptors: Mutant mice provide new insights for drug development. Nat Rev Drug Discov. 2007;6:721–33.

Wessler I, Kirkpatrick CJ. Acetylcholine beyond neurons: The non-neuronal cholinergic system in humans. Br J Pharmacol. 2008;154:1558–71.

Witzemann V, Boustead C. Changes in acetylcholinesterase molecular forms during the embryonic development of *Torpedo marmorata*. J Neurochem. 1982;39:747–55.

Woolf NJ. Cholinergic systems in mammalian brain and spinal cord. Prog Neurobiol. 1991;37:475–524.

Yamada M, Lamping KG, Duttaroy A, Zhang W, Cui Y, Bymaster FP, et al. Cholinergic dilation of cerebral blood vessels is abolished in M5 muscarinic acetylcholine receptor knockout mice. Proc Natl Acad Sci U S A. 2001a;98:14096–101.

Yamada M, Miyakawa T, Duttaroy A, Yamanaka A, Moriguchi T, Makita R, et al. Mice lacking the M3 muscarinic acetylcholine receptor are hypophagic and lean. Nature. 2001b;410:207–12.

Yarnall A, Rochester L, Burn DJ. The interplay of cholinergic function, attention, and falls in Parkinson's disease. Mov Disord. 2011;26:2496–503.

Young J, Inouye SK. Delirium in older people. Br Med J. 2007;334:842–6.

Zaborszky L. The modular organization of brain systems. Basal forebrain: The last frontier. Prog Brain Res. 2002;136:359–72.

Zaborszky L, Hoemke L, Mohlberg H, Schleicher A, Amunts K, Zilles K. Stereotaxic probabilistic maps of the magnocellular cell groups in human basal forebrain. Neuroimage. 2008;42:1127–41.

Zare-Shahabadi A, Ataei A, Rezaei N. Proteins brighten the brain. Life Sci. 2016;167:1–5.

Zhang W, Yamada M, Gomeza J, Basile AS. Striatal dopamine release, as studied with M 1 – M 5 muscarinic receptor knock-out mice. J Neurosci. 2002;22:6347–52.

Zhang ZX, Roman GC, Hong Z, Wu CB, Qu QM, Huang JB, et al. Parkinson's disease in China: Prevalence in Beijing, Xian, and Shanghai. Lancet. 2005;365:595–7.

Zhou FM, Wilson CJ, Dani JA. Cholinergic interneuron characteristics and nicotinic properties in the striatum. J Neurobiol. 2002;53:590–605.

7 Neurochemical Alterations in Parkinson's Disease

Syed Arman Rabbani, Boshra Khaled, and Huda Khaled
Department of Clinical Pharmacy and Pharmacology, RAK College of Pharmacy, RAK Medical and Health Sciences University, Ras Al Khaimah, UAE

Shrestha Sharma
Amity Institute of Pharmacy, Amity University, Gurgaon, Haryana, India

Faheem Hyder Pottoo
Department of Pharmacology, College of Clinical Pharmacy, Imam Abdulrahman Bin Faisal University, Dammam, Saudi Arabia

Asiya Mahtab
Department of Pharmaceutics, School of Pharmaceutical Sciences and Research, Jamia Hamdard, New Delhi, India

CONTENTS

DOI: 10.1201/9780429265198-9

7.1 INTRODUCTION

Dopamine (DA; 3,4-dihydroxy-phenylethylamine) is a monoamine catecholamine neurotransmitter that has various roles in the functioning of locomotor activity, emotion, cognition, and positive reinforcement (Bhardwaj and Deshmukh, 2018). Dopaminergic neurons are found mainly in the ventral tegmental area (VTA) and the substantia nigra pars compacta (SNpc) in the rostral midbrain and the striatum (over 80% of the total dopamine is present in the basal ganglia) (Ayano, 2016; Ikonen, 2001). DA is known to be the most plentiful catecholamine neurotransmitter in the brain (Bhardwaj and Deshmukh, 2018). However, dopaminergic neurons account for > 1% of neurons in the brain. Its time of action ranges from hundreds of milliseconds to several hours.

Dopaminergic neurons have four main pathways in the basal ganglia of the brain (the nigrostriatal pathway, the mesolimbic pathway, the mesocortical pathway, and the tuberoinfundibular pathway) as they project into different areas (Ayano, 2016). The nigrostriatal pathway includes dopaminergic neuron projections from the substantia nigra to the striatum (the striatum is formed by the caudate and putamen). The loss of those dopaminergic neurons primarily affects the nigrostriatal pathway which is linked to Parkinson's disease (PD) (Marsden, 2006).

PD is a neurodegenerative disorder which happens due to the loss of DA-producing neurons of the SNpc due to an endogenous or exogenous influence (Bhardwaj and Deshmukh, 2018). However, in a healthy individual, the nigrostriatal pathway of DA modulates the striatal pathways by exciting the D1 receptor-mediated pathway and suppressing the D2 receptor-mediated pathway. Therefore, motor functions and movements are well coordinated in a healthy individual (Zisapel, 2001).

Moreover, the dopaminergic system has a crucial function in behavior activation, motivated behavior, and reward process (Everitt and Robbins, 2000). By modulating aggressive behavior, it has consequently been interrelated with the experience of aggression and recognition (Lawrence et al., 2002). One study revealed that elevated levels of DA are associated with impulsive behavior in humans (Bergh et al., 1993). Therefore, antipsychotic drugs acting on D_2 dopamine receptors have been found to be effective in lowering anger levels in aggressive patients (Brizer, 1988). DA interacts with many neurotransmitters, including serotonin [5-hydroxytryptamine (5-HT)]. The serotonergic system has potent interactions with the dopaminergic system both on the anatomic, as well as functional, level (Kapur and Remington, 1996).

The activity of the serotonergic system is inversely proportional to that of DA. One study showed that DA activity was inhibited by 5-HT_2 receptors and enhanced by 5-HT_2 antagonists (Shi et al., 1995). DA acts also on the peripheral blood vessels. It can be given as low- or immediate-rate infusions. As a low-rate infusion (0.5–2 μg/kg/min) it causes vasodilation (by enhancing renal blood flow) in the kidneys, which therefore increases urine output. Immediate-rate infusions (2–10 μg/kg/min) increase heart contraction and thereby increase electrical conductivity, which increases cardiac output (Sonne and Lopez, 2020). DA is often indicated in congestive heart failure, renal failure, and trauma; low doses of dopamine may be effective in managing hypotension (De Backer D et al., 2010). In other words, DA modulates cardiovascular functions; it enhances the contraction of the cardiac muscle and promote vasodilation required for proper blood flow.

Since DA cannot cross the blood–brain barrier (BBB), it is not given directly to compensate for the deficiency in PD; instead, L-dihydroxyphenylalanine (L-DOPA), which crosses the BBB, is given to help improve motor symptoms, but it also known to cause motor side effects and addiction-related behavioral issues (Ghadery et al., 2018). To prevent the occurrence of peripheral effects of L-DOPA, it is usually co-administered with carbidopa (a decarboxylase inhibitor), as this prevents the common side effects, including emesis and nausea (Gilbert et al., 2000). This chapter aims to discuss primarily the neurochemical changes occurring in PD, highlighting the various neurotransmitters involved in the pathophysiology of the disease, along with their receptors and actions. The chapter also covers the dopaminergic and non-dopaminergic pathways involved and how their imbalance contributes to the occurrence of the disease symptoms. Furthermore, the chapter gives an insight into the latest developments and challenges in the treatment of PD.

7.2 SYNTHESIS OF DOPAMINE

Tyrosine is an amino acid that is basically the precursor for DA synthesis. Phenylalanine is converted into tyrosine by phenylalanine hydroxylase in the liver. Later, via an active transport, tyrosine is taken up into the brain where it is converted into DA in dopaminergic neurons. Tyrosine undergoes a hydroxylation reaction by tyrosine hydroxylase in the presence of oxygen to form L-DOPA, which undergoes decarboxylation to produce DA (in order to oxidize tyrosine to levodopa, iron and tetrahydropteridine are required) (Ayano, 2016). The rate-limiting step in this process is the synthesis of DOPA from tyrosine; this step is based on the activity of tyrosine hydroxylase, which is influenced by many factors. Figure 7.1 shows the process of dopamine synthesis. (Cooper et al., 2003). However, one factor is the competitive inhibition of the inhibitory end-product catecholamines for the binding site on tyrosine hydroxylase. Another factor is the flow of an impulse through the dopaminergic neurons; the tyrosine hydroxylation rate is increased as a result of an increase in the impulse flow rate and, thus, the biosynthesis of catecholamine is increased (Feldman et al., 1997). It has been found that the synthesis of DA is higher in the limbic regions than in the striatal regions (Andén et al., 1983). However, this conversion reaction generates CO_2 due to the catalyzing effect of the aromatic amino acid decarboxylase. DA is then preserved in monoaminergic vesicles, from where it can be released later into the synaptic junction and bind to its postsynaptic receptors (Ayano, 2016; Mishra et al., 2018).

DA is a monoamine neurotransmitter which has a few characteristics that make it different from other classes of hormones or peptides. First, DA synthesis occurs in the neuron terminals; second, the synthesis of DA takes place near its releasing site; and, third, DA has a distinguished reuptake process, where remaining DA is restored back into the previous secretory vesicles. However, other classes, like neuropeptide, cannot be restored. Nevertheless, after release, DA is transported back into its original vesicles (1) to avoid the enzymatic degradation of DA by monoamine oxidase (MAO); (2) to reduce its constitutive release via diffusion; (3) to promote a controlled release; and (4) to maintain a replenished store, which is depleted by DA release (Ben-Jonathan and Hnasko, 2001).

7.3 TRANSPORTERS OF DOPAMINE

Reuptake is a process that facilitates the restoration of released neurotransmitters, bringing them back into their presynaptic terminals. This process is very effective if released transmitters produce an action that has to be terminated. The reuptake of transmitters will conserve the energy spent in their biosynthesis by reusing the same molecule. Two types of transporters facilitate the reuptake of DA: DA transporter (DAT) and vesicular monoamine transporter (VMAT; type 1 and type 2). These two transporters vary in structure, cellular localization, selectivity to substrate and antagonist specificity, and their energy requirements. DAT deliver DA from the extracellular to the intracellular space, while VMAT restores DA into the vesicles from the intracellular space (Ben-Jonathan and Hnasko, 2001; German et al., 2015), mediating the mobility of DA by directing it toward the storage vesicles (Mignini et al., 2009).

The availability of DA is monitored by the DAT. It is a plasma membrane protein that transports the released DA from the extracellular space in the synaptic gap into the presynaptic nerve terminal (influx of DA). The function of DAT is usually affected in two ways: first, some drugs may bind to the protein and prevent the transport of DA (e.g., cocaine); second, some drugs are capable of reversing the direction of transportation when they are transported; in this case, they cause efflux of DA from the intracellular space [e.g., amphetamine and methamphetamine (METH)]. DAT

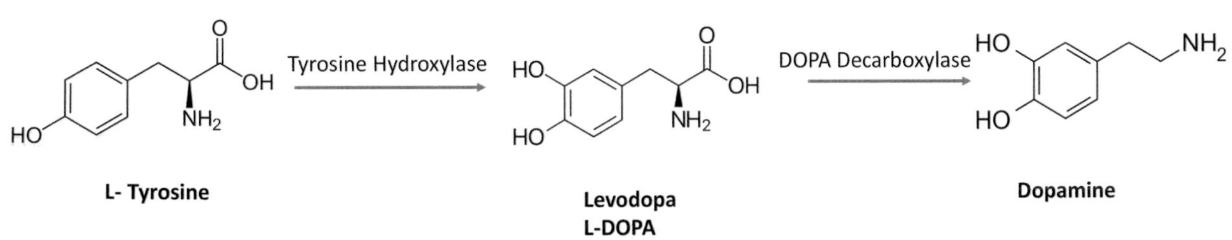

FIGURE 7.1 Dopamine synthesis. L-tyrosine is the precursor for dopamine synthesis. Tyrosine hydroxylase converts L-tyrosine into L-DOPA, which, using the DOPA decarboxylase, results in dopamine (Daubner et al., 2011).

is classified under the SLC6 group of transporters (Vaughan and Foster, 2013). This transmembrane DAT transports the solute in a downhill movement of Na^+ and Cl (sodium chloride-dependent protein) (Manepalli et al., 2012; Pramod et al., 2013). DAT is present on the presynaptic neuron of nigrostriatal cells (Kaufman and Madras, 1991). It is specific for its corresponding neurotransmitter; moreover, it can translocate synthetic or natural structural analogues of DA (Storch et al., 2004).

7.3.1 Regulation of Dopamine Transporter

Three different mechanisms are involved in the control of DAT: post-translational modification; protein–protein interactions; and intracellular localization. These three mechanisms are highly interconnected (German et al., 2015.)

7.3.1.1 Post-Translation Modification

DAT undergoes four types of post-translational modification: phosphorylation; ubiquitination; glycosylation; and palmitoylation. Oxidation and nitrosylation may also regulate DAT. However, these changes mediate DAT–protein interactions, altering DAT transport kinetics by changing its distribution within the plasma membrane. Most post-translational modifications of DAT take place within the N- and C-terminal cytoplasmic domains, and the extracellular sections in the midst of the third and fourth transmembrane domains (German et al., 2015).

7.3.1.1.1 Phosphorylation

Phosphorylation is the most well characterized mechanism, but the functional side is still not fully understood in the context of post-translational modifications of DAT; its N-terminus has many residues of threonine and serine that may undergo phosphorylation by the action of several enzymes, including protein kinases C, A, and G, and Ca^{2+}/calmodulin protein kinase (Gorentla et al., 2009).

7.3.1.1.2 Ubiquitination

Ubiquitination of DAT's N-terminus will determine its fate; if the internalization is temporary, reuptake by the membrane will occur, but if the internalization is more permanent, DAT will degrade. (German et al., 2015).

7.3.1.1.3 Glycosylation

Glycosylation takes place in the extracellular domain, which is located between the third and fourth transmembrane domains, and involves 'asparagines' (Vaughan and Kuhar, 1996); however, the glycosylation of DAT seems to stabilize its localization to the plasma membrane and thus more DA is transported. (Vaughan and Kuhar, 1996; Li et al., 2004).

7.3.1.1.4 Nitrosylation

Nitrosylation is mediated via nitric oxide and can control DAT activity (Volz and Schenk, 2004). The function of DAT is decreased when it is exposed to reactive oxygen species (Berman et al., 1996; Fleckenstein et al., 1997), which results in DAT complexes with higher molecular weights, and is considered to contribute to METH-induced dopaminergic deficits (Baucum et al., 2004).

7.3.1.2 Protein–Protein Interactions

In order for the DAT to be localized to the plasma membrane domains, it needs to interact with several synaptic scaffolding proteins (Eriksen et al., 2010). The PDZ domain protein, which interacts with C kinase 1 has been found to bind the transporter's C-terminal in both vitro and vivo; thus, it is thought to have a role in the regulation of plasma protein expression, despite having divergent functions (Torres et al., 2001; Bjerggaard et al., 2004).

7.3.1.3 Transporter Localization

When DA is expressed at the plasma membrane it can be transported. The function of DAT is based on two aspects of localization; one is the localization of DAT within the plasma membrane lipids; however, protein interactions and relative mobility on the membrane is dictated by the localization within the lipid rafts. The other is the distribution of DAT between the cell membrane and intracellular membrane-bound structures (German et al., 2015).

7.3.2 Regulation of Vesicular Monoamine Transporter-2

As with DAT, vesicular monoamine transporter-2 (VMAT2) is also controlled by three mechanisms. It was assumed that VMAT2 follows the stages of the synaptic vesicle cycle. Vesicles are translocated to the plasma membrane, where the release of neurotransmitter into the synapse takes place, after which there is reformation of vesicles via internalization. Then, through membrane-integrated transporters, the neurotransmitter is transported back to these vesicles from the cytoplasm. (Bjerggaard et al., 2004). Internalization of VMAT2 and reformation of vesicles occurs at a faster rate in the midbrain, where the separation into different terminal and axonal sites takes place. Compared to other vesicles associated with the vesicular glutamate transporter, VMAT2 is released at a faster rate (Onoa et al., 2010).

7.4 DOPAMINE RECEPTORS

DA binds to its receptors in order to modify motor and non-motor functions. The DA receptors are seven transmembrane G protein-coupled receptors (GPCRs). The presence of DA receptors in the brain was proved through experiments that stimulated the activity of adenylyl cyclase using DA in 1972. DA receptors are classified into two distinguished types as per pharmacologic and biochemical evidence that revealed the presence of multiple binding sites for DA. The first type of receptors is positively coupled to adenylyl cyclase, whereas the second type is independent of the adenosine 3′,5′-cyclic monophosphate (cAMP)-generating system.

TABLE 7.1
Locations and Functions of Dopamine Receptors

Dopaminergic Pathways	Function
Nigrostriatal pathway	Involved in movement and sensory stimulus
Mesolimbic pathway	Involved in reward-seeking behavior, pleasure, emotion, and perception
Mesocortical pathway	Cognition, memory, attention, and emotional behavior, in addition to learning
Tuberoinfundibular pathway	Control of the hypothalamic pituitary endocrine system, and inhibiting the secretion of prolactin

Thus, DA receptors are categorized into D1-like and D2-like receptors based upon their ability to regulate cAMP and their pharmacology (Snyder, 2011; Magrinelli et al., 2016).

These receptors are subdivided into D1, D2, D3, D4, and D5 receptors. D1-like receptors (the D1 and D5 receptors) exhibit a similarity in structure and sensitivity to drugs. D2, D3, and D4 are D2-like receptors. The prevalence of D1 receptors (D1R) in the brain is more than the other receptors, and is followed by D2 (D2R), D3, D5, and D4 receptors. D1Rs are the most abundant in the striatum of the brain and they have various roles such as regulating motor functions (Groenewegen, 2003; Benazzouz et al., 2000). However, D1Rs are involved in postsynaptic inhibition, while D2-like receptors are involved in pre- and postsynaptic inhibition. Activation of D1Rs can cause an excitation or inhibition effect (the opening of sodium channels leads to exciting effect, while the opening of potassium channels can lead to an inhibitory effect). Through the activation of the G protein, the D1R works on stimulating the activity of adenylyl cyclase and thus cAMP is produced as a second messenger; the activation of D2Rs is not related to adenylate cyclase (Ayano, 2016; Santens et al., 2003). In contrast, the activation of D2-like receptors can cause an inhibitory postsynaptic effect; whether the effect is inhibition or stimulation depends basically upon the type of receptors present on the postsynaptic neuron (Ayano, 2016).

Table 7.1 illustrates the locations of the subtypes of DA receptors and their main function in the central nervous system and other body parts (Ayano, 2016; Santens et al., 2003). The D1Rs modulate intracellular calcium ions by activating phospholipase C and stimulating the release of intracellular calcium ions (Missale et al., 1998). The D2-like receptors act by inhibiting calcium channels and activating potassium ion channels, which leads to an alteration in membrane potential and hyperpolarization. Binding to D2Rs causes an inhibitory effect. Why the binding of DA to D1Rs causes a stimulation of the direct pathway will be explained later in this chapter (Lester et al., 2010).

7.4.1 D1 Receptors

D1Rs are expressed in the striatum, nucleus accumbens, substantia nigra pars reticulate, and olfactory bulb in higher concentrations; in moderate levels in the entopeduncular nucleus; and in the dorsolateral prefrontal cortex, cingulate cortex, and hippocampus in lower concentrations (Missale et al., 1998). The D1R family (D1 and D5) has two effects: an activating effect via the opening of Na^+ channels, or an inhibitory effect via the opening of K^+ channels. However, activating D2-like receptors causes an inhibitory effect in the target neuron. The type of effect produced (wither a stimulatory or an inhibitory effect) depends upon the DA receptor type present on the neuron's membrane and the neuron's internal response to the second messenger, cAMP. When D1Rs are activated, cyclic AMP is formed. Activating D2Rs yields the opposite effect. D1Rs are highly concentrated in the hippocampus, striatum, nucleus accumbens, hypothalamus, substantia nigra pars reticulata, and olfactory tubercle, in addition to the frontal and temporal cortex. D1Rs and D5R are classified under the same family (D1-like receptors); these receptors possess a high structural similarity (up to a 50% degree of homology). The main unique difference between these two types is that DA has a greater binding affinity for the D5R than the D1R (Ayano, 2016).

7.4.2 D2 Receptors

D2Rs are the second major abundant receptors after D1Rs; the remaining receptors are found in lesser concentrations (Ayano, 2016). D2Rs are GPCRs made up of seven transmembrane domains, where the extracellular part is made by the amino (N) terminal and the intracellular one is formed by the carboxyl terminal. Compared to D1-like receptors, the carboxyl terminal is shorter (Hung and Schwarzschild, 2014). However, those receptors are coupled with G proteins, which consist of three protein subunits (α,β, and γ); the longer intracellular loop is usually the one that interacts with the G protein. Basically, D2Rs are coupled to $G_{\alpha i}$ and $G_{\alpha 0}$ proteins. Guanosine diphosphate is substituted by guanosine triphosphate upon ligand binding. and the α-subunit is detached from the βγ complex. Thus, the α-subunit and the βγ complex are now capable of conducting a signal and stimulating other pathways. Both subgroups ($G_{\alpha i}$ and $G_{\alpha 0}$) work on decreasing the synthesis of cAMP by inhibiting adenyl cyclase, resulting in decreased protein kinase A activity. D2Rs have two subtypes, which are distributed in various patterns. A high concentration of D2Rs is found in the striatum (caudate and putamen), basal ganglia, nucleus accumbens, VTA, and substantia nigra, and are present in lower concentrations in the septal region, amygdala, hippocampus, thalamus, and cerebral cortex (Cazorla et al., 2015).

7.4.3 D3 Receptors

The D3 receptors (D3R) are similar in their structure and pharmacology to the D2Rs. These receptors encode 446 amino acids and have five introns. This type of receptor is basically found in the subcortical limbic regions (nucleus accumbens septic and olfactory tubercle) and present in low concentrations in the basal ganglia. D3Rs are capable of inhibiting the activity of adenylyl cyclase due to their structural similarity to D2Rs. However, studies indicate that D3Rs might facilitate a positive regulatory influence of DA on neurotensin peptide production (Fornai et al., 2007). D3Rs are distinguishable from other receptors in the D2-like family due to their continuous high affinity for DA binding (D3R affinity is <20 times higher than that of D2Rs), suggesting that endogenous DA, upon binding to D3R for long periods, may result in high spontaneous D3R activation. Alterations in the abundance or function of D3R may cause intense effects on synaptic transmission; thus, it is assumed that D3Rs could be critical modulators of dopaminergic function (Richtand et al., 2001). Tonic levels of DA can stimulate D3Rs, unlike D2Rs, as a result of their strong affinity for DA (Sokoloff et al., 1990).

7.4.4 D4 Receptors

D4 receptors (D4Rs) have 41% homology with D2Rs and 39% with D3Rs. However, D1 and D5 receptors exhibit overall 50% homology, which is higher than that of D2 and D4 receptors (Ayano, 2016). The distribution of D3Rs and D4Rs is restricted to the limbic areas. D4Rs are expressed in the hippocampus and frontal cerebral cortex. Stimulating the D4Rs inhibits the activity of adenylyl cyclase and stimulates the release of arachidonic acid in brain neurons, just like the D2Rs (Fornai et al., 2007). D4Rs are located in pre- and postsynaptic neurons; they are postsynaptic in the terminals of glutamatergic neurons, and presynaptic in the dendrites of the GABAergic efferent neurons (Woods, 2010).

7.4.5 D5 Receptors

D5 receptors (D5Rs) encode 477 amino acids and they have no introns ('intron-less'). As mentioned previously, D5Rs have 50% homology with D1Rs and, if we consider the seven transmembrane domains, the homology reaches 80%. Like D1Rs, the interaction between the D5$s and G proteins is capable of stimulating the activity of adenylyl cyclase (Fornai et al., 2007). Pharmacologic and physiologic studies are not able to recognize the individual role of D1Rs and D5Rs due to absence of definite agonists and antagonists. One main difference is that D1R subtypes show lesser affinity for DA than D5Rs, as they display higher affinity; thus, it suggested that D5 subtypes must have more effective dopaminergic functions (Khan et al., 2000). The blockage of D5Rs alone suppresses reproductive behavior associated with D1 agonist stimulation (non-selective for D1Rs and D5Rs) (Dziewczapolski et al., 1998).

7.5 METABOLISM OF DOPAMINE

When DA is released into the synaptic cleft, it can be bound to its receptors in the next neuron, or subjected to reuptake to be recycled later or degraded via two pathways to its inactive metabolites. DA undergoes O-methylation and/or oxidative deamination by the action of MAO. DA is degraded to produce 3,4-dihydroxyphenylacetaldehyde (DOPAL) and hydrogen peroxide. DOPAL can be further reduced to yield 3,4-dihydroxyphenylethanol, or oxidized by alcohol dehydrogenase to 3,4-dihydroxyphenylacetic acid (DOPAC; German et al., 2015).

In general, MAO has two isoforms (isozymes): MAO-A and MAO-B. They have 70% similarity in their primary structural identity. MAO-A is found in catecholaminergic neurons and their axons, and it is sensitive to inhibition by clorgyline. MAO-B is inhibited by deprenyl and is expressed in serotonergic and histaminergic neurons, and in astrocytes (Mishra et al., 2018). MAO-A is more sensitive and has a higher affinity for serotonin, DA, epinephrine, and norepinephrine. Thus, it maintains lower concentrations of DA in the cytosol. MAO-B has high affinity for tyramine, 1-methyl-4-phenyl-1,2,3,6-tetrahydropyridine (MPTP), and phenyl ethylamine (Hung and Schwarzschild, 2014).

Moreover, DA degrades to produce 3-methoxytyramine (3-MT) by the action of catechol-O-methyltransferase (COMT), and 3-MT is further reduced to homovanillic acid (HVA) by MAO (Ayano, 2016). The metabolism of DA takes place in glial cells and in the cytoplasm of the nerve terminal (Janhunen et al., 2005). The level of DOPAC, HVA, and 3-MT present in the brain or cerebrospinal fluid may indicate the functional activity of dopaminergic neurons; in other words, their production is increased due to an increased metabolism of DA or enhanced dopaminergic neuron stimulation (Cooper et al., 2003). COMT is present in two forms – soluble and membrane-bound form – in glial cells and postsynaptic neurons (Rivett et al., 1983). Since COMT degrades DA, COMT inhibitors prevent degradation and yield higher concentrations of DA in the synaptic gap, extending the action on DA receptors and prolonging their activation. In patients with PD, the dopaminergic function can be enhanced via using COMT inhibitors along with levodopa (Janhunen S, 2005).

7.6 THE DOPAMINERGIC PATHWAYS IN THE BRAIN

The dopaminergic system includes four major pathways: the nigrostriatal pathway, the mesolimbic pathway, the mesocortical pathway, and the tuberoinfundibular pathway (Marsden, 2006; Ayano, 2016). In the nigrostriatal pathway, dopaminergic fibers originate from the SNpc into the basal nuclei, where DA plays a critical role in regulating voluntary motor movement. However, the neurodegeneration of DA-producing neurons in this pathway is the hallmark of PD. Table 7.2 illustrates the different DA pathways in the brain and their functions (Groenewegen, 2003; Tarazi, 2001; Lester et al., 2010).

TABLE 7.2
Dopaminergic Pathways and Their Functions

Receptor Type	Location	Function
D1	Present in high concentrations in the striatum, substantia nigra, olfactory bulb, nucleus accumbens, hypothalamus, and frontal cortex	Regulate motor function, including voluntary movements, and developmental processes, control impulse, and involved in emotional process (Ayan, 2016; Santens et al., 2003).
D2	Found in high concentrations in the striatum, ventral tegmental area, olfactory bulb, substantia nigra, and cerebral cortex	Involved in locomotion, working memory, and learning, in addition to blood pressure regulation (Ayan, 2016; Santens et al., 2003)
D3	Present mainly in the CNS in the olfactory bulb, cerebral cortex, striatum, islands of Calleja, and nucleus accumbens	Involved in modulating the endocrine functions, regulating locomotion, food intake, and cognition (Ayan, 2016; Santens et al., 2003)
D4	Substantia nigra, frontal cortex, thalamus, hypothalamus, amygdala, and nucleus accumbens	Control impulse, modulate cognitive functions, and regulate functions of kidney (Ayan, 2016; Santens et al., 2003)
D5	Substantia nigra, hippocampus, hypothalamus, and cortex	Involved in cognition, pain processes, attention, and endocrinal functions of dopamine (Ayan, 2016; Santens et al., 2003)

(Groenewegen, 2005; Tarazi, 2000; Lester et al., 2010).

7.6.1 Nigrostriatal Pathway

In this pathway, dopaminergic neurons originate from the pars compacta region of the substantia nigra in the midbrain and project into the dorsal striatum (Grace et al., 2010; Snyder, 2011). In the striatum, both D1 and D2 dopaminergic receptors are found, where DA binding to the D1R results in a stimulatory effect in the postsynaptic neuron, while binding the latter leads to an inhibitory effect (Marsden, 2006). However, D2Rs are present on the pre- and postsynapse, while the D1Rs are present only on the postsynapse.

These two receptors work as tuners that modulate the non-dopaminergic pathway functions (the direct and the indirect pathways), which are involved in initiating and inhibiting motor movements (Marsden, 2006; Benazzouz et al., 2000). A small part of the nigrostriatal pathway is made up of cell bodies axons in the A8 area (German et al., 2015) that project into the ventral putamen. The axons of the nigrostriatal pathway are present along with noradrenaline and 5-HT (i.e., serotonin) containing fibers throughout the medial forebrain bundle and internal capsule to the striatum. This pathway incorporates 75% of the total DA in the brain, which is responsible for controlling motor behavior, habits, and posture (Janhunen, 2005).

7.6.2 Mesolimbic Pathway

In this pathway, the dopaminergic neurons originate in the VTA and project into the limbic areas of the brain (the amygdala, pyriform cortex, lateral septal nuclei, and ventral striatum) (Marsden, 2006; Ayano, 2016). The mesolimbic pathway mediates happiness in the brain as DA is released during pleasurable conditions; thus, it functions in the emotion and reward system, and encourages a person to seek pleasurable activities, such as drugs, food, and sex, which stimulate the release of DA in specific areas, including the prefrontal cortex and the nucleus accumbens. It also has a role in schizophrenia as the antipsychotic drugs used to decrease the positive symptoms of schizophrenia work as DA receptor blockers in this pathway (Ayano, 2016). DA in the mesolimbic pathway is involved in addiction to many drugs (e.g., nicotine, opiates, psychostimulants, and cannabinoids), and also in human romantic relationships. Using the blood oxygen level-dependent method in functional magnetic resonance imaging, different brain regions of people who are in love have been observed while watching random pictures on a scanner (Acevedo et al., 2012; Aron et al., 2005; Bartels and Zeki, 2000; Xu et al., 2011). The images included facial photographs of the loved ones of the participants and another person, who was the same age and sex as the loved person. The studies revealed that the brain areas rich in DA and the mesolimbic dopaminergic pathway were more active when the participants were watching their loved one's images than they were when watching images of the other, similar people (De Boer et al., 2012; Diamond and Dickenson, 2012; Hatfield and Rapson, 2009; Ortigue et al., 2010; Zeki, 2007). The most stable activation was observed in the caudate and VTA (Acevedo et al., 2012; Bartels and Zeki, 2000; Xu et al., 2011; Ortigue et al., 2007).

7.6.3 Mesocortical Pathway

In the mesocortical pathway, the dopaminergic neurons originate in the VTA region and project mainly to the frontal cortex and septo-hippocampal regions. The hippocampal region is a part of the limbic cortex, which is involved in forming new memories via recollecting daily events. However, these dopaminergic neurons promote learning, memory, attention, and the cognitive emotional behavior (Ayano, 2016; Galvan et al., 2006). Due to the overlapping of the mesocortical and mesolimbic pathways, they are known as the mesocorticolimbic system (Wise, 2005). The mesolimbic pathway differs from the mesocortical pathway in a few functional properties, including physiologic activity and firing mode. While the mesolimbic pathway is involved in controlling motor behavior, emotion, and motivation; the mesocortical is involved in higher cognitive functions such as learning and rewarding behavior (Janhunen, 2005). Dopaminergic neurons account for 30%–40% of the total mesofrontal pathway neurons, unlike the nigrostriatal pathway which is almost entirely composed of dopaminergic neurons (Swanson, 1982).The release of DA in this pathway can be upregulated by some medications (e.g., amphetamines); thus, both cognition and activity in the prefrontal cortex are increased (Prakash et al., 2014).

7.6.4 Tuberoinfundibular Pathway

The fourth pathway is the tuberoinfundibular pathway in which the dopaminergic fibers originate in the hypothalamic arcuate nucleus (arcuate and paraventricular nuclei) and project to the median eminence of the hypothalamus and pituitary gland (Marsden, 2006). The anterior part of the pituitary gland is involved in prolactin secretion and, as DA is known to be the main neuroendocrine inhibitor, it inhibits the release of prolactin. Thus, blockers of dopaminergic receptors, such as antipsychotic drugs, inhibit DA and facilitate the release of prolactin (Ayano, 2016). Therefore, DA is termed prolactin-inhibiting factor or prolactin-inhibiting hormone, or prolactostatin (Ayano, 2016). The secretion of prolactin is monitored by a short-loop feedback mechanism, where the activity of the tuberoinfundibular system is increased due to the increase in prolactin levels (Moore, 1987). However, the D2Rs are found in this pathway (Marsden, 2006).

7.7 THE BASIC STRUCTURES AND PROJECTIONS OF BASAL GANGLIA

Basal ganglia/nuclei are a group of structures present deep within the brainstem. Structures forming the basal nuclei include the caudate, putamen, globus pallidus (internus [GPi] and externus [GPe]), substantia nigra, and subthalamic nucleus; the caudate and putamen together are known as striatum, and both the globus pallidus and putamen are known as lentiform nuclei. However, there are two substantia nigra, one on each side of mid-line of the mid-brain. Each substantia nigra is divided into two parts, one is the SNpc and the other is substantia nigra pars reticulata (SNpr). The dark color of neuromelanin pigment in the SNpc refers to the large number of DA neurons. Thus, most of the DA-producing neurons are present in the SNpc; however, most of the GABA neurons are present in the SNpr area. The basal nuclei have an important role in coordinating body movement, and are involved in processing emotional-, motivational-, and cognitive-related information. Hence, if dysfunction occurs in this part, it will result in a number of disorders. Separated structures form the basal ganglia, but these parts are interrelated and connected via neurons between which various neurotransmitters are transferred. The main input region of the basal nuclei is the striatum. It receives signals from three major areas: the cerebral motor cortex, the thalamus, and the brainstem (Segura-Aguilar et al., 2014)

7.8 THE EFFECT OF DOPAMINE ON THE BASAL GANGLIA CIRCUITRY

D1Rs and D2Rs are located in the striatum, which is the primary input region, and, via the action of other neurotransmitters, striatal neurons are either inhibited or stimulated. DA regulates the action of those neurotransmitters by acting directly on their pathways in the basal nuclei; however, PD occurs when the dopaminergic pathways that innervate the direct and the indirect pathways of the basal nuclei are degenerated. How dopaminergic loss results in the motor symptoms of PD will be discussed later.

Dopaminergic striatal neurons in the direct pathway act on D1Rs, while those in the indirect pathway act on D2Rs. When D1Rs are activated they enhance the activity of adenylate cyclase, which, in turn, stimulates the release of γ-aminobutyric acid (GABA) and substance P from the medium spiny neurons. However, the activation of D2Rs inhibits the activity of adenylate cyclase, which further inhibits the release of GABA. The direct and indirect pathways have opposite actions. However, activating the direct pathway will result in stimulation of the motor cortex, and inhibiting the direct pathway will cause the motor cortex to be released from inhibition 'stimulation' as the pathways carry out opposite actions but yield the same effect (Frank, 2005). Figure 7.2 illustrates the direct and indirect non-dopaminergic pathways.

7.9 NON-DOPAMINERGIC DIRECT NEUROTRANSMITTER PATHWAYS IN BASAL NUCLEI

Neurotransmitters act through various pathways in order to achieve their functions; and their role in the basal nuclei is to stimulate or inhibit the body movements through excitatory and inhibitory signals. However, these neurotransmitters act through direct and indirect pathways. The main neurotransmitters in basal nuclei are DA, GABA, glutamate, substance P, and acetylcholine. These neurotransmitters have a crucial role in the pathophysiology of PD. Figure 7.2 illustrates the direct pathway of some neurotransmitters in the basal ganglia. To initiate a physical activity, the motor cortex, including motor and premotor areas, sends stimulatory signals to the striatum. The stimulation of the corticostriatal glutaminergic fibers causes the release of glutamate into the striatum, and, as glutamate is a stimulatory neurotransmitter, by binding to its postsynaptic *N*-methyl-D-aspartate (NMDA) receptors, it enhances the influx of Ca^{+2} ions into the neuron, resulting in an excitatory postsynaptic potential, which, in turn, stimulates the generation of an action potential via striatopallidal inhibitory GABAergic neurons. Stimulation of GABAergic neurons results in the release of GABA in the GPi.

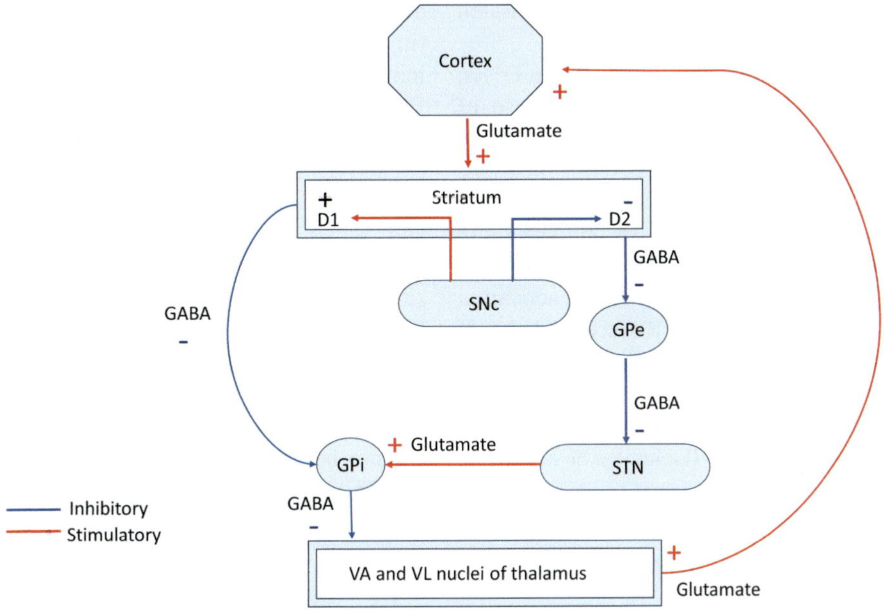

FIGURE 7.2 Direct and indirect non-dopaminergic pathways and their effects on the basal ganglia circuity. GABA: γ-aminobutyric acid; pc: pars compacta; STN: subthalamic nucleus. *SNc: Substantia Nigra pars compacta, STN: Subthalamic Nucleus, VA: Ventro-anterior, VL: Ventro-lateral, GPe: Lateral globus pallidus, GPi: Medial globus pallidus.*

As these neurons release GABA they are an inhibitory neuron; thus, released glutamate stimulates inhibitory GABAergic neurons. However, GABA is known to be the chief inhibitory neurotransmitter in the central nervous system (CNS) and will stimulate the influx of Cl^- ions or the efflux of K^+ ions in the GPi, resulting in an inhibitory postsynaptic potential (IPSP), and will inhibit pallidothalamic GABAergic fibers from releasing GABA in the thalamus. Thus, less production of GABA means fewer IPSPs, or 'releasing the postsynaptic neuron from inhibition', more stimulation of the glutaminergic cerebral motor cortex takes place, and therefore stimulation of skeletal muscle to enable an individual to initiate a movement through the direct stimulatory/excitatory pathway (Jankovic, 2008; Delaville et al., 2011; Hung and Schwarzschild, 2014).

7.10 NON-DOPAMINERGIC INDIRECT NEUROTRANSMITTER PATHWAYS IN BASAL NUCLEI

The indirect pathway is known to be an inhibitory pathway. Both pathways are required to work simultaneously because while some muscles contract, other muscles need to be relaxed and their muscle tone must be maintained (e.g., bending the elbow requires the biceps to contract and the triceps to relax). The indirect pathway starts when the motor cortex stimulates the striatum through the activation of the corticostriatal glutaminergic fibers and causes the release of glutamate into the striatum, which stimulates postsynaptic GABAergic neurons and releases more GABA. Here lies the difference between the two pathways: the neuronal GABAergic fibers extend into the GPe only and not the internus, and thus the release of GABA takes place in the GPe. These GABAergic neurons project from the GPe to the subthalamic nucleus; however, GABA production inhibits these neurons in the subthalamic nucleus, resulting in less GABA production, which means less inhibition (fewer IPSPs; more stimulation) for the next glutaminergic neuron, releasing it from inhibition. Here the glutaminergic neurons extend from the subthalamic nucleus until it goes back into the GPi, and stimulates glutamate release, which, in turn, stimulates pallidothalamic GABAergic fibers and releases more GABA in the thalamus. More GABA means more inhibition; thus, the cerebral cortex is inhibited from initiating a movement (Jankovic, 2008; Delaville et al., 2011; Hung and Schwarzschild, 2014).

7.11 THE EFFECT OF DOPAMINE ON THE DIRECT AND INDIRECT PATHWAYS IN THE BRAIN

PD is characterized by the loss of dopaminergic neurons in SNpc. Normally, the dopaminergic neurons that arise from the SNpc influence both the direct and the indirect pathways; they act as a modulator. The direct pathway contains GABA, glutamate neurotransmitters, and substance P, and expresses D1Rs in the striatum. Those dopaminergic neurons provide an excitatory input to the direct pathway and thus enhance the input from the cortex by enhancing transmission. When DA binds to its D1R in the striatum it overactivates the inhibitory GABAergic neurons, resulting in more production of GABA in the GPi. Increased release of GABA inhibits pallidothalamic neurons and releases less of GABA in the thalamus. When less GABA is produced in the thalamus, thalamocortical neurons are released from inhibition, and their stimulatory effect will increase as compared to the effect produced by the action of glutamate alone. The indirect pathway, which has GABA and glutamate neurotransmitters, expresses D2Rs and results in an inhibitory input to the striatum, which reduces transmission over the indirect pathway.

The effect of DA on D2Rs is antagonistic to that on D1Rs. When DA binds to D2Rs located on the postsynapse of GABAergic dendrites, it will antagonize the effect of glutamate and suppress the inhibitory effect of GABAergic neurons. Reduced release of GABA in the GPe will stimulate (less inhibited) the pallidosubthalamic neurons to produce more of GABA in the subthalamic nucleus, which will inhibit the release of glutamate in the GPi, and will lead to less activation of pallidothalamic GABAergic neurons. Reduced release of GABA in the thalamus will release the thalamocortical fibers from inhibition (stimulatory effect). Thereby, DA activates the direct pathway and inhibits the indirect one, and results in a stimulatory effect on both pathways (Delaville et al., 2011; Magrinelli et al., 2016).

7.12 THE IMBALANCE OF DOPAMINE IN PARKINSON'S DISEASE

In PD, dopaminergic axons which are projected into the striatum gradually disappear or degenerate over many years; thus, the amount of DA inhibiting the indirect pathway and stimulating the direct pathway is gradually diminished. However, the degeneration of dopaminergic neurons projected to the putamen occurs before that of striatal limbic neurons. When neuronal degeneration starts, motor symptoms and signs start to develop; thereafter, non-motor signs occur. Both motor and non-motor signs are detected after significant nigrostriatal neurons have been degenerated; in other words, when 70% of the nigrostriatal neurons are affected (Bernheimer et al., 1973). The loss of DA results in a reduction in medium spiny neurons density in the putamen (Ingham CA, et al., 1989; Villalba et al., 2009; Zaja et al., 2005), which may affect corticostriatal transmission. Studies have recommended that medium spiny neurons with D2Rs are expressed in the indirect pathway could be affected by the loss of the spine (Day et al., 2006). Medium spiny neurons (MSNs) contribute toward many important functions in the brain, which is indicated by the consecutive behavioral relationship between Huntington's disease and PD (Glass et al., 2000).

MSNs can be divided into two groups, one of which expresses D1Rs (known as D1-MSNs), and the other group expresses D2Rs (D2-MSNs). The earlier project into the GPi and SNpr and thus forms the direct pathway; while the latter project into the GPe, forming the indirect pathway (Gerfen and Bolam, 2016).

In the direct pathway, the reduction in the DA-producing neurons will activate GABAergic neurons less, which, in turn, produce less GABA in the GPi. This will stimulate the release of GABA in the thalamus and therefore the thalamocortical neurons are inhibited (less stimulated). In contrast, when DA is not inhibiting the indirect pathway, the glutaminergic effect on the striatopallidal fibers is increased, and the release of GABA is increased in the GPe, inhibiting the release of GABA in the subthalamic nucleus and stimulating the production of glutamate in the GPi. The stimulatory neurotransmitter activates the pallidothalamic neurons to release more GABA (in the thalamus) after all the thalamocortical fibers are inhibited. However, the indirect pathway, which expresses D2Rs, will be overactive and the thalamus will remain inhibited. Therefore, the motor cortex is unable to become excited (Obeso et al., 2000).

However, the depletion of DA can alter the density and sensitivity of dopaminergic receptors. An increase (upregulation) in mRNA expression for D2Rs and binding sites in striatal indirect pathway-containing MSNs has been found in patients with PD, along with a reduction in the expression of D1R mRNA in direct pathway-containing MSNs (Gerfen and Bolam, 2016). It is now clear that the neurodegeneration of striatal DA in PD interrupts the balance between the direct and the indirect pathways, which in turn produce a hypokinetic (parkinsonian) state (Obeso et al., 2000).

7.13 NEUROCHEMISTRY OF OTHER NEUROTRANSMITTERS IN PARKINSON'S DISEASE

7.13.1 Classification of Neurotransmitters

Specific criteria are required to consider a substance as a neurotransmitter: first, localization – a substance must be present within the neuron from which the presynaptic terminal arises; second, release – a substance must be shown to be released from the presynapse on activation of the axon terminal; third, identity – a substance must act on the target cells and produce the same biologic effect. Thus, when neurotransmitters bind to their specific receptors on the postsynapse they initiate an action potential that is described as an electrical change in the membrane potential of a neuron, either causing a stimulation effect in the next neuron or an inhibition effect, depending upon the action on the neurotransmitter. However, neurotransmitters are classified according to their chemical nature, and their effects. (Haines, 2013). Table 7.3 illustrates the main neurotransmitters and neuromodulators that act in the basal ganglia of the brain.

7.13.1.1 GABA

GABA is the major inhibitory neurotransmitter in the CNS; it is an amino acid made from the precursor glutamate (when glutamic acid decarboxylase converts glutamate to GABA) in brain cells. GABA causes an inhibitory effect by binding to its receptors, both on the pre and postsynapse. Receptors of GABA are GABA-**A**, GABA-B, and GABA-C. GABA-A and GABA-C receptors are (ionotropic) ligand-gated ion channel receptors, while GABA-B is (metabotropic) a GPCR. When GABA-A receptors are activated, the opening of Cl^- channels causes an influx of chloride ions into the neuron, resulting in hyperpolarization due to an increase in the negative charge inside the neuron. Thus, the generation of an action potential will be less in the next neuron, leading to an inhibitory effect known as an inhibitory postsynaptic potential (IPSP). In addition to the primary binding sites for GABA-A receptors, it

TABLE 7.3
Classification of Various Neurotransmitters and their Site of Action

Chemical Nature	Neurotransmitters	Action	Site of Action
Amino acids	GABA (γ -amino butyric acid)	Inhibitory	Basal ganglia, cerebral cortex, cerebellum
	Glutamate	Excitatory	Brainstem, cerebral cortex, cerebellum
Neuropeptide	Substance P	Excitatory	Mediate nociception within spinal cord
Esters	Acetylcholine	Excitatory	Basal ganglia, cerebral cortex, pre- and postganglionic sympathetic and parasympathetic nerve endings

(Sheffler et at., 2020; Galvan et al., 2006).

has an allosteric binding site to which other GABA modulatory substances can bind; the binding to this allosteric site may cause an increase or decrease in the action of GABAergic receptors. Thus, they may cause a negative or a positive modulatory effect. GABA-B receptors are present in both the pre- and postsynapse. Those receptors exhibit an auto receptor-like mechanism as they activate K^+ channels and inhibit Ca^{+2} channels; thus, the efflux of K^+ ions from the neuron will make the neuron less positive, which will mediate the inhibition of neurotransmitter release. However, the activation of GABA-B receptors will produce a slower hyperpolarization, termed a slow IPSP (Ayano, 2016; Grace et al., 2010). Hence, it is clearly seen that GABA controls calcium influx by inhibiting calcium channels via GABAergic receptors (Yamakage and Namiki, 2002). During the hyperpolarized phase, neurons will remove Ca^{2+} ions with the help of adenosine triphosphate. Mitochondrial dysfunction leads to the pathogenesis of PD: long-lasting calcium concentrations will occur due to the reduction in calcium-buffering capacity, leading to mitochondrial oxidative stress, which, in turn, is responsible for neuronal loss in the substantia nigra (Hurley et al., 2013; Surmeier and Schumacker, 2013).

7.13.1.2 Glutamate

Glutamate is the main stimulatory neurotransmitter in the CNS. Glutamate is a nonessential amino acid which needs to be synthesized in the neurons as it does not cross the BBB. The prevalent precursor is glutamine, which is released by glial cells and then taken up into the presynaptic terminal. Glutamine is metabolized by the glutaminase into glutamate, which it is then packaged into synaptic vesicles. In the presence of any stimuli, synaptic vesicles release their components into the synaptic junction. Glutamate provides the excitatory signals required to balance the large number of inhibitory signals produced by other neurotransmitters like GABA, DA, and serotonin. Glutamate receptors are mainly of two types: ionotropic (voltage-sensitive) receptors such as AMPA (α-amino-3-hydroxy-5-methyl-4-isoxazole propionate) and NMDA; and metabotropic (ligand-sensitive) and kainate (KA) receptors. Despite the fact that AMPA and NMDA receptors are the primary mediators of glutamatergic transmission in the brain, KA, or kainic acid, receptors also participate in providing neuronal excitability. However, both AMPA and NMDA receptors are stimulated during synaptic transmission. The main difference is that AMPA receptors promote a faster neurotransmission, whereas NMDA receptors mediate slower and a longer-lasting responses. These glutamergic receptors are gathered as homo- or hetero-oligomeric complexes of protein subunits. The AMPA receptor family consists of two subunits GluR1 and GluR4. NMDA receptors consist of NR1, NR2A–NR2D, and NR3A–NR3B subunits, while KA receptors include GluR5–GluR7 and KA1/KA2 subunits (Fox SH et al., 2009). In PD, there will be an imbalanced GABAergic–glutamatergic neurotransmitter system seen as glutamergic hyperactivity which, via NMDA receptors, will be capable of inhibiting the presynaptic D2R, leading to an enhanced DA deficiency (Werner and Coveñas, 2015).

7.13.1.3 Acetylcholine

Acetylcholine (ACh) is a widespread neurotransmitter. It has a large number of interventions in the cerebral cortex, cerebellum, and subcortical structures. ACh has two types of receptors: muscarinic (mAChR) and nicotinic (nAChR). Muscarinic receptors are found primarily in the striatum in high concentrations. Nicotinic receptors are present in the cerebral cortex and have a high binding property in the striatum, thalamus, and SNpc. In fact, cholinergic activity has been found to affect the release of DA and vice versa. Activation of nAChR by ACh controls the action potential on which DA depends to be released. However, vesamicol (ACh uptake inhibitor) is used as a pretreatment, to prevent the release of ACh. Thus, the stimulus-induced DA release decreases by 90%. Cholinergic interneurons modulate the activity of spiny projection neurons, which affects the output of the striatum. These spiny neurons innervate the output nuclei of the basal ganglia, and the synapse between the cholinergic neurons and the spiny neurons acts in a modulatory way. However, when presynaptic muscarinic neurons, which are located on fast-spiking interneurons are stimulated, the release of GABA is decreased. Thus, the spiny projection neurons are no longer inhibited by GABA's inhibitory effect. Muscarinic activity acts on a slower scale, while the nicotinic influence/activity acts rapidly on the striatal output (German et al., 2005).

However, the cholinergic system in the brain can be categorized into two subsystems: the interneuron system (where ACh is provided via its own cholinergic interneurons) and the projection neuronal system (in which ACh is provided through the axons of cholinergic neurons). The cerebral cortex contains the neuronal projection type. Unlike other projecting cholinergic neurons present throughout the brain, striatal cholinergic neurons are interneurons capable of providing very rich innervation at the local level; thus, the striatum contains the interneuron system (intrinsic cholinergic neurons). However, 2% of striatal neurons are cholinergic. (Zhou et al., 2002). In PD, it is reported that there is a 50% loss in the large cholinergic neurons of the lateral part of the pedunculopontine tegmental nucleus pars compacta is seen where the neurons are degenerated, according to neuropathologic studies (Müller and Bohnen, 2013).

7.13.1.4 Serotonin

Serotonin, also known as 5-hydroxytryptamine (5-HT), is synthesized from tryptophan by the aromatic amino acid decarboxylase and tryptophan hydroxylase. It has two types of receptors: serotonin-gated ion channels (5-HT_3 receptors) or GPCRs (5-HT_1, 5-HT_2, and 5-HT_{4-7} receptors). The raphe nuclei in the brainstem contain most of the serotonin neuron bodies. Some serotonergic projections ascend to brain regions such as the cortex, hippocampus, and hypothalamus. Neurons in the dorsal raphe nuclei (DRN) project mainly to the basal ganglia, specifically the striatum. Normally, there are dense serotonergic innervations from these DRN into the basal ganglia. And, as the striatum and other output regions of the basal ganglia (substantia nigra pars reticulata, and globus pallidum), receive a serotonergic input, it is suggested that serotonin has a crucial role in PD.

Serotonin is involved in the regulation of many functions, including mood, attention, appetite, and sleep. 5-HT-producing neurons are affected in PD, which could be the reason for the high incidence of depression in patients with PD. Basically, serotonergic projections of the basal ganglia and frontal cortex originate in the DRN, as the output regions of the basal ganglia (striatum, substantia nigra, and medial globus pallidus) are receiving serotonergic input; thus, the involvement of serotonin in the pathology of PD has been suggested. Studies have revealed the loss of serotonin in the caudate, hypothalamus, and frontal cortex, but at a lower level than DA loss: one pathologic study showed a 66% loss in serotonin and 98% loss in DA (Zhou et al., 2002).

7.13.1.5 Adenosine

Adenosine is a CNS neurotransmitter. Adenosine receptors are categorized into four groups (A1, A2A, A2B, and A3). They are GPCRs and are classified based on their ability to either enhance or suppress the activity of adenylate cyclase. A1 and A2A receptors are responsible for the physiologic effects of adenosine in the striatum of the brain, where they are widely distributed. However, both A1 and A2A receptors are present in the striatum, but A2A is more concentrated. Those A2A receptors are co-localized with D2Rs of the indirect pathway; A1 receptors are found with D1Rs in the direct pathway, particularly in the substantia nigra. A1 receptors are the most plentiful adenosine receptors in the brain and inhibits neurotransmission by blocking adenylyl cyclase activity Thus, adenosine receptors have an essential role in modulating the effect of DA receptors in the basal ganglia. Interactions between A1 receptors and D1R facilitate the release of GABA in substantia nigra and DA in the striatum. This interaction promotes the release of GABA in the substantia nigra and the release of DA in the striatum. However, adenosine A2A receptors and dopamine D2Rs have an antagonist effect; activation of the A2A receptor has the opposite of D2Rs. Thus, adenosine receptors play an important role in modulating the effect of DA in the basal ganglia. Adenosine can modulate the process of neurotransmission by modulating the basal levels of several neurotransmitters, including DA, serotonin, glutamate, and GABA via A1 and A2A receptors. It was found that patients with PD had altered adenosine A2A receptor expression in the basal ganglia. In a postmortem study of patients with PD, the density of A2A receptors was increased up to 2.95-fold in the striatum versus healthy individuals. (Kulisevsky and Poyurovsky, 2012).

7.13.1.6 Noradrenaline

The release of noradrenaline has significant neuromodulator effects, which include a change in membrane potential, stimulation of neurons, plasticity of synapses, and transmission across synapses. The locus coeruleus is the main source of noradrenaline in the CNS. The input signals from various areas affect the activity of the neurons present in the locus coeruleus. The function of the noradrenergic system is to modulate the brainstem, and cortical and subcortical pathways through its widely spread efferent neurons, which release the noradrenaline into the different regions, including the subthalamic nucleus, thalamus, and substantia nigra. However, noradrenaline works by binding to its adrenergic receptors (ARs). There are two types of AR: alpha (α) and beta (β). The α receptor has two subtypes (α1 and α2), and the β receptor has three subtypes (β1, β2, and β3). As these receptors are GPCRs, they are coupled to different G proteins. Usually, the α1 and β postsynaptic ARs promote a stimulatory effect, while α2 presynaptic receptors promote an inhibitory effect. In fact, the ARs are found in the striatum and substantia nigra (Galvan and Wichmann, 2008). The loss of NAergic cells has been related to PD because L-DOPA is the metabolic precursor of noradrenaline and also DA. The loss of NAergic neurons in the locus coeruleus in PD has been noted, despite a poor anatomic connection between the substantia nigra and the locus coeruleus (Ben-Jonathan and Hnasko, 2001; Galvan et al., 2006). It was found in the 1980s that there is a relationship between PD and various types of AR. It was revealed that the density of adrenergic α1 and β1 receptors was elevated, and that of α2 adrenergic receptors is decreased in patients with dementia and patients with PD. Other studies have shown that, through the activation of ARs (α1 and α2) in the subthalamic nucleus in 6-hydroxydopamine (6-OHDA)-lesioned rats, the firing activity of the subthalamic nucleus was noticeably controlled and therefore the locomotor activity was decreased, proving the role of ARs in controlling the activity of the subthalamic nucleus (Rinne et al., 2002; Macphee and Stewart, 2010; McGeer et al., 1984).

7.13.1.7 Histamine

Histamine is widely spread neurotransmitter in the brain. The stomata of the histaminergic neurons (cell bodies) are limited to the tuberomammillary nucleus of the posterior hypothalamus; from these nuclei the histaminergic fibers project to the cerebral cortex and cerebellar cortex. The neurodegeneration of DA-producing neurons in the substantia nigra affects other neurotransmitter systems, including the histaminergic system. There is evidence of an increase in histamine concentrations in the blood in patients with PD, and higher levels of pros-methylimidazoleacetic acid (an isomer of a histamine metabolite called tele-methylimidazoleacetic acid) have been reported. Furthermore, H2 blockers like famotidine have been given to patients with PD, who subsequently showed improved motor function. Despite the fact that plasma histamine levels were altered, the activity of the enzyme responsible for producing histamine (histidine decarboxylase) remained the same in the frontal cortex, caudate nucleus, hypothalamus, and substantia nigra. However, a noticeable increase was observed in histaminergic innervation in the regions of degeneration in PD (Hung and Schwarzschild, 2014).

7.14 PATHOPHYSIOLOGY OF MOTOR SYMPTOMS IN PARKINSON'S DISEASE

7.14.1 Bradykinesia

The neurodegeneration of dopaminergic neurons in PD causes multiple motor symptoms arising from the disturbance in the basal ganglia pathways, leading to disruption in the activity of motor cortex (Missale et al., 1998). These motor symptoms include bradykinesia, hypokinesia, and akinesia, and these symptoms exhibit a disturbance in the voluntary movement primarily associated with slowness. Bradykinesia refers to slowness of movement; hypokinesia refers to repetitive movements, which are less than the desired movement; and akinesia refers to the failure of voluntary spontaneous movement (Magrinelli et al., 2016). Bradykinesia is the most salient clinical feature of PD; however, this symptom is common to other disorders like depression. Neurodegeneration of dopaminergic neurons is the hallmark of PD and bradykinesia is the hallmark of basal ganglia disorders; thus, difficulties in initiating movements can occur. Other features of bradykinesia are the loss of spontaneous movement and excess salivation due to impaired swallowing, loss of facial expression (hypomimia), and reduced blinking (Jankovic, 2008).

7.14.2 Rigidity

Rigidity occurs in patients with PD due to alteration in the activity of the joints, muscles, and tendons, and is characterized by resistance to stretching (Mishra et al., 2018). Rigidity is usually associated with pain. One of the most common manifestations of PD is painful shoulder; it is more prominent in the flexor muscle, and is more easily seen during slow stretching movements. When rigidity coexists with tremor, it is called the 'cogwheel phenomenon' (Magrinelli et al., 2016).

7.14.3 Tremor

Various types of tremor can be seen in patients with PD: resting tremor, action tremor, essential tremor, exaggerated tremor, and dystonic tremor. When the patient shifts from rest to posture, action tremor occurs after few seconds. Rest tremor, essential tremor, and dystonic tremor have the same frequency as action tremor. Rest tremor manifestations are a pill-rolling movement, and/or a yes-yes or no-no motion with the head (Magrinelli et al., 2016).

7.14.4 Motor Fluctuations

Levodopa has long been used as the first choice treatment of PD. However, previously stable responses to treatment can change and the development of motor fluctuations can occur. Various mechanisms can lead to motor fluctuations and dyskinesia. The main mechanisms are: (1) an increase in neurodegeneration of the nigrostriatal pathway containing DA, resulting in a decreased amount of DA in storage vesicles; (2) the stimulated conversion of levodopa into dopamine; (3) a change in dopaminergic receptors undergoing plastic alterations; and (4) elevated glutamergic activity in the striatum. Peripheral mechanisms are (1) competition between levodopa and protein-rich food containing amino acids (as levodopa is a neutral amino acid that requires a transporter to cross the BBB and the gut–blood barrier); and (ii) decreased gastric emptying linked to PD (Magrinelli et al., 2016)

7.14.5 Gait Disturbance and Balance

Other common motor symptoms in PD are posture, balance, and gait disturbances, which significantly contribute to motor impairment and risk of falls. Features seen include flexed hips and knees, and postural deformities. Disorders in balance and gait are significant obstacles for patients with PD. Drugs intended to improve symptoms may not work as expected, and in some cases symptoms may worsen (Magrinelli et al., 2016).

7.14.6 Postural Instability

Postural instability usually happens after the occurrence of other features. This is a late-stage manifestation characterized by the loss of postural reflexes. Postural instability associated with freezing gait is the primary reason for falls, which could lead to bone fractures. Other factors affecting postural instability include orthostatic hypotension, age, and related sensory alterations. Patients who are afraid of falling may lose their balance and fall. However, one study showed that the frequency of falls is proportionally increased with disease severity (Mishra et al., 2018).

7.15 STAGES OF PARKINSON'S DISEASE

Stage I: The mildest stage of PD is stage I. In this stage, the symptoms are unknown and unusual, and they are minor in severity. However, these symptoms do not affect the patient's lifestyle or quality of life. The person may not require medical care, and the physician is unable to give a definite diagnosis. There may be a wait to check whether the symptoms worsen until a formal diagnosis is given (Richtand et al., 2001). In stage I, tremors and other motor disturbances generally occur on one side of the body (unilateral) and can be improved with prescribed medications (Bhardwaj and Deshmukh, 2018). Other symptoms are intermittent tremor of one hand, rigidity, and one hand or leg may feel clumsier than the other. (Richtand et al., 2001).

Stage II: The moderate stage of PD is stage II, this stage is considered as an early disease state. In stage II, symptoms are more obvious and experienced than stage I. Tremors, rigidity, and difficulties in movements are increased in comparison to Stage I. Stage II is also associated with changes in the patient facial expressions. The time taken to complete tasks is prolonged due to muscle stiffness; stage II does not affect balance. In this stage motor disabilities are bilateral – both body sides are affected (Bhardwaj and Deshmukh, 2018). The symptoms are accompanied by speech problems, including slurring and a fading volume after starting to speak loudly. Muscle stiffness in the truck result in backache or neck pain, and general slowness in performing daily life activities, and a stooped posture (Richtand et al., 2001). The PD may be misinterpreted as the normal aging process if stage II symptoms are only slowness or a lack of normal movement. The progression from stage I to stage II may take months or years (Bhardwaj and Deshmukh, 2018).

Stage III: This stage is the middle stage of the disease, and is the main turning point in the progression of PD. Both signs and symptoms are similar to those seen in stage II, with the exception of the loss of balance and reduced reflexes, which can also occur. In this stage, daily tasks and the quality of life of the patient can be influenced by the progression of symptoms. Patients will require more time to achieve a task. A combination of both occupational therapy and prescribed medication may be useful in reducing the symptoms. Nowadays, artificial intelligence can help to improve the quality of care of patients with PD (Bhardwaj and Deshmukh, 2018). In this stage, balance is compromised by lack of initiation of the rapid, automatic, and involuntary adjustments required to prevent a fall, and therefore falls are common. The diagnosis is made by the physician who checks for impairments in reflexes. The physician stands behind the patient and slightly pulls the patient backward to determine whether the patient falls backward due to an inability to maintain balance (the patient will not fall as the physician supports the patient from behind). One important feature of stage III is that the patient is still independent and is capable of carrying out activities of daily living (e.g. dressing, hygiene, and eating; Prakash et al., 2014).

Stage IV: In this stage of PD, the remarkable reduction in motor movement and other cognitive abnormalities make it difficult for the patient to survive; thus, they will need assistance or a walker to carry out tasks. For a patient who is in stage IV or V it is impossible to carry out basic daily tasks, which increases the risk of dangerous events, especially if the patient lives alone (Bhardwaj and Deshmukh, 2018).

Stage V: This is a very severe stage and the overall strength of the patient PD is poor. Patients need wheelchairs or assisters due to the risk of falling when standing, because of severe stiffness in legs. PD symptoms may worsen, and hallucinations and occasional delusions may be experience. Side effects arising from medications may increase versus any benefits (Bhardwaj and Deshmukh, 2018). Over time, symptoms will worsen, meaning that some patients may not reach stage V, as the duration of disease progression varies from an individual to another. Different treatments are given to patients with PD depending on the disease stage; however, the earlier the diagnosis of the stage, the more efficient the treatment is in improving symptoms (Prakash et al., 2014).

7.16 RECENT DEVELOPMENTS AND CHALLENGES IN THE TREATMENT OF PARKINSON'S DISEASE

Levodopa has been the golden agent since it was developed in the 1960s to normalize motor functions and it proved to be the most effective drug in the treatment of PD. However, it is frequently associated with cognitive problems, dystonia, levodopa-induced dyskinesia, and on-off fluctuations. (Stoker and Barker, 2020). Hence, the need for newer developments and treatments is crucial.

Alpha-synuclein is a key protein involved in the pathology of PD. One experimental method proposed the use of antibodies to limit the propagation of alpha-synuclein; therefore, targeting and degrading extracellular alpha-synuclein and hindering it from infecting adjacent cells may be an option.

Another approach is the humanized monoclonal antibody, where the C-terminal of alpha-synuclein (prasinezumab or PRX002; Prothena) is targeted and results in a 97% reduction in free alpha-synuclein in serum, has been found to be well-tolerated in phase I trials. Immunization techniques – both active and passive – have shown neuroprotective benefits in animal models. Some preclinical approaches aimed at reducing the synthesis of alpha-synuclein also use antisense oligonucleotide and RNA interference techniques.

Targeting the non-dopaminergic neurotransmitter system in PD treatment is a new avenue. Safinamide is a recently approved novel drug used in PD that has multimodal actions; it is a potent MAO-B inhibitor and glutamate transmission modulator. Safinamide reduced and improved depression in PD, and is now being used as an adjunct therapy to levodopa-based treatment in cases where dyskinesia and fluctuations are experienced (Stoker and Barker, 2020)

In addition to the abovementioned pharmacologic treatments, regenerative medicines are gaining more interest. Gene therapies are introduced where genes producing DA are synthesized. Two gene therapies include the genes encoding both tyrosine hydroxylase and DOPA decarboxylase (Stoker and Barker, 2020)

7.17 CONCLUSION

PD is a lifelong progressive neurodegenerative disorder and this chapter has covered the neurochemistry, synthesis, metabolism, and receptors of many neurotransmitters involved in the pathophysiology of PD, including DA, serotonin, noradrenaline, GABA, glutamate, histamine, adenosine, and ACh, and also the dopaminergic and non-dopaminergic direct and indirect pathways in the basal ganglia of the brain. The progression of PD goes through various stages, where motor symptoms may worsen with time. Levodopa has long been recognized as the golden treatment of PD, but the emergence of levodopa-induced dyskinesia and other side effects associated with long-term use has led to the search for alternatives such as immunization techniques, targeting non-dopaminergic systems via novel drugs like safinamide and regenerative medicines, including cell and gene therapy.

REFERENCES

Acevedo BP, Aron A, Fisher HE, Brown LL. Neural correlates of long-term intense romantic love. Social Cognitive and Affective Neuroscience. 2012;7:145–59.

Andén NE, Grabowska-Andén M, Lindgren S, Thornström U. Synthesis rate of dopamine: difference between corpus striatum and limbic system as a possible explanation of variations in reactions to drugs. Naunyn-Schmiedeberg's Archives of Pharmacology. 1983;323:193–8.

Aron A, Fisher H, Mashek DJ, Strong G, Li H, Brown LL. Reward, motivation, and emotion systems associated with early-stage intense romantic love. Journal of Neurophysiology. 2005;94:327–37.

Ayano G. Dopamine: receptors, functions, synthesis, pathways, locations and mental disorders: review of literatures. Journal of Mental Disorders and Treatment. 2016;2:2.

Bartels A, Zeki S. The neural basis of romantic love. Neuroreport. 2000;11:3829–34.

Baucum AJ, Rau KS, Riddle EL, Hanson GR, Fleckenstein AE. Methamphetamine increases dopamine transporter higher molecular weight complex formation via a dopamine-and hyperthermia-associated mechanism. Journal of Neuroscience. 2004;24:3436–43.

Benazzouz A, Piallat B, Ni ZG, Koudsie A, Pollak P, Benabid AL. Implication of the subthalamic nucleus in the pathophysiology and pathogenesis of Parkinson's disease. Cell Transplantation. 2000;9:215–21.

Ben-Jonathan N, Hnasko R. Dopamine as a prolactin (PRL) inhibitor. Endocrine Reviews 2001;22:724–63.

Bergh C, Eklund T, Sodersten P, Nordin C. Altered dopamine function in pathological gambling. Neuropsychobiology. 1993;28:30–6.

Berman SB, Zigmond MJ, Hastings TG. Modification of dopamine transporter function: effect of reactive oxygen species and dopamine. Journal of Neurochemistry. 1996;67:593–600.

Bernheimer H, Birkmayer W, Hornykiewicz O, Jellinger K, Seitelberger F. Brain dopamine and the syndromes of Parkinson and Huntington clinical, morphological and neurochemical correlations. Journal of the Neurological Sciences. 1973;20:415–55.

Bhardwaj R, Deshmukh R. Parkinson's disease: an insight into mechanisms and model systems. International Journal of Medical Research & Health Sciences. 2018;7:38–51.

Bjerggaard C, Fog JU, Hastrup H, Madsen K, Loland CJ, Javitch JA, Gether U. Surface targeting of the dopamine transporter involves discrete epitopes in the distal C terminus but does not require canonical PDZ domain interactions. Journal of Neuroscience. 2004;24:7024–36.

Brizer DA. Psychopharmacology and the management of violent patients. Psychiatric Clinics of North America. 1988;11:551–68.

Cazorla M, Kang UJ, Kellendonk C. Balancing the basal ganglia circuitry: a possible new role for dopamine D2 receptors in health and disease. Movement Disorders. 2015;30: 895–903.

Cooper JR, Bloom FE, Roth RH. The Biochemical Basis of Neuropharmacology. Oxford University Press, 2003.

Daubner SC, Le T, Wang S. Tyrosine hydroxylase and regulation of dopamine synthesis. Archives of Biochemistry and Biophysics. 2011;508:1–2.

Day M, Wang Z, Ding J, An X, Ingham CA, Shering AF, et al. Selective elimination of glutamatergic synapses on striatopallidal neurons in Parkinson disease models. Nature Neuroscience. 2006;9:251–9.

De Backer D, Biston P, Devriendt J, Madl C, Chochrad D, Aldecoa C, et al., SOAP II Investigators. Comparison of dopamine and norepinephrine in the treatment of shock. New England Journal of Medicine. 2010;362:779–89.

De Boer A, Van Buel EM, Ter Horst GJ. Love is more than just a kiss: a neurobiological perspective on love and affection. Neuroscience. 2012;201:114–24.

Delaville C, De Deurwaerdère P, Benazzouz A. Noradrenaline and Parkinson's disease. Frontiers in Systems Neuroscience. 2011; 5:31.

Diamond LM, Dickenson JA. The neuroimaging of love and desire: review and future directions. Clinical Neuropsychiatry. 2012;9.

Dziewczapolski G, Menalled LB, García MC, Mora MA, Gershanik OS, Rubinstein M. Opposite roles of D1 and D5 dopamine receptors in locomotion revealed by selective antisense oligonucleotides. Neuroreport. 1998;9:1–5.

Eriksen J, Jørgensen TN, Gether U. Regulation of dopamine transporter function by protein-protein interactions: new discoveries and methodological challenges. Journal of Neurochemistry. 2010;113:27–41.

Everitt BJ, Robbins TW. Second-order schedules of drug reinforcement in rats and monkeys: measurement of reinforcing efficacy and drug-seeking behavior. Psychopharmacology 2000;153:12–30.

Feldman RS, Meyer JS, Quenzer LF. Principles in Neuropsychopharmacology. Sinauer Associates, 1997.

Fleckenstein AE, Metzger RR, Beyeler ML, Gibb JW, Hanson GR. Oxygen radicals diminish dopamine transporter function in rat striatum. European Journal of Pharmacology. 1997;334:111–14.

Fornai F, di Poggio AB, Pellegrini A, Ruggieri S, Paparelli A. Noradrenaline in Parkinson's disease: from disease progression to current therapeutics. Current Medicinal Chemistry. 2007;14:2330–4.

Fox SH, Chuang R, Brotchie JM. Serotonin and Parkinson's disease: on movement, mood, and madness. Movement Disorders. 2009;24:1255–66.

Frank MJ. Dynamic dopamine modulation in the basal ganglia: a neurocomputational account of cognitive deficits in medicated and nonmedicated Parkinsonism. Journal of Cognitive Neuroscience. 2005;17:51–72.

Galvan A, Wichmann T. Pathophysiology of parkinsonism. Clinical Neurophysiology, 2008;119:1459–74.

Galvan A, Kuwajima M, Smith Y. Glutamate and GABA receptors and transporters in the basal ganglia: what does their subsynaptic localization reveal about their function? Neuroscience. 2006;143:351–75.

Gerfen CR, Bolam JP. The neuroanatomical organization of the basal ganglia. In: Handbook of Basal Ganglia Structure and Function (2nd ed.), edited by Steiner H, Tseng KY. Academic, 2016, pp. 3–30.

German CL, Baladi MG, McFadden LM, Hanson GR, Fleckenstein AE. Regulation of the dopamine and vesicular monoamine transporters: pharmacological targets and implications for disease. Pharmacological Reviews. 2015;67:1005–24.

Ghadery C, Valli M, Mihaescu A, Strafella R, Navalpotro I, Kim J, Strafella AP. Molecular imaging of addictive behavior in idiopathic Parkinson's disease. International Review of Neurobiology. 2018;141:365–404.

Gilbert JA, Frederick LM, Ames MM. The aromatic-L-amino acid decarboxylase inhibitor carbidopa is selectively cytotoxic to human pulmonary carcinoid and small cell lung carcinoma cells. Clinical Cancer Research. 2000;6:4365–72.

Glass M, Dragunow M, Faull RL. The pattern of neurodegeneration in Huntington's disease: a comparative study of cannabinoid, dopamine, adenosine and GABAA receptor alterations in the human basal ganglia in Huntington's disease. Neuroscience. 2000;97:505–19.

Gorentla BK, Moritz AE, Foster JD, Vaughan RA. Proline-directed phosphorylation of the dopamine transporter N-terminal domain. Biochemistry. 2009;48:1067–76.

Grace AA, Lodge DJ, Buffalari DM. Dopamine-CNS pathways and neurophysiology. In: Encyclopedia of neuroscience. Elsevier, 2010, pp. 549–55.

Groenewegen HJ. The basal ganglia and motor control. Neural Plasticity. 2003;10:107–20.

Haber SN. The place of dopamine in the cortico-basal ganglia circuit. Neuroscience. 2014;282:248–57.

Haines DE (editor). Fundamental Neuroscience for Basic and Clinical Applications (4th ed.), Elsevier, 2013, pp. 52–4.

Hatfield E, Rapson RL (editors). The neuropsychology of passionate love. In: Psychology of Relationships. Nova Science Publishers, 2009.

Hurley MJ, Brandon B, Gentleman SM, Dexter DT. Parkinson's disease is associated with altered expression of CaV1 channels and calcium-binding proteins. Brain 2013;136:2077–97.

Hung AY, Schwarzschild MA. Treatment of Parkinson's disease: what's in the non-dopaminergic pipeline? Neurotherapeutics. 2014;11:34–46.

Ikonen S. The Role of the Septohippocampal Cholinergic System in Cognitive Functions. University of Kuopio, 2001.

Ingham CA, Hood SH, Arbuthnott GW. Spine density on neostriatal neurones changes with 6-hydroxydopamine lesions and with age. Brain Research. 1989;503:334–8.

Janhunen S. Different responses of the nigrostriatal and mesolimbic dopaminergic pathways to nicotinic receptor agonists. Academic dissertation at University of Helsinki. 2005; 9–20. https://core.ac.uk/download/pdf/14914635.pdf.

Jankovic, J. Parkinson's disease: clinical features and diagnosis. Journal of Neurology, Neurosurgery & Psychiatry. 2008;79:368–76.

Kapur S, Remington G. Serotonin-dopamine interaction and its relevance to schizophrenia. American Journal of Psychiatry. 1996;153:436–76.

Kaufman MJ, Madras BK. Severe depletion of cocaine recognition sites associated with the dopamine transporter in Parkinson's-diseased striatum. Synapse. 1991;9:43–9.

Khan ZU, Gutierrez A, Martın R, Penafiel A, Rivera A, De La Calle A. Dopamine D5 receptors of rat and human brain. Neuroscience. 2000;100:689–99.

Kulisevsky J, Poyurovsky M. Adenosine A2A-receptor antagonism and pathophysiology of Parkinson's disease and drug-induced movement disorders. European Neurology. 2012;67:4–11.

Lawrence AD, Calder AJ, McGowan SW, Grasby PM. Selective disruption of the recognition of facial expressions of anger. Neuroreport. 2002;13:881–884.

Lester DB, Rogers TD, Blaha CD. Acetylcholine–dopamine interactions in the pathophysiology and treatment of CNS disorders. CNS Neuroscience & Therapeutics. 2010;16:137–62.

Li LB, Chen N, Ramamoorthy S, Chi L, Cui XN, Wang LC, Reith ME. The role of N-glycosylation in function and surface trafficking of the human dopamine transporter. Journal of Biological Chemistry. 2004;279:21012–20.

McGeer EG, Staines WA, McGeer PL. Neurotransmitters in the basal ganglia. Canadian Journal of Neurological Sciences. 1984;11(S1):89–99.

Macphee GJ, Stewart DA. Parkinson's disease–pathology, aetiology and diagnosis. Reviews in Clinical Gerontology. 2010; 22:165–78.

Magrinelli F, Picelli A, Tocco P, Federico A, Roncari L, Smania N, et al. Pathophysiology of motor dysfunction in Parkinson's disease as the rationale for drug treatment and rehabilitation. Parkinson's Disease. 2016;2016.

Manepalli S, Surratt CK, Madura JD, Nolan TL. Monoamine transporter structure, function, dynamics, and drug discovery: a computational perspective. The AAPS Journal. 2012;14:820–31.

Marsden CA. Dopamine: the rewarding years. British Journal of Pharmacology. 2006;147(S1):S136–44.

Mignini F, Tomassoni D, Traini E, Amenta F. Dopamine, vesicular transporters and dopamine receptor expression and localization in rat thymus and spleen. Journal of Neuroimmunology. 2009;206:5–13.

Mishra A, Singh S, Shukla S. Physiological and functional basis of dopamine receptors and their role in neurogenesis: possible implication for Parkinson's disease. Journal of Experimental Neuroscience. 2018;12:1179069518779829.

Missale C, Nash SR, Robinson SW, Jaber M, Caron MG. Dopamine receptors: from structure to function. Physiological Reviews. 1998;78:189–225.

Moore KE. Interactions between prolactin and dopaminergic neurons. Biology of Reproduction. 1987;36:47–58.

Müller ML, Bohnen NI. Cholinergic dysfunction in Parkinson's disease. Current Neurology and Neuroscience Reports. 2013;13:377.

Obeso JA, Rodriguez-Oroz MC, Rodriguez M, Lanciego JL, Artieda J, Gonzalo N, Olanow CW. Pathophysiology of the basal ganglia in Parkinson's disease. Trends in Neurosciences. 2000;23:S8–19.

Onoa B, Li H, Gagnon-Bartsch JA, Elias LA, Edwards RH. Vesicular monoamine and glutamate transporters select distinct synaptic vesicle recycling pathways. Journal of Neuroscience. 2010;30:7917–27.

Ortigue S, Bianchi-Demicheli F, de C. Hamilton AF, Grafton ST. The neural basis of love as a subliminal prime: an event-related functional magnetic resonance imaging study. Journal of Cognitive Neuroscience. 2007;19:1218–30.

Ortigue S, Bianchi-Demicheli F, Patel N, Frum C, Lewis JW. Neuroimaging of love: fMRI meta-analysis evidence toward new perspectives in sexual medicine. The Journal of Sexual Medicine. 2010;7:3541–52.

Prakash J, Chouhan S, Yadav SK, Westfall S, Rai SN, Singh SP. *Withania somnifera* alleviates parkinsonian phenotypes by inhibiting apoptotic pathways in dopaminergic neurons. Neurochemical Research. 2014;39:2527–2536.

Pramod AB, Foster J, Carvelli L, Henry LK. SLC6 transporters: structure, function, regulation, disease association and therapeutics. Molecular Aspects of Medicine. 2013;34:197–219.

Richtand NM, Woods SC, Berger SP, Strakowski SM. D3 dopamine receptor, behavioral sensitization, and psychosis. Neuroscience & Biobehavioral Reviews. 2001;25:427–43

Rinne JO, Anichtchik OV, Eriksson KS, Kaslin J, Tuomisto L, Kalimo H, et al. Increased brain histamine levels in Parkinson's disease but not in multiple system atrophy. Journal of Neurochemistry. 2002;81:954–60.

Rivett AJ, Francis A, Roth JA. Localization of membrane-bound Catechol-O-methyltransferase. Journal of Neurochemistry. 1983;40:1494–6.

Roshan MH, Tambo A, Pace NP. Potential role of caffeine in the treatment of Parkinson's disease. The Open Neurology Journal. 2016;10:42.

Santens P, Boon P, Van Roost D, Caemaert J. The pathophysiology of motors symptoms in Parkinson's disease. Acta Neurologica Belgica. 2003;103:129–34.

Segura-Aguilar J, Paris I, Muñoz P, Ferrari E, Zecca L, Zucca FA. Protective and toxic roles of dopamine in Parkinson's disease. Journal of Neurochemistry. 2014;129:898–915.

Sheffler ZM, Reddy V, Pillarisetty LS. Physiology, neurotransmitters. In StatPearls. StatPearls Publishing, 2020.

Shi WX, Nathaniel P, Bunney BS. Ritanserine, a 5-HT2A/2C antagonist, reverses direct dopamine agonist-induced inhibition of midbrain dopamine neurons. Journal of Pharmacology Experimental Therapeutics. 1995; 274:735–40.

Snyder SH. What dopamine does in the brain. Proceedings of the National Academy of Sciences. 2011;108:18869–71.

Sokoloff P, Giros B, Martres MP, Bouthenet ML, Schwartz JC. Molecular cloning and characterization of a novel dopamine receptor (D3) as a target for neuroleptics. Nature. 1990;347:146–51.

Sonne J, Lopez-Ojeda W. Dopamine. In StatPearls. StatPearls Publishing, 2020.

Storch A, Ludolph AC, Schwarz J. Dopamine transporter: involvement in selective dopaminergic neurotoxicity and degeneration. Journal of Neural Transmission. 2004;111:1267–86.

Stoker TB, Barker RA. Recent developments in the treatment of Parkinson's Disease. F1000Research. 2020;9.

Surmeier DJ, Schumacker PT. Calcium, bioenergetics, and neuronal vulnerability in Parkinson's disease. Journal of Biological Chemistry. 2013;288:10736–41.

Swanson LW. The projections of the ventral tegmental area and adjacent regions: a combined fluorescent retrograde tracer and immunofluorescence study in the rat. Brain Research Bulletin. 1982;9:321–53.

Tarazi FI. Neuropharmacology of dopamine receptors: Implications in neuropsychiatric diseases. Journal for Scientific Research. 2001;3:93.

Torres GE, Yao WD, Mohn AR, Quan H, Kim KM, Levey AI, et al. Functional interaction between monoamine plasma membrane transporters and the synaptic PDZ domain–containing protein PICK1. Neuron. 2001;30:121–34.

Vaughan RA, Foster JD. Mechanisms of dopamine transporter regulation in normal and disease states. Trends in Pharmacological Sciences. 2013;34:489–96.

Vaughan RA, Kuhar MJ. Dopamine transporter ligand binding domains structural and functional properties revealed by limited proteolysis. Journal of Biological Chemistry. 1996;271:21672.

Villalba RM, Lee H, Smith Y. Dopaminergic denervation and spine loss in the striatum of MPTP-treated monkeys. Exp Neurol. 2009 Feb;215(2):220–7.

Volz TJ, Schenk JO. L-arginine increases dopamine transporter activity in rat striatum via a nitric oxide synthase-dependent mechanism. Synapse. 2004;54:173–82.

Werner FM, Coveñas R. Classical neurotransmitters and neuropeptides involved in Parkinson's disease: focus on anti-Parkinsonian drugs. Current Drug Therapy. 2015;10:66–81.

Wise RA. Forebrain substrates of reward and motivation. Journal of Comparative Neurology. 2005;493:115–21.

Woods AS. The dopamine D4 receptor, the ultimate disordered protein. Journal of Receptors and Signal Transduction. 2010;30:331–6.

Xu X, Aron A, Brown L, Cao G, Feng T, Weng X. Reward and motivation systems: A brain mapping study of early-stage intense romantic love in Chinese participants. Human Brain Mapping. 2011;3:249–57.

Yamakage M, Namiki A. Calcium channels – basic aspects of their structure, function and gene encoding; anesthetic action on the channels – a review. Canadian Journal of Anesthesia. 2002;49:151–64.

Zaja-Milatovic S, Milatovic D, Schantz AM, Zhang J, Montine KS, Samii A, et al. Dendritic degeneration in neostriatal medium spiny neurons in Parkinson disease. Neurology. 2005;64:545–7.

Zeki S. The neurobiology of love. FEBS Letters. 2007;581: 2575–9.

Zhou FM, Wilson CJ, Dani JA. Cholinergic interneuron characteristics and nicotinic properties in the striatum. Journal of Neurobiology. 2002;53:590–605.

Zisapel N. Melatonin–dopamine interactions: from basic neurochemistry to a clinical setting. Cellular and Molecular Neurobiology. 2001;21:605–16.

8 Therapeutic Targets and Neurochemical Signaling in Huntington's Disease

Newman Osafo and Oduro Kofi Yeboah
Kwame Nkrumah University of Science and Technology, Kumasi, Ghana

CONTENTS

8.1 INTRODUCTION

Huntington's disease (HD) is the commonest genetic neurodegenerative disorder, and mostly surfaces in middle age. The disorder is marked by characteristic chorea, dystonia, Parkinsonism, cognitive deterioration, and behavioral changes. Clinical presentation of the disorder arises from gradual neural cell loss in the cortico-striato-thalamo-cortical circuits, as well as predominant cell loss in the neostriatum (Novak and Tabrizi, 2011; Martino et al., 2013).

The causative mutation of HD is a prolongation of the CAG trinucleotide within *IT15*. As in similar trinucleotide repeat-associated disorders, there exists a reciprocal association between the length of the repeat and the onset of the disease (Martino et al., 2013). The time of onset of HD is linked with the clinical outlook of both motor and cognitive features of the disease (Novak and Tabrizi, 2011).

The condition was initially referred to as Huntington's chorea as a result of the characteristic movement of patients identified

DOI: 10.1201/9780429265198-10

with it. The pathophysiology of HD entails a default in a range of biologic mechanisms, which results in gliosis, loss of astrocytes and oligodendrocytes, and neuronal death, as well as the shriveling of tissues of the brain (Vonsattel, 2008). Dysfunction in the neurotransmission in patients with HD is also marked by cellular inclusions of protein aggregates (Cattaneo et al., 2005), altered signaling pathways (Tourette et al., 2014), oxidative balance, inflammatory mechanism, energy metabolism, and altered neurotrophic factors (Kumar and Ratan, 2016).

The dysfunction in neurotransmission seen in HD is intricately linked by numerous neurotransmitters such as dopamine (DA), gamma-aminobutyric acid (GABA), glutamate, brain-derived neurotrophic factor (BDNF), adenosine, and endocannabinoids, which modulate synaptic activity and plasticity, especially in the cortical synapse (Veldman and Yang, 2018). Alterations in the signaling of these neurochemicals are also reported in the hippocampus and the cortex (Dargaei et al., 2018; Estrada-Sánchez and Rebec, 2013). This chapter discuses HD in depth, focusing on the role of mutant Huntingtin (mHTT) in the modulation of the signaling pathways of these neurotransmitters in the HD brain. The authors also explore the possible direction of future research on this neurodegenerative disorder.

Currently, there is no treatment that can alter the course of disease progression in HD. However, education and symptomatic management can be effective approach for attending to individuals with HD and their families. This chapter discuses HD in depth and also explores the possible direction of future research on this neurodegenerative disorder.

8.2 SIGNS AND SYMPTOMS OF HUNTINGTON'S DISEASE

The onset of symptoms in HD can occur anytime between the age of 1 and 80 years, with predominant clinical presentations including cognitive, seizure, and oropharyngeal dysfunctions, and gait, fine motor, and behavioral disorders (Gonzalez-Alegre and Afifi, 2006; Myers, 2004). Prior to the appearance of clinically discernible symptoms, patients exhibit minute variations in cognition, personality, and motor control, a stage known as the prediagnostic stage (Walker, 2007). During this stage, patients may or may not be aware of the changes, and can thus delay diagnosis, as the diagnosis is made when signs and symptoms become clinically observable and specific (Paulsen et al., 2005). Although the spectrum of motor manifestations is easily discernible, and might be the source of anxiety and social withdrawal, it is often better handled by the sufferer and family alike, in contrast to the cognitive and behavioral symptoms, which frequently affect family, social, and occupational well-being (Bachoud-Lévi et al., 2019).

8.2.1 Motor Disorders

The biphasic course of the loss of medium spiny neurons in the striatum and the initial loss of neurons of the indirect pathway, followed by neuronal loss in the direct pathway, results in two disease phenotypes: hyperkinetic and hypokinetic (McColgan and Tabrizi, 2018). The former is characterized predominantly by chorea at an earlier disease phase, which then plateaus as the condition advances (Dorsey et al., 2013). While chorea is relevant in diagnosing HD, it does not correlate with disease severity, as some patients may not develop it, develop it but only transiently, or chorea may become less prominent due to the later onset of dystonia and rigidity (Mahant et al., 2003; Walker, 2007). However, the hypokinetic phenotype is characterized by the disturbance of gait and balance, dystonia, and bradykinesia (McColgan and Tabrizi, 2018).

The inability to steadily maintain voluntary muscle contraction, described as motor impersistence, such as in milkmaid's grip, contributes to overactivity in HD and sometimes results in compensatory repositioning (Walker, 2007; Gordon et al., 2000). In the early stages of the disease, fine motor skills, such as rhythmic finger tapping, may be present, which may help in early diagnosis. Postural maintenance and gait, however, deteriorate later in the course of the disease (Walker, 2007). Chorea manifests as non-rhythmic, purposeless movements involving the face, trunk, and limbs, and is abnormal, involuntary, spontaneous, and intermittent in onset (Bachoud-Lévi et al., 2019). In contrast to chorea, motor impersistence is linearly progressive and may serve as a better surrogate marker for monitoring disease progression (Reilmann et al., 2001).

In HD progression, dystonia – characterized by an abnormal posture which may affect all body segments – varies in intensity from slight intermittent abnormal postures to severe muscles twitches that may impair function in the patient (Bachoud-Lévi et al., 2019). It is often associated with rigidity, which may be induced or worsened by the use of pharmacological agents such as antipsychotics (van de Zande et al., 2017). Slowed saccadic movements of the eye are usually prominent in juvenile-onset HD (younger than 20 years of age; Kremer, 2002). Other motor disorders associated with HD are bruxism, akathisia, manual dexterity, myoclonus, and swallowing disorders (Nash et al., 2004; Bonelli et al., 2002).

8.2.2 Cognitive and Behavioral Disorders

Cognitive dysfunction often appears before the onset of motor disturbances and follows a subcortical pattern, characterized by impairment in executive function, visuospatial and visual perceptual disturbances, and impaired emotion recognition (Paulsen et al., 2017; Papoutsi et al., 2014). Long-term memory is, however, frequently spared in patients with HD with cognitive deficits (Craufurd and Snowden, 2002). The development of cognitive impairments in the course of the disease leads to intense psychological discomfort, which leads to psychiatric and behavioral symptoms (Bachoud-Lévi et al., 2019). Unlike cognitive impairments, psychiatric and behavioral symptoms develop with some frequency but do not show linear progression with disease severity (Craufurd and Snowden, 2002).

During progression of the disease, slowing of cognition and extension of reaction time subject to the complexity of the cognitive task – termed bradyphrenia – becomes more apparent. The individual's ability to perform tasks such as the acquisition of new motor skills and planning/organizing becomes impaired (Papoutsi et al., 2014). The onset of language and communication disorders also occurs in the initial phase of the disease and may be characterized by slurred and slowed speech resulting in dysarthria and progressive loss of verbal fluency. Difficulties in understanding and producing complex sentences occurs earlier, with the onset of lexical capacity deficit happening later in the course of the disease (Ludlow et al., 1987; Bachoud-Lévi et al., 2019).

The onset and progression of HD is marked by an array of neuropsychiatric symptoms, including depression, psychosis, visuospatial disorders, suicidal ideation and attempts, apathy, anxiety, irritability, and obsession and compulsivity (Craufurd et al., 2001; Martinez-Horta et al., 2016). Suicidal ideation and attempts are frequent findings in patients with HD (5–10 times that of general population), with and increased risk in patients with a positive personal and/or family history of suicide, and the presence of depression, particularly during disease stages where symptoms are non-specific (Fiedorowicz et al., 2011). Apathy is the most frequent psychological and behavioral manifestation in HD, and may alternate with irritability – often at different times of the day (Bouwens et al., 2015). Other neuropsychiatric symptoms that may manifest in the course of the disease include decreased libido and hallucinations (Bachoud-Lévi et al., 2019).

8.2.3 Other Symptoms and Cause of Mortality

In addition to motor and cognitive dysfunction, other deficits contribute to the clinical spectrum of HD. These include sleep disorders in about two-thirds of patients with HD, pain, urinary incontinence, weight loss, excessive perspiration, and hypersalivation (Cochen et al., 2000; Sciacca et al., 2017).

As motor and cognitive impairments become severe and debilitating, death usually occurs due to complications from falls, aspiration, and extreme lethargy due to malnourishment. Typically, death occurs about 20 years after diagnosis (Walker, 2007).

8.3 PATHOPHYSIOLOGIC MECHANISM OF HUNTINGTON'S DISEASE

The pathological bases of HD originate from mutation in the huntingtin (HTT) protein, although other factors play significant roles, and will be discussed later. Neurodegeneration can result in about a 30% loss of brain mass, depending on its severity. The significant loss of brain mass occurs in the neostriata, which is made up of the caudate nucleus and putamen (Waldvogel et al., 2015). Although studies have proposed that the neuronal losses begin after the onset of the motor symptoms of HD, recent studies have pointed to changes in volume and shape of the basal nuclei and other brain regions before the onset of motor symptoms (Reading et al., 2005; Rosas et al., 2005). This establishes the origin of cognitive impairment before the development of motor symptoms in HD.

The striatal circuit is made up of projection neurons, with GABA as their main neurotransmitter. The striatum receives projections from almost all cortical regions but significantly from the sensory and motor cortices (Rüb et al., 2015). Mutation in *HTT* not only causes changes in GABA signaling, but also changes DA, glutamate, adenosine, BDNF, and endocannabinoid signaling in the brain, and alters synaptic functioning, as summarized in Figure 8.1.

8.3.1 The Role of Mutant Huntingtin and Huntington's Disease Genes

The expression of *HTT* is key to the survival of humans. Studies in mice have shown that deletion of *HD* results in embryonic lethality between days 8.5 and 9.5 (Zeitlin et al., 1995). Further studies have pointed to the realization that eliminating HD expression postnatally leads to the loss of *HTT* with deleterious effects inside and outside the brain (Dietrich et al., 2017). However, the sole loss of *HTT* function is unlikely to cause HD, with studies presenting sufficient data to establish that the symptoms of the disease are dependent on the level of mutant or normal allele.

HTT is ubiquitously expressed by all cells but is found abundantly in the neocortex and the cerebellar cortex, as well as the striatum and hippocampus (Fusco et al., 1999).

8.3.2 Dopamine Signaling in Huntington's Disease

In HD, the first cells to degenerate is spiny projection neurons (SPNs) in the striatum, with each SPN demonstrating varying susceptibility to the disease (Cepeda et al., 2007; Plotkin and Surmeier, 2015). During neurotransmission, the main neurochemicals received by SPNs are DA from the substantia nigra pars compacta, and glutamate from the cortex and thalamus (Cepeda et al., 2007). The released DA from the substantia nigra pars compacta acts on the G_s-coupled D_1 and G_i-coupled D_2 receptors on SPNs, cholinergic and GABAergic interneurons, and D_2 receptors expressed on cortical afferents and terminals of substantia nigra pars compacta, to modulate striatal glutamatergic, GABAergic, cholinergic, and dopaminergic neurotransmission (Yan and Surmeier, 1997; André et al., 2010; Rangel-Barajas and Rebec, 2016) (Figure 8.2).

According to Bird (1980), increased levels of DA induces chorea, whereas low levels result in akinesia. This explains the biphasic nature of motor disturbances in early- and late-stage HD (Chen et al., 2013). Moreover, a reduction in the release of DA during late-stage HD is believed be the cause of hypoactivity and the demonstration of apathy in patients with late-stage HD (Dallérac et al., 2011; Covey et al., 2016).

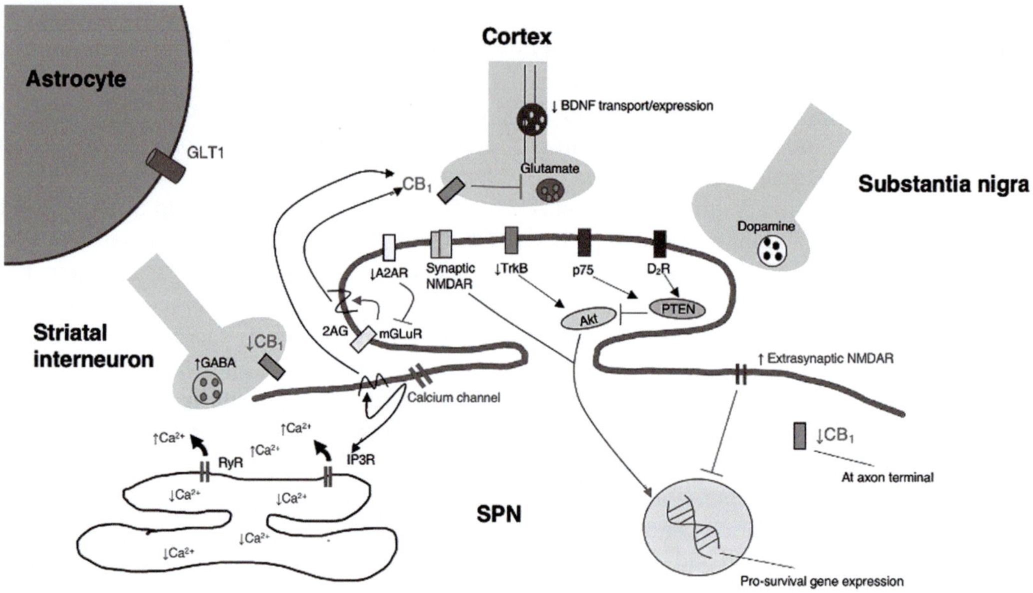

FIGURE 8.1 Altered synaptic function in spiny projection neurons (SPN) following changes in various neurochemical signaling pathways (adapted from Smith-Dijak et al., 2019). BDNF, brain-derived neurotrophic factor; CB, cannabinoid; NMDAR, *N*-methyl-D-aspartate receptor; PTEN, phosphatase and tensin homolog; GABA, γ-aminobutyric acid; A2AR, adenosine A2A receptor; mGLuR, metabotropic glutamate receptor; 2AG, 2-arachidonoylglycerol; RyR, ryanodine receptor; IP3R, inositol triphosphate receptor; GLT1, glutamate transporter 1.

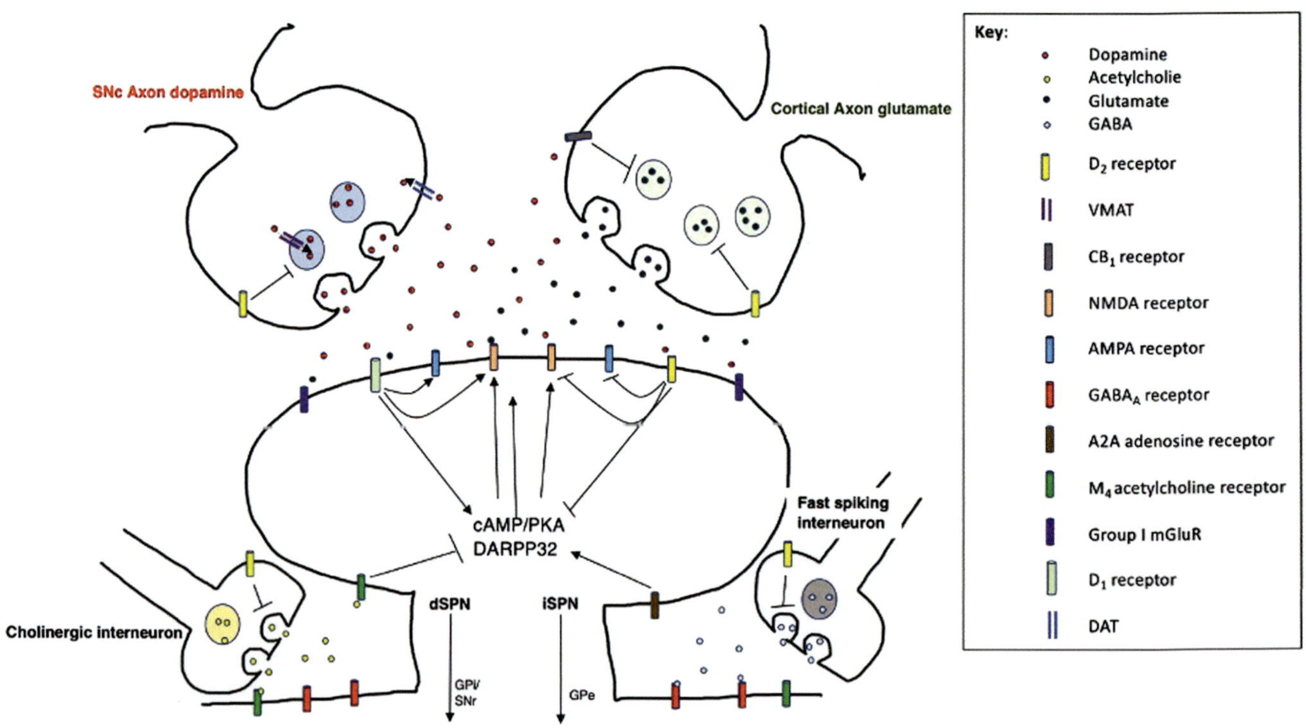

FIGURE 8.2 Dopamine effects at D_1 and D_2 receptors in the healthy striatum. SNc, substantia nigra pars compacta; cAMP, cyclic adenosine monophosphate; PKA, protein kinase A; DARPP-32, dopamine and cAMP-regulated neuronal phosphoprotein; dSPN, direct spiny projection neuron; iSPN, indirect spiny projection neuron; GPi, globus pallidus internus; SNr, substantia nigra; GPe, globus pallidus externus; GABA, γ-aminobutyric acid; VMAT2, vesicular monoamine transporter 2; CB_1, cannabinoid receptor 1; NMDAR, *N*-methyl-D-aspartate; AMPA, α-amino-3-hydroxy-5-methyl-4-isoxazolepropionic acid; mGluR, metabotropic glutamate receptor; DAT, dopamine transporter (adapted from Koch and Raymond, 2019).

An additional factor that contributes to the role of DA in the pathogenesis of HD is DA transporter activity. The knockout of DA transporter genes in mice is associated with increased DA levels in the striatum, hyperactivity, and cognitive and behavioral deficits (Zhuang et al., 2001). As evidenced from the crossing of DA transporter knockout mice with knock-in Hdh(Q92) HD mice, increased extracellular DA levels predispose mice to early aggregation of mHTT, and increased motor deficits and stereotypical behaviors (Cyr et al., 2006).

Another neurochemical pathway that contributes to the role of DA in the neuropathology of HD is the D_1-regulated DA and cyclic adenosine monophosphate (cAMP)-regulated neuronal phosphoprotein (DARPP-32) signaling pathway. D_1-mediated enhancement of striatal *N*-methyl-D-aspartate (NMDA) receptor effects is dependent, in part, on DARPP-32 expression, which, in turn, is affected by mHTT dose (Flores-Hernández et al., 2002; Young et al., 2013). However, the levels of these neurochemicals are reduced in early stages of HD (Dau et al., 2014; Gladding et al., 2014). The phosphorylation of extracellular regulated kinase (ERK)1/2 and cAMP response element-binding protein (CREB) can be stimulated by D_1 receptor activation and glutamatergic transmission via NMDA receptor activation or by striatal-enriched protein tyrosine phosphatase (Figure 8.2). These modulators of ERK1/2 and CREB phosphorylation have been shown to be altered in HD (Gladding et al., 2014).

8.3.3 Glutamate Signaling in Huntington's Disease

Aside from DA, enhanced glutamatergic inputs to the striatum may also cause excitotoxicity owing to the subsequent impairment of calcium-buffering capacity and resultant neurodegeneration of the matrix compartment (Crittenden and Graybiel, 2001; Bezprozvanny and Hayden, 2004; Tang et al., 2007). In HD, this glutamate-induced striatal neurotoxicity is believed to be the underlying cause of the vulnerability of medium spiny neurons (MSNs) in patients with HD and the reduction in striatal glutamate release and receptors in HD animal models (Morigaki and Goto, 2017; Li et al., 2003; Calabresi et al., 1999). Studies have demonstrated a reduction in amino-3-hydroxy-5-methyl-4-isoxazolepropionic acid (AMPA) and NMDA receptors in the caudate nucleus and cerebral cortex of symptomatic patients and late stage presymptomatic HD animal models (Starling et al., 2005).

Moreover, in HD, the increase in expression of the calcium-binding proteins parvalbumin and calretinin positively correlates with the survival of interneurons (Gerfen et al., 1992; Morigaki and Goto, 2017). During calcium-induced excitotoxicity, these proteins exert calcium-buffering effects and protect interneurons against neurotoxicity (Sun et al., 2001).

8.3.4 Dopamine and Glutamate Receptor Interactions in Huntington's Disease

DA and glutamate neurotransmission are closely linked. In HD, early increases in corticostriatal glutamate are associated with hyperexcitability of the cortex, whereas the loss of D_2 receptors enhances synaptic activity (Spampanato et al., 2008; Cummings et al., 2009). Stimulation of corticostriatal neurons activates striatal DA release. Considering that frontal cortical neuron-regulated neurotransmission along DA neurons is upregulated in early HD, it can be inferred that the biphasic changes in DA levels during early and late human HD correlates with the biphasic changes in the extent of glutamate release by cortical afferents (Chen et al., 2013).

In forebrain neurons, a reduction in DA levels decreases the strength of the glutamate signal above background activity, thus constraining the propensity to regulate neural noise with subsequent impairment of motor and cognitive functions (Kiyatkin and Rebec, 1996). This is common in neurodegenerative conditions, including HD and ageing (Chen et al., 2013). Synergistic signaling of DA and glutamate can enhance the sensitivity of MSNs to mHTT toxicity via D_1, but not D_2, receptor activation (Tang et al., 2007; Paoletti et al., 2008). This is supported by the finding that the pretreatment of knock-in HD mice with NMDA increases D_1 receptor-induced cell death in MSNs versus cells from wild-type mice (Paoletti et al., 2008).

8.3.5 Gamma-Aminobutyric Acid

During neural plasticity, inhibitory interneurons mark the boundary of learned motor activity sequences (Martiros et al., 2018). This implies that, during HD, increased activity at the synapses of these inhibitory interneurons onto SPNs may be a compensatory mechanism to suppress the onset and severity of chorea (Smith-Dijak et al., 2019). This, in addition to increased glutamate excitotoxicity in SPNs, suggests that increased GABAergic neurotransmission is a form of homeostatic plasticity that can counter the enhanced activation of SPNs. This hypothesis is based on findings from animal studies that have demonstrated an increase in GABAergic activity in striatal inhibitory interneurons that most probably synapse onto indirect pathway SPNs (iSPNs) (Dvorzhak et al., 2016; Cepeda et al., 2013).

Aside from the striatum, the release of GABA in the cortex, onto layer II/III cortical pyramidal neurons, is also altered in various HD mouse models. Early-stage disease is characterized by enhanced inhibitory synaptic activity, whereas symptomatic stage is characterized by reduced inhibitory synaptic activity HD in murine models (Cummings et al., 2009; Raymond et al., 2011). In humans with HD, both interneurons and pyramidal neurons in the cortex degenerate, with the former correlating with the appearance and severity of motor symptoms, as discussed earlier (Mehrabi et al., 2016). Moreover, the activation of $GABA_A$ receptors on cortical pyramidal neurons changes their depolarizing membrane potential by altering the expression of the potassium and chloride co-transporters, NKCC1 and KCC2. The resultant increase in GABAergic signaling can be effectively inhibited by administration of a NKCC1 inhibitor (Dargaei et al., 2018). However, clinical trials with agents that target this mechanism did not show positive results, and some patients involved in the trial developed worse motor deficits, such as dystonia (Shoulson et al., 1978; Foster et al., 1983).

8.3.6 Brain-Derived Neurotrophic Factor

Another synaptic function-modulating neurochemical that is impaired in the HD brain is the neurotrophin, BDNF. With varying affinities, BDNF binds to two neurotrophin receptors, TrkB and p75 (Chao and Hempstead, 1995). Stimulation of the high-affinity TrkB turns on various pathways associated with synaptic plasticity and cell survival, whereas stimulation of the low-affinity p75 results in apoptotic cell death (Smith-Dijak et al., 2019). It is therefore unsurprising that BDNF signaling through TrkB is impaired at multiple pre- and postsynaptic points in the HD brain (Plotkin and Surmeier, 2015). Increasing BDNF production, release, and/or signaling through the TrkB receptor improves HD phenotypes in HD mouse models (Silva et al., 2015; Zhao et al., 2016).

Enhancement of BDNF signaling in HD, through increased expression and transport of BDNF and its TrkB receptor, subverts the effects of neuronal loss in the corticostriatal synapse and beyond. Normalizing BDNF signaling in the HD brain improves SPN survival, and rescues SPN morphology, advance motor function, and practical learning (Xie et al., 2010; Zhao et al., 2016). Procedural learning may be enhanced by improvement of neuronal function and long-term potentiation at the CA3–CA1 hippocampal synapse, following normalization of BDNF signaling (Kolodziejczyk et al., 2014; Giralt et al., 2017). The beneficial effect of fingolimod in HD, is also attributed to its ability to normalize the balance in TrkB and p75 signaling, by reducing expression of p75 (Miguez et al., 2015). Pridopidine and amitriptyline also increase BDNF signaling.

8.3.7 Endocannabinoids

The endocannabinoid system comprises cannabinoid receptors (CB) 1 and 2, their endogenous ligands, and enzymes responsible for the synthesis and degradation of these ligands (Argenziano et al., 2019). Of the two main receptors, CB_1 is mainly distributed in the central nervous system (CNS), and is one of the most densely distributed G protein-coupled receptors (GPCR) within the brain (Sharkey and Wiley, 2016). Endocannabinoid signaling via CB_1 plays a relevant role in the striatum and at the corticostriatal synapse, where they contribute to goal-oriented behavior (Augustin et al., 2018). A number of findings have established a consistent reduction in the distribution of CB_1 receptors in the striatum of HD mouse models and patients (Laere et al., 2010; Horne et al., 2013). Studies in rodents have showed that deletion of CB_1 worsens the HD phenotype, and enhances mHTT-induced morphological and behavioral changes in the brain (Blázquez et al., 2015; Mievis et al., 2011).

In HD models, there is an observed increase in activity at GABAergic synapses, most likely due to the less effective inhibitory action of CB_1 agonists on the release of GABA in the striatum (Chiodi et al., 2012). During the early stage of the disease, the expression and function of CB_1 appears to be preserved on cortical afferents to striatal SPNs, though endocannabinoid signaling at GABAergic synapses is reduced during this stage in HD mouse models (Chiodi et al., 2012; Chiarlone et al., 2014). It has been observed that selective expression of mHTT in the cortex but not in the striatum increases the sensitivity of dSPNs to CB_1 receptor loss at the corticostriatal synapse. This may imply that mHTT action in the striatum accounts for the vulnerability of iSPNs in HD, as supported by findings of improvement in morphology and survival of R6/2 SPNs but not motor symptoms following the rescue of CB_1 expression (Naydenov et al., 2014; Ruiz-Calvo et al., 2018).

8.3.8 Adenosine

In addition to the abovementioned neurochemical alterations, one other mode of signaling altered in the early stage of HD is adenosinergic signaling via the A_{2A} receptor. The A_{2A} receptor is enriched in iSPNs and is dramatically reduced in the globus pallidus following the loss of iSPNs in HD. iSPNs are also expressed on cholinergic interneurons and glutamatergic terminals (Blum et al., 2018; Reiner and Deng, 2018).

Both the stimulation of the A_{2A} receptor in cultured, inducible pluripotent stem cells and overexpression of A_{2A} in patients with HD and rats injected with 3-nitropropionic acid, respectively, have been reported to be neuroprotective (Chiu et al., 2015; Domenici et al., 2018). Unsurprisingly, a daily provision of 190 mg caffeine – an adenosine receptor antagonist – is associated with the early onset of HD (Simonin et al., 2013). This further shows that decreased A_{2A} receptor activity is not a compensatory mechanism for other primary causes, but rather a deleterious trigger of HD onset (Smith-Dijak et al., 2019). Findings of reduced extracellular adenosine levels in mouse models and in patients with HD may suggest that, in addition to reduced expression, adenosine receptors receive less stimulation in HD (Guitart et al., 2016; Kao et al., 2017).

The activity of striatal A_{2A} receptors functionally antagonizes D_2 receptor activity in the striatum by inhibiting the induction of endocannabinoid-mediated long-term potentiation at corticostriatal synapses (Figure 8.1). The activation of A_{2A} receptors stimulates acetylcholine release from cholinergic interneurons, which subsequently inhibits the synthesis of endocannabinoid in all SPNs via the activation of M_1 acetylcholine receptors (Tozzi et al., 2011). Given that antagonism of A_{2A} receptors by caffeine produces psychomotor stimulation, the loss of A_{2A} receptors may be thought to favor the inhibition of iSPNs, thus facilitating the lowering of the threshold for the activation of motor programs in HD chorea (Smith-Dijak et al., 2019).

Although stimulation and overexpression of A_{2A} receptors appear to be neuroprotective, some level of evidence suggests that inactivation of these receptors can rescue behavioral deficits in R6/2 mice (Li et al., 2015). The reason for this phenomenon is currently unknown, and further studies of how signaling through the A_{2A} receptor contributes to either plasticity or neuroprotection are needed to fully understand the effect of A_{2A} signaling in HD.

8.3.9 Impaired Autophagy

The accumulation of damaged organelles and toxic protein aggregates due to abnormally low autophagic flux characterizes neuronal death in many neurodegenerative disorders (Menzies et al., 2011). In these samples, autophagic pathways are activated by mHTT-induced sequestration and inactivation of mammalian target of rapamycin (mTOR; Ravikumar et al., 2004). Although the number of autophagosomes increases, autophagy substrates, such as HTT, build up in the cytoplasm due to deficits in cargo recognition, resulting in the formation of empty autophagic vacuoles (Martinez-Vicente et al., 2010).

Studies in the fruit fly *Drosophila* and mouse CNS have shown that the loss of function or depletion of HTT impairs retrograde transport of autophagosomes containing non-degraded selective cargo, resulting in the buildup of toxic proteins (Wong and Holzbau, 2014). HTT facilitates this selective autophagy by enhancing cargo recognition and promoting autophagy initiation, in both *Drosophila* and mammalian cells. Thus, HTT mediates binding between p62 and Unc-51-like autophagy activating kinase (ULK1), an autophagy-initiating enzyme (Rui et al., 2015).

The defects in the autophagy pathway may result in a negative feedback loop, where the presence of mHTT leads to defective autophagy-mediated degradation and the accumulation of toxic protein aggregates, with subsequent activation of further compensatory autophagic pathways by the cell. This inevitably results not only in persistent upregulation of autophagy, but also accumulation of mHTT and subsequent neurotoxicity (Martin et al., 2015).

8.3.10 Mitochondrial Dysfunction

Mitochondria dysfunction is implicated in a number of cellular processes, associating altered mitochondria function to a wide array of human diseases, including neurological disorders (Bhatti et al., 2017). In HD, increased mitochondrial fragmentation and inefficient cargo loading of mitochondria into autophagosomes, termed as mitophagy, is observed (Costa et al., 2010; Wong and Holzbaur, 2014). This may be, in part, as a result of decreased PINK1 (PTEN-induced putative kinase 1) expression (Martin et al., 2015). It has been proposed that mHTT interferes with mitochondrial function, directly and indirectly, resulting in increased oxidative damage and compromised energy metabolism through decreased oxygen utilization, electron transport chain activity, Ca^{2+} buffering, and decreased nicotinamide adenine dinucleotide (NAD^+) and adenosine triphosphate (ATP) production (Ross and Tabrizi, 2011; Oliveira, 2010). In support of this, studies employing the use of positron emission tomography (PET) and proton nuclear magnetic resonance have showed decreased striatal glucose uptake and pyruvate utilization in presymptomatic patients with HD, years before the onset of motor symptom (Feigin et al., 2001; Intihar et al., 2019).

Mitochondrial dysfunction in HD has been partly attributed to dysregulation and crosstalk between the transcriptional factors p53, peroxisome proliferator-activated receptor γ coactivator 1α (PGC-1α), and heat shock transcription factor 1 (HSF1) (Intihar et al., 2019). In HD, it is believed that mHTT strongly interacts with the tumor suppressor and p53, and impairs E3 ligase Mdm2 recruitment with resultant stabilization of p53 (Bae et al., 2005). Other than decreasing the level of tumor incidence in patients with HD, increased levels of p53 induce the expression of mitochondrial proteins such as Puma and Bax, which decrease cell viability and activate apoptosis (La Spada and Morrison, 2005; Bae et al., 2005). This effect can be blocked by pifithrin-α, a p53 inhibitor (Bae et al., 2005).

PGC-1α regulates nuclear-encoded mitochondrial gene expression, and downregulation of it in HD results in decreased mitochondrial biogenesis, inhibition of autophagy, and a decrease in turnover of mHTT aggregates with subsequent neuronal death (Tsunemi et al., 2012). A number of studies have suggested that PGC-1α may act jointly with p53 to control mitochondrial function, depending on the physiological state (Villeneuve et al., 2013; Aquilano et al., 2013). The stress-responsive transcription factor HSF1 has been shown to be fundamental in the development of HD, at least in part, due to modulation of mitochondrial function and regulation of PGC-1α and p53 levels (Intihar et al., 2019). It is believed that the depletion of striatal levels of HSF1 and a reduction in its activity lead to strong mitochondrial depolarization, resulting in neuronal dysfunction in HD (Neef et al., 2011). The depletion may occur as a result of the inappropriate upregulation of protein kinase CK2α' and E3 ligase Fbxw7, which marks the protein for proteasomal degradation. The latter ubiquitylates, while the former phosphorylates, HSF1 (Gomez-Pastor et al., 2017). It has been further shown that HSF1 regulates the expression of PGC-1α. Thus, depletion of HSF1 leads to reduction in PGC-1α levels, which also worsens neuronal function (Gomez-Pastor et al., 2017).

8.3.11 Alpha-Synuclein and Huntington's Disease

The occurrence of expanded polyglutamine sequences in HD compromises the cell's ability to correctly fold proteins, leading to protein misfolding and the formation of fibrillar aggregates (Monsellier et al., 2016; Orr and Zoghbi, 2007). Emerging evidence that implicates mixed proteinopathies in the pathogenesis of HD suggests a possible crosstalk between α-synuclein and mHTT (Jimenez-sanchez et al., 2017).

Studies utilizing the R6/1 mouse model of HD, or postmortem brain samples, have indicated the buildup of α-synuclein in polyglutamine residues in HD brains. It is also known from these studies that N-terminal mHTT and α-synuclein form independent filamentous microaggregates in the mouse brain (Tomás-Zapico et al., 2012). Upregulation of the expression of N-terminal mHTT thus increases the load of α-synuclein filaments. This is supported by the finding that α-synuclein knockout in mice leads to a reduced number of N-terminal mHTT inclusions in transfected neurons and also in vivo in the brain. Unsurprisingly, α-synuclein knockout mitigates the loss of body weight and early motor symptoms of HD mice (Abeliovich et al., 2000). Moreover, significant levels of α-synuclein are realized in patients with HD, regardless of disease stage (Breza et al., 2020).

8.4 DIAGNOSIS OF HUNTINGTON'S DISEASE

In addition to the establishment of motor, cognitive, and behavioral symptoms, HD can be diagnosed with the use of genetic testing and imaging techniques (Craufurd et al., 2015). Complications such as the development of seizures can be monitored using electroencephalography, electromyography, and nerve-conduction studies. Neuropsychological tests, including the Wechsler Intelligence Scale for Children-Revised (WISC-R) testing, have been used to assess the level of cognitive decline in patients with HD (Gonzalez-Alegre and Afifi, 2006).

8.4.1 Genetic Testing

In HD, the gene responsible for the pathology is located on the short arm of chromosome four and is associated with an expanded trinucleotide repeat. However, normal alleles containing CAG repeats also occur at this point (Rubinsztein, 2003). HD becomes fully penetrant with 41 or more repeats, while 36–40 repeats are described as incomplete penetrance. Generally, 35 or fewer CAG repeats are not characteristic of HD (McNeil et al., 1997). CAG trinucleotide expansion mutation in *HTT*, which encodes the ubiquitously expressed HTT, results in the production of mHTT, leading to the development of HD (HDCRG, 1993; Craufurd et al., 2015). Since the discovery of the mutation, genetic testing by direct mutation analysis has been utilized in the diagnosis of the disorder (Craufurd et al., 2015).

Currently, guidelines developed for genetic testing are formulated to exclude children or those with suicidal ideation, and also to inform individuals of the impact of test results on both their lives and those of relatives. Genetic testing for mHTT mutation can be either diagnostic, where it is done to confirm clinical diagnosis in symptomatic patients, or predictive, in determining the possibility of development of HD in at-risk individuals. In the case of diagnostic testing, the individual must present with unequivocal features of motor dysfunction (Craufurd et al., 2015). As such, behavioral and cognitive impairments, and the presence of a positive family history alone, is not sufficient justification for diagnostic genetic testing. In such an instance, predictive genetic testing is an option and the patient must receive appropriate genetic counseling before and after undergoing the test (Craufurd et al., 2015).

A predictive genetic test is an alternative for individuals at risk of inheriting the mutant *HTT* gene. It is done before the onset of unequivocal motor symptoms and helps to determine the likelihood of an at-risk person developing symptoms in the future. As updated in 2013, internationally agreed guidelines state that predictive genetic counseling can only be requested by a genetic counseling unit (MacLeod et al., 2013). Thus, the individual must undergo pretest counseling to allow the patient ample chance to reflect on what will be an unalterable resolution. The individual must therefore be adequately informed on HD and its hereditary nature, as a positive test will impact significantly on both the patient and their family.

Pretest counseling should be followed by neurological and psychological assessments. Neurological examinations will ensure that the individual is not already symptomatic, while psychological assessments will enable the identification of individuals at high risk of suicide in the event the predictive test is positive. It is important that follow-up is carried out post-test, to monitor the impact of the test result on patients and also to assess the need for further support (Hersch, 1994; McColgan and Tabrizi, 2018).

8.4.2 Imaging

Although imaging techniques have not been useful in diagnosing HD due to the lack of specificity, magnetic resonance imaging (MRI), PET, and computed tomography (CT) offer an opportunity to view the pathology macroscopically. Routine MRI and CT have been used to show striatal atrophy and increased size of the frontal horns of the lateral ventricles in moderate-to-severe HD (Stober et al., 1984; Walker, 2007). Evidence that pathological changes in affected brains occur before symptom onset have also been shown with PET and functional MRI (Burggren and Brown, 2014). Some improved MRI techniques, such as 3-Tesla MRI and diffusion tensor imaging, can precisely measure the cortex and striatum of affected brains (Dumas et al., 2012). These imaging techniques have the advantages of identifying putaminal and caudate atrophy as early as 9 and 11 years, respectively, before the estimated symptom onset (Rodrigues et al., 2016).

8.4.3 Other Biomarkers

Testing for mHTT has received much research attention. However, a number of potential blood biomarkers are being studied for use in diagnosing HD. One such biomarker is neurofilament light (NFL), which has shown to be very promising (McColgan and Tabrizi, 2018). It has been shown that the baseline plasma concentration of NFL correlates with progression in brain atrophy over time. During the disease course, the baseline levels also correlate with motor and cognitive measures, and are associated with disease onset over a 3-year period (Byrne et al., 2017). Since plasma NFL highly correlates with cerebrospinal fluid (CSF) NFL, the use of peripheral blood may be sufficient to detect NFL accurately, reducing the need for the more invasive approach used in CSF sampling. The use of a proteomics approach in determining the plasma levels of immune proteins have also demonstrated correlation with disease progression (Dalrymple et al., 2007).

8.4.4 Oxidative Stress and Neuroinflammation in Huntington's Disease

Neuroinflammation and oxidative stress are two intertwined pathological processes in the development of neurodegenerative disorders (Leszek et al., 2016; Kumar and Ratan, 2016). Neuroinflammation has its onset prior to the development of pathologically relevant neuronal depletion in the disease profile (Leszek et al., 2016). However, oxidative stress serves as a principal culprit in multifaceted CNS mechanistic processes leading to the development of neurodegenerative disorders (Shukla et al., 2011; Moylan et al., 2013).

Mitochondrial dysfunction is important in the development of neurodegenerative disorders. Findings by Federico et al. (2012) have established a reciprocal relationship between elevated oxidative and nitrosative stress and mitochondrial dysfunction. Hence, there is an impaired mitochondrial respiratory chain complex and oxidative damages in the striatal structures when there is elevated oxidative stress (Browne et al., 1997).

When there are significant intracellular aggregates from mHTT, there is also an observable increase in reactive oxygen species (ROS) levels in the neuronal and non-neuronal cells. This points to a causal relationship between the two processes (Hands et al., 2011). Also, elevated ROS levels inhibit the formation of autophagosomes, leading to the loss of dopaminergic cells from human neuroblastomas expressing transgenic mHTT (Vidoni et al., 2016).

Iron plays a significant role, via the Fenton reaction, in the mechanism of oxidative damage (Ignácio et al., 2019). Hence, the exposure of biomolecules to elevated iron levels leads to oxidative damage, contributing to the development of neurodegeneration (Rosas et al., 2012). However, there is no established causative relationship between high iron levels and neurodegeneration.

A number of studies have proposed the possible underpinning of neuroinflammation in all stages of HD. mHTT also interferes with key mechanistic processes directed by microglia and astrocytes. These processes include the release of cytokines, cell signaling, and transcription mechanisms (Crotti and Glass, 2015). mHTT expression tends to lead to peripheral immune response activation. This has been identified to result in excessive expression of inflammatory cytokines in myeloid cells in HD (Träger et al., 2014). Some studies suggest that mHTT itself may be identified as an inflammatory marker of HD since its mutated protein expression in monocytes and T cells has a corresponding relationship with striatal atrophy (Weiss et al., 2012). Proinflammatory cytokines have also been identified to be highly expressed in later stages of HD (Björkqvist et al., 2008), in both microglia and serum. However, there is a decreased expression of anti-inflammatory cytokines in microglia and CSF (Valekova et al., 2016), with studies identifying inflammatory markers expressed at various stages of HD as markers of the disease (Chang et al., 2015).

8.5 DIFFERENTIATING HUNTINGTON'S DISEASE-LIKE 2 FROM HUNTINGTON'S DISEASE

Huntington disease-like 2 (HDL2) has been identified as the most common HD phenocopy in patients with African ancestry (Krause et al., 2015). In effect, HDL2 should be suspected in all patients of South African ancestry presenting with symptoms consistent with HD since they account for up to 50% of HDL2 cases (Margolis et al., 2005; Martino et al., 2013). Both disorders are marked by progressive subcortical dementia and psychiatric symptoms that normally progress to death in about 20 years after disease onset (Margolis et al., 2001; Walker et al., 2003). The HDL2 phenotype has been identified to have high parkinsonism, less dysarthria, and quite unaltered oculomotor function when compared to HD (Walker et al., 2003; Schneider et al., 2012). Some of these findings were confirmed through a systematic review of 69 cases of HDL2 (Anderson et al., 2017).

In a study by Anderson et al. (2019), the clinical phenotypes of HDL2 and HD were compared using UHDRS® (Unified Huntington's Disease Rating Scale). In neurocognitive assessment, the median scores on the Stroop Color and Word Test for patients with HD were significantly higher compared with those of patients with HDL2. Patients with HD also performed better than those with HDL2 in the Luria Test, although this test has limitations in differentiating dementia. Although most movement abnormalities are similar in HD and HDL2, the extent of dysarthria and dystonia were worse in HDL2.

Evidence currently available points to the fact that HD expansion mutation leads to disease mainly through the expression of a protein with an overly long stretch of polyglutamine residues. This is coupled with the toxic role of mutant RNA and loss of protein function (Romo et al., 2018). In HDL2, data point to the loss of *JPH3* expression and toxic property of *JPH3* RNA in the neurodegeneration. There are also data establishing the role of cryptic antisense expression of a transcript encoding a protein with an expanded stretch of polyglutamine residues (Seixas et al., 2012; Rudnicki et al., 2007; Wilburn et al., 2011). This similarity between the two conditions points to their common pathogenesis (Anderson et al., 2019). However, the subtle divergency in the pathogenesis of the two conditions leads to phenotypic differences such as dysarthria and dystonia.

8.6 PARKINSONISM IN HUNTINGTON'S DISEASE

Atypical parkinsonism can be a key symptom in late-onset HD (Ciammola et al., 2011). With the development of parkinsonism in HD, there is an associated progressive functional disability (Garcia Ruiz et al., 2002), which makes in-depth knowledge of the pathological mechanism of parkinsonism very relevant in the management of HD.

8.7 MANAGEMENT OF HUNTINGTON'S DISEASE

More than a century after its first description, no curative treatment has been discovered for HD, and management remains largely symptomatic. However, these symptomatic treatments are thought to be efficacious in managing some of the disease's worrisome symptoms (Bachoud-Lévi et al., 2019). The primary goals of therapy therefore remain improving patients' quality of life, assisting with the changing needs of patients, and reducing mortality from complications.

8.7.1 Pharmacologic Management

8.7.1.1 Managing Motor Features

Abnormal movements such as chorea and dystonia in HD are due to changes in neural circuitry and neurotransmitters such as DA. D_2 receptor antagonists such as risperidone and olanzapine have been used to suppress the extent and incidence of these abnormal movements (Loi et al., 2018). Consensus states that pharmacologic intervention should be considered if chorea causes distress to the patient (Bachoud-Lévi et al., 2019). Unless the patient has a history of poorly managed depression and suicidal ideation, tetrabenazine – an inhibitor of synaptic vesicular monoamine transporter 2 – should be considered as one of the first-line agents (Huang et al., 1976). A longer-acting derivative, deutetrabenazine, is available and has been shown in randomized controlled trial to significantly reduce chorea (Loi et al., 2018). Tetrabenazine depletes DA stores and helps rebalance iSPN and direct pathway SPN (dSPN) activity by decreasing D_1-mediated activation of dSPNs (Smith-Dijak et al., 2019). In addition, tetrabenazine may also reduce DA tone and restore normal modulation of glutamatergic activity at the corticostriatal synapse (André et al., 2010).

Another drug that has been studied in clinical trials for the improvement of motor symptoms, including hand movements, gait, and balance, in HD is pridopidine (de Yebenes et al., 2011). Pridopidine binds to D_2 receptors and acts as a partial agonist to enhance receptor activity when the levels are low. However, in the presence of high levels of DA, the drug acts as an antagonist to inhibit full activation of the D_2 receptor (Dyhring et al., 2010). As a partial agonist, pridopidine attenuates dSPN activation by increasing stimulation of D_2 receptors on presynaptic glutamatergic inputs to SPNs. As an antagonist, the drug reduces D_2 receptor oversight on cholinergic neurons and thus increases the release of acetylcholine and inhibition of dSPNs. Through these mechanisms pridopidine could improve motor symptoms in early HD (Smith-Dijak et al., 2019). However, in clinical trials, the ability of pridopidine to improve motor symptoms was not significant (de Yebenes et al., 2011).

In the event that personality, behavioral, and/or psychotic symptoms are present, second-generation narcoleptics can be used in the first-line management of chorea (Bachoud-Lévi et al., 2019). Drugs such as olanzapine, sulpiride, quetiapine, and risperidone have been successfully used in the treatment of HD chorea. However, side effects such as weight gain and, most commonly, sedation, limit their use (Coppen and Roos, 2017). Pharmacologic management of chorea may also improve gait and balance (Bachoud-Lévi et al., 2019).

In the treatment of focal dystonia or prevention of secondary deformities, botulinum toxin injections have been used. Injection of botulin toxin A into the masseter muscles is also the proposed first-line treatment for bruxism. However, treatment should be administered by a trained professional to minimize the risk of paralysis (Ross and Tabrizi, 2011; Bachoud-Lévi et al., 2019). It is worth noting that bruxism and akathisia may occur as a side effect of therapy involving antipsychotics and serotonin reuptake inhibitors, and thus a dose reduction should be considered (Nash et al., 2004).

In situations where myoclonus impacts on the functional capacities of patients, treatment can be initiated with sodium valproate or clonazepam, as mono-/bi-therapy. The doses can be carefully titrated until symptom control is achieved or the patient cannot tolerate further increments (Vogel et al., 1991). An alternative is the anticonvulsant, levetiracetam. Another pyrrolidone derivative, piracetam, is indicated for the treatment of myoclonus of cortical origin that is not associated with epileptic seizures. Clonazepam, a benzodiazepine, may also be used in the management of myoclonus; however, the incidence of adverse events such as somnolence, increasing falls, and the risk of drug-dependence must be carefully monitored (Bachoud-Lévi et al., 2019). A summary of pharmacological agents used in the management of motor symptoms in HD and their common side effects are presented in Table 8.1.

8.7.1.2 Management of Cognitive Dysfunction

Cognitive dysfunction is functionally disabling and often appears before the onset of motor symptoms (Beglinger et al., 2010). Currently, no pharmacologic treatments are recommended for the management of these deficits. Citalopram was studied and discovered to have no effect on cognitive function in HD (Beglinger et al., 2014). Two other randomized controlled trials assessed the benefit of other anti-cholinesterase inhibitors in cognitive dysfunction in HD; however, the studies were not adequately powered, and the results were conflicting (Li et al., 2015; Han et al., 2019). Atomoxetine, a norepinephrine reuptake inhibitor, also failed to enhance cognitive function in small groups of patients with HD in the early stage of the disease, despite showing benefits in preclinical studies (Beglinger et al., 2009).

Alternate approaches to indirectly improving cognition by enhancing alertness with agents such as modafinil have also not been proven to be effective in HD (Blackwell et al., 2008). Sedatives, antipsychotics, and tetrabenazine may impact negatively on memory (Bachoud-Lévi et al., 2019). With

TABLE 8.1
Drug Therapy for Controlling Motor Symptoms in Huntington's Disease

Symptom	Drug	Class	Common Side Effects
Chorea	Tetrabenazine	Dopamine-depleting agents	Sedation, depression
	Olanzapine	Dopamine receptor antagonist	Sedation, extrapyramidal effects, weight gain, parkinsonism, and akathisia
Dystonia	Botulinum toxin	Acetylcholine release inhibitor at neuromuscular junction	Paralysis in adjacent muscles
Rigidity	Levodopa	Amino acid precursor of dopamine	Postural hypotension, insomnia, gastrointestinal disturbances, agitation, and psychiatric symptoms; may worsen chorea
	Baclofen	Skeletal muscle relaxant	Sedation, hypotension, drowsiness, confusion, gastrointestinal disturbances
Myoclonus	Sodium valproate	Anticonvulsant	Gastrointestinal disturbance, weight gain, blood dyscrasia, liver dysfunction
	Clonazepam	Benzodiazepine	Somnolence and increasing falls, risk of drug dependence
	Levetiracetam	Anticonvulsant	Gastrointestinal disturbance, mood changes, rash, myalgia
Gait and balance	Tetrabenazine	Dopamine-depleting agents	Sedation, depression, and can worsen walking disorders
Bruxism	Botulinum toxin A	Acetylcholine release inhibitor at neuromuscular junction	Paralysis in adjacent muscles
Manual dexterity	Tetrabenazine	Dopamine-depleting agents	Sedation, depression, and can worsen walking disorders

cognitive dysfunction in patients with HD going unmanaged, it is important for patient-management teams to ensure that medications chosen for the management of other dysfunctions or comorbid conditions do not exacerbate any cognitive deficits (Mason and Barker, 2015).

8.7.1.3 Management of Affective Disorders

Affective disorders are common in HD and multifactorial in etiology, and focusing on their management may have the greatest impact on the quality of life of patients and their families (Loi et al., 2018). Depression is the most commonly encountered psychiatric disorder, and evidence from case studies shows its responsiveness to antidepressant medications such as fluoxetine, mirtazapine, and monoamine oxidase inhibitors (Bonelli, 2003). The serotonin norepinephrine reuptake inhibitor venlafaxine significantly reduces depression, but it has been associated with a high incidence of side effects such as nausea and irritability (Holl et al., 2010).

A review of international prescribing practices for affective disorder management in HD has shown serotonin reuptake inhibitors to be the most commonly prescribed class of drugs in the management of mild-to-moderate irritability and obsessive–compulsive behaviors, whereas antipsychotics are the preferred class in the treatment of severe aggressive behaviors (Anderson et al., 2011; Groves et al., 2011). Evidence supporting these practices is however weak (Mason and Barker, 2015). Psychotic symptoms such as hallucinations and delusions often respond to higher doses of antipsychotics, and risperidone may be used as a first-line agent since it also suppresses movement deficits (Loi et al., 2018).

Frontal dysexecutive syndrome, characterized by irritability, agitation, impulsivity, apathy, disinhibition, and perseveration, is extremely challenging to manage. Thus, agents and events that predispose patients to developing these symptoms should be avoided as much as possible (Loi et al., 2018). Mood stabilizers such as carbamazepine and valproate, and selective serotonin reuptake inhibitors (SSRIs) can be used to manage irritability and aggression. Obsessiveness and compulsivity, repetitive behaviors, inflexibility, and rigidity can be managed with the tricyclic antidepressant clomipramine and SSRIs (Loi et al., 2018). Per current treatment guidelines, impulsivity associated with depression or personality disorders should be managed with an antipsychotic in combination with a SSRI. This is due to the risk of auto- or heteroaggressiveness in affected patients (Bachoud-Lévi et al., 2019). Table 8.2 presents a summary of the pharmacologic interventions available for the management of affective disorders in HD.

8.7.1.4 Management of Other Features of Huntington's Disease

Aside the classic triad (motor, cognitive, and psychiatric features), other symptoms are often present and may reduce the quality of life of patients. These commonly include weight loss, sleep disturbance, and dysphagia (Bachoud-Lévi et al., 2019). Weight loss responds well to a combination of dietary supplements and olanzapine (Loi et al., 2018). The use of pharmacologic intervention in sleep disturbance should be considered only after lifestyle strategies fail. A hypnotic, for a short duration, may be used to avoid the risk of drug dependence. Agents such as mianserin, mirtazapine, and antihistamines may be used for long-term therapy, as they are associated with a reduced tendency to cause dependency. Melatonin may also be considered in the treatment of sleep phase inversion. In patient with sleep disorders associated with chorea or psychiatric symptoms, an antipsychotic should be prescribed in the evening to manage the symptoms (Bachoud-Lévi et al., 2019).

8.7.2 Gene Therapy

Although the current management of HD focuses on symptomatic management and does not slow disease

TABLE 8.2
Drug Therapy for Controlling Affective Symptoms in Huntington's Disease

Symptom	Drug	Class	Common Side Effects
Depression	Fluoxetine, paroxetine	SSRI	Sedation, drowsiness, gastrointestinal disturbance, and postural hypotension
	Venlafaxine	SNRI	Nausea, irritability, hypertension, and gastrointestinal disturbance
Irritability	Fluoxetine	SSRI	Sedation, drowsiness, gastrointestinal disturbance, and postural hypotension
	Carbamazepine	Benzodiapine	Dizziness, dry mouth, fluid imbalance, movement disorders, and thrombocytopenia
	Mirtazapine	Tricyclic antidepressant	Sedation, tremors, edema, headache, and weight gain
Apathy	Fluoxetine	SSRI	As for SSRIs
	Venlafaxine	SNRI	Nausea, irritability, hypertension, and gastrointestinal disturbance
Anxiety	Citalopram	SSRI	As for SSRIs
	Venlafaxine	SNRI	As for SNRIs
	Lorazepam	Benzodiazepine	Dizziness, dry mouth, fluid imbalance, movement disorders, and thrombocytopenia
Psychosis	Olanzapine	Atypical antipsychotic	Sedation, parkinsonism, tardive dyskinesia, and neuroleptic malignant syndrome
	Sulpiride, amisulpride	First-generation antipsychotic	Dry mouth, constipation, sedation, dystonia, agitation, dystonia, akathisia, and hypotension
Obsession and compulsivity	Olanzapine and risperidone	Atypical antipsychotic	As above
Sexual disorders	Risperidone	Atypical antipsychotic	As for atypical antipsychotics
		SSRI	As for SSRIs
Hallucinations	Risperidone	Atypical antipsychotic	Atypical antipsychotic

SSRIs, selective serotonin reuptake inhibitors; SNRIs, serotonin norepinephrine reuptake inhibitors.

progression, emerging therapies aimed at lowering neurotoxic mHTT perhaps serve as the most promising disease-modifying therapeutic options. These approaches targets DNA, RNA, and mutant *HTT*, and involve the use of antisense oligonucleotides (ASOs) and RNA interference compounds (RNAi) that target mRNA, and zinc-finger transcriptional repressors (ZFTR) and clustered inter-spaced short palindromic repeats (CRISPR/Cas9) methods that target mHTT DNA (McColgan and Tabrizi, 2018; Wild and Tabrizi, 2017). Additionally, the conversion of striatal astrocytes into GABAergic neurons in mouse models has also been achieved with gene therapy (Wu et al., 2020).

Several ASO therapies have made it into human trials for HD. IONIS-HTT$_{Rx}$ – a synthetic 20-nucleotide sequence phosphorothioate – is an ASO and the first mHTT-lowering compound to enter into clinical trials (Wild and Tabrizi, 2017). It is safe and well tolerated in patients with early-stage HD, and reduces mHTT levels in CSF by up to 40% (Tabrizi et al., 2019). Because ASOs cannot cross the blood–brain barrier, they are delivered intrathecally, following which they target and bind specific mRNA sequences. They then catalyze the degradation of HTT mRNA by ribonuclease H (McColgan and Tabrizi, 2018). ASOs can be designed to be either allele specific, where they interact only with mutant *HTT* allele, or allele non-specific, where they bind to both mHTT mRNA and normal HTT mRNA (Stahl and Feigin, 2020). As opposed to ASOs, RNAi utilizes double-stranded RNA sequences that are complementary to HTT mRNA. However, they bind mRNA and initiate their degradation and subsequent reduction in the target protein similar to ASOs. As opposed to ASOs, current RNAi therapies in development utilize viral vectors, such as adeno-associated virus, as delivery systems to target regions in the brain – most often the striatum – through stereotactic neurosurgery (Bashir, 2019).

Editing mHTT DNA has been achieved using ZFTR and CRISPR/Cas9 approaches. In recent years, the system has been harnessed as a tool for genome editing with a multitude of applications to human disease. This CRISPR/Cas9 system, originally derived from genome-editing technique in bacteria, has been used to achieve permanent and mutant allele-specific inactivation of the mutant *HTT* gene in patients with HD (Shin et al., 2016).

Another prominent agent still in clinical trial (SIGNAL; NCT02481674) is pepinemab, a monoclonal antibody that blocks semaphoring-4D, a transmembrane signaling protein involved in the processes of neuroinflammation and neurodegeneration (Stahl and Feigin, 2020).

8.8 RECENT DEVELOPMENTS AND CHALLENGES

It is apparent that the aggregation of mHTT produces alterations in the activity of and communication between neurons, which results in the characteristic presentation of HD. Changes in the function of synaptic signaling molecules such as DA and BDNF are relatively well documented. However, there remains much to be understood about the exact roles of endocannabinoids, adenosine, and GABA, and their respective receptors in the striatum, cortex, and hippocampus of the HD brain. Further studies are needed to enable researchers understand whether the apparent increase in corticostriatal GABAergic signaling is a result of a pathogenic effect of mHTT upon inhibitory interneurons, or a homeostatic response to other primary disease-causing factors in HD. Sufficiently elucidating these signaling pathways will aid in the identification of potential targets for drug development in HD treatment.

Therapy that focuses on lowering mHTT protein by targeting RNA, DNA, or the mutant *HTT* gene currently remains the most promising in altering the disease course of HD. Future studies may thus focus on expanding the sample size in randomized controlled trials, to enable clear descriptions of the efficacy of these interventions. Studies are also needed to establish the long-term safety profiles of these therapies, as data remain sparse. Although gene therapy holds promise in modifying the course of HD, better therapies are also needed to manage cognitive and psychiatric disturbances. Some natural products such as cannabinoids, luteolin, and the flavonoids myricetin, hesperidin, quercetin, naringin, and fisetin have been shown to be neuroprotective in HD (Khan et al., 2020). Further studies are needed to investigate these natural products as alternatives to conventional therapies in the symptomatic treatment of HD. The translation of results from animal studies to clinical settings remains one of the most challenging hurdles in HD research. More robust models are therefore needed to maximize safety and the chance of successful translation from the bench to the clinic.

8.9 CONCLUSION

This chapter has expansively discussed the most relevant neurochemical signaling pathways in HD, and has suggested a possible path of future research into this neurodegenerative disorder. With extensive studies still ongoing on HD, the far-reaching consequences of no definitive cure for the condition remain palpable. This opens avenues for more to be done in understanding this genetic disorder and in directing research towards the development of a treatment for HD. With advances in gene therapy, there is an auspicious dimension to this quest.

REFERENCES

Abeliovich A, Schmitz Y, Farinas I, et al. Mice lacking alpha-synuclein display functional deficits in the nigrostriatal dopamine system. Neuron. 2000;25:239–252.

Anderson DG, Ferreira-Correia A, Rodrigues FB, et al. Comparison of the Huntington's disease like 2 and Huntington's disease clinical phenotype. Mov Disord Clin Pract 2019;6(4):302–311.

Anderson DG, Walker RH, Connor M, et al. A systematic review of the Huntington disease-Like 2 phenotype. J Huntingtons Dis. 2017;6:37–46.

Anderson K, Craufurd D, Edmondson MC, et al. An international survey-based algorithm for the pharmacologic treatment of obsessive-compulsive behaviors in Huntington's disease. PLoS Curr. 2011;3:RRN1261.

André VM, Cepeda C, Levine MS. Dopamine and glutamate in Huntington's disease: a balancing act. CNS Neurosci Ther. 2010;16:163–178.

Aquilano K, Baldelli S, Pagliei B, et al. p53 orchestrates the PGC-1a-mediated antioxidant response upon mild redox and metabolic imbalance. Antioxid Redox Signal. 2013;18:386–399.

Argenziano M, Tortora C, Bellini G, et al. The endocannabinoid system in pediatric inflammatory and immune diseases. Int J Mol Sci. 2019; 20:5875.

Augustin SM, Chancey JH, Lovinger DM. Dual dopaminergic regulation of corticostriatal plasticity by cholinergic interneurons and indirect pathway medium spiny neurons. Cell Rep. 2018; 24:2883–2893.

Bachoud-Lévi A-C, Ferreira J, Massart R, et al. International guidelines for the treatment of Huntington's disease. Front Neurol. 2019;10:710.

Bae BI, Xu H, Igarashi S, et al. p53 mediates cellular dysfunction and behavioral abnormalities in Huntington's disease. Neuron. 2005;47:29–41.

Bashir H. Emerging therapies in Huntington's disease. Expert Rev Neurother. 2019;19:983–995.

Beglinger LJ, Adams WH, Langbehn D, et al. Results of the citalopram to enhance cognition in Huntington disease trial. Mov Disord. 2014; 29: 401–405.

Beglinger LJ, Adams WH, Paulson H, et al. Randomized controlled trial of atomoxetine for cognitive dysfunction in early Huntington disease. J Clin Psychopharmacol. 2009;29:484–487.

Beglinger LJ, O'Rourke JJ, Wang C, et al. Earliest functional declines in Huntington disease. Psychiatry Res. 2010;178:414–418.

Bezprozvanny I, Hayden MR. Deranged neuronal calcium signaling and Huntington disease. Biochem Biophys Res Commun. 2004;322:1310–1317.

Bhatti JS, Bhatti GK, Reddy H. Mitochondrial dysfunction and oxidative stress in metabolic disorders—a step towards mitochondria based therapeutic strategies. Biochim Biophys Acta Mol Basis Dis. 2017;1863:1066–1077.

Bird ED. Chemical pathology of Huntington's disease. Annu Rev Pharmacol Toxicol. 1980;20:533–551.

Björkqvist M, Wild EJ, Thiele J, et al. A novel pathogenic pathway of immune activation detectable before clinical onset in Huntington's disease. J Exp Med. 2008;205:1869–1877.

Blackwell AD, Paterson NS, Barker RA, et al. The effects of modafinil on mood and cognition in Huntington's disease. Psychopharmacology. 2008;199:29–36.

Blázquez C, Chiarlone A, Bellocchio L, et al. The CB1 cannabinoid receptor signals striatal neuroprotection via a PI3K/Akt/mTORC1/BDNF pathway. Cell Death Differ. 2015;22:1618–1629.

Blum D, Chern Y, Rosaria D, et al. The role of adenosine tone and adenosine receptors in Huntington's disease. J Caffeine Adenosine Res. 2018;8:43–58.

Bonelli RM. Mirtazapine in suicidal Huntington's disease. Ann Pharmacother. 2003;37:452.

Bonelli RM, Mahnert FA, Niederwieser G. Olanzapine for Huntington's disease: an open label study. Clin Neuropharmacol. 2002;25:263–265.

Bouwens JA, van Duijn E, van der Mast RC, et al. Irritability in a prospective cohort of Huntington's disease mutation carriers. J Neuropsychiatry Clin Neurosci. 2015;27:206–12.

Breza M, Emmanouilidou E, Leandrou E, et al. Elevated serum α-synuclein levels in Huntington's disease patients. Neuroscience. 2020;431:34–39

Browne SE, Bowling AC, MacGarvey U, et al. Oxidative damage and metabolic dysfunction in Huntington's disease: selective vulnerability of the basal ganglia. Ann Neurol. 1997;41:646–653.

Burggren A, Brown J. Imaging markers of structural and functional brain changes that precede cognitive symptoms in risk for Alzheimer's disease. Brain Imaging Behav 2014;8(2):251–261.

Byrne LM, Rodrigues FB, Blennow K, et al. Neurofilament light protein in blood as a potential biomarker of neurodegeneration in Huntington's disease: a retrospective cohort analysis. Lancet Neurol. 2017;16:601–609.

Calabresi P, Centonze D, Gubellini P, et al. Glutamate-triggered events inducing corticostriatal long-term depression. J Neurosci. 1999;19:6102–6110.

Cattaneo E, Zuccato C, Tartari M. Normal huntingtin function: an alternative approach to Huntington's disease. Nat Rev Neurosci. 2005;6:919–930.

Cepeda C, Galvan L, Holley SM, et al. Multiple sources of striatal inhibition are differentially affected in Huntington's disease mouse models. J Neurosci. 2013; 33:7393–7406.

Cepeda C, Wu N, Andre VM, et al. The corticostriatal pathway in Huntington's disease. Prog Neurobiol. 2007;81:253–271.

Chang KH, Wu YR, Chen YC, Chen CM. Plasma inflammatory biomarkers for Huntington's disease patients and mouse model. Brain Behav Immun. 2015;44:121–127.

Chao MV, Hempstead BL. p75 and Trk: a two-receptor system. Trends Neurosci. 1995;18:321–326.

Chen KC, Yang YK, Howes O, et al. Striatal dopamine transporter availability in drug-naive patients with schizophrenia: a case-control SPECT study with [(99m)Tc]-TRODAT-1 and a meta-analysis. Schizophr Bull. 2013;39:378–386.

Chiarlone A, Bellocchio L, Blázquez C, et al. A restricted population of CB_1 cannabinoid receptors with neuroprotective activity. Proc Natl Acad Sci U S A. 2014;111:8257–8262.

Chiodi V, Uchigashima M, Beggiato S, et al. Unbalance of CB1 receptors expressed in GABAergic and glutamatergic neurons in a transgenic mouse model of Huntington's disease. Neurobiol Dis. 2012;45:983–991.

Chiu F-L, Lin J-T, Chuang C-Y, et al. Elucidating the role of the A2A adenosine receptor in neurodegeneration using neurons derived from Huntington's disease iPSCs. Hum Mol Genet. 2015;24:6066–6079.

Ciammola A, Sassone J, Poletti B, et al. Atypical parkinsonism revealing a late onset, rigid and akinetic form of Huntington's disease. Case Rep Neurol Med. 2015;2011:696953.

Cochen V, Degos JD, Bachoud-Lévi AC. Efficiency of carbamazepine in the treatment of micturitional disturbances in Huntington disease. Neurology. 2000;55:1934.

Coppen EM, Roos RA. Current pharmacological approaches to reduce chorea in Huntington's disease. Drugs. 2017;77:29–46.

Costa V, Giacomello M, Hudec R, et al. Mitochondrial fission and cristae disruption increase the response of cell models of Huntington's disease to apoptotic stimuli. EMBO Mol Med. 2010;2:490–503.

Covey DP, Dantrassy HM, Zlebnik NE, et al. Compromised dopaminergic encoding of reward accompanying suppressed willingness to overcome high effort costs is a prominent prodromal characteristic of the Q175 mouse model of Huntington's disease. J Neurosci. 2016;36:4993–5002.

Craufurd D, MacLeod R, Frontali M, et al. Diagnostic genetic testing for Huntington's disease. Pract Neurol. 2015;15:80–84.

Craufurd D, Snowden J. Neuropsychological and neuropsychiatric aspects of Huntington's disease. In: Bates G, Harper P, Jones L, eds. Huntington's Disease. New York: Oxford University Press, 2002:62–94.

Craufurd D, Thompson JC, Snowden JS. Behavioral changes in Huntington disease. Neuropsychiatry Neuropsychol Behav Neurol. 2001;14:219–226.

Crittenden JR, Graybiel AM. Basal ganglia disorders associated with imbalances in the striatal striosome and matrix compartments. Front Neuroanat. 2001;5.

Crotti A, Glass CK. The choreography of neuroinflammation in Huntington's disease. Trends Immunol. 2015;36:364–373.

Cummings DM, Andre VM, Uzgil BO, et al. Alterations in cortical excitation and inhibition in genetic mouse models of Huntington's disease. J Neurosci. 2009;29:10371–10386.

Cyr M, Sotnikova TD, Gainetdinov RR, Caron MG. Dopamine enhances motor and neuropathological consequences of polyglutamine expanded huntingtin. FASEB J. 2006;20:2541–2543.

Dallerac GM, Vatsavayai SC, Cummings DM, et al. Impaired long-term potentiation in the prefrontal cortex of Huntington's disease mouse models: rescue by D1 dopamine receptor activation. Neurodegener Dis. 2011;8:230–239.

Dalrymple A, Wild EJ, Joubert R, et al. Proteomic profiling of plasma in Huntington's disease reveals neuroinflammatory activation and biomarker candidates. J Proteome Res. 2007;6:2833–2840.

Dargaei Z, Bang JY, Mahadevan V, et al. Restoring GABAergic inhibition rescues memory deficits in a Huntington's disease mouse model. Proc Natl Acad Sci U S A. 2018;115:E1618–E1626.

Dau A, Gladding CM, Sepers MD, Raymond LA. Chronic blockade of extrasynaptic NMDA receptors ameliorates synaptic dysfunction and pro-death signaling in Huntington disease transgenic mice. Neurobiol Dis. 2014;62:533–542.

de Yebenes JG, Landwehrmeyer B, Squitieri F. et al. Pridopidine for the treatment of motor function in patients with Huntington's disease (MermaiHD): a phase 3, randomised, double-blind, placebo-controlled trial. Lancet Neurol 2011;10(12):1049–1057.

Dietrich P, Johnson IM, Alli S, Dragatsis I. Elimination of huntingtin in the adult mouse leads to progressive behavioral deficits, bilateral thalamic calcification, and altered brain iron homeostasis. PLOS Genet. 2017; https://doi.org/10.1371/journal.pgen.1006846.

Domenici MR, Chiodi V, Averna M, et al. Neuronal adenosine A2A receptor overexpression is neuroprotective towards 3-nitropropionic acid-induced striatal toxicity: a rat model of Huntington's disease. Purinergic Signal. 2018;14:235–243.

Dorsey ER, Beck CA, Darwin K, et al. Natural history of Huntington disease. JAMA Neurol. 2013;70:1520–1530.

Dumas EM, van den Bogaard SJ, Ruber ME, et al. Early changes in white matter pathways of the sensorimotor cortex in premanifest Huntington's disease. Hum Brain Mapp. 2012;33:203–212.

Dvorzhak A, Vagner T, Kirmse K, Grantyn R. Functional indicators of glutamate transport in single striatal astrocytes and the influence of Kir4.1 in normal and Huntington mice. J Neurosci. 2016;36:4959–4975.

Dyhring T, Nielsen EØ, Sonesson C, et al. The dopaminergic stabilizers pridopidine (ACR16) and (-)-OSU6162 display dopamine D2 receptor antagonism and fast receptor dissociation properties. Eur J Pharmacol2010;628:19–26.

Estrada-Sánchez AM, Rebec GV. Role of cerebral cortex in the neuropathology of Huntington's disease. Front Neural Circuits. 2013;7:19.

Federico A, Cardaioli E, Da Pozzo P, et al. Mitochondria, oxidative stress and neurodegeneration. J Neurol Sci. 2012;322:254–262.

Feigin A, Leenders KL, Moeller JR, et al. Metabolic network abnormalities in early Huntington's disease: an [(18)F]FDG PET study. J Nucl Med. 2001;42:1591–1595.

Fiedorowicz JG, Mills JA, Ruggle A, et al. Suicidal behavior in prodromal Huntington disease. Neurodegener Dis. 2011;8:483–490.

Flores-Hernandez J, Cepeda C, Hernandez-Echeagaray E, et al. Dopamine enhancement of NMDA currents in dissociated medium-sized striatal neurons: role of D1 receptors and DARPP-32. J Neurophysiol. 2002;88:3010–3020.

Foster NL, Chase TN, Denaro A, et al. THIP treatment of Huntington's disease. Neurology. 1983;33:637–639.

Fusco FR, Chen Q, Lamoreaux WJ, et al. Cellular localization of huntingtin in striatal and cortical neurons in rats: lack of correlation with neuronal vulnerability in Huntington's disease. J Neurosci. 1999;19:1189–1202.

Garcia Ruiz PJ, Hernández J, Cantarero S, et al. Bradykinesia in Huntington's disease. A prospective, follow-up study. J Neurol. 2002;249:437–440.

Gerfen CR. The neostriatal mosaic: multiple levels of compartmental organization. Trends Neurosci. 1992;15:133–139.

Giralt A, Brito V, Chevy Q, et al. Pyk2 modulates hippocampal excitatory synapses and contributes to cognitive deficits in a Huntington's disease model. Nat Commun. 2017;8:15592.

Gladding CM, Fan J, Zhang LYJ, et al. Alterations in striatal-enriched protein tyrosine phosphatase (STEP) expression, activation and downstream signaling in early and late stages of the YAC128 Huntington's disease mouse model. J Neurochem. 2014;130:145–159.

Gomez-Pastor R, Burchfiel ET, Neef DW, et al. Abnormal degradation of the neuronal stress-protective transcription factor HSF1 in Huntington's disease. Nat Commun. 2017;8:14405.

Gonzalez-Alegre P, Afifi AK. Clinical characteristics of childhood-onset (juvenile) Huntington disease: report of 12 patients and review of the literature. J Child Neurol. 2006;21:223–229.

Gordon AM, Quinn L, Reilmann R, Marder K. Coordination of prehensile forces during precision grip in Huntington's disease. Exp Neurol. 2000;163:136–148.

Groves M, Van Duijn E, Anderson K, et al. An international survey-based algorithm for the pharmacologic treatment of irritability in Huntington's disease. PLoS Curr. 2011;3:RRN1259.

Guitart X, Bonaventura J, Rea W, et al. Equilibrative nucleoside transporter ENT1 as a biomarker of Huntington disease. Neurobiol Dis. 2016; 96:47–53.

Han J-Y, Besser LM, Xiong C, et al. Cholinesterase inhibitors may not benefit mild cognitive impairment and mild Alzheimer disease dementia. Alzheimer Dis Assoc Disord 2019;33(2):87–94.

Hands S, Sajjad MU, Newton MJ, Wyttenbach A. In vitro and in vivo aggregation of a fragment of huntingtin protein directly causes free radical production. J Biol Chem. 2011;286:44512–44520.

HDCRG (The Huntington's Disease Collaborative Research Group). A novel gene containing a trinucleotide repeat that is expanded and unstable on Huntington's disease chromosome. Cell. 1993;72:971–983.

Hersch S, Jones R, Koroshetz W, Quaid K. The neurogenetics genie: testing for the Huntington's disease mutation. Neurology. 1994;44:1369–1373.

Holl AK, Wilkinson L, Painold A, et al. Combating depression in Huntington's disease: effective antidepressive treatment with venlafaxine XR. Int Clin Psychopharmacol. 2010;25:46–50.

Horne EA, Coy J, Swinney K, et al. Downregulation of cannabinoid receptor 1 from neuropeptide Y interneurons in the basal ganglia of patients with Huntington's disease and mouse models. Eur J Neurosci. 2013;37:429–440.

Huang CY, McLeod JG, Holland RT, Elliot C. Tetrabenazine in the treatment of Huntington's chorea. Med J Aust. 1976;1:583–584.

Ignácio ZM, Quevedo J, Réus GZ. Pathophysiological mechanisms of Huntington's disease. In: S Singh, N Joshi (eds). Pathology, Prevention and Therapeutics of Neurodegenerative Disease. Singapore: Springer Nature, 2019;49–59.

Intihar TA, Martinez EA, Gomez-Pastor R. Mitochondrial dysfunction in Huntington's disease; interplay between HSF1, p53 and PGC-1α. Transcription Factors Front Cell Neurosci. 2019;13:103.

Jimenez-Sanchez M, Licitra F, Underwood BR, Rubinsztein DC. Huntington's disease: mechanisms of pathogenesis and therapeutic strategies. Cold Spring Harb Perspect Med. 2017;7:a024240.

Kao Y-H, Lin M-S, Chen C-M, et al. Targeting ENT1 and adenosine tone for the treatment of Huntington's disease. Hum Mol Genet. 2017;26:467–478.

Khan H, Ullah H, Tundis R, et al. Dietary flavonoids in the management of Huntington's disease: mechanism and clinical perspective. eFood. 2020;1:38–52.

Kiyatkin EA, Rebec GV. Dopaminergic modulation of glutamate-induced excitation of neurons in the neostriatum and nucleus accumbens of awake, unrestrained rats. J Neurophysiol. 1996;75:142–153.

Koch ET, Raymond LA. Dysfunctional striatal dopamine signaling in Huntington's disease. J Neurosci Res. 2019; 97:1636–1654.

Kolodziejczyk K, Parsons MP, Southwell AL, et al. Striatal synaptic dysfunction and hippocampal plasticity deficits in the Hu97/18 mouse model of Huntington disease. PLoS ONE. 2014;9:e94562.

Krause A, Mitchell C, Essop F, et al. Junctophilin 3 (JPH3) expansion mutations causing Huntington disease like 2 (HDL2) are common in South African patients with African ancestry and a Huntington disease phenotype. Am J Med Genet B Neuropsychiatr Genet. 2015;168:573–585.

Kremer B. Clinical neurology of Huntington's disease. In: Bates G, Harper P, Jones L (eds). Huntington's Disease. New York: Oxford University Press, 2002;3–27.

Kumar A, Ratan RR. Oxidative stress and Huntington's disease: the good, the bad, and the ugly. J Huntingtons Dis. 2016;5:217–217.

La Spada AR., Morrison RS. The power of the dark side: Huntington's disease protein and p53 form a deadly alliance. Neuron. 2005;47:1–3.

Laere KV, Casteels C, Dhollander I, et al. Widespread decrease of type 1 cannabinoid receptor availability in Huntington disease in vivo. J Nucl Med. 2010;51:1413–1417.

Leszek J, Barreto GE, Gąsiorowski K, et al. Inflammatory mechanisms and oxidative stress as key factors responsible for progression of neurodegeneration: role of brain innate immune system. CNS Neurol Disord Drug Targets. 2016;15:329–336.

Li H, Wyman T, Yu ZX, et al. Abnormal association of mutant huntingtin with synaptic vesicles inhibits glutamate release. Hum Mol Genet. 2003;12:2021–2030.

Li W, Silva HB, Real J, et al. Inactivation of adenosine A2A receptors reverses working memory deficits at early stages of Huntington's disease models. Neurobiol Dis. 2015;79:70–80.

Li Y, Hai S, Zhou Y, Dong BR. Cholinesterase inhibitors for rarer dementias associated with neurological conditions. Cochrane Database Syst Rev. 2015;CD009444.

Loi SM, Walterfang M, Velakoulis D, Looi JCL. Huntington's disease: managing neuropsychiatric symptoms in Huntington's disease. Australas Psychiatry. 2018;26:376–380.

Ludlow CL, Connor NP, Bassich CJ. Speech timing in Parkinson's and Huntington's disease. Brain Lang. 1987;32:195–214.

McColgan P, Tabrizi SJ. Huntington's disease: a clinical review. Eur J Neurol. 2018;25:24–34.

McNeil SM, Novelletto A, Srinidhi J, et al. Reduced penetrance of the Huntington's disease mutation. Hum Mol Genet. 1997;6:775–79.

Mahant N, McCusker EA, Byth K, Graham S. Huntington's disease: clinical correlates of disability and progression. Neurology. 2003;61:1085–1092.

Margolis RL, O'Hearn E, Rosenblatt A, et al. A disorder similar to Huntington's disease is associated with a novel CAG repeat expansion. Ann Neurol. 2001;50:373–380.

Margolis RL, Rudnicki DD, Holmes SE. Huntington's disease like-2: review and update. Acta Neurol Taiwan. 2005;14:1–8.

Martin DDO, Ladha S, Ehrnhoefer DE, Hayden MR. Autophagy in Huntington disease and huntingtin in autophagy. Trends Neurosci. 2015;38:26–35.

Martinez-Horta S, Perez-Perez J, van Duijn E, et al. Neuropsychiatric symptoms are very common in premanifest and early stage Huntington's disease. Parkinsonism Relat Disord. 2016;25:58–64.

Martinez-Vicente M, Talloczy Z, Wong E, et al. Cargo recognition failure is responsible for inefficient autophagy in Huntington's disease. Nat Neurosci. 2010;13:567–576.

Martino D, Stamelou M, Bhatia KP. The differential diagnosis of Huntington's disease like syndromes: 'red flags' for the clinician. J Neurol Neurosurg Psychiatry. 2013;84:650–656.

Martiros N, Burgess AA, Graybiel AM. Inversely active striatal projection neurons and interneurons selectively delimit useful behavioral sequences. Curr Biol. 2018;28:560–573.

Mason SL, Barker RA. Advancing pharmacotherapy for treating Huntington's disease: a review of the existing literature. Expert Opin Pharmacother. 2015;17:1–12.

MacLeod R, Tibben A, Frontali M, et al. Recommendations for the predictive genetic test in Huntington's disease. Clin Genet 2013;83(3):221–231.

Mehrabi NF, Waldvogel HJ, Tippett LJ, et al. Symptom heterogeneity in Huntington's disease correlates with neuronal degeneration in the cerebral cortex. Neurobiol Dis. 2016;96:67–74.

Menzies FM, Moreau K, Rubinsztein DC. Protein misfolding disorders and macroautophagy. Curr Opin Cell Biol. 2011;23:190–197.

Mievis S, Blum D, Ledent C. Worsening of Huntington disease phenotype in CB1 receptor knockout mice. Neurobiol Dis. 2011;42:524–529.

Miguez A, García-Díaz Barriga G, Brito V, et al. Fingolimod (FTY720) enhances hippocampal synaptic plasticity and memory in Huntington's disease by preventing p75NTR up-regulation and astrocyte-mediated inflammation. Hum Mol Genet. 2015;24:4958–4970.

Monsellier E, Bousset L, Melki R. α-Synuclein and Huntingtin exon 1 amyloid fibrils bind laterally to the cellular membrane. Sci Rep. 2016;6:19180.

Morigaki R, Goto S. Striatal vulnerability in Huntington's disease: neuroprotection versus neurotoxicity. Brain Sci. 2017;7:2–25.

Moylan S, Maes M, Wray NR, Berk M. The neuroprogressive nature of major depressive disorder: pathways to disease evolution and resistance, and therapeutic implications. Mol Psychiatry. 2013;18:595–606.

Myers RH. Huntington's disease genetics. NeuroRx. 2004;1:255–162.

Nash MC, Ferrell RB, Lombardo MA, Williams RB. Treatment of bruxism in Huntington's disease with botulinum toxin. J Neuropsychiatry Clin Neurosci. 2004;16:381–382.

Naydenov AV, Sepers MD, Swinney K, et al. Genetic rescue of CB1 receptors on medium spiny neurons prevents loss of excitatory striatal synapses but not motor impairment in HD mice. Neurobiol Dis. 2014;71:140–150.

Neef DW, Jaeger AM, Thiele DJ. Heat shock transcription factor 1 as a therapeutic target in neurodegenerative diseases. Nat Rev Drug Discov. 2011;10:930–944.

Novak MJ, Tabrizi SJ. Huntington's disease: clinical presentation and treatment. Int Rev Neurobiol. 2011;98:297–323.

Oliveira JM. Nature and cause of mitochondrial dysfunction in Huntington's disease: focusing on huntingtin and the striatum. J Neurochem. 2010;114:1–12.

Orr HT, Zoghbi HY. Trinucleotide repeat disorders. Annu Rev Neurosci. 2007;30:575–621.

Paoletti P, Vila I, Rifé M, et al. Dopaminergic and glutamatergic signaling crosstalk in Huntington's disease neurodegeneration: the role of p25/cyclin-dependent kinase 5. J Neurosci. 2008;28:10090–10101.

Papoutsi M, Labuschagne I, Tabrizi SJ, Stout JC. The cognitive burden in Huntington's disease: pathology, phenotype, and mechanisms of compensation. Mov Disord. 2014;29:673–683.

Paulsen JS, Hoth KF, Nehl C, Stierman L. Critical periods of suicide risk in Huntington's disease. Am J Psychiatry. 2005;162:725–31.

Paulsen JS, Miller AC, Hayes T, Shaw E. Cognitive and behavioral changes in Huntington disease before diagnosis. Handb Clin Neurol. 2017;144:69–91.

Plotkin JL, Surmeier DJ. Corticostriatal synaptic adaptations in Huntington's disease. Curr Opin Neurobiol. 2015; 33:53–62.

Rangel-Barajas C, Rebec GV. Dysregulation of corticostriatal connectivity in Huntington's disease: a role for dopamine modulation. J Huntingtons Dis. 2016;5:303–331.

Ravikumar B, Vacher C, Berger Z, et al. Inhibition of mTOR induces autophagy and reduces toxicity of polyglutamine expansions in fly and mouse models of Huntington disease. Nat Genet. 2004;36:585–595.

Raymond LA, André VM, Cepeda C, et al. Pathophysiology of Huntington's disease: time-dependent alterations in synaptic and receptor function. Neuroscience. 2011;198:252–273.

Reading SA, Yassa MA, Bakker A, et al. Regional white matter change in pre-symptomatic Huntington's disease: a diffusion tensor imaging study. Psychiatry Res. 2005;140:55–62.

Reilmann R, Kirsten F, Quinn L, et al. Objective assessment of progression in Huntington's disease: a 3-year follow-up study. Neurology. 2001;57:920–924.

Reiner A, Deng Y-P. Disrupted striatal neuron inputs and outputs in Huntington's disease. CNS Neurosci Ther. 2018;24:250–280.

Rodrigues FB, Byrne LM, McColgan P, et al. Cerebrospinal fluid inflammatory biomarkers reflect clinical severity in Huntington's disease. PLoS One. 2016;11:e0163479.

Romo L, Mohn ES, Aronin N. A fresh look at Huntingtin mRNA processing in Huntington's disease. J Huntingtons Dis. 2018;7:101–108.

Rosas HD, Chen YI, Doros G, et al. Alterations in brain transition metals in Huntington disease: an evolving and intricate story. Arch Neurol. 2012;69:887–893.

Rosas HD, Hevelone ND, Zaleta AK, et al. Regional cortical thinning in preclinical Huntington disease and its relationship to cognition. Neurology. 2005;65:745–747.

Ross CA, Tabrizi SJ. Huntington's disease: from molecular pathogenesis to clinical treatment. Lancet Neurol. 2011;10:83–98.

Rüb U, Vonsattel JP, Heinsen H, Korf HW. The neuropathology of Huntington's disease: classical findings, recent developments and correlation to functional neuroanatomy. Adv Anat Embryol Cell Biol. 2015;217:1–146.

Rubinsztein DC. Molecular biology of Huntington's disease (HD) and HD-like disorders. In: Pulst S (ed.). Genetics of Movement Disorders. San Diego, CA: Academic Press, 2003:365–367.

Rudnicki DD, Holmes SE, Lin MW, et al. Huntington's disease-like 2 is associated with CUG repeat-containing RNA foci. Ann Neurol. 2007;61:272–282.

Rui YN, Xu Z, Patel B, et al. Huntingtin functions as a scaffold for selective macroautophagy. Nat Cell Biol. 2015;17:262–275.

Ruiz-Calvo A, Maroto IB, Bajo-Grañeras R, et al. Pathway-specific control of striatal neuron vulnerability by corticostriatal cannabinoid CB1 receptors. Cereb Cortex. 2018;28:307–322.

Schneider SA, Marshall KE, Xiao J, LeDoux MS. JPH3 repeat expansions cause a progressive akinetic-rigid syndrome with severe dementia and putaminal rim in a five-generation African-American family. Neurogenetics. 2012;13:133–140.

Sciacca S, Favellato M, Madonna M, et al. Early enteric neuron dysfunction in mouse and human Huntington disease. Parkinsonism Relat Disord. 2017;34:73–74.

Seixas AI, Holmes SE, Takeshima H, et al. Loss of junctophilin-3 contributes to Huntington disease-like 2 pathogenesis. Ann Neurol. 2012;71:245–257.

Sharkey KA, Wiley JW. The role of the endocannabinoid system in the brain-gut axis. Gastroenterol. 2016;151:252–266.

Shin JW, Kim KH, Chao MJ, et al. Permanent inactivation of Huntington's disease mutation by personalized allele-specific CRISPR/Cas9. Hum Mol Genet. 2016;25:4566–4576.

Shoulson I, Goldblatt D, Charlton M, Joynt RJ. Huntington's disease: treatment with muscimol, a GABA-mimetic drug. Ann Neurol. 1978;4:279–284.

Shukla V, Mishra SK, Pant HC. Oxidative stress in neurodegeneration. Adv Pharmacol Sci. 2011;2011:572634.

Silva A, Naia L, Dominguez A, et al. Overexpression of BDF and full-length TrkB receptor ameliorate striatal neural survival in Huntington's disease. Neurodegener Dis. 2015;15:207–218.

Simonin C, Duru C, Salleron J, et al. Association between caffeine intake and age at onset in Huntington's disease. Neurobiol Dis. 2013;58:179–182.

Smith-Dijak AI, Nassrallah WB, Zhang LYJ, et al. Impairment and restoration of homeostatic plasticity in cultured cortical neurons from a mouse model of Huntington disease. Front Cell, Neurosci. 2019;13:209.

Spampanato J, Gu X, Yang XW, Mody I. Progressive synaptic pathology of motor cortical neurons in a BAC transgenic mouse model of Huntington's disease. Neuroscience. 2008;157:606–620.

Stahl CM, Feigin A. Medical, surgical, and genetic treatment of Huntington disease. Neurol Clin. 2020;38:367–378.

Starling AJ, André VM, Cepeda C, et al. Alterations in N-methyl-D-aspartate receptor sensitivity and magnesium blockade occur early in development in the R6/2 mouse model of Huntington's disease. J Neurosci Res. 2005;82:377–386.

Stober T, Wussow W, Schimrigk K. Bicaudate diameter: the most specific and simple CT parameter in the diagnosis of Huntington's disease. Neuroradiology. 1984;26:25–28.

Sun Y, Savanenin A, Reddy PH, Liu YF. Polyglutamine-expanded huntingtin promotes sensitization of N-methyl-D- aspartate receptors via post-synaptic density 95. J Biol Chem. 2001;276:24713–24718.

Tabrizi SJ, Leavitt BR, Landwehrmeyer GB, et al. Targeting huntingtin expression in patients with Huntington's disease. N Engl J Med. 2019;380:2307–2316.

Tang T-S, Chen X, Liu J, Bezprozvanny I. Dopaminergic signaling and striatal neurodegeneration in Huntington's disease. J Neurosci. 2007;27:7899–7910.

Tomás-Zapico C, Díez-Zaera M, Ferrer I, et al. α-Synuclein accumulates in huntingtin inclusions but forms independent filaments and its deficiency attenuates early phenotype in a mouse model of Huntington's disease. Hum Mol Genet. 2012;21:495–510.

Tourette C, Li B, Bell R, et al. A large-scale Huntingtin protein interaction network implicates Rho GTPase signaling pathways in Huntington disease. J Biol Chem. 2014;289:6709–6726.

Tozzi A. de Iure A, Filippo MD, et al. The distinct role of medium spiny neurons and cholinergic interneurons in the D2/A2A receptor interaction in the striatum: implications for Parkinson's disease. J Neurosci 2011;31:1850–1862.

Träger U, Andre R, Lahiri N, et al. HTT-lowering reverses Huntington's disease immune dysfunction caused by NFκB pathway dysregulation. Brain. 2014;137:819–833.

Tsunemi T, Ashe TD, Morrison BE, et al. PGC-1a rescues Huntington's disease proteotoxicity by preventing oxidative stress and promoting TFEB function. Sci Transl Med. 2012;4:142–197.

Valekova I, Jarkovska K, Kotrcova E, et al. Revelation of the IFNα, IL-10, IL-8 and IL-1β as promising biomarkers reflecting immuno- pathological mechanisms in porcine Huntington's disease model. J Neuroimmunol. 2016;293:71–81.

van de Zande NA, Massey TH, McLauchlan D, et al. Clinical characterization of dystonia in adult patients with Huntington's disease. Eur J Neurol. 2017;24:1140–1147.

Veldman MB, Yang XW. Molecular insights into cortico-striatal miscommunications in Huntington's disease. Curr Opin Neurobiol. 2018;48:79–89.

Vidoni C, Castiglioni A, Seca C, et al. Dopamine exacerbates mutant Huntingtin toxicity via oxidative-mediated inhibition of autophagy in SH-SY5Y neuroblastoma cells: beneficial effects of anti-oxidant therapeutics. Neurochem Int. 2016;101:132–143.

Villeneuve C, Guilbeau-Frugier C, Sicard P, et al. p53-PGC-1a pathway mediates oxidative mitochondrial damage and cardiomyocyte necrosis induced by monoamine oxidase-A upregulation: role in chronic left ventricular dysfunction in mice. Antioxid Redox Signal. 2013;18:5–18.

Vogel CM, Drury I, Terry LC, Young AB. Myoclonus in adult Huntington's disease. Ann Neurol. 1991;29:213–215.

Vonsattel JP. Huntington disease models and human neuropathology: similarities and differences. Acta Neuropathol. 2008;115:55–69.

Waldvogel HJ, Kim EH, Tippett LJ, et al. The neuropathology of Huntington's disease. Curr Top Behav Neurosci. 2015;22:33–80.

Walker FO. Huntington's disease. Lancet. 2007;369:218–228.
Walker RH, Jankovic J, O'Hearn E, Margolis RL. Phenotypic features of Huntington's disease-like 2. Mov Disord. 2003;18:1527–1530.
Weiss A, Träger U, Wild EJ, et al. Mutant huntingtin fragmentation in immune cells tracks Huntington's disease progression. J Clin Invest. 2012;122:3731–3736.
Wilburn B, Rudnicki DD, Zhao J, et al. An antisense cag repeat transcript at JPH3 locus mediates expanded polyglutamine protein toxicity in Huntington's disease-like 2 Mice. Neuron. 2011;70:427–440.
Wild EJ, Tabrizi S. Therapies targeting DNA and RNA in Huntington's disease. Lancet Neurol. 2017;16:837–847.
Wong YC, Holzbaur EL. The regulation of autophagosome dynamics by huntingtin and HAP1 is disrupted by expression of mutant huntingtin, leading to defective cargo degradation. J Neurosci. 2014;34:1293–1305.
Wu Z, Parry M, Hou X-Y, et al. Gene therapy conversion of striatal astrocytes into GABAergic neurons in mouse models of Huntington's disease. Nat Commun. 2020;11:1105.
Xie Y, Hayden MR, Xu B. BDNF overexpression in the forebrain rescues Huntington's disease phenotypes in YAC128 mice. J Neurosci. 2010;30:14708–14718.
Yan Z, Surmeier DJ. D5 dopamine receptors enhance Zn^{2+}-sensitive GABA(A) currents in striatal cholinergic interneurons through a PKA/PP1 cascade. Neuron. 1997;19:1115–1126.
Young D, Mayer F, Vidotto N, et al. Mutant huntingtin gene-dose impacts on aggregate deposition, DARPP32 expression and neuroinflammation in HdhQ150 mice. PLoS ONE. 2013;8:e75108.
Zeitlin S, Liu JP, Chapman DL, et al. Increased apoptosis and early embryonic lethality in mice nullizygous for the Huntington's disease gene homologue. Nat Genet. 1995;11:155–163.
Zhao X, Chen X-Q, Han E, et al. TRiC subunits enhance BDNF axonal transport and rescue striatal atrophy in Huntington's disease. Proc Natl Acad Sci U S A. 2016;113:E5655–E5664.
Zhuang X, Oosting RS, Jones SR, et al. Hyperactivity and impaired response habituation in hyperdopaminergic mice. Proc Natl Acad Sci U S A. 2001;98:1982–1987.

9 Schizophrenia

Neurochemical Insight into a Mind's Faulty Dimension

Md Zeeshan Ali, Anushree, and Jawaid Ahsan
Department of Biotechnology, Central University of South Bihar, Panchanpur, Gaya, Bihar, India

CONTENTS

9.1 INTRODUCTION

Schizophrenia is a complex psychiatric syndrome that disrupts the social and mental well-being of an individual (Jablensky, 2010). According to the World Health Organization, schizophrenia is ranked eighth as the prime reason of disability in people aged 15–44 years globally (Haddad and Correll, 2018; Theodoridou and Rossler, 2010). Statistically, this accounts for 0.5–1% of the world's population (He *et al.*, 2012; Tsuang, 2000). A major problem with schizophrenia is the way it progresses. The onset of illness – known as the prodromal phase – is difficult to decipher. Mostly, it begins in adolescence and exhibits various non-specific symptoms such as depression, mood swings, anxiety, anger, and so on (Podea *et al.*, 2019). This episode becomes glaringly prominent in the next 10 years, with the unfolding of a complete range of symptoms (Table 9.1). These symptoms can be divided into three classes: positive, negative, and cognitive (Aleman, 2014; Patel *et al.*, 2014). The positive symptomatic signs include hallucinations, delusions, chaotic behavior, and muddled speech, whereas negative ones encompass anhedonia, asociality, reduced energy, and unwillingness to take the initiative (Correll and Schooler, 2020; Kaneko, 2018). Lastly, cognitive indicators express a broad set of cerebral dysfunctions that form the central feature of schizophrenia (Avramopoulos, 2018; Bowie and Harvey, 2006; Tripathi *et al.*, 2018). Now, what is interesting at this point is that different schizophrenic patients exhibit different sets of warning signs. The pattern is not uniform and this ambiguous nature of schizophrenia makes it a very complex and heterogeneous syndrome (Lewis and Lewitt, 2002).

DOI: 10.1201/9780429265198-11

TABLE 9.1
Outline of the Various Symptoms Observed in Schizophrenia

Positive Symptoms	Negative Symptoms	Cognitive Symptoms
Hallucinations	Inability to feel pleasure (anhedonia)	Deficit in attention
Delusions	Social withdrawal (asociality)	Lack of general intelligence
Disorganized behavior	Reduced energy	Working memory dysfunction
Muddled/distorted speech	Loss of motivation, desires, and lack of taking initiatives (avolition)	Poor verbal learning and memory
Abnormal motor behavior	Lack of emotional expression (affective flattening)	Disorganized executive functioning
Racing thoughts/confused thoughts	Reduced amount of speech and drop of eloquence in conversation (alogia)	

Data from: Aleman, 2014; Bowie and Harvey, 2006; Correll and Schooler, 2020; Kaneko, 2018, and Tripathi *et al.*, 2018.

What actually causes schizophrenia is still unclear. The various factors that amalgamate into this disorder remain elusive (Jablensky, 2006; Mathalon and Ford, 2012). Numerous studies have suggested that many genetic and environmental factors – either singly or in groups – may be causative of the ailment (Moran *et al.*, 2016; van Os *et al.*, 2008). Several high-risk genes, such as those encoding neuregulin, dysbindin, D-amino acid oxidase (DAAO), proline dehydrogenase, catechol-O-methyltransferase, along with functional candidate genes of dopamine D-3 receptor and serotonin receptor (5-HT_{2A}), have been found to be associated with schizophrenia (Elkin *et al.*, 2004; Salleh 2004). In addition, environmental factors also play a catastrophic role. Pre-/postnatal infection, maternal stress, child abuse, drug abuse, social adversity, and so on, augment the chances of developing schizophrenia (Dean and Murray, 2005; Stilo *et al.*, 2011) (Figure 9.1).

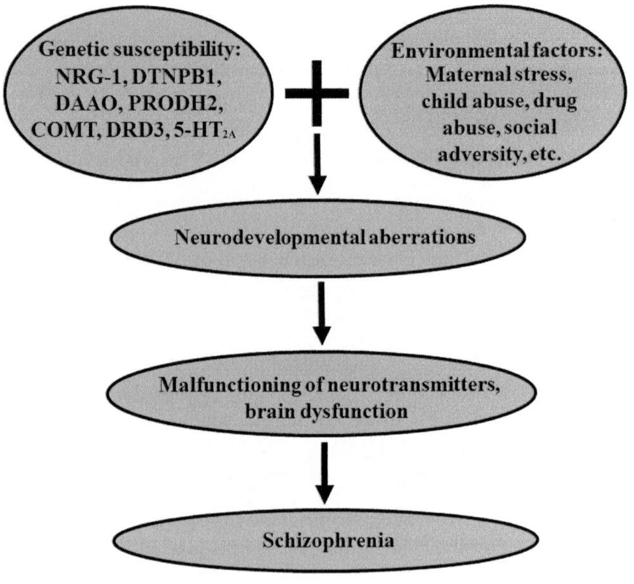

FIGURE 9.1 Genes encoding neuregulin (NRG-1), dysbindin (DTNPB1), D-amino acid oxidase (DAAO), proline dehydrogenase (PRODH2), catechol-O-methyltransferase (COMT), dopamine D-3 receptor (DRD3), and serotonin receptor (5-HT_{2A}) are high-risk genes for schizophrenia. Environmental factors such as maternal stress, child abuse, drug abuse, social adversity, and so on, augment the chances of developing this mental disorder. Data from Dean and Murray (2005); Elkin *et al.* (2004); Salleh (2004), and Stilo *et al.* (2011).

Furthermore, in the last few years, research in the field of schizophrenia has sped up. With improved structural and functional neuroimaging and post-mortem studies, the scientific community has ascertained a few biological mechanisms of this disorder and the environmental factors inducing them (Kahn *et al.*, 2015; Schultz *et al.*, 2007). The altered functioning of neurotransmitters such as γ-aminobutyric acid (GABA), dopamine, glutamate, and serotonin in the pathology of schizophrenia is a step in this direction (Brisch *et al.*, 2014; Dean, 2002). As far as treatment is considered, antipsychotic drugs do help in checking the psychotic symptoms and have small benefits on the social and professional well-being of patients (Kucharska-Pietura and Mortimer, 2013; Roberts *et al.*, 2010). Thus, a parallel delivery of medications accompanied by social and psychological interventions can be of much greater help (Chien *et al.*, 2013). In this chapter, the authors enlighten the neurochemical aspects of schizophrenia and the ways to tackle it in an improved way.

9.2 NEUROCHEMISTRY OF SCHIZOPHRENIA

Neurotransmitters perform a vital part in the normal functioning of our bodies. They transmit signals from neurons to the target cells and are key to the physical and psychological well-being of an individual (Valenzuela *et al.*, 2011). In schizophrenia, the deterioration in the psychological condition of patients is well documented, which has led the scientific community to shift its focus on finding altered neurotransmissions involved in this cognitive syndrome. Much work has been done in this direction in the last 10–12 years. The analyses of living brain cells of schizophrenic patients using magnetic resonance spectroscopy (MRS), positron emission tomography (PET), and single-photon emission computed tomography (SPECT) has validated the presence of neurochemical asymmetry (Reynolds, 2008). This has highlighted the role of dopamine, glutamate, GABA, acetylcholine, and serotonin in the pathology of schizophrenia (Brisch *et al.*, 2014; Howes

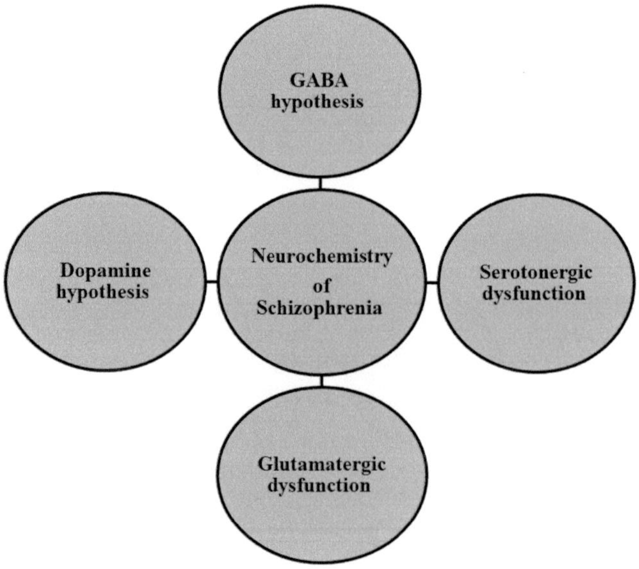

FIGURE 9.2 Various hypotheses, such as dopamine hypothesis, γ-aminobutyric acid (GABA) hypothesis, serotonergic dysfunction, and glutamatergic dysfunction, have been proposed suggesting the misbehavior of neurotransmitter systems in schizophrenia.

et al., 2015). A few psychotic drugs such as lysergic acid diethylamide (LSD), amphetamine, phencyclidine, and so on, can trigger schizophrenic symptoms in healthy people (Paparelli *et al.*, 2011). These chemical formulations are known to disrupt specific neurotransmitter systems. As a result of all these findings, a number of hypotheses have been put forward suggesting the malfunctioning of neurotransmitters in schizophrenia (Figure 9.2).

9.2.1 Altered Dopaminergic Neurotransmission (Dopamine Hypothesis)

Dopamine (3, 4-dihydroxytyramine) is an important neurotransmitter that relays signals between nerve endings in the brain (Lovinger, 2008). Being both excitatory and inhibitory in nature, it is synthesized and released by dopaminergic neurons located at the substantia nigra, ventral tegmental area (VTA), and hypothalamus of the brain (Olguin *et al.*, 2016). After being released by nerve cells, it binds to a number of receptors. These G protein-coupled receptors (GPCRs) can be separated into D1-like (D1 and D5) and D2-like (D2, D3, and D4) classes (Hasbi *et al.*, 2011; Neve *et al.*, 2004). Dopamine is known to be critical to the normal functioning of the human mind. It regulates varied physical and psychological responses in our daily lives. Enough scientific literature is available that points toward its involvement in the inflection of behavior, perception, enthusiasm, and making a choice between reward or punishment. In addition, the neurotransmitter is also known to modulate mood, the sleep–wake cycle, dreams, learning, and working memory in all of us (Best *et al.*, 2009; Olguin *et al.*, 2016).

The emergence of the dopamine hypothesis stems from the fact that most of the responses regulated by dopamine seem to be altered in a schizophrenic mind. As per the hypothesis, subcortical hyperactivity in dopamine transmission leads to the positive symptoms of schizophrenia. At the same time, the cortical deficit of dopamine manifests as the negative and cognitive symptoms (Brisch *et al.*, 2014; Shen *et al.*, 2012) (Figure 9.3). Now, let us try to understand the crux of this theory. It is based on the following two assumptions. First, a normal individual subjected to continuous exposure of D2 receptor agonists exhibits positive schizophrenic symptoms; second, most antipsychotic drugs (chlorpromazine, haloperidol, etc.) function by blocking the D2 receptor to varying degrees (Abi-Dargham, 2004; Toda and Abi-Dargham, 2007). To clarify, it has been found that there is amplified D2 mRNA ($D2_{Longer}$, $D2_{Long}$, and $D2_{Short}$) production in the frontal cortex of the brain in schizophrenic patients (Brisch *et al.*, 2014; Tallerico *et al.* 2001). This leads to augmented D2 receptor activity resulting from the binding of dopamine released in excess from the subcortical region of the brain. The enhanced receptor–ligand commotion then triggers the manifestation of positive symptoms in schizophrenic patients (Shen *et al.*, 2012). Various research with drugs such as amphetamines, that are known to enhance dopamine production, has shown that they lead to psychotic symptoms in normal people, if administered repeatedly (Lyon *et al.*, 2011; Yui *et al.*, 2002). Finally, a few studies using PET scans have concluded that radiolabeled L-DOPA (L-[β-^{11}C] DOPA), a precursor of dopamine, when administered to schizophrenic patients is taken up more quickly than in normal people (Lindstrom *et al.*, 1999; Miyake *et al.*, 2011; Nozaki *et al.*, 2009). Similar studies using ^{18}F-DOPA have also pointed toward enhanced dopamine production in patients (Howes *et al.*, 2009). To sum up, these findings suggest that dopamine D2 hyperactivity pre-dates the onset of positive symptoms in people suffering from schizophrenia.

Contrary to this, reduced cortical dopamine transmission has been found to be associated with the negative and cognitive symptoms of schizophrenia. There are a few reports that hint at the involvement of dysfunction or reduced density of D1 receptors in the prefrontal cortex in the manifestation of these symptomatic signs (Avery and Krichmar, 2015; Brisch *et al.*, 2014; Goldman-Rakic *et al.*, 2004; Shen *et al.*, 2012; Stenkrona *et al.*, 2019 Walter *et al.*, 2009). Some studies have demonstrated the correlation of a lower plasma homovanillic acid – the chief metabolite of dopamine – to a higher negative symptom score in schizophrenic patients (Mitra *et al.*, 2016; Ribeyre *et al.*, 1994). Therefore, given these points, it can be said that malfunctioning in the dopamine neurotransmission is intricately involved in the manifestation of schizophrenia.

9.2.2 Abnormal Gabaergic Neurotransmission in Schizophrenia (Gaba Hypothesis)

GABA is the chief inhibitory neurotransmitter in the mammalian brain (Boonstra *et al.*, 2015; Petroff, 2002).

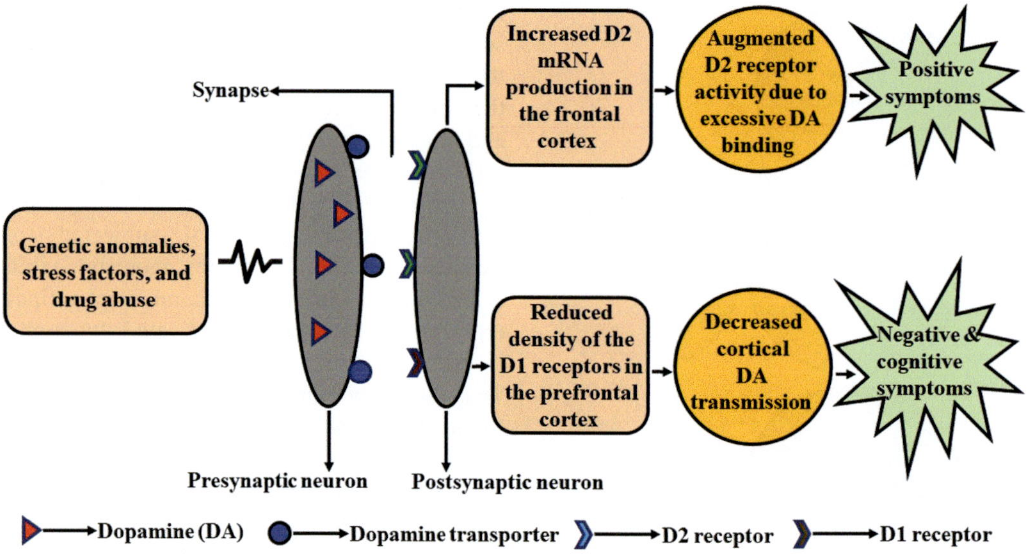

FIGURE 9.3 Dopamine (DA) hypothesis of schizophrenia: enhanced D2 receptor activity in the frontal cortex of the brain triggers the positive symptoms of schizophrenia, whereas reduced D1 receptor activity in the prefrontal cortex manifests into the negative symptomatic signs. Adapted from Howes and Kapur (2009). Data from Brisch *et al.* (2014) and Shen *et al.* (2012).

Synthesized and secreted by the GABAergic neurons, GABA intricately regulates the development of cognition, behavior, and the body's response to stress factors (Fogaca and Duman, 2019). GABA is basically created by the decarboxylation of glutamate by the enzyme glutamic acid decarboxylase (GAD) (de Jonge *et al.*, 2017; Petroff, 2002). This enzyme exists in two isotypic forms – GAD65 and GAD67 (encoded by *GAD1* and *GAD2*, respectively) – both of which regulate different aspects of GABAergic neurotransmission (Bu *et al.*, 1992; Kass *et al.*, 2014; Pinal and Tobin, 1998; Thompson *et al.*, 2009). GAD65 (65-kDa isotype positioned on axon terminals) is involved in the packaging and release of GABA. It comes into action when a quick synthesis of GABA is needed, resulting from a high synaptic demand (Patel *et al.*, 2006; Tian *et al.*, 1999). Contrary to this, GAD67 (67-kDa isotype located in the cytosol) is responsible for the majority of GABA synthesis, accounting for 80–90% of the total amount (Asada *et al.*, 1997). Apart from this, it also plays a key role in the maintenance of basal GABA levels in the central nervous system (CNS) (Curley *et al.*, 2011; Hashimoto *et al.*, 2003). To put it differently, GABA works in tandem with glutamate (an excitatory neurotransmitter), and thus maintains an inhibitory–excitatory balance for the proper functioning of the normal human brain (Wu and Sun, 2015; Xu *et al.*, 2011).

Furthermore, the implication of GABA in the pathophysiology of schizophrenia was first put forward by Eugene Roberts in 1972. He proposed that the behavioral and cognitive abnormalities in schizophrenic patients might result from a defect in the inhibitory GABAergic neuronal circuit (Blum and Mann, 2002; Roberts, 1972). These assumptions gave rise to the GABA hypothesis. To have a clearer picture, let us consider the details of this proposition. As per the GABA hypothesis, there is a decreased concentration of GABA and neurotransmission in patients with schizophrenia (Chang *et al.*, 2016; Chattopadhyaya *et al.*, 2007; Senagolage and Sullivan, 2013; Yoon *et al.*, 2010) (Figure 9.4). A number of key physiologic processes, such as handling of information (Kann *et al.*, 2014), cognition-associated gamma oscillations (Moran and Hong, 2011), modulation of the activity of hippocampus and cortex, (Benes and Berretta, 2001; Kelsom and Lu, 2013), and so on, involve GABAergic neurotransmission (Xu and Wong, 2018). These processes are intricately involved in the maintenance of behavioral/cognitive functions and are found to be severely hampered in schizophrenic patients. Consequently, such findings have led to a flurry of research in this direction, unearthing some interesting facts.

Postmortem analyses and multiple imaging techniques have revealed a reduction in the mRNA and protein levels of GAD67 in patients with schizophrenia. In one of the studies, scientists quantified GAD67 mRNA levels in postmortem specimens of the dorsolateral prefrontal cortex (DLPFC) from schizophrenic patients. The quantitative polymerase chain reaction data they obtained indicated a reduction of 15% in GAD67 mRNA levels in diseased subjects versus the control. Also, GAD67 protein levels were reduced by 10% in total gray matter and by 49% in parvalbumin axon terminals (Curley *et al.*, 2011). Likewise, an additional study also reported an approximately 10% reduction in GABA concentration in patients with schizophrenia. The researchers used high-field MRS to measure visual cortical GABA levels and reported a drop in neurotransmission (Yoon *et al.*, 2010). Consequently, these studies have led to the conclusion that such drops result in lower levels of GABA in the DLPFC, with ensuing hampered working memory and selective attention. In addition, a few reports have hinted at the role of parvalbumin in the manifestation of schizophrenia (Taylor and Tso, 2015). This calcium-binding protein acts as a buffer, influencing GABA release.

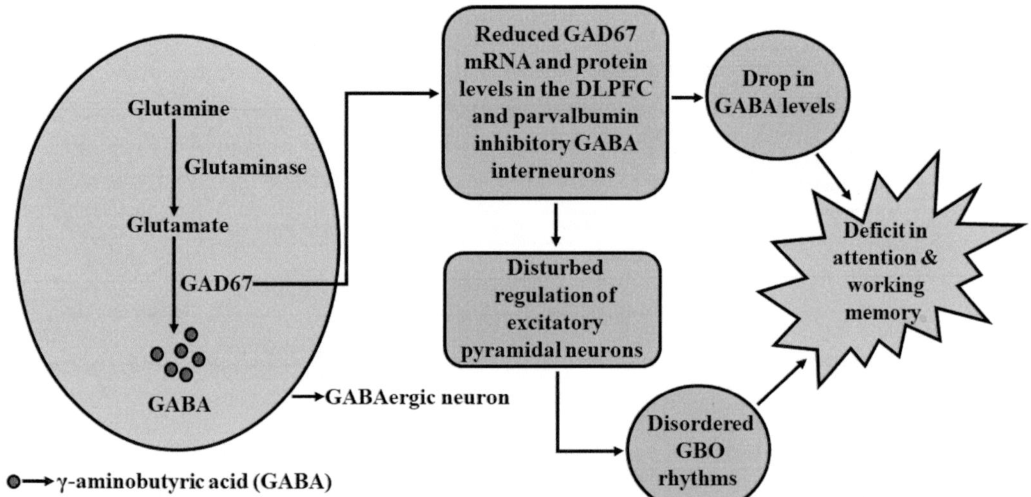

FIGURE 9.4 γ-Aminobutyric acid (GABA) hypothesis of schizophrenia: reduced GAD67 (glutamic acid decarboxylase, 67-kDa isoform) mRNA and protein levels in the dorsolateral prefrontal cortex (DLPFC) and parvalbumin axon terminals result in a decreased amount of GABA in the brain. The reduction also hampers the rhythmic pattern of γ-band oscillations (GBO). Ultimately, both these circumstances lead to the manifestation of cognitive symptoms in schizophrenic patients. Data from Chang *et al.* (2016); Curley *et al.* (2011); and Hashimoto *et al.* (2003).

It has been found that genetic elimination of this protein alters GABA neurotransmission, which develops into schizophrenic symptoms (Lewis, 2014; Xu and Wong, 2018). Lastly, some scientists have found an association between impaired gamma-band oscillations (GBOs) and schizophrenia. These oscillations are controlled by complex chemistry between the excitatory pyramidal cells and inhibitory GABAergic interneurons (McNally and McCarley, 2016). An important point at this juncture is that neuronal oscillations in the gamma frequency spectrum (30–80 Hz) in the prefrontal cortex (PFC) is linked to selective attention and working memory (Benchenane *et al.*, 2011; Womelsdorf and Fries, 2007). It has also been observed that there is a deficit in gamma activity in the frontal area of the brain during mathematical tasks (Kissler *et al.*, 2000) and the recovery phase of working memories in schizophrenic patients (Haenschel *et al.*, 2009; Shin *et al.*, 2011). These reports clearly indicate the involvement of GBO anomalies in the progression of schizophrenia. These abnormalities in GBO have been attributed to a reduced expression of GAD67 and parvalbumin mRNA in cortical parvalbumin GABA interneurons (Hashimoto *et al.*, 2003; Woo *et al.*, 2010). Accordingly, this weakening leads to disturbed GABAergic regulation of the pyramidal neurons, upsetting GBO rhythms. Therefore, all things considered, the role of faulty GABA neurotransmission in the development of schizophrenia cannot be ruled out.

9.2.3 Serotonergic Dysfunction in Schizophrenia (Serotonin Hypothesis)

Serotonin [5-hydroxytryptamine (5-HT)] plays a vital role in regulating numerous behavioral and somatic functions controlled by the brain (Frazer and Hensler, 1999). To put it differently, this inhibitory neurotransmitter is intricately involved in the synchronization of both the neuropsychological, as well as the biological, processes of our body (Berger *et al.*, 2009). The modulation of stress, anxiety, anger, reward, sexuality, and perception, as well as peristalsis, blood pressure regulation, radial pulse management, bladder rheostat, and blood clotting, all involve serotonin at one or more steps (Berger *et al.*, 2009). As far as the synthesis of serotonin goes, the serotonergic neurons in our brain secrete only about 1–2% of the total amount of serotonin produced. A much larger quantity, constituting nearly 90%, is discharged from enterochromaffin (EC) cells of the gastrointestinal (GI) tract (Gershon and Tack, 2007; Lv and Liu, 2017). In both the brain and the GI tract, the synthesis of serotonin takes place via a multifactorial pathway. Initially, L-tryptophan is transformed into L-5-hydroxytryptophan (L-5-HTP) by an enzyme called tryptophan hydroxylase. This enzyme exists in two isoforms encoded by *TPH1* and *TPH2* expressed, respectively, in the EC cells and neurons (Cote *et al.*, 2003; Gershon and Tack, 2007; Walther *et al.*, 2003). This is followed by the conversion of L-5-HTP into serotonin by an aromatic L-amino acid decarboxylase (Ichiyama *et al.*, 1970; Walther *et al.*, 2003).

Furthermore, in schizophrenic patients, there is a disturbance in cognition, mood, sexual drive, appetite, sleep, and pain sensitivity, all involving serotonergic neurotransmission (Ohuoha *et al.*, 1993). This relationship has made researchers consider the possibility of serotonergic dysfunction in the manifestation of schizophrenia (Abi-Dargham *et al.*, 1997; Nudmamud-Thanoi, 2005) (Figure 9.5). Interestingly, the serotonin hypothesis has originated more than a few decades ago. A few scientists proposed that the psychedelic drug LSD acted as a peripheral 5-HT antagonist

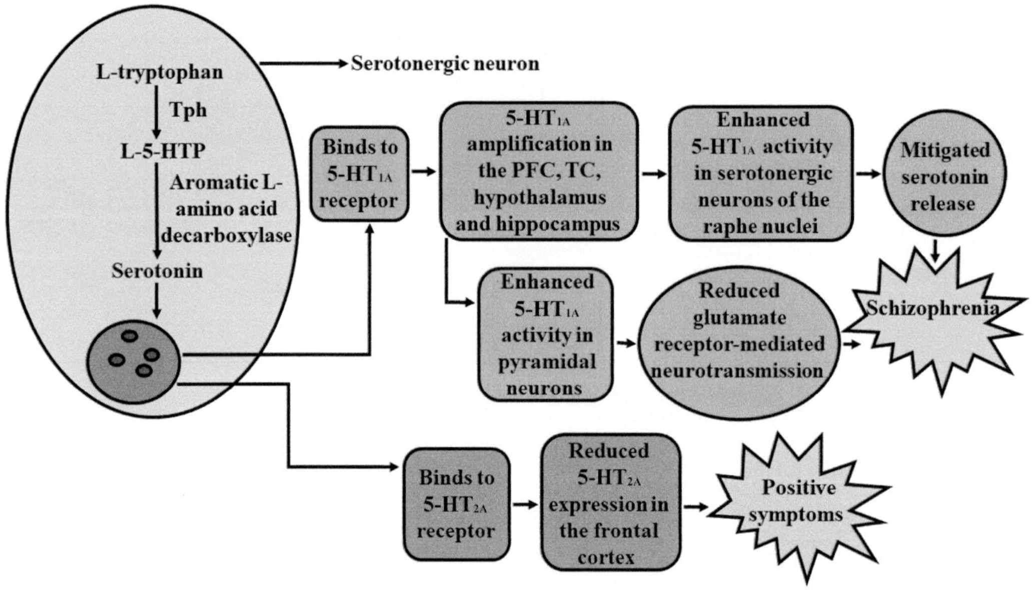

FIGURE 9.5 Serotonin (5-HT) hypothesis of schizophrenia: a reduced serotonergic and glutamatergic neurotransmission due to augmented 5-HT_{1A} synthesis leads to the unfolding of schizophrenia. A reduced 5-HT_{2A} receptor density might be responsible for triggering the positive symptoms in patients. Tph, tryptophan hydroxylase; PFC, prefrontal cortex; TC, temporal cortex; L-5HTP, L-5-hydroxytryptophan. Data from Costa *et al.* (2012); Meltzer *et al.* (2003); Polter and Li (2010); and Pralong *et al.* (2000).

and conceived that a deficit in serotonin might be responsible for inducing schizophrenia (Bleich *et al.*, 1988; Gaddum, 1954; Woolley and Shaw, 1954). Going into a bit of detail will help with understanding the proposition in a much better way. First, neurotransmission via serotonin involves a number of receptors, a few of which are important, namely 5-HT_{1A} and 5-HT_{2A} (Carhart-Harris and Nutt, 2017). 5-HT_{1A} receptors are mainly expressed in the limbic-cortical and medial regions of the brain (Varnas *et al.*, 2004). Here they are involved in the coordination of anxiety, stress, and emotions, exhibiting a high affinity for serotonin (Carhart-Harris and Nutt, 2017; Peroutka and Snyder, 1979). Contrary to this, the 5-HT_{2A} receptors are predominantly expressed throughout the cortex of the brain. They come into play when conditions are extreme and there is increased release of serotonin in the CNS (Andrade, 2011).

In this case, postmortem studies and PET scans have revealed that there is an altered manifestation of these receptors in different areas of the brain. The 5-HT_{1A} receptors (inhibitory GPCRs) have been found to be amplified by approximately 15–80% in the prefrontal cortex, temporal cortex, hippocampus, and thalamus of schizophrenic patients. (Gurevich and Joyce, 1997; Hashimoto *et al.*, 1991; Meltzer *et al.*, 2003; Millan, 2000; Ohno, 2011; Tauscher *et al.*, 2002). What electrophysiologic studies have highlighted is also interesting. As per these analyses, activation of 5-HT_{1A} receptors in serotonergic neurons of the raphe nuclei (autoreceptors) triggers membrane hyperpolarization and a mitigated serotonin release (Polter and Li, 2010). In addition, the 5-HT_{1A} receptors located on pyramidal neurons (heteroreceptors) induce a truncated glutamate receptor-mediated neurotransmission (Bantick *et al.*, 2001; Costa *et al.*, 2012; Rojas and Fiedler, 2016). Given these arguments, it can be said that there is reduced serotonergic and glutamatergic neurotransmission in schizophrenic patients owing to augmented 5-HT_{1A} synthesis. Another key point at this juncture is to understand the role of 5-HT_{2A} receptors (excitatory GPCRs) in the progression of schizophrenia. Postmortem analyses of cerebral tissues have revealed reduced 5-HT_{2A} receptor expression in the frontal cortex of schizophrenic patients (Matsumoto *et al.*, 2005; Pralong *et al.*, 2000). Also, there has been a report of a truncated cortical 5-HT_{2A} mRNA translation in people suffering from schizophrenia (Hernandez and Sokolov, 2000). In the same fashion, in vivo studies using PET scans have exhibited that there is an overall dip in 5-HT_{2A} receptor binding in the cortex of male schizophrenic patients (first episode), triggering positive symptoms (Hurlemann *et al.*, 2005; Rasmussen *et al.*, 2010). Consequently, these findings have led scientists to believe that a reduced 5-HT_{2A} receptor density might represent a primary pathophysiologic commotion in the unfolding of schizophrenia (Ebdrup *et al.*, 2011). Therefore, on the whole, a faulty serotonergic neurotransmission seems play one or more parts in schizophrenia, though further validation studies need to be done.

9.2.4 Flawed Glutamatergic Neurotransmission in Schizophrenia (Glutamate Hypothesis)

The sensorimotor and cognitive circuitry is involved in a number of neuronal activities, such as neuroprotection, glial neuronal communication, and synaptic plasticity. Glutamate, being the main excitatory neurotransmitter in the brain, is indispensable in their normal functioning (Gaspar *et al.*, 2009; Zhou and Danbolt, 2014). Its importance can be understood from the fact that glutamatergic neurons make use of 60–80% of the total metabolic activity in the brain (Howes *et al.*, 2015; Rothman *et al.*, 2003). As far as the synthesis of glutamate is considered, it is well known that it is a non-essential amino acid that cannot breach the blood–brain barrier enveloping the CNS (Hawkins, 2009; Hawkins and Vina, 2016). To overcome this hurdle, glutamatergic neurons synthesize glutamate by the enzymatic conversion of local precursor molecules, the most dominant being glutamine (Kvamme, 1998; Newsholme *et al.*, 2003). In other words, it is actually a cycling of events between neurons and glial cells. Glutamine synthetase present in the glial cells converts glutamate into glutamine. It is then shipped out of the glia to enter into the presynaptic terminal of the neurons where it is transformed into glutamate by the mitochondrial enzyme glutaminase. The glutamate is then packed into the synaptic vesicles to be released into the synaptic cleft. As the uptake of glutamate by postsynaptic neurons is slow, there is an active reuptake of it into the presynaptic compartments and glial cells to keep the glutamate–glutamine cycle going (Daikhin and Yudkoff, 2000; Purves *et al.*, 2001; Shen, 2013). In addition, the neurons sometimes synthesize glutamate by transamination of 2-oxoglutarate (a tricarboxylic cycle intermediate metabolite) via glutamate dehydrogenase, with ammonium acting as the nitrogen source (Kaneko, 2000; Torgner and Kvamme, 1990; Walker and van der Donk, 2016).

Furthermore, there are two significant groups of receptors on which glutamate acts: ionotropic glutamatergic receptors, namely *N*-methyl-D-aspartate (NMDA), α-amino-3-hydroxy-5-methyl-4-isoxazolepropionic acid (AMPA), and kainite; and metabotropic glutamate receptors (mGluRs) (Masu *et al.*, 1993; Rousseaux, 2008; Sherman, 2014). Among these, the role of NMDA has been found to be critical in the manifestation of schizophrenia, giving rise to the glutamate hypothesis (Olney *et al.*, 1999; Stone *et al.*, 2007). However, to be more precise, the proposition of faulty glutamatergic neurotransmission in schizophrenia dates back almost 50 years. There were early records outlining the treatment of schizophrenic patients with glutamic acid (Kitzinger and Arnold, 1949). In addition, depleted levels of glutamate in cerebrospinal fluid of patients was also reported (Kim *et al.*, 1980). However, with time, a flurry of research has changed the scenario to a considerable extent. Now, the hypothesis does not just involve a straightforward reduction in the levels of glutamate progressing to schizophrenia. At present, a number of glutamate receptors have been implicated in the manifestation of this cognitive syndrome, with NMDA being the key one (Balu, 2016; Stone *et al.*, 2007) (Figure 9.6).

One of the hallmarks of schizophrenia is the escalating decline in the cognitive abilities of patients. With the passage of time, they find it difficult to concentrate on things, which takes a toll on working memory and behavior (Bowie and Harvey, 2006; Tripathi *et al.*, 2018). Therefore, what the glutamate hypothesis proposes is that an anomalous relationship between the glutamatergic and GABAergic neurotransmitter systems ultimately progresses into the cerebral deficit. To put it differently, the fast-spiking parvalbumin GABAergic interneurons are stimulated by NMDA. Therefore, reduced functioning of the receptor

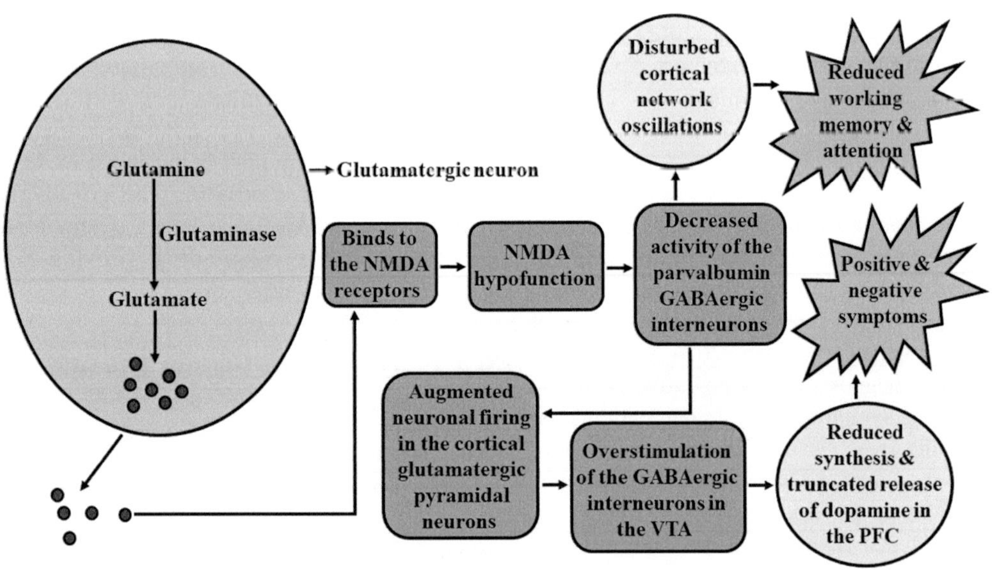

FIGURE 9.6 Glutamate hypothesis of schizophrenia: *N*-methyl-D-aspartate (NMDA) hypofunction results in decreased parvalbumin GABAergic interneural activity, leading to disturbed cortical network oscillations. This progresses into cognitive symptoms. The hypoactive parvalbumin γ-aminobutyric acid (GABA) interneurons also cause augmented neuronal firing in the cortical glutamatergic pyramidal neurons. This leads to hyperactivity in the GABAergic interneurons of the ventral tegmental area (VTA) causing reduced dopamine release in the prefrontal cortex (PFC). This sequence of events then progresses into the positive and negative symptoms of schizophrenia. Data from Ellaithy *et al.* (2015); Gaspar *et al.* (2009); and Lisman *et al.* (2008).

(NMDA hypofunction) results in decreased activity of these GABA interneurons. In turn, this leads to modulations in the cortical network oscillations of the brain, ultimately hampering working memory and attention in schizophrenic patients (Balu, 2016; Gaspar *et al.*, 2009; Jadi *et al.*, 2016; Lisman *et al.*, 2008). In addition, the reduction in the work of the parvalbumin GABAergic interneurons leads to augmented neuronal firing in the glutamatergic pyramidal neurons of the cerebral cortex. As a result, the GABAergic interneurons in the VTA become overstimulated, dampening the mesocortical pathway of the dopamine synthesis. This ultimately results in a truncated dopamine release in the PFC that might be conducive to the positive and negative symptoms of schizophrenia (Balu, 2016; Ellaithy *et al.*, 2015).

With this in mind, a number of scientific reports authenticate the legitimacy of the glutamate hypothesis. First, a few NMDA competitors such as ketamine, phencyclidine, and dizocilpine (MK-801) are known to induce psychological effects in humans that closely resemble the symptoms of schizophrenia (Carlsson *et al.*, 2004; Javitt, 2007; Krystal *et al.*, 1994; Lisman *et al.*, 2008; Morgan and Curran, 2006). Second, a handful of drugs known as positive allosteric modulators have been found to help alleviate psychotic conditions in schizophrenic patients. They are known to act as mGluR agonists and might revolutionize the treatment of schizophrenia in the future (Conn *et al.*, 2009). In addition, in vivo imaging studies have also highlighted the hypofunction of NMDA in the manifestation of schizophrenia. Using SPECT, reduced NMDA activity has been observed in the left hippocampus of unmedicated schizophrenic patients (Pilowsky *et al.*, 2006). Also, a study using proton MRS (1H-MRS) has shown that there is an increased glutamine-to-glutamate ratio in patients with chronic schizophrenia, hinting at reduced NMDA activity (Bustillo *et al.*, 2014; Howes *et al.*, 2015). Thus, the disturbed functioning of the glutamate receptors is visible in the progression of schizophrenia in patients.

9.3 DIAGNOSIS

Before discussing the pharmacologic treatments available for schizophrenia, a brief understanding of how to diagnose it in patients will be of great help. As already stated, schizophrenia is a serious mental disorder that progresses through an extensive range of symptoms (Aleman, 2014; Patel *et al.*, 2014). What makes the diagnosis of this cognitive syndrome a difficult task is that the symptoms vary among patients. No two patients exhibit the same set of warning signs, making it tough to diagnose the disorder (Jablensky, 2010). In order to make things a bit easier for doctors, the Diagnostic and Statistical Manual of Mental Disorders, Fifth Edition (DSM-5) – an American Psychiatric Association manual – has laid down certain criteria to establish that a person suffers from schizophrenia (Patel *et al.*, 2014). The measures described in the manual are as follows:

1. A person must exhibit a minimum of two or multiple active-phase symptoms such as muddled speech, fallacies, hallucinations, chaotic behavior, and negative symptoms, each persisting for a period of at least 1 month. In other words, for a person to be considered schizophrenic, he/she must exhibit at least one of the symptomatic signs from hallucinations, muddled speech, and fallacies besides the other symptoms mentioned above.
2. Certain drugs, such as methamphetamines or LSD, are known to induce schizophrenic symptoms in people consuming them. Therefore, it is necessary for doctors to ascertain that the symptoms are not arising from illegal drug use or prescribed medications.
3. There must be a reduced level of interest in work, self-care, social relationships, and everyday tasks, hampering the normal well-being of an individual.
4. Lastly, in addition to the 1-month period of active-phase symptoms, a person must display persistence of the symptomatic signs for a duration of 6 months or more (Mattila *et al.*, 2015; Tandon *et al.*, 2013).

Given these points, with no single physical or laboratory test available, careful evaluation of the symptoms and the course of illness for at least 6 months can ensure a correct diagnosis of schizophrenia.

9.4 PHARMACOLOGIC TREATMENTS OF SCHIZOPHRENIA

As there is no cure for schizophrenia, for now, getting proper treatment early in life is the best way of managing this mental disorder. The involvement of virtually all the neurotransmitter systems in the etiology of schizophrenia has made researchers develop a number of drug therapies for its management. These medications help in alleviating positive symptoms in schizophrenic patients. Though psychosocial interventions are also being implemented in parallel, the use of medicines remains prevalent. In addition, efforts are also being made to design such drugs that can help in controlling the negative and cognitive symptoms, too. Now let us have a detailed outlook on a few prominent pharmacologic drugs proving to be effective in treating schizophrenia.

9.4.1 Drugs Targeting the Dopamine System

9.4.1.1 Chlorpromazine

Also known as thorazine, chlorpromazine is one of the most common drugs used worldwide to treat people with schizophrenia (Adams *et al.*, 2005). It belongs to the group of typical antipsychotics, also known as the first-generation antipsychotics (FGAs). Chlorpromazine aids in alleviating positive symptoms in schizophrenic patients by blocking dopamine D2 receptors in the mesolimbic pathway. However, its administration sometimes results in drug-induced movement disorders in patients owing to the D2 receptor blockade in the nigrostriatal pathway (Mann and Marwaha, 2020).

A few other significant FGAs that act as D2 antagonists are fluphenazine, haloperidol, loxapine, perphenazine, pimozide, thioridazine, thiothixene, and trifluoperazine (Guzman, 2019).

9.4.1.2 Phosphodiesterase 10A Inhibitors

Papaverine, MP-10, TP-10, and TAK-063 are some of the most potent phosphodiesterase (PDE) 10A inhibitors known to date. PDE 10A acts by disabling the intracellular second messengers cyclic guanosine monophosphate (cGMP) and cyclic adenosine monophosphate (cAMP). Therefore, its inhibition by the drugs mentioned above results in reduced D2 receptor activity. This mitigation in the conditioning of the D2 receptor thus helps in alleviating the psychotic symptoms in schizophrenic patients (Chappie *et al.*, 2009; Grauer *et al.*, 2009; Gresack *et al.*, 2014; Kehler and Nielsen, 2011; Nishi *et al.*, 2008; Schmidt *et al.*, 2008; Snyder and Vanover, 2017; Suzuki *et al.*, 2016).

9.4.1.3 Cariprazine

Cariprazine is a D2/D3 receptor partial agonist with a higher D3 potency (Citrome, 2013). Its affinity for the D3 receptors is so high that it results in a unique blockade of them when administered at physiologic doses. This dopamine receptor is primarily located in the hypothalamus, limbic system, VTA, and substantia nigra (Stahl, 2017). Therefore, a limited agonist action of cariprazine on presynaptic D3 autoreceptors leads to a disinhibited dopamine release in the PFC. Subsequently, this augmented discharge of dopamine progresses into improvement in cognition, mood, and negative symptoms in schizophrenic patients (Gyertyan *et al.*, 2011; Leggio *et al.*, 2013; Stahl, 2016; Stahl *et al.*, 2020; Zimnisky *et al.*, 2013).

9.4.1.4 Stepholidine

Stepholidine is a tetrahydroberberine alkaloid isolated from the Chinese herb *Stephania intermedia*. It acts as a partial agonist for the dopamine D1 receptor and as an antagonist for the D2 receptor (Yang *et al.*, 2007; Zhu *et al.*, 2000). The proposed role of the dysfunction of the D1 receptors in the medial PFC and hyperactivity of the D2 receptors in the subcortical regions of the brain in schizophrenia makes it an ideal drug (Jin *et al.*, 2002). So, in this case, stepholidine reduces the enhanced functioning of D2 receptors and helps alleviate the positive symptoms in schizophrenic patients. Sometimes the D2 receptor blockade also leads to a reduced negative feedback mechanism of dopaminergic neurons, resulting in increased dopamine release. In turn, this may result in deranged overstimulation of D1 receptors. Here, the partial agonist feature of stepholidine comes into play by ensuring a submaximal stimulation of the D1 receptors. In turn, this helps in lessening the negative and cognitive symptoms in schizophrenic patients (Abi-Dargham, 2004; Jin *et al.*, 2002; Mo *et al.*, 2005; Natesan et al., 2008; Yang *et al.*, 2007).

9.4.1.5 L-DOPA

Also known as 3,4-L-Dihydroxyphenylalanine, L-DOPA is a precursor of dopamine (Abbott, 2010). For a long time, its role as a therapeutic agent has been under study by scientists worldwide. A few scientific reports do shed light on it. It has been observed that, in schizophrenic patients already on antipsychotic drugs, the use of L-DOPA as an adjunctive treatment has proved beneficial (Jaskiw and Popli, 2004; Kay and Opler, 1985).

9.4.2 Drugs Targeting the Glutamate System

9.4.2.1 Bitopertin

The reduced functioning of the NMDA receptor is proposed to underlie many symptomatic signs of schizophrenia, especially the negative symptoms (Stone *et al.*, 2007). Therefore, efforts have been made to design specific drugs that can enhance NMDA functioning. Bitopertin is one such drug that augments its functioning by blocking the glycine transporter type 1, a major glycine reuptake receptor. Subsequently, this leads to an inhibited glycine reuptake, resulting in an increased concentration of synaptic glycine. Now, as glycine is an NMDA co-agonist, the final result is enhanced NMDA transmission (Chue, 2013; Umbricht *et al.*, 2014).

9.4.2.2 D-Amino Acid Oxidase Inhibitors

The functioning of NMDA demands the presence of both the agonist (glutamate) and the co-agonist (D-serine, D-glycine, or D-alanine) simultaneously (Mothet *et al.*, 2000; Smith *et al.*, 2010). The breakdown of these D-amino acids by the flavoenzyme DAAO results in a reduced NMDA transmission. Therefore, DAAO inhibitors have been tested as therapeutic agents to alleviate the degradation of the co-agonist D-amino acids in schizophrenic patients. The idea behind this line of therapy is that increasing NMDA functioning by enhancing the levels of the co-agonist D-amino acids might improve the positive and negative symptoms of schizophrenia (Sacchi *et al.*, 2013; Smith *et al.*, 2010).

9.4.3 Drugs Targeting the Serotonin System

9.4.3.1 Clozapine

Clozapine is an atypical antipsychotic drug belonging to the second generation of antipsychotics (SGAs). It is a monoaminergic antagonist of the serotonin receptor 5-HT_{2A} and dopamine receptor D2 (Fakra and Azorin, 2012). It has also been found that the affinity of clozapine is much greater for the 5-HT_{2A} receptor than the D2 receptor (Elkis and Meltzer, 2010; Seeman, 2014). As far as the use of clozapine as a drug is considered, it has been found to be quite beneficial in managing treatment-resistant schizophrenic patients with minimal risk of death (Conley *et al.*, 1988; Elkis and Meltzer, 2010; Fakra and Azorin, 2012; Haddad and Correll, 2018; Naber *et al.*, 1989).

Other prominent SGAs that act as serotonin (5-HT_{2A}) and dopamine (D2) antagonists are asenapine, brexpiprazole, iloperidone, lurasidone, olanzapine, paliperidone, quetiapine, risperidone, and ziprasidone (Chokhawala and Stevens, 2020; Divac *et al.*, 2014; Haddad and Correll, 2018). There is an added advantage of SGAs in that, in general, they induce fewer extrapyramidal symptoms in patients in than FGAs (Divac *et al.*, 2014).

9.4.3.2 Ondansetron

Ondansetron acts as an antagonist for the serotonin receptor 5-HT_3. Its use as an adjunctive to antipsychotic drugs in the treatment of schizophrenia has given encouraging results. Reports have hinted at the alleviation of negative symptoms (Bennett and Vila, 2010), cognitive symptoms (Kulkarni *et al.*, 2018), and psychotic symptoms (Sirota *et al.*, 2000) in schizophrenic patients administered ondansetron in addition to antipsychotic medications. However, the exact mechanism of the action of ondansetron is still unclear and requires further understanding.

9.4.3.3 Tropisetron

Tropisetron is also an antagonist of the serotonin receptor 5-HT_3. The use of this drug as an add on to risperidone is known to improve the primary negative symptoms in schizophrenic patients (Noroozian *et al.*, 2013). In addition, some reports hint at its involvement as an agonist of the α7 nicotinic acetylcholine receptor to help improve the cognitive symptomatic signs of schizophrenia (Shiina *et al.*, 2010; Xia *et al.*, 2020; Zhang *et al.*, 2012).

9.4.4 Drugs Targeting the Gabaergic System

Dysfunctional GABAergic neurotransmission is proposed to be involved in the pathophysiology of schizophrenia (Blum and Mann, 2002; Roberts, 1972). Therefore, efforts have been made to design therapeutic drugs that can target the GABAergic system. However, the use of GABAergic medications in the treatment of schizophrenia has been unconvincing owing to multiple methodologic restrictions and the toxicity of the drugs (Wassef *et al.*, 1999), though, over time, some improvements have been made. Benzodiazepines – a class of drugs acting as allosteric modulators of the GABA-A receptor subtypes (α5GABA-A receptor) – have exhibited positive signs in ameliorating cognitive deficits in schizophrenic patients (Gill and Grace, 2014; Yang and Tsai, 2017).

These are some of the potent pharmacologic drugs known to date that help in managing the various symptomatic signs of schizophrenia. However, all these drugs come with the liability of one or more side effects. Therefore, researchers worldwide are working extensively in this direction to design better drug therapies. On the whole, a more advanced and focused approach in the long run will surely ensure the availability of much-improved therapeutic drugs in the future.

9.5 RECENT DEVELOPMENTS AND CHALLENGES

With advancements in neurochemical science, genetics, and behavioral science, people suffering from schizophrenia can imagine a better life. However, a lot remains to be understood and done. Every year, many young people worldwide are diagnosed with schizophrenia (Haddad and Correll, 2018; Theodoridou and Rossler, 2010). This chronic psychiatric disorder leads them to a confusing dimension that hampers their personal well-being (Podea *et al.*, 2019). Though not a deadly disease, schizophrenia is devastating to patients and their families. Therefore, the mental issues and their associated disorders must not be taken lightly. Researchers worldwide should continue to find the root cause of schizophrenia and develop methods to treat this mental disorder in a better way. Designing much-improved drug therapies with minimal side effects should be the priority. Governments and non-governmental organizations should also come forward to fund research agencies working on schizophrenia and associated syndromes. A few organizations like the National Institute of Mental Health (NIMH) and Recovery After an Initial Schizophrenia Episode (RAISE) are extensively working in this direction. They are trying to develop a better prognosis for schizophrenia through synchronized treatment in the prodromal phase of the disorder (Dixon *et al.*, 2018), which will help to reduce the chance of schizophrenic patients developing long-term disabilities. An improved prognosis will go a long way in assisting patients to live independent and productive lives. Putting it all together, until medical science can completely prevent or cure schizophrenia, work remains to be done.

9.6 CONCLUSION

Schizophrenia continues to present a stiff challenge to researchers worldwide. Its pathophysiology and etiology still remain to be completely decoded. However, with advancements in medical science and imaging technologies, there has been a shift in the focus to find the core of schizophrenia at the molecular grade. Many hypotheses, such as the dopamine hypothesis and the GABA hypothesis, serotonergic dysfunctions, and glutamatergic abnormalities have been put forward, hinting at a role of disturbed neurochemical transmissions in schizophrenia. These propositions have their fair share of scientific backing, too. Numerous research reports point at dysfunctions at the molecular level and misexpression of neurotransmitter receptors in schizophrenic patients. These studies have subsequently led to the development of the current batch of FGAs and SGAs. Postmortem investigations of brain tissues have revealed the combinatorial effects of environment and genetic mutations in the manifestation of schizophrenia. All things considered, the future direction should be more focused on understanding the neurochemistry and molecular genetics of schizophrenia in a much-improved way so that better therapeutic interventions can be developed.

REFERENCES

Abi-Dargham A. Do we still believe in the dopamine hypothesis? New data bring new evidence. Int J Neuropsychopharmacol. 2004;7(Suppl 1):S1–S5.

Abi-Dargham A, Laruelle M, Aghajanian GK, et al. The role of serotonin in the pathophysiology and treatment of schizophrenia. J Neuropsychiatry Clin Neurosci. 1997; 9:1–17.

Abbott A. Levodopa: The story so far. Nature. 2010;466:S6–S7.

Adams CE, Rathbone J, Thornley B, et al. Chlorpromazine for schizophrenia: A Cochrane systematic review of 50 years of randomised controlled trials. BMC Med. 2005;3:15.

Aleman A. Neurocognitive basis of schizophrenia: Information processing abnormalities and clues for treatment. Adv Neurosci. 2014;6:1–15.

Andrade R. Serotonergic regulation of neuronal excitability in the prefrontal cortex. Neuropharmacology. 2011;61:382–386.

Asada H, Kawamura Y, Maruyama K, et al. Cleft palate and decreased brain gamma-aminobutyric acid in mice lacking the 67-kDa isoform of glutamic acid decarboxylase. Proc Natl Acad Sci U S A. 1997;94:6496–6499.

Avery MC, Krichmar JL. Improper activation of D1 and D2 receptors leads to excess noise in prefrontal cortex. Front Comput Neurosci. 2015;9:31.

Avramopoulos D. Recent advances in the genetics of schizophrenia. Mol Neuropsychiatry 2018;4:35–51.

Balu DT. The NMDA receptor and schizophrenia: From pathophysiology to treatment. Adv Pharmacol. 2016;76: 351–382.

Bantick RA, Deakin JF, Grasby PM. The 5-HT1A receptor in schizophrenia: A promising target for novel atypical neuroleptics? J Psychopharmacol. 2001;15:37–46.

Benchenane K, Tiesinga PH, Battaglia FP. Oscillations in the prefrontal cortex: A gateway to memory and attention. Curr Opin Neurobiol. 2011;21:475–485.

Benes FM, Berretta S. GABAergic interneurons: Implications for understanding schizophrenia and bipolar disorder. Neuropsychopharmacology. 2001;25:1–27.

Bennett AC, Vila TM. The role of ondansetron in the treatment of schizophrenia. Ann Pharmacother. 2010;44:1301–1306.

Berger M, Gray JA, Roth BL. The expanded biology of serotonin. Annu Rev Med. 2009;60:355–366.

Best JA, Nijhout HF, Reed MC. Homeostatic mechanisms in dopamine synthesis and release: A mathematical model. Theor Biol Med Model. 2009;6:21.

Bleich A, Brown SL, Kahn R, van Praag HM. The role of serotonin in schizophrenia. Schizophr Bull. 1988;14:297–315.

Blum BP, Mann JJ. The GABAergic system in schizophrenia. Int J Neuropsychopharmacol. 2002;5:159–179.

Boonstra E, de Kleijn R, Colzato LS, et al. Neurotransmitters as food supplements: The effects of GABA on brain and behavior. Front Psychol. 2015;6:1520.

Bowie CR, Harvey PD. Cognitive deficits and functional outcome in schizophrenia. Neuropsychiatr Dis Treat. 2006;2:531–536.

Brisch R, Saniotis A, Wolf R, et al. The role of dopamine in schizophrenia from a neurobiological and evolutionary perspective: Old fashioned, but still in vogue. Front Psychiatry. 2014;5:47.

Bu DF, Erlander MG, Hitz BC, et al. Two human glutamate decarboxylases, 65-kDa GAD and 67-kDa GAD, are each encoded by a single gene. Proc Natl Acad Sci U S A. 1992;89:2115–2119.

Bustillo JR, Chen H, Jones T, et al. Increased glutamine in patients undergoing long-term treatment for schizophrenia: A proton magnetic resonance spectroscopy study at 3 T. JAMA Psychiatry. 2014;71:265–272.

Carhart-Harris RL, Nutt DJ. Serotonin and brain function: A tale of two receptors. J Psychopharmacol. 2017;31:1091–1120.

Carlsson ML, Carlsson A, Nilsson M. Schizophrenia: From dopamine to glutamate and back. Curr Med Chem. 2004;11:267–277.

Chang CY, Chen YW, Wang TW, Lai WS. Akting up in the GABA hypothesis of schizophrenia: Akt1 deficiency modulates GABAergic functions and hippocampus-dependent functions. Sci Rep. 2016;6:33095.

Chappie T, Humphrey J, Menniti F, Schmidt C. PDE10A inhibitors: An assessment of the current CNS drug discovery landscape. Curr Opin Drug Discov Devel. 2009;12:458–467.

Chattopadhyaya B, Di Cristo G, Wu CZ, et al. GAD67-mediated GABA synthesis and signaling regulate inhibitory synaptic innervation in the visual cortex. Neuron. 2007;54:889–903.

Chien WT, Leung SF, Yeung FK, Wong WK. Current approaches to treatments for schizophrenia spectrum disorders, part II: Psychosocial interventions and patient-focused perspectives in psychiatric care. Neuropsychiatr Dis Treat. 2013;9:1463–1481.

Chokhawala K, Stevens L. Antipsychotic Medications. Treasure Island (FL): StatPearls Publishing; 2020.

Chue P. Glycine reuptake inhibition as a new therapeutic approach in schizophrenia: Focus on the glycine transporter 1 (GlyT1). Curr Pharm Des. 2013;19:1311–1320.

Citrome L. Cariprazine in schizophrenia: Clinical efficacy, tolerability, and place in therapy. Adv Ther. 2013;30:114–126.

Conley RR, Schulz SC, Baker RW, et al. Clozapine efficacy in schizophrenic nonresponders. Psychopharmacol Bull. 1988;24:269–274.

Conn PJ, Lindsley CW, Jones CK. Activation of metabotropic glutamate receptors as a novel approach for the treatment of schizophrenia. Trends Pharmacol Sci. 2009;30:25–31.

Correll CU, Schooler NR. Negative symptoms in schizophrenia: A review and clinical guide for recognition, assessment, and treatment. Neuropsychiatr Dis Treat. 2020;16:519–534.

Costa L, Trovato C, Musumeci SA, et al. 5-HT(1A) and 5-HT(7) receptors differently modulate AMPA receptor-mediated hippocampal synaptic transmission. Hippocampus. 2012;22:790–801.

Cote F, Thevenot E, Fligny C, et al. Disruption of the nonneuronal *tph1* gene demonstrates the importance of peripheral serotonin in cardiac function. Proc Natl Acad Sci U S A. 2003;100:13525–13530.

Curley AA, Arion D, Volk DW, et al. Cortical deficits of glutamic acid decarboxylase 67 expression in schizophrenia: Clinical, protein, and cell type-specific features. Am J Psychiatry. 2011;168:921–929.

Daikhin Y, Yudkoff M. Compartmentation of brain glutamate metabolism in neurons and glia. J Nutr. 2000;130(4S Suppl):1026S–1031S.

Dean B. Understanding the pathology of schizophrenia: Recent advances from the study of the molecular architecture of postmortem CNS tissue. Postgrad Med J. 2002;78:142–148.

Dean K, Murray RM. Environmental risk factors for psychosis. Dialogues Clin Neurosci. 2005;7:69–80.

de Jonge JC, Vinkers CH, Hulshoff Pol HE, Marsman A. GABAergic mechanisms in schizophrenia: Linking postmortem and in vivo studies. Front Psychiatry. 2017;8:118.

Divac N, Prostran M, Jakovcevski I, Cerovac N. Second-generation antipsychotics and extrapyramidal adverse effects. Biomed Res Int. 2014;2014:656370.

Dixon LB, Goldman H, Srihari VH, Kane J. Transforming the treatment of schizophrenia in the United States: The RAISE initiative. Annu Rev Clin Psychol. 2018;14:6.1–6.22.

Ebdrup BH, Rasmussen H, Arnt J, Glenthoj B. Serotonin 2A receptor antagonists for treatment of schizophrenia. Expert Opin Investig Drugs. 2011;20:1211–1223.

Ellaithy A, Younkin J, González-Maeso J, Logothetis DE. Positive allosteric modulators of metabotropic glutamate 2 receptors in schizophrenia treatment. Trends Neurosci. 2015;38:506–516.

Elkin A, Kalidindi S, McGuffin P. Have schizophrenia genes been found? Curr Opin Psychiatry. 2004;17:107–113.

Fakra E, Azorin JM. Clozapine for the treatment of schizophrenia. Expert Opin Pharmacother. 2012;13:1923–1935.

Fogaca MV, Duman RS. Cortical GABAergic dysfunction in stress and depression: New insights for therapeutic interventions. Front Cell Neurosci. 2019;13:87.

Frazer A, Hensler JG. Serotonin involvement in physiological function and behavior. In: Siegel GJ, Agranoff BW, Albers RW, et al., editors. Basic Neurochemistry: Molecular, Cellular and Medical Aspects. 6th ed. Philadelphia, PA: Lippincott-Raven; 1999.

Gaddum JH. Drugs antagonistic to 5-hydroxytryptamine. In: Wolstenholme GE, Cameron MP, editors. Ciba Foundation Symposium – Hypertension: Humoral and Neurogenic Factors. Boston, MA: Little Brown and Company; 1954.

Gaspar PA, Bustamante ML, Silva H, Aboitiz F. Molecular mechanisms underlying glutamatergic dysfunction in schizophrenia: Therapeutic implications. J Neurochem. 2009;111:891–900.

Gershon MD, Tack J. The serotonin signaling system: From basic understanding to drug development for functional GI disorders. Gastroenterology. 2007;132:397–414.

Gill KM, Grace AA. The role of α5 GABAA receptor agonists in the treatment of cognitive deficits in schizophrenia. Curr Pharm Des. 2014;20:5069–5076.

Goldman-Rakic PS, Castner SA, Svensson TH, et al. Targeting the dopamine D1 receptor in schizophrenia: Insights for cognitive dysfunction. Psychopharmacology. 2004;174:3–16.

Grauer SM, Pulito VL, Navarra RL, et al. Phosphodiesterase 10A inhibitor activity in preclinical models of the positive, cognitive, and negative symptoms of schizophrenia. J Pharmacol Exp Ther. 2009;331:574–590.

Gresack JE, Seymour PA, Schmidt CJ, Risbrough VB. Inhibition of phosphodiesterase 10A has differential effects on dopamine D1 and D2 receptor modulation of sensorimotor gating. Psychopharmacology (Berl). 2014;231:2189–2197.

Gurevich EV, Joyce JN. Alterations in the cortical serotonergic system in schizophrenia: A postmortem study. Biol Psychiatry. 1997;42:529–545.

Guzman F. First-generation antipsychotics: An Introduction. Psychopharmacology Institute. 2019. Available at: https://psychopharmacologyinstitute.com/publication/first-generation-antipsychotics-an-introduction-2110 (accessed 18 July 2020).

Gyertyan I, Kiss B, Saghy K, et al. Cariprazine (RGH-188), a potent D3/D2 dopamine receptor partial agonist, binds to dopamine D3 receptors in vivo and shows antipsychotic-like and procognitive effects in rodents. Neurochem Int 2011;59:925–935.

Haddad PM, Correll CU. The acute efficacy of antipsychotics in schizophrenia: A review of recent meta-analyses. Ther Adv Psychopharmacol. 2018;8:303–318.

Haenschel C, Bittner RA, Waltz J, et al. Cortical oscillatory activity is critical for working memory as revealed by deficits in early-onset schizophrenia. J Neurosci. 2009;29:9481–9489.

Hasbi A, O'Dowd BF, George SR. Dopamine D1-D2 receptor heteromer signaling pathway in the brain: Emerging physiological relevance. Mol Brain. 2011;4:26.

Hashimoto T, Nishino N, Nakai H, Tanaka C. Increase in serotonin 5-HT1A receptors in prefrontal and temporal cortices of brains from patients with chronic schizophrenia. Life Sci. 1991;48:355–363.

Hashimoto T, Volk DW, Eggan SM, et al. Gene expression deficits in a subclass of GABA neurons in the prefrontal cortex of subjects with schizophrenia. J Neurosci. 2003;23:6315–6326.

Hawkins RA. The blood–brain barrier and glutamate. Am J Clin Nutr. 2009;90:867S–874S.

Hawkins RA, Vina JR. How glutamate is managed by the blood–brain barrier. Biology (Basel). 2016;5:37.

Hernandez I, Sokolov BP. Abnormalities in 5-HT2A receptor mRNA expression in frontal cortex of chronic elderly schizophrenics with varying histories of neuroleptic treatment. J Neurosci Res. 2000;59:218–225.

He Y, Yu Z, Giegling I, et al. Schizophrenia shows a unique metabolomics signature in plasma. Transl Psychiatry. 2012;2:e149.

Howes OD, Kapur S. The dopamine hypothesis of schizophrenia: Version III – The final common pathway. Schizophr Bull. 2009;35:549–562.

Howes OD, Montgomery AJ, Asselin MC, et al. Elevated striatal dopamine function linked to prodromal signs of schizophrenia. Arch Gen Psychiatry. 2009;66:13–20.

Howes O, McCutcheon R, Stone J. Glutamate and dopamine in schizophrenia: An update for the 21st century. J Psychopharmacol. 2015;29:97–115.

Hurlemann R, Boy C, Meyer PT, et al. Decreased prefrontal 5-HT2A receptor binding in subjects at enhanced risk for schizophrenia. Anat Embryol (Berl). 2005;210:519–523.

Elkis H, Meltzer HY. Role of clozapine in treatment-resistant schizophrenia. Therapy-resistant schizophrenia. Adv Biol Psychiatry. 2010;26:114–128.

Ichiyama A, Nakamura S, Nishizuka Y, Hayaishi O. Enzymic studies on the biosynthesis of serotonin in mammalian brain. J Biol Chem. 1970;245:1699–1709.

Jablensky A. Subtyping schizophrenia: Implications for genetic research. Mol Psychiatry. 2006;11:815–836.

Jablensky A. The diagnostic concept of schizophrenia: Its history, evolution, and future prospects. Dialogues Clin Neurosci. 2010;12:271–287.

Jadi MP, Behrens MM, Sejnowski TJ. Abnormal gamma oscillations in NMDAR hypofunction models of schizophrenia. Biol Psychiatry. 2016;79:716–726.

Jaskiw GE, Popli AP. A meta-analysis of the response to chronic L-dopa in patients with schizophrenia: Therapeutic and heuristic implications. Psychopharmacology (Berl). 2004;171:365–374.

Javitt DC. Glutamate and schizophrenia: Phencyclidine, N-methyl-D-aspartate receptors, and dopamine-glutamate interactions. Int Rev Neurobiol. 2007;78:69–108.

Jin G, Zhu Z, Fu Y. (−)-Stepholidine: A potential novel antipsychotic drug with dual D1 receptor agonist and D2 receptor antagonist actions. Trends Pharmacol. Sci. 2002;23:4–7.

Kahn RS, Sommer IE, Murray RM, et al. Schizophrenia. Nat Rev Dis Primers. 2015;1:15067.

Kaneko K. Negative symptoms and cognitive impairments in schizophrenia: Two key symptoms negatively influencing social functioning. Yonago Acta Med. 2018;61:91–102.

Kaneko K. Enzymes responsible for glutamate synthesis and degradation. In: Otterson OP, Storm-Mathisen J, editors. Handbook of Chemical Neuroanatomy. 18th vol. Amsterdam: Elsevier; 2000.

Kann O, Papageorgiou IE, Draguhn A. Highly energized inhibitory interneurons are a central element for information processing in cortical networks. J Cereb Blood Flow Metab. 2014;34:1270–1282.

Kass I, Hoke DE, Costa MG, et al. Cofactor-dependent conformational heterogeneity of GAD65 and its role in autoimmunity and neurotransmitter homeostasis. Proc Natl Acad Sci U S A. 2014;111:E2524–E2529.

Kay SR, Opler LA. L-dopa in the treatment of negative schizophrenic symptoms: A single-subject experimental study. Int J Psychiatry Med. 1985;15:293–298.

Kehler J, Nielsen J. PDE10A inhibitors: Novel therapeutic drugs for schizophrenia. Curr Pharm Des. 2011;17:137–150.

Kelsom C, Lu W. Development and specification of GABAergic cortical interneurons. Cell Biosci. 2013;3:19.

Kim JS, Kornhuber HH, Schmid-Burgk W, Holzmuller B. Low cerebrospinal fluid glutamate in schizophrenic patients and a new hypothesis on schizophrenia. Neurosci Lett. 1980;20:379–382.

Kissler J, Muller MM, Fehr T, et al. MEG gamma band activity in schizophrenia patients and healthy subjects in a mental arithmetic task and at rest. Clin Neurophysiol. 2000;111:2079–2087.

Kitzinger H, Arnold DG. A preliminary study of the effects of glutamic acid on catatonic schizophrenics. Rorschach Res Exch J Proj Tech. 1949;13:210–218.

Krystal JH, Karper LP, Seibyl JP, et al. Subanesthetic effects of the noncompetitive NMDA antagonist, ketamine, in humans. Psychotomimetic, perceptual, cognitive, and neuroendocrine responses. Arch Gen Psychiatry. 1994;51:199–214.

Kucharska-Pietura K, Mortimer A. Can antipsychotics improve social cognition in patients with schizophrenia? CNS Drugs. 2013;27:335–343.

Kulkarni J, Thomas N, Hudaib AR, et al. Ondansetron – A promising adjunctive treatment for persistent schizophrenia. J Psychopharmacol. 2018;32:1204–1211.

Kvamme E. Synthesis of glutamate and its regulation. Prog Brain Res. 1998;116:73–85.

Leggio GM, Salomone S, Bucolo C, et al. Dopamine D(3) receptor as a new pharmacological target for the treatment of depression. Eur J Pharmacol. 2013;719:25–33.

Lewis DA. Inhibitory neurons in human cortical circuits: Substrate for cognitive dysfunction in schizophrenia. Curr Opin Neurobiol. 2014;26:22–26.

Lewis DA, Levitt P. Schizophrenia as a disorder of neurodevelopment. Annu Rev Neurosci. 2002;25:409–432.

Lindstrom LH, Gefvert O, Hagberg G, et al. Increased dopamine synthesis rate in medial prefrontal cortex and striatum in schizophrenia indicated by L-(beta-11C) DOPA and PET. Biol Psychiatry. 1999;46:681–688.

Lisman JE, Coyle JT, Green RW, et al. Circuit-based framework for understanding neurotransmitter and risk gene interactions in schizophrenia. Trends Neurosci. 2008;31:234–242.

Lovinger DM. Communication networks in the brain: Neurons, receptors, neurotransmitters, and alcohol. Alcohol Res Health. 2008;31:196–214.

Lv J, Liu F. The role of serotonin beyond the central nervous system during embryogenesis. Front Cell Neurosci. 2017;11:74.

Lyon GJ, Abi-Dargham A, Moore H, et al. Presynaptic regulation of dopamine transmission in schizophrenia. Schizophr Bull. 2011;37:108–117.

McNally JM, McCarley RW. Gamma band oscillations: A key to understanding schizophrenia symptoms and neural circuit abnormalities. Curr Opin Psychiatry. 2016;29:202–210.

Mann SK, Marwaha R. Chlorpromazine. Treasure Island (FL): StatPearls Publishing; 2020.

Masu M, Nakajima Y, Moriyoshi K, et al. Molecular characterization of NMDA and metabotropic glutamate receptors. Ann N Y Acad Sci. 1993;707:153–164.

Mathalon DH, Ford JM. Neurobiology of schizophrenia: Search for the elusive correlation with symptoms. Front Hum Neurosci. 2012;6:136.

Matsumoto I, Inoue Y, Iwazaki T, et al. 5-HT2A and muscarinic receptors in schizophrenia: A postmortem study. Neurosci Lett. 2005;379:164–168.

Mattila T, Koeter M, Wohlfarth T, et al. Impact of DSM-5 changes on the diagnosis and acute treatment of schizophrenia. Schizophr Bull. 2015;41:637–643.

Meltzer HY, Li Z, Kaneda Y, Ichikawa J. Serotonin receptors: Their key role in drugs to treat schizophrenia. Prog Neuropsychopharmacol Biol Psychiatry. 2003;27: 1159–1172.

Millan MJ. Improving the treatment of schizophrenia: Focus on serotonin (5-HT)(1A) receptors. J Pharmacol Exp Ther. 2000;295:853–861.

Mitra S, Mahintamani T, Kavoor AR, Nizamie SH. Negative symptoms in schizophrenia. Ind Psychiatry J. 2016;25:135–144.

Miyake N, Thompson J, Skinbjerg M, Abi-Dargham A. Presynaptic dopamine in schizophrenia. CNS Neurosci Ther. 2011;17:104–109.

Mo YQ, Jin XL, Chen YT, et al. Effects of l-stepholidine on forebrain Fos expression: Comparison with clozapine and haloperidol. Neuropsychopharmacology. 2005;30:261–267.

Moran LV, Hong LE. High vs low frequency neural oscillations in schizophrenia. Schizophr Bull. 2011;37:659–663.

Moran P, Stokes J, Marr J, et al. Gene × environment interactions in schizophrenia: Evidence from genetic mouse models. Neural Plast. 2016;2016:2173748.

Morgan CJ, Curran HV. Acute and chronic effects of ketamine upon human memory: A review. Psychopharmacology (Berl). 2006;188:408–424.

Mothet JP, Parent AT, Wolosker H, et al. D-serine is an endogenous ligand for the glycine site of the N-methyl-D-aspartate receptor. Proc Natl Acad Sci U S A. 2000;97:4926–4931.

Naber D, Leppig M, Grohmann R, Hippius H. Efficacy and adverse effects of clozapine in the treatment of schizophrenia and tardive dyskinesia--A retrospective study of 387 patients. Psychopharmacology (Berl). 1989;99(Suppl.):S73–S76.

Natesan S, Reckless GE, Barlow KB, et al. The antipsychotic potential of l-stepholidine – A naturally occurring dopamine receptor D1 agonist and D2 antagonist. Psychopharmacology (Berl). 2008;199:275–289.

Neve KA, Seamans JK, Tranthum-Davidson H. Dopamine receptor signaling. J Recept Sig Transd. 2004;24:165–205.

Newsholme P, Lima MM, Procopio J, et al. Glutamine and glutamate as vital metabolites. Braz J Med Biol Res. 2003;36:153–163.

Nishi A, Kuroiwa M, Miller DB, et al. Distinct roles of PDE4 and PDE10A in the regulation of cAMP/PKA signaling in the striatum. J Neurosci. 2008;28:10460–10471.

Noroozian M, Ghasemi S, Hosseini SM, et al. A placebo-controlled study of tropisetron added to risperidone for the treatment of negative symptoms in chronic and stable schizophrenia. Psychopharmacology (Berl). 2013;228:595–602.

Nozaki S, Kato M, Takano H, et al. Regional dopamine synthesis in patients with schizophrenia using L-[beta-11C]DOPA PET. Schizophr Res. 2009;108:78–84.

Nudmamud-Thanoi S. Neurochemical abnormalities in schizophrenia. NUJ. 2005;13:61–72.

Ohno Y. Therapeutic role of 5-HT1A receptors in the treatment of schizophrenia and Parkinson's disease. CNS Neurosci Ther. 2011;17:58–65.

Ohuoha, DC, Hyde TM, Kleinman JE. The role of serotonin in schizophrenia: An overview of the nomenclature, distribution and alterations of serotonin receptors in the central nervous system. Psychopharmacology. 1993;112(S1):S5–S15.

Olguin HJ, Guzman DC, Garcia EH, Mejia GB. The role of dopamine and its dysfunction as a consequence of oxidative stress. Oxid Med Cell Longev. 2016;2016:9730467.

Olney JW, Newcomer JW, Farber NB. NMDA receptor hypofunction model of schizophrenia. J Psychiatr Res. 1999;33:523–533.

Paparelli A, Di Forti M, Morrison PD, Murray RM. Drug-induced psychosis: How to avoid star gazing in schizophrenia research by looking at more obvious sources of light. Front Behav Neurosci. 2011;5:1.

Patel AB, de Graaf RA, Martin DL, et al. Evidence that GAD65 mediates increased GABA synthesis during intense neuronal activity in vivo. J Neurochem. 2006;97:385–396.

Patel KR, Cherian J, Gohil K, Atkinson D. Schizophrenia: Overview and treatment options. P T. 2014;39:638–645.

Peroutka SJ, Snyder SH. Multiple serotonin receptors: Differential binding of [3H]5-hydroxytryptamine, [3H]lysergic acid diethylamide and [3H]spiroperidol. Mol Pharmacol. 1979;16:687–699.

Petroff OA. GABA and glutamate in the human brain. Neuroscientist. 2002;8:562–573.

Pilowsky LS, Bressan RA, Stone JM, et al. First in vivo evidence of an NMDA receptor deficit in medication-free schizophrenic patients. Mol Psychiatry. 2006;11:118–119.

Pinal CS, Tobin AJ. Uniqueness and redundancy in GABA production. Perspect Dev Neurobiol. 1998;5:109–118.

Podea DM, Moldovan RT, Popa LC. Schizophrenia: Early recognition and prevention. In: Fitzgerald M, editor. Neurodevelopment and Neurodevelopmental Disorder. London: IntechOpen; 2019.

Polter AM, Li X. 5-HT1A receptor-regulated signal transduction pathways in brain. Cell Signal. 2010;22:1406–1412.

Pralong D, Tomaskovic-Crook E, Opeskin K, et al. Serotonin(2A) receptors are reduced in the planum temporale from subjects with schizophrenia. Schizophr Res. 2000;44:35–45.

Purves D, Augustine GJ, Fitzpatrick D, et al., editors. Neuroscience. 2nd ed. Sunderland, MA: Sinauer Associates; 2001.

Rasmussen H, Erritzoe D, Andersen R, et al. Decreased frontal serotonin2A receptor binding in antipsychotic-naive patients with first-episode schizophrenia. Arch Gen Psychiatry. 2010;67:9–16.

Reynolds GP. The neurochemistry of schizophrenia. Psychiatry. 2008;7:425–429.

Ribeyre JM, Lesieur P, Varoquaux O, et al. A comparison of plasma homovanillic acid in the deficit and nondeficit subtypes of schizophrenia. Biol Psychiatry. 1994;36:230–236.

Roberts DL, Penn DL, Corrigan P, et al. Antipsychotic medication and social cue recognition in chronic schizophrenia. Psychiatry Res. 2010;178:46–50.

Roberts E. Prospects for research on schizophrenia. An hypotheses suggesting that there is a defect in the GABA system in schizophrenia. Neurosci Res Program Bull. 1972;10:468–482.

Rojas PS, Fiedler JL. What do we really know about 5-HT1A receptor signaling in neuronal cells? Front Cell Neurosci. 2016;10:272.

Rothman DL, Behar KL, Hyder F, Shulman RG. In vivo NMR studies of the glutamate neurotransmitter flux and neuroenergetics: implications for brain function. Annu Rev Physiol. 2003;65:401–427.

Rousseaux CG. A review of glutamate receptors II: Pathophysiology and pathology. J Toxicol Pathol. 2008;21:133–173.

Sacchi S, Rosini E, Pollegioni L, Molla G. D-amino acid oxidase inhibitors as a novel class of drugs for schizophrenia therapy. Curr Pharm Des. 2013;19:2499–2511.

Salleh MR. The genetics of schizophrenia. Malays J Med Sci. 2004;11:3–11.

Schmidt CJ, Chapin DS, Cianfrogna J, et al. Preclinical characterization of selective phosphodiesterase 10A inhibitors: A new therapeutic approach to the treatment of schizophrenia. J Pharmacol Exp Ther. 2008;325:681–690.

Schultz SH, North SW, Shields CG. Schizophrenia: A review. Am Fam Physician. 2007;75:1821–1829.

Seeman P. Clozapine, a fast-off-D2 antipsychotic. ACS Chem Neurosci. 2014;5:24–29.

Senagolage M, Sullivan P. Firings of a schizophrenic mind: Faulty neurotransmission and genetics. Eukaryon. 2013;9.

Shen J. Modeling the glutamate-glutamine neurotransmitter cycle. Front Neuroenergetics. 2013;5:1.

Shen LH, Liao MH, Tseng YC. Recent advances in imaging of dopaminergic neurons for evaluation of neuropsychiatric disorders. J Biomed Biotechnol. 2012;2012:259349.

Sherman SM. The function of metabotropic glutamate receptors in thalamus and cortex. Neuroscientist. 2014;20:136–149.

Shiina A, Shirayama Y, Niitsu T, et al. A randomised, double-blind, placebo-controlled trial of tropisetron in patients with schizophrenia. Ann Gen Psychiatry. 2010;9:27.

Shin YW, O'Donnell BF, Youn S, Kwon JS. Gamma oscillation in schizophrenia. Psychiatry Investig. 2011;8:288–296.

Sirota P, Mosheva T, Shabtay H, et al. Use of the selective serotonin 3 receptor antagonist ondansetron in the treatment of neuroleptic-induced tardive dyskinesia. Am J Psychiatry. 2000;157:287–289.

Smith SM, Uslaner JM, Hutson PH. The therapeutic potential of D-amino acid oxidase (DAAO) inhibitors. Open Med Chem J. 2010;4:3–9.

Snyder GL, Vanover KE. PDE inhibitors for the treatment of schizophrenia. Adv Neurobiol. 2017;17:385–409.

Stahl SM. Drugs for psychosis and mood: Unique actions at D3, D2, and D1 dopamine receptor subtypes. CNS Spectr. 2017;22:375–384.

Stahl SM, Laredo S, Morrissette DA. Cariprazine as a treatment across the bipolar I spectrum from depression to mania: mechanism of action and review of clinical data. Ther Adv Psychopharmacol. 2020;10:2045125320905752.

Stahl SM. Mechanism of action of cariprazine. CNS Spectr. 2016;21:123–127.

Stenkrona P, Matheson GJ, Halldin C, et al. D1-Dopamine receptor availability in first-episode neuroleptic naive psychosis patients. Int J Neuropsychopharmacol. 2019;22:415–425.

Stilo SA, Forti MD, Murray R. Environmental risk factors for schizophrenia: Implications for prevention. Neuropsychiatry. 2011;1:457–466.

Stone JM, Morrison PD, Pilowsky LS. Glutamate and dopamine dysregulation in schizophrenia – A synthesis and selective review. J Psychopharmacol. 2007;21:440–452.

Suzuki K, Harada A, Suzuki H, et al. TAK-063, a PDE10A inhibitor with balanced activation of direct and indirect pathways, provides potent antipsychotic-like effects in multiple paradigms. Neuropsychopharmacology. 2016;41:2252–2262.

Tallerico T, Novak G, Liu IS, et al. Schizophrenia: Elevated mRNA for dopamine D2(Longer) receptors in frontal cortex. Brain Res Mol Brain Res. 2001;87:160–165.

Tandon R, Gaebel W, Barch DM, et al. Definition and description of schizophrenia in the DSM-5. Schizophr Res. 2013;150:1.

Tauscher J, Kapur S, Verhoeff NP, et al. Brain serotonin 5-HT(1A) receptor binding in schizophrenia measured by positron emission tomography and [11C]WAY-100635. Arch Gen Psychiatry. 2002;59:514–520.

Taylor SF, Tso IF. GABA abnormalities in schizophrenia: A methodological review of in vivo studies. Schizophr Res. 2015;167:84–90.

Theodoridou A, Rossler W. Disease burden and disability-adjusted life years due to schizophrenia and psychotic disorders. In: Preedy VR, Watson RR. Handbook of Disease Burdens and Quality of Life Measures, Pt. 2, 2.6. New York: Springer; 2010, pp. 1493–1507.

Thompson M, Weickert CS, Wyatt E, Webster MJ. Decreased glutamic acid decarboxylase(67) mRNA expression in multiple brain areas of patients with schizophrenia and mood disorders. J Psychiatr Res. 2009;43:970–977.

Tian N, Petersen C, Kash S, et al. The role of the synthetic enzyme GAD65 in the control of neuronal gamma-aminobutyric acid release. Proc Natl Acad Sci U S A. 1999;96:12911–12916.

Toda M, Abi-Dargham A. Dopamine hypothesis of schizophrenia: Making sense of it all. Curr Psychiatry Rep. 2007;9:329–336.

Torgner I, Kvamme E. Synthesis of transmitter glutamate and the glial-neuron interrelationship. Mol Chem Neuropathol. 1990;12:11–17.

Tripathi A, Kar SK, Shukla R. Cognitive deficits in schizophrenia: Understanding the biological correlates and remediation strategies. Clin Psychopharmacol Neurosci. 2018;16:7–17.

Tsuang M. Schizophrenia: genes and environment. Biol Psychiatry. 2000;47:210–220.

Umbricht D, Alberati D, Martin-Facklam M, et al. Effect of bitopertin, a glycine reuptake inhibitor, on negative symptoms of schizophrenia: A randomized, double-blind, proof-of-concept study. JAMA Psychiatry. 2014;71:637–646.

Valenzuela CF, Puglia MP, Zucca S. Focus on: Neurotransmitter systems. Alcohol Res Health. 2011;34:106–120.

van Os J, Rutten BP, Poulton R. Gene-environment interactions in schizophrenia: Review of epidemiological findings and future directions. Schizophr Bull. 2008;34:1066–1082.

Varnas K, Halldin C, Hall H. Autoradiographic distribution of serotonin transporters and receptor subtypes in human brain. Hum Brain Mapp. 2004;22:246–260.

Walker MC, van der Donk WA. The many roles of glutamate in metabolism. J Ind Microbiol Biotechnol. 2016;43:419–430.

Walter H, Kammerer H, Frasch K, et al. Altered reward functions in patients on atypical antipsychotic medication in line with the revised dopamine hypothesis of schizophrenia. Psychopharmacology. 2009;206:121–132.

Walther DJ, Peter JU, Bashammakh S, et al. Synthesis of serotonin by a second tryptophan hydroxylase isoform. Science. 2003;299:76.

Wassef AA, Dott SG, Harris A, et al. Critical review of GABA-ergic drugs in the treatment of schizophrenia. J Clin Psychopharmacol. 1999;19:222–232.

Womelsdorf T, Fries P. The role of neuronal synchronization in selective attention. Curr Opin Neurobiol. 2007;17:154–160.

Woolley DW, Shaw E. A biochemical and pharmacological suggestion about certain mental disorders. Proc Natl Acad Sci U S A. 1954;40:228–231.

Woo TU, Spencer K, McCarley RW. Gamma oscillation deficits and the onset and early progression of schizophrenia. Harv Rev Psychiatry. 2010;18:173–189.

Wu C, Sun D. GABA receptors in brain development, function, and injury. Metab Brain Dis. 2015;30:367–379.

Xia L, Liu L, Hong X, et al. One-day tropisetron treatment improves cognitive deficits and P50 inhibition deficits in schizophrenia. Neuropsychopharmacology. 2020;45:1362–1368.

Xu G, Broadbelt KG, Haynes RL, et al. Late development of the GABAergic system in the human cerebral cortex and white matter. J Neuropathol Exp Neurol. 2011;70:841–858.

Xu MY, Wong AH. GABAergic inhibitory neurons as therapeutic targets for cognitive impairment in schizophrenia. Acta Pharmacol Sin. 2018;39:733–753.

Yang AC, Tsai SJ. New targets for schizophrenia treatment beyond the dopamine hypothesis. Int J Mol Sci. 2017;18:1689.

Yang K, Jin G, Wu J. The neuropharmacology of (-)-stepholidine and its potential applications. Curr Neuropharmacol. 2007;5:289–294.

Yoon JH, Maddock RJ, Rokem A, et al. GABA concentration is reduced in visual cortex in schizophrenia and correlates with orientation-specific surround suppression. J Neurosci. 2010;30:3777–3781.

Yui K, Ikemoto S, Goto K, et al. Spontaneous recurrence of methamphetamine induced paranoid-hallucinatory states in female subjects: Susceptibility to psychotic states and implications for relapse of schizophrenia. Pharmacopsychiatry. 2002;35:62–71.

Zhang XY, Liu L, Liu S, et al. Short-term tropisetron treatment and cognitive and P50 auditory gating deficits in schizophrenia. Am J Psychiatry. 2012;169:974–981.

Zhou Y, Danbolt NC. Glutamate as a neurotransmitter in the healthy brain. J Neural Transm (Vienna). 2014;121:799–817.

Zhu ZT, Fu Y, Hu GY, Jin GZ. Electrophysiological study on biphasic firing activity elicited by D(1) agonistic-D(2) antagonistic action of (-)-stepholidine in nucleus accumbens. Sheng Li Xue Bao. 2000;52:123–130.

Zimnisky R, Chang G, Gyertyan I, et al. Cariprazine, a dopamine D(3)-receptor-preferring partial agonist, blocks phencyclidine-induced impairments of working memory, attention set-shifting, and recognition memory in the mouse. Psychopharmacology (Berl). 2013;226:91–100.

10 Role of Endocannabinoids in Neurocognitive Dysfunctions

Swathi Putta and Eswar Kumar Kilari
A.U. College of Pharmaceutical Sciences, Andhra University, Visakhapatnam, India

Chandrasekar SB, Hemavathi G, and Lokesh Prasad
Hi-Tech Lab, Drugs Testing Laboratory, Drugs Control Department, Bangalore, India

Anil T. Pawar
School of Pharmacy, Dr.Vishwanath Karad MIT World Peace University, Pune, India

Aman Upaganlawar
SNJB's Shriman Suresh Dada Jain College of Pharmacy, Nashik, India

CONTENTS

10.1 INTRODUCTION

The use of cannabis plants has been described in several pharmacopoeias (Indian, Arab, and Chinese), and has been well documented over the centuries, with the plants being used for their therapeutic and recreational properties (Mechoulam, 1986; Samuelsson and Bohlin, 2010). Extensive research over more than 50 years with respect to the underlying mechanism of action of the biologically active chemical constituents of cannabinoids (CBDs) led to their approval by European medicinal authorities in 2010 (Kmietowicz, 2010). A treasure of outcomes acquired from various preclinical and clinical studies from more than two decades of research has immensely enhanced the description of the mechanisms underlying the effects of endocannabinoids (eCBs) and their synthetic equivalents, and how they control the CBD receptors. This scientific research has been the beacon in directing the eCBs signaling pathway to the treatment of numerous distressing illnesses such as neurocognitive dysfunctions, as well as chronic inflammatory diseases (Zou and Kumar, 2018).

DOI: 10.1201/9780429265198-12

The endocannabinoid system (ECS) regulates an enormous variety of physiological activities, involving pain, mood, motor regulation, feeding behavior, and mood (Mechoulam, 1986; Kmietowicz, 2010; Pertwee, 2006; Gorzo, 2012; Velayudhan et al., 2014; Maccarrone et al., 2015). In recent times, several studies have revealed the role played by eCBs in various pathological conditions.

Almost 105 phytocannabinoids have been identified to date, including the most important, cannabidiol and Δ^9-tetrahydrocannabinol (THC) (Devane et al., 1988; Matsuda et al., 1990; Russo, 2011). Numerous medicines which are CBD-associated are officially permitted to treat pathological conditions such as spasticity, epileptogenic encephalopathy associated with infancy and childhood, nausea induced by chemotherapy, anorexia, and pain (Velayudhan et al., 2014; Devane et al., 1988; Matsuda et al., 1990; Munro et al., 1993).

After alcohol, cannabis is ranked the second most consumed drug. Approximately 20% of college students and about 23% of high school students are thought to use it (Johnston et al., 2011). Investigation of the neurocognitive effects of cannabis in adolescents and young adults is a great health concern.

Research on the neurocognitive effects of cannabis continues to grow rapidly. It is an especially pertinent area of research because cannabis continues to be the most widely used illicit substance worldwide (United Nations Office on Drugs and Crimes, 2011).

When animals are exposed to THC at doses evoking a psychoactive outcome, cognitive dysfunctions appear that are analogous to those observed in humans following cannabis consumption (Solowij et al., 2011a). These cognitive shortfalls continue, without change, even after a prolonged time of self-restraint, and it has been proposed that cannabis consumption during the initial phase of life might quicken brain disorders connected with aging and psychosis (Bolla et al., 2002; Mata et al., 2010; Solowij et al., 2011b).

This study emphasizes on eCB signaling pathways in cognitive dysfunctions, which is evident by its diverse molecular mechanisms in this signaling pathway (enzymes and receptors creating and deactivating eCBS) by considerably modulating central and peripheral tissues with regard to aging. It is proposed that the effect of eCB based medicines vary with respect to age, sex and disease severity of the patients. Hence, therapy involving eCBs essentially needs to be customized according to this patient (Velayudhan et al., 2014; Hollister, 1986; Ben Amar, 2006).

10.2 COMPOSITION AND SYNTHESIS OF THE ENDOCANNABINOID SYSTEM

There are two classes of CBDs: synthetic CBDs (*WIN55,212-2*) and eCBs (Toth et al., 2019). The eCBs are produced through biosynthetic pathways occurring at lipid bilayers, and exert biological effects through cannabinoid receptor type 1 (CB_1) and cannabinoid receptor type 2 (CB_2). The synthesis of eCBs is represented in (Figure 10.1). 2-Arachidonylglycerol (2-AG) functions as a true ligand to CBD receptors,

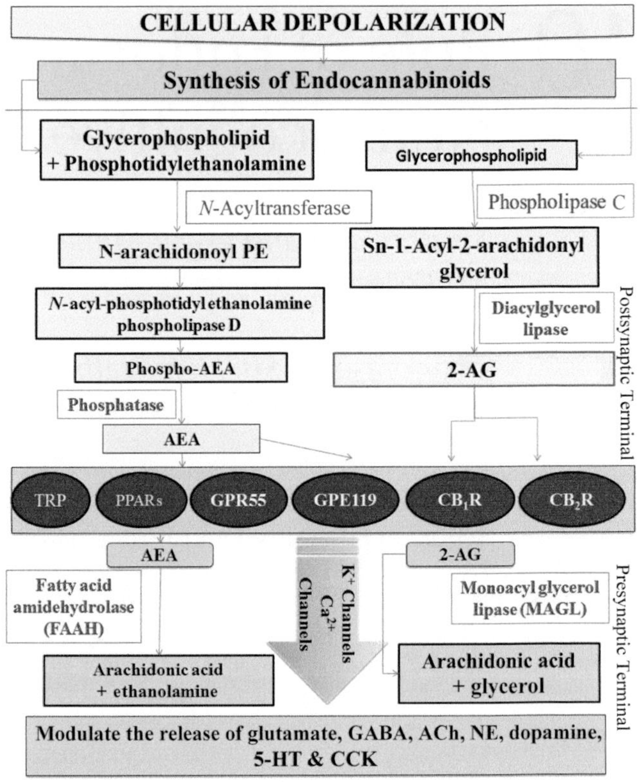

FIGURE 10.1 Synthesis of endocannabinoids (eCBs). PE, phosphotidylethanolamine AEA, anandamide; 2-AG, 2-arachidonoylglycerol; TRP, transient receptor potential; PPAR, peroxisome proliferator-activated receptor; GPR55, G protein-coupled receptor 55; GPE119, globus pallidus external 119; CB_1R, cannabinoid receptor type 1; CB_2R, cannabinoid receptor type 2; ACh, acetylcholine; NE, norepinephrine; 5-HT, 5-hydroxytryptamine; CCK, cholecystokinin.

since 2-AG is abundantly expressed in the brain (Sugiura et al, 2002). The on-demand production of eCBs stimulates neuronal membrane depolarization or activation of immune cells. The concentration of eCBs escalates in response to trauma and other pathological events, leading to regulatory or compensatory mechanisms in their production. The activation of both macrophages and dendritic cells stimulates eCBs, which reduces the expression of degrading enzymes for eCBs (Klein, 2005).

10.2.1 Endogenous Ligands

The first endogenous ligands, anandamide (AEA) and 2-AG, were obtained from porcine brain to evaluate their interaction with CBD receptors (Devane et al., 1992). Of these, 2-AG is a high-efficacy ligand due its agonistic activity on CBD receptors. AEA is known to have little efficiency for CB_1 and negligible efficiency for CB_2 (Luk et al., 2004). AEA produces antagonistic effects and agonistic actions, with low receptor expression or weak signaling actions (Mackie et al., 1993). The endogenous ligands that are synthesized from arachidonic acid are virodhamine and 2-AG. These are also produced, when required, from the lipid membrane. The end products all elicit their local effects with a very short half-life. These endogenous ligands are degraded by means of fatty acid amide hydrolase (FAAH) along with mono-acylglycerol lipase (Howlett et al., 2002). Cyclooxygenase-2 (COX-2) is selective for AEA, compared to other acyl ethanolamides. It elicits more specificity for the inhibition of AEA than FAAH (Hermanson et al., 2014).

AEA is converted to prostaglandin-ethanolamides by means of COX-2. Lipoxygenase (LOX) converts AEA into hydroxy-anandamides. Cytochrome P450s convert AEA to epoxyeicosatrienoic acid ethanolamides. Diacylglycerol lipase (DAGL) alpha and beta synthesize 2-AG and *N*-acyl-phosphatidylethanolamine (NAPE) phospholipase D (PLD). They are involved in the biosynthesis of eCBs complemented by complete characterization of AEA (Howlett et al., 2011). The transportation of these eCBs is limited due to the lipid medium; therefore, their movement in the cytosol is obstructed. Thus, intracellular AEA transporters assist the transit of AEA to accurate ultimate targets, which include (Hillard et al., 2017):

(i) Metabolism in the endoplasmic reticulum (ER) through adiposomes (FAAH-1/FAAH-2) for deposition (Meccariello et al., 2020).
(ii) Metabolism by adiposomes or oxidation via COX-2 or LOXs (Chan et al., 2019).
(iii) Oxidation at mitochondria and likely stimulation of CB_1 (Bénard et al., 2012).
(iv) Metabolism at lysosomes by P450 or *N*-acylethanolamine acid amide hydrolase.
(v) Nucleus for stimulation of peroxisome proliferator-activated receptors (PPARs; Maccarrone, 2017).

10.2.2 Receptors

CBD receptors are found in abundance all over the body, distributed in cell membranes. Various CBD receptors, including CB_1, CB_2, transient receptor potential (TRP) channels, and PPARs, are stimulated and take part in physiological processes (Mackie and Lu, 2016). The central nervous system (CNS), peripheral tissues, and other organs all contain types of CBD receptor. The activation of receptors is based on cell type, dose administered to produce bipolar actions, relaxation of muscles, appetite, dry mouth, and analgesic actions (Gonclaves et al., 2019). Blockage or antagonism of CBD receptors, modulation of eCBs, and inhibition of the degradation of eCBs occur during the activation of CBD receptors (Grotenhermen, 2006). CB_1 and CB_2 are G protein-coupled receptors (GPCRs), which are similar to the orphan receptors GPR55 and GPR119 (Barzegar et al., 2015). The abundance of CB_1 and the scarcity of CB_2 in the CNS might be the result of the antipsychotic activity of exogenous and endogenous CBDs (Wang and Ueda, 2009). Figure 10.2 summarizes the types of receptors and their distribution patterns in the CNS and peripheral nervous system.

10.2.2.1 CB_1

CB_1 belongs to the rhodopsin member subfamily of GPCRs. The activation of GPCRs occurs at three cytosolic loops and at the palmitoylation sites (at the C-terminal region of the juxta membrane, where it is activated by the fourth loop (Demuth and Molleman, 2006).

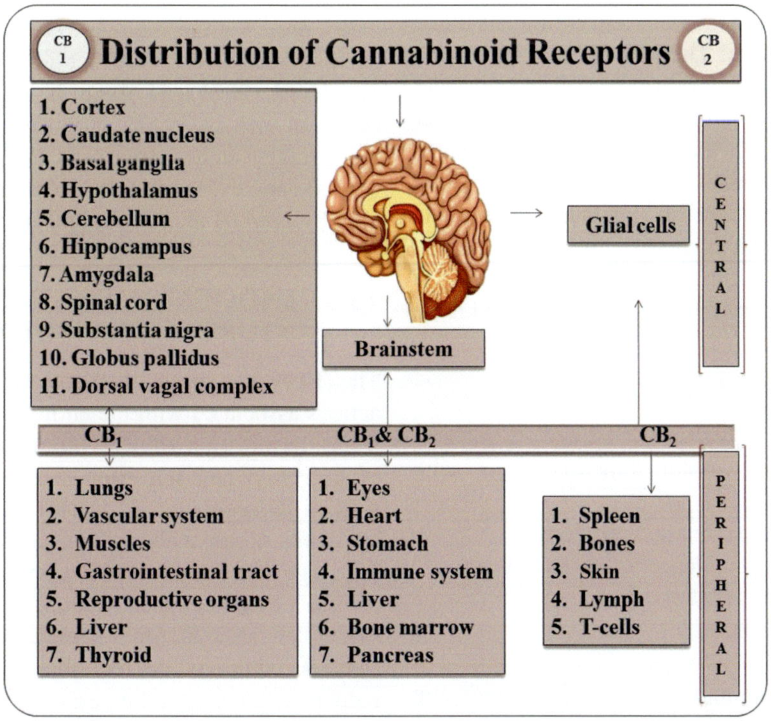

FIGURE 10.2 Distribution of cannabinoid receptors. CB_1, cannabinoid receptor type 1; CB_2, cannabinoids receptor type 2.

Amino acids 401–417, present at the intracellular portion of the C-terminal tail domain of CB_1, are considered to be vital for G protein (GP) coupling. Amino acids 418–472, present at the distal C-terminal tail domain of the CB_1, are considered to regulate the extent of signal transduction (Nie and Lewis, 2001).

CB_1 is found in abundance in the brain region and the area of expression is responsible for its nature of action (Kendall and Yudowski, 2017), for example the basal ganglia and cerebellum for movement, the hippocampus and cerebral cortex for memory processing, and the spinal cord and some parts of the gray matter, particularly the periaqueductal gray, for pain modulation. CB_1 is also found in peripheral organs. The secretion of neurotransmitters (inhibitory and excitatory) is modulated by central and peripheral neurons, and its activity is confined by CB_1 (Szabo and Schlicker, 2005). CB_1 binding and mRNA levels are decreased in basal ganglia during aging.

The eCBs maintain synaptic plasticity by:

1. Depolarization-induced suppressed inhibition / depolarization-induced suppressed excitation (Straiker and Mackie, 2005).
2. Metabotropic-induced suppressed inhibition / metabotropic-induced suppressed excitation (Straiker and Mackie, 2007).
3. eCB-induced or mediated long-term depression (Heifets and Castillo, 2009).

CB_1 is capable of enhancing the tropomyosin receptor kinase B (TrkB) signaling pathway by enhancing the activity of MAP kinase/extracellular regulated kinase (ERK) by direct action on the TrkB receptor (Jin, 2020). The decreased expression of brain-derived neurotrophic factor (BDNF) is a result of CB_1 genetic deletion, and eCBs exert their controlling action over BDNF expression. The lack of CB_1 expression in the brainstem is the reason for cannabis-related fatalities (Pertwee and Ross, 2002). Stimulation of CB_1 modulates the synthesis of cytokines, which are majorly associated with neurodegenerative processes by increasing the levels of interleukin (IL)-10 and trophic factors (Fernandez-Ruiz et al., 2005). All the mentioned actions of CB_1 might be due to the phosphorylation of GPs and the activation of ERK1/2, p38 mitogen-activated protein kinase and C-Jun N terminal kinase (JNK) (Howlett, 2005).

CB_1 agonists act by:

1. Decreasing tremor.
2. Improving impaired motor function seen with dopaminergic agonists.
3. Protecting against dopaminergic cell death.
4. Slowing down the incidence of L-DOPA-induced dyskinetic movements (LID).
5. Reducing bradykinesia.

As a result, CB_1 antagonists may be beneficial in reducing bradykinesia in cases of Parkinsonism or LID, as they reduce the CB_1 signaling in either GP_e or GP_i (Kunos and Tam, 2011).

10.2.2.2 CB_2

CB_2 receptors are present in the immune system on all types of immune cells; they are peripherally present in the pancreas, bones, spleen, skin, and tonsils; and centrally present in the microglial cells of the CNS (Goodman and Packard, 2015). Schizophrenia, depression, and bipolar disorder are associated with polymorphisms of CB_2 and genes encoding CB_2 (Tong et al., 2013). These diseases might also occur due to changes in hippocampal synaptic transmissions. Activation of CB_2 reduces pain perception, impulse behavior, vomiting, and locomotor activity triggered by the stimulation of CB_2. Stimulation of CB_2 has also shown a decrease in excitability in the ventral tegmental area, prominently at sensory, pyramidal, and dopaminergic neurons (Kim and Li, 2015). Based on their action in eCBs, multimodal therapeutic approaches in drug design have been developed. Due to their inducible nature, CB_2 exhibits mainly anti-inflammatory and neuroprotective effects in the treatment of Alzheimer's disease. Activity against inflammatory mediators is linked to enhanced proliferation of immune cells, as well as the production of immune cells involved in repair, mediated via immunity to neuronal damage (Habtemariam, 2019).

10.2.3 Enzymes

AEA is produced by phosphatidylethanolamine activation in the presence of two enzymes – PLD and *N*-acyltransferase – in the cell membrane (Leung et al., 2006). 2-AG, initiated by phospholipase C and calcium, involved in the transformation of membrane phosphoinositides into diacylglycerol synthesized in the presence of DAGL, also takes part in eCB synthesis (Muccioli, 2010). This discovery enabled the identification of new inhibitors of the metabolism of eCBs used for the modulation of AEA or 2-AG and their accumulation for site-specific action (Pamplona and Takahashi, 2011). If there is a deficiency in metabolic enzymes such as FAAH, attenuation of aging and associated dysregulated cardiac functions, and decreased inflammation-related gene expression due to higher levels of anandamide, are observed (Batkai et al., 2007).

10.3 ENDOCANNABINOIDS IN NEUROCOGNITIVE DYSFUNCTION

The ECS has been implicated in various physiological actions associated with the entire nervous system. Moreover, dense localization of ECS components in various regions of the brain and has been seen to control movement, as well as dysregulation (Concannon et al., 2015). Traditional medicine, along with preclinical and clinical interventions, have suggested that modulation of the ECS is useful in diverse pharmacological actions under different conditions (Yuan et al., 2016). Pharmacological actions such as sedation, anxiety, depression, memory impairment, pain reduction, hunger, and emesis control are modulated by the ECS. All reported actions are correlated and linked with the structure–activity relationships of functional groups present in phytocannabinoids and the cloning properties of cannabinoid receptors (Manzanares et al., 2006). Based on the concept of the role of eCB in the ECS, the development of new therapeutic molecules to elicit different intrinsic activities has been possible. The neuronal toxicity caused by the induction of beta amyloid (Aβ), and the consequent cellular events required for the activation of eCBs in the brain, lead to the blockage of neurodegenerative process and result in neuronal death. The ECS plays a neuroprotective role in both neuronal acute injuries and chronic neurodegenerative problems (Adams and Martin, 1996; Chandrasekar et al. 2011). All the actions might be due to the following eCB events:

1. Altered glutamatergic transmission due to excitatory neurotransmission and synaptic plasticity through presynaptic CB_1.
2. Altered immune response by the release of inflammatory mediators via eCBs along with astrocytes, neurons, microglia, macrophages, and lymphocytes.
3. Activation of protein kinase (A and B) cytoprotective signaling pathways.
4. Altered calcium homeostasis through its effect on ion channels, macula/zonula communicans, *N*-methyl-D-aspartate receptors, and intracellular stocks of calcium during excitation.
5. Antioxidant properties of CBD.
6. Controlling the metabolic rate and demand on oxygen during CB_1-mediated hypothermia.

The role of the ECS in different neurological dysfunctions such as Alzheimer's disease, Parkinson's disease, cognition, Huntington's disease, psychosis, anxiety, depression, and aging is explained based on physiological and pathological aspects (Baquero and Martin, 2015).

10.3.1 Alzheimer's Disease

Alzheimer's disease (AD) involves an innate, adaptive, central, and peripheral pathophysiology (Eldik et al., 2016). eCBs play a leading role in AD therapy owing to their neuroprotective, antioxidant, and anti-inflammatory characteristics. Notably, two areas of the brain influence the ECS: (i) the area where there is control over memory and learning (hippocampus); and (ii) the plural cortices (Graham et al., 2017).

The ECS processes are composed of various physiological functions which regulate neuronal damage and neuron production, free radical production, the activation of glial cells, and the elimination of cellular elements and macromolecules. These processes chiefly participate in the pathogenesis of AD (Molina-Holgado and Molina-Holgado, 2010; Bilkei-Gorzo, 2012). In microglia, the innate response of AD is associated with the triggering of TNF-α, IL-β, pro-inflammatory mediators, chemokines such as nitric oxide, and other components of the complement cascade (Monsonego and Weiner, 2003). Increased survival of PC12 neuronal cells has been observed by the scavenging of free radicals and lipid peroxidation upon exposure to Aβ. Cannabidiol also inhibits hyperphosphorylation of tau proteins, inducible nitric oxide synthase, and IL-β production, and glial fibrillary acidic protein expression during Aβ induction, which might cause neuroinflammatory processes to be evoked (Esposito et al., 2006). All the events counteracting these consequences happen during the inflammatory process in response to the levels of neurochemicals during the deposition of Aβ, hyperphosphorylation of Aβ, and the imbalance of acetylcholine (ACh) and glutamate in the neuronal system.

The abundant CB_2 expression in the brain in AD is a chief factor for its immune activation in cells. Areas of the cortex and parahippocampus are filled with neuritic plaques, and this is accompanied by the overexpression of CB_2 and its polyclonal antibodies. CB_2 produces beneficial actions on immune-mediated neurotoxicity, as it protects the AD brain from Aβ deposition by downregulating microglial activation, and also by producing cytokines (Aso et al., 2016).

10.3.2 Parkinson's Disease

Parkinson's disease (PD) involves a series of biochemical, anatomic, and electrophysiologic events in the brain. It involves different components of eCBs, which are present in different regions of the brain, such as the basal ganglia neuronal circuit, to modulate motor function. It has been found that reduced secretion of glutamate is caused by the stimulation of presynaptic CB_1 on the termini of corticostriatal regions (Brotchie, 2003). The inhibition of glutamate and gamma-aminobutyric acid (GABA) from afferent and striatal afferents respectively, is due to the activation of CB_1 and the basal ganglia nuclei output (globus pallidus internus and substantia nigra pars compacta). For example, AEA levels increase when CB_1 is upregulated in the brain after neuronal damage in response to neuronal injury and inflammation (Sagredo et al., 2007).

The role of eCBs in PD is bidirectional. CB_1 and D_2 receptors are GPCRs and elicit a common pool of signal transduction mechanisms (Meschler and Howlett, 2001). However, D_1 receptor-mediated adenyl cyclase activation is blocked completely by the stimulation of CB_1. eCBs show a protective role against excitotoxicity, cell damage, and death – the cascade of events leading to pars compacta neuronal death in PD (Monory et al., 2006).

The overall role of eCBs in PD might be in neuronal survival and their influence on physiologic synaptic functions via CB_1 neuronal activity and CB_2 immune-mediated activity. The actions of both CB_1 and CB_2 in PD are within the boundaries of the CNS (Behl et. al., 2020).

10.3.3 Cognition

Human memory mainly depends on receptor density, which is most highly expressed in the temporal lobe and frontal lobe, along with asymmetric expression in the left hemisphere. The eCB system has the capability to impair and enhance memory processes in different stages of memory by both receptor-mediated (direct) and non-receptor-mediated (indirect) mechanisms (Rasch and Born, 2013). Therefore, they facilitate the extinction or forgetting process of memory. Many studies have reported that eCBs are able to promote neuronal changes related to memory by enhancing long-term synaptic plasticity (Pape and Pare, 2010). Chronic administration of cannabis leads to cognitive impairments in attention, memory functions, and verbal learning, associated with lack of mental flexibility, uncontrolled preservation, and loss of ability to concentrate (Zanettini et al., 2011). All the effects of intoxication and cognitive dysfunction caused by CBDs are due to changes in cholinergic, glutamatergic, and GABAergic neurochemicals expressed in the hippocampus (Lichtman et al., 2002).

The ECS acts on various paradigms of behavior such as acquisition, consolidation, retrieval, and extinction. Activation of the ECS system can interfere with working and long-term memory; inhibition of the ECS system can enhance the same phases of memory, except memory retrieval, due to higher resistance over CBD alteration. In cognition, eCBs act via contradictory mechanisms when compared to THC (Gerdeman and Lovinger, 2003).

10.3.4 Huntington's Disease

Huntington's disease (HD) is an advanced, consistently disastrous, degenerative neurologic disorder triggered by a mutation in genes, creating an extension of glutamine repeats. It is also accompanied by a typical movement resembling dancing (chorea) (Chen et al., 2013). The chief pathology is damage to the striatal and cortical projection neurons. The initial stimulation of CB_1 does not control the onset and symptoms of HD. The CBD receptors provide neuroprotective activity by reducing the inflammatory processes exerted by the activation of CB_1 in glial cells, limiting excitotoxicity by normalizing glucose homeostasis and its antioxidant potential (Zou and Kumar, 2018). All changes are produced dose dependently, and the striatal degeneration and up-regulatory mechanisms also support a neuroprotective role on eCBs in HD by recruiting CB_2 at the site of lesions (Pazos et al., 2008).

With regard to pathophysiology, neurotransmitters such as GABA and enkephalin, initially generated at efferent nerve terminals of the globus pallidus, are affected. Loss of CB_1 at external and internal segments is present in co-localized GABA receptors and at neuronal ends with substance P. However, certain studies have reported that CB_2 is expressed and up-regulated in the striatum in HD (Allen, 2009). The CAG trinucleotide is repeated in the Huntingtin gene (*HTT*) with the outcome of producing HTT protein expanded in the glutamine tract leading to HD.

10.3.5 Psychosis and Anxiety

Three important characteristics – localization, basal activation, and sensitivity of anxiety – are associated with the biphasic effect of the CB_1 regulatory mechanism. CB_1 elicits anxiogenic activation on CBDs due to the GABAergic localization with high sensitivity and higher basal activation. However, the GABAergic localization with low sensitive basal activation plays an essential role in the anxiolytic responses of CBDs stimulated by CB_1. Emotional homeostasis is also dose-dependently regulated by monoaminergic transmission-related cannabinoid action. The alpha-2A adrenergic receptor (α2A-AR) acts as an autoreceptor for the inhibition of norepinephrine (NE) release from the presynaptic nerve terminal and its molecular mechanism down-regulates increased NE release. This mechanism is supported by previous studies, which linked CB_2 activities with anxiety, emesis, schizophrenia-related behavior, alcohol addiction, and impulsive behaviors. All these functions are mediated via CB_2 cellular mechanisms. The first episode of psychosis is characterized by the dysregulation of peripheral eCBs, with significant reduction in eCB synthesis, a reduction in enzymes, and also by the increased expression of degrading enzymes (Bioque et al., 2013).

AEA levels are enhanced in cerebrospinal fluid in acute schizophrenic conditions. Schizophrenic conditions show a decline in, or unmodified, mRNA levels, and elevated CB_1 binding. They also exhibit altered trafficking and high expression of membrane-bound CB_1 receptors with a high affinity for CB_1 (Desfosses, et al. 2010).

The typical indicators of psychosis include a noteworthy decrease in AEA, accompanied by CB_2 mRNA transcription occurring in peripheral blood mononuclear cells. When patients with schizophrenia underwent brain imaging procedures in vivo, increased binding of CB_1 was found in the nucleus accumbens, pons, cingulate, and insular cortex. Susceptibility to a single nucleotide polymorphism in the CB_2 gene was increased in schizophrenia. AEA hydrolysis was inhibited by CBDs. CBDs stimulated vanilloid type receptors. The cortex regions of the brain displayed elevated, as well as reduced, CB_1 density associated with schizophrenia. The organs involved in such conditions are dorsolateral PFC, as well as the anterior cingulate cortex. Decreased corticotropin-releasing hormone levels were found in stress conditions (Rodriguez de Fonseca et al., 1997). During anxiety and fear, elevated CB_1, as well as other elevated neurochemical transmitters such as serotonin, cholecystokinin, and opioid, was observed. It has been reported that 130–140% of GABA is released after treatment with a CB_1 antagonist. GABAergic synaptic tonic inhibition is more often found on glutamatergic synapses, mediated by CB_1 (Nuss, 2015).

10.3.6 Depression

The incidence of the depression has been found to be associated with cannabis use. However, THC – the major psychoactive major of cannabis – has been reported to reduce depressive behavior (Scherma, et al. 2018). Based on certain studies, CB_1 blockers have been suggested to treat depressive conditions. A high density of CB_1 and elevated levels of mRNA have been found in the prefrontal cortex (dorsolateral part). There is increased CB_1 expression with a lack of alternatives to the $G_{i/o}$ proteins in the brain's prefrontal cortex region. However, cortical mRNA concentration has been found to be unchanged in the depressive condition (Basavarajappa, 2007).

Genetic polymorphisms also occur in CBD-associated depression. During antidepressant therapy, there is modulation of the *CNR1* gene polymorphism (Kong et al., 2019). The vulnerability that *CNR1* carriers influences the severity of mental disorders such as depression. Reduced cellular expression of functional, non-synonymous single nucleotide polymorphisms in FAAH (*C385A* and *rs324420*) are observed in the depressive condition in humans. Of these two polymorphisms, *C385A* forms allele carriers in depression and bipolar disorder. In such cases, the levels of AEA are high due to reduced allele carriers for FAAH activity and the induction of CB1 desensitization (Hariri et al., 2009).

The activity of α-glutamatergic neurons, together with high cortisol levels in childhood trauma, is due to a reduction in CB_1 and tends toward anxiety and depression (Schoenfeld and Cameron, 2015). The CBDs appear to play a negative role in controlling NMDA receptor function in the absence of CB_1, and NMDA receptor-induced excitotoxicity is increased. GPCRs, μ-opioid receptors, and dopamine D_2 receptors are altered, with no changes in CB_1 or NMDA receptor subunits (Munoz et al., 2017).

10.3.7 Aging

The ECS system is actively involved in the molecular and cellular events of the aging process and acts against age-related alterations, thereby delaying the aging of the brain. This activity might be due to its metal-chelating capability and related neuroprotective activity on brain. The eCBs can alter astrocyte and microglial actions. During aging an increased number of activated astrocytes and microglia results in the elevated production of proinflammatory cytokines (Vincenzo et al., 2015).

CB_1 receptors are found to be active, whereas CB_2 receptors are inactive, in age-related defense. They act through GABAergic regulation of glial activity and intracellular lysosomal and endosomal CB_1 receptor clearance of damaged macromolecules in the brain (Bilkei-Gorzo, 2012). eCBs activate glutamatergic CB_1 receptor-mediated excitotoxicity. Astrocyte and microglial activities do not occur during aging due to a reduction in ECS activity. The functional defect also leads to a decline in the expression of BDNF during aging and affects synaptic plasticity. Many CB_1 receptor agonists exert neuroprotective activity through improved synaptic plasticity with enhanced BDNF expression (Moore, 2010).

10.4 PRECLINICAL AND CLINICAL INTERVENTIONS WITH ENDOCANNABINOIDS

A wide range of research has been carried out, and is still underway, to identify suitable agonist and antagonists for eCBs. Figure 10.3 shows the available drugs and their targets

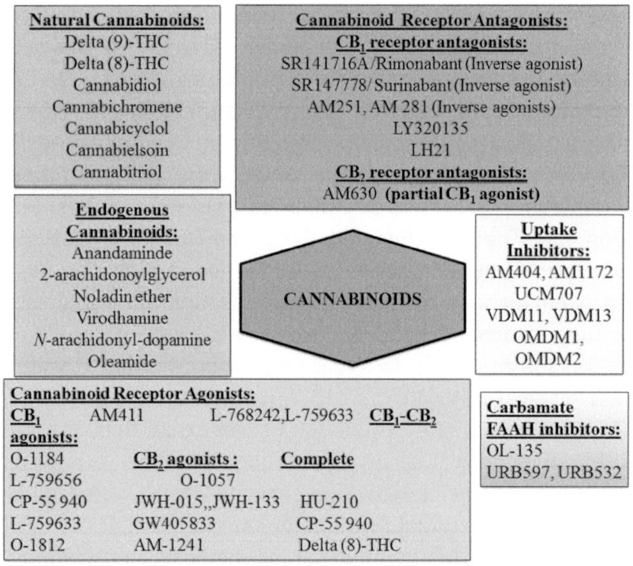

FIGURE 10.3 Types of cannabinoid (CBD) and drugs acting on CBD receptors. THC, Δ^9-tetrahyrocannabinol; CB_1, cannabinoids receptor type 1; CB_2, cannabinoid receptor type 2; FAAH, fatty acid amidehydrolase.

of action. There is still a large scope for further preclinical and clinical studies to explore the exact mechanism of eCBs and their analogs on neurocognitive dysfunctions.

10.4.1 Preclinical Studies

Rodent and animal models have demonstrated that the up-regulation of eCBs in the striatum was followed by depletion of dopamine, which tends to modulate the eCB signaling pathway, with improved PD symptoms (Van Vliet et al., 2008). Glutamate release and induced synaptic plasticity is specifically observed at corticostriatal synapses, which can be the result of CB_1 stimulation; the events were counteracted by overactivity of the glutamatergic system in an animal study of PD (Kreitzer and Malenka, 2007). Long term depression (LTD) pathogenesis also revealed elevated dopamine receptor signaling and transmission of glutamatergic signals caused by an alteration in CB_1 signaling (Benarroch, 2007). There is a loss of corticostriatal eCB-dependent LTD, and synaptic depotentiation occurs as a result of reduced signaling by CB_1 in a direct way in experimentally induced PD models (Picconi et al., 2003).

In auditory fear conditioned tests conducted in CB_1 knockout mice, unaffected memory acquisition and consolidation impairment has been observed. It has been reported that CB_1 is required for memory extinction in a recombinant mimicked phenotype of CB_1-deficient mice versus control mice (Marsicano et al., 2002). The systemic injection of the antagonist SR141716 showed impaired extinction in mice before trial, indicating that CB_1 is essential for movement in extinction training. The intraperitoneal administration of THC at doses of 3.0, 5.0, and 6.0 mg/kg resulted in impairment in acquiring memory during passive avoidance tests in rats. Pretreatment with 1 mg/kg of the recombinant CB_1 antagonist SR141716A showed reversed deterioration in mice (Campolongo et al., 2012). Pretreatment with SR141716A or SR147778 at 1.0 mg/kg intraperitoneally prevented the impairment in animal models (Phillips and LeDoux, 1992). Acute drug administration of WIN55,212-2 showed impaired memory acquisition using the Cognitive-Functional Composite (CFC) test; it was reported that the action might be due to hippocampus-dependent learning and memory, and its action was not caused by hippocampus-dependent fear conditioning (Pamplona and Takahashi, 2006). In a fear-conditioning task and water maze reversal task, WIN55,212-2, administered at a dose of 0.25 mg/kg, facilitated extinction of contextual fear, as well as spatial memory. The FAAH inhibitor OL-135 (30 mg/kg) resulted in enhanced extinction capacity on the water maze test. Using a contextual fear memory model, AM281 – a CB_1 antagonist – exerted its influence on memory acquisition at a dose of 2.5 mg/kg in adult mice, administered via both the intraperitoneal and intrahippocampal routes (Vaseghi et al., 2013).

AM251, a CB_1 antagonist, exhibited better social cognition than controls during a social discrimination experiment, indicating the critical role performed by CB_1 in memory building. Rats treated with reserpine and a 6-hydroxydopamine lesion in the basal ganglia showed decreased transportation rate of eCBs and FAAH degradation with increased tissue eCBs and CB_1 in induced Parkinson's disease (PD) (Lin et al., 2011). Mice deficient in anandamide-degrading enzyme (i.e., FAAH) showed anxiolytic activity when compared with control mice. Anxiolytic activity was reversed by the administration of rimonabant to the circulatory system (Moreira and Lutz, 2008).

Microinjection of methanandamide at a dose of 0.1 μg administered into the prefrontal cortex of rats resulted in anxiolytic responses. A dose of 10 μg methanandamide revealed an anxiety-like action. A study based on experimentation of the basal ganglia, as well as the β-amyloid peptide (BAP_{1-42}), reported a weakened cognitive response of $D3R^{-/-}$ mice, including a significance of eCBs/endovanilloids. These studies revealed the participation of TRP vanilloid-type 1 (TRPV1) channels and D_3 receptors (D_3R) in cognition, and the contribution of eCBs to beta peptides inhibiting memory retention in mice (Micale et al., 2010).

Based on the experiment conducted by Gajadro-Gomez et al. (2017), WIN, 2-AG or methanandamide completely prohibited the activities of hemichannels, as well as inflammatory activities triggered by Aβ present in the astrocytes. C57BL/6J and APP/PS1 mice showed inhibition of prostaglandin production and the accumulation of Aβ42 with exposure to nitrogen dioxide upon JZL 184 treatment (Yan et al., 2016). Some drugs, like JZL 184 and GW9662, are reported to the enhance formation, activity, and expression of MiR118-3, beta secretase 1 (BACE-1), and Aβ, with improved synaptic and cognition functions in both preclinical (5XFAD transgenic mice) and clinical studies (AD brains) (Zhang et al., 2014). There was improved cognitive performance with reduced neuroinflammation and Aβ levels following treatment with WIN55,212-2 and JWH-123 in transgenic APP2576 mice. Treatment with THC reversed the accumulation of Aβ in human neuron and MC65 cell lines (Currais et al., 2016). When treated with anandamide (0.1 mg/kg), the mice exhibited an upsurge in the time spent in the open arms of the elevated plus maze model. In AD, treatment with JWH133 lead to diminished levels of glycogen synthase kinase 3 beta (GSK3β), p38, and stress-activated protein kinase/c-Jun N-terminal kinases, with decreased activity and phosphorylation of tau protein at the phospho-tau (Thr181) site. It was also noted that the performance of superoxide dismutase was decreased in AD plaques with controlled cognitive activity (Aso et al., 2016). There was no significant behavioral change observed in BALB/c mice treated with CB_2 antagonist (AM630) at 1 mg/kg and 3 mg/kg, once daily for a period of 4 weeks, for sucrose consumption in an anhedonia test. AM630, when administered at doses of 1 mg/kg, 2 mg/kg, and 3 mg/kg, twice daily for 4 weeks, showed antidepressant action in a chronic mild stress study in Swiss ICR mice; at the same doses, it elicited anxiolytic activity in a light/dark box model, as well as an elevated plus maze for a period of 1 week with twice daily dosing. Treatment with a CB_2 agonist, namely GW405833 at 10 and 30 mg/kg, elicited no noteworthy observations in a marble-burying test for anxiety; at 100 mg/kg, it also did not cause any response in a rota rod apparatus. Another CB_2 agonist, JWH015, resulted in the increased consumption of sucrose (anhedonia test) at 20 mg/kg once daily dose for a period of 4 weeks in BALB/c mice.

AM251 at 5 mg/kg intraperitoneally showed enhanced freezing behavior in a hippocampal- and amygdala-dependent manner in trace fear conditioning and delay fear conditioning testing respectively during a tone footshock conditioning test (Reich et al., 2008). The stopping of increased conduction in the central nucleus of the amygdala was observed with CB_1 inverse agonists (AM251 and SR141716) by the stimulation of CB_1 on GABAergic neurons bound by the influence of basal conditions (Roberto et al., 2010). There was a controlled response to anxiety observed with the CB_2 agonist, JWH015, at 1–20 mg/kg in an elevated plus maze test using C57BL/6 mice; at a similar dose, the same drug showed enhanced anxiety in a light/dark box study. The CB_2 agonist (JWH133) was not successful in producing any effect against the changed behavior at 0.5 mg/kg, 1 mg/kg, and 2 mg/kg using a light/dark box test, as well as elevated plus maze test. CB_2 antagonists such as SR144528 also did not show any change in behavior at 1–20 mg/kg using BALB/c mice in a light/dark box study (Onaivi et al., 2008).

An age-related alteration of cognitive abilities, such as learning and memory, was observed during genetic removal of CB_1. No significant enhanced accumulation of lipofuscin or the activation of brain glial cells with the genetic deletion of CB_2 has been observed (Bilkei-Gorzo, 2012).

A THC-induced anxiolytic response was inhibited by SR141716A (CB_1 antagonist) at 0.5 mg/kg, indicating that the activity was mediated by CBD receptors. Parenteral administration of AM251 showed enhanced reconsolidation and AEA hyperactivation with ECS hypoactivation. WIN55,212-2 or HU-210 showed reduced reconsolidation in fear and memory in rats. Twelve-month-old CB_1 2/2 mice showed atrophy of the subdermal fat layer but no significant signs of aging in peripheral organs, even in wild-type mice (Ruehle, 2012).

One study has reported that diminished microglial performance, as well as reduced cortical Aβ levels, were improved with long-lasting treatment of CB_1 agonists in transgenic mice (APP/PS1 and APP2576). In a genetic model of overexpressed amyloid precursor protein transgenic mice (TgAPP-2576), reduced levels of AEA in hippocampal regions were observed (Kofalvi et al., 2016). WIN, 2-AG or methanandamide in Aβ-prompted astrocytes revealed inhibition of hemichannel activity, as well as inflammation in hippocampal slices. Treatment with CBDs also showed reduced excitotoxic glutamate release and neuronal damage (Gajardo-Gomez et al. 2017). CB_1 knockout mice showed an increased response to elevated basal plasma adrenocorticotropic hormone (ACTH) versus normal mice, indicating that the signaling of eCBs modifies hypothalamus–pituitary–adrenal (HPA) axis activity in the opposite way. Increased time spent on the open arm model was seen with AM404 treatment and eCB uptake/catabolism inhibitor at doses of 0.3–10 mg/kg doses, but did not show any significant alteration at higher doses.

A dose-dependent anxiolytic activity was observed in animal models when treated with AM404, acting on both the FAAH and eCBs systems. The elevated plus maze model, as well as the defensive withdrawal test, were used for this purpose (Bortolato et al., 2006). CB_1 antagonists in CB_1 knockout mice, when compared to normal mice, showed ACTH and corticosterone in stress-induced conditions. CB_1 knockout mice also exhibited increased levels of response to anxiety under basal conditions, compared to normal animals, indicating that loss of CB_1 causes anxiety under adverse conditions.

Conditional mutant mice showed anxiogenic responses, evidenced by a biphasic response due to the presence of CB_1 on specific subtypes of neurons and GABAergic neurons at CB_1. SR141716A (recombinant CB_1 antagonist), used to treat obesity, induced a depressed mood. CB_1 knockout mice exhibited enhanced expression of histidine triad nucleotide-binding protein 1 (HINT-1) and *N*-methyl-D-aspartate receptors within the prefrontal cortex at the NR1 subunit. In a cued conditioning test, CB_1 knockout mice showed impairment in short-term freezing reduction with activity after 200 s within the session, and showed long-term extinction between sessions.

WIN55,212-2 – a CB_1 agonist – showed enhanced secretion of extracellular NE in the prefrontal cortex. Increased NE was due to the elevated expression of enzymes such as tyrosine hydroxylase in catecholamine and the biosynthesis of other neurotransmitters, with associated anxiety-like behaviors (Boundy et al., 1998), indicating a role of NE release and transmission in eliciting anxiogenic responses.

A potent CB_1 agonist (Hu-210) exhibited significant activation of the HPA axis during stress, as well as reduced activation, with both prazosin (an α1-adrenoceptor antagonist) and propranolol (a β-adrenoceptor antagonist) (Mclaughlin et al., 2009). The FAAH inhibitor URB597 was shown to have enhanced NE and serotonergic activity in association with antidepressant behavior. There was reduced anxiety and increased social response with a clear emotional phenotypic activity observed in FAAH knockout mice. With antisense oligonucleotide induction, an anxiolytic effect was produced in an elevated plus maze test due to the blocking of CB_2 expression (Ravi et al., 2014). Microinjection of cannabidiol at lower doses in the area of periaqueductal gray showed anxiolytic effects due to the involvement of 5-HT_{1A} receptor and the actions elicited were inhibited by their inhibitors (WAY-100635) but not with CB_1 antagonists (AM251) (Campos and Guimaraes, 2008).

A whole-cell voltage clamp test was performed to study the effects of CB_1 agonists using slices of rat corpus striatum. Synaptic transmission was measured (Gerdeman and Lovinger, 2001). EPSCs prompted by the electric activation in afferent fibers were considerably decreased due to application of a CB_1 agonist. The decrease observed was in a dose-dependent manner (Gerdeman et al., 2002).

10.4.2 Clinical Studies (Table 10.1)

Currently available drugs include nabilone and dronabinol (CBD receptor agonists) and Sativex®, a formulation prescribed as an adjunct to treat pain (neuropathic) in patients suffering with multiple sclerosis (MS). Sativex is a spray formulation applied to the mucosal area (Turri et al., 2018). In the UK and Spain, its use is limited and it is awaiting approval for use in MS as a therapeutic agent. Rimonabant is approved in 38 countries including European Union, Mexico and Brazil. But it was rejected for approval for use in United States, as it may increase suicidal tendencies and depression (dyslipidemia and diabetes; Waterlow and Chrisp, 2007). However, its use has been limited and is not currently approved by the U.S. Food and Drugs Administration (FDA) due to their adverse psychotropic events and suicidal tendencies. The major psychoactive compound, THC, has been marketed with the international non-proprietary name of dronabinol. Synthetic dronabinol at doses of 2.5 mg, 5 mg, and 10 mg was approved by the FDA, under the name of Marinol® in capsule form, for treating motion sickness and vomiting prompted by drug therapy; Cesamet™ capsules are used for nausea and vomiting induced by chemotherapy. Rimonabant 20 mg has been marketed by Sanofi Aventis (Grotenhermen and Vahl, 2012).

Endogenous cannabinoids (AEA) have been found in the cerebrospinal fluid (CSF) of patients with PD (Pisani et al., 2005). Patients with PD are shown to have increased CB_1 binding in the striatum with enhanced activation of CB_1 (Lastres-Becker et al., 2001). In a double-blind, randomized, controlled study it was observed that blockage of CB_1 in patients with PD does not show any improvement in motor function or on levodopa-induced dyskinesia (Mesnage et al.,

2004). Patients with post-traumatic stress disorder (PSTD), particularly in a skin conductance test, showed extinction recall for 2 days by using extinction procedures with fear conditioning (Milad et al., 2008).

There is up-regulation of 2-AG and its biosynthesizing enzyme, DAGL, in specific tissues of the hippocampal region in patients with AD. Cognitive impairment and neurotoxic Aβ42 (one of the chief isoforms of Aβ, with 42 residues) levels are correlated with lower levels of cortical anandamide, but the levels of anandamide do not have any correlation with amyloid plaque load or formation of Aβ40, or tau hyperphosphorylation (Jung et al., 2012). The AD brain has

TABLE 10.1
Clinical Studies of Drugs that Act on Endocannabinoid Receptors

Type of Study	Indication	Dose/Patient Number	Outcomes	Reference
Combined meta-analysis	Psychosis	Observational study/256	Impaired global and prospective memory; verbal, immediate, and delayed recall; visual recognition in healthy individuals; better global, visual, and immediate recall and recognition in psychotic patients	Schoeler et al., 2016
Correlation	Schizophrenia	600 mg/15	Frequent cannabis exposure may down-regulate anandamide signaling in the CNS of schizophrenic patients	Leweke et al., 2007
Preliminary prospective study	Depression	Depressed patients selected for study/12	Electroconvulsive therapy enhances the AEA and 2-AG levels in depressive patients	Kranaster et al., 2017
Preliminary study	Depression	Depressed patients selected for study/109	Plasma concentrations of both AEA and 2-AG were also found to be elevated in depressed patients treated with SSRI-type antidepressants	Romero-Sanchiz et al., 2019
Sociodemographic and clinical data	Depression	10–320 mg daily/85	CB_1R C allele of *rs2023239* showed an independent protective effect against lifetime major depressive disorder in opiate-dependent outpatients	Icick et al., 2015
Power analysis	Schizophrenia	619 mg daily/90	The specific binding of [^{3}H]MePPEP to CB_1 receptors was 20% higher in patients with schizophrenia than in controls.	Jenko et al., 2012
Investigation	Schizophrenia	Schizophrenia patients selected for study/67	CB_1 receptor binding is altered in the mesocorticolimbic circuitry of schizophrenia both with and without antipsychotic therapy, especially in the nucleus accumbens	Ceccarini et al., 2013
Cross-sectional, case–control study	Psychosis	82.78 mg daily/58	Using the Brief Psychiatric Rating Scale for fluoride 18-labeled FMPEP-d_2 1 and the Positive and Negative Syndrome Scale for carbon 11-labeled MePPEP indicated the availability of CB_1R was lowered, and indicated that greater reductions in CB_1R levels were associated with greater symptom severity and poorer cognitive functioning in male patients	Borgan et al., 2019
Correlation	Psychosis	Correlation study/159	Association of cannabinoid receptor *CNR1* genotypes (*rs7766029* and *rs12720071*) with changes in neurocognitive performance in psychotic patients. There was higher improvement in verbal memory with *rs7766029*, and improvement in executive functions with *rs12720071*.	Rojnic Kuzman et al., 2019
Cohort	Psychosis	Cohort study/407	Prevalence of metabolic syndrome is associated with *CNR1* (*rs806377*, *rs1049353*, *rs6454674*, and *rs806379*) in patients with schizophrenia during long-term antipsychotic treatment	Yu et al., 2013
Preliminary study	Bipolar disorder	Preliminary study/352	*rs6454674*, *rs806368*, and *rs1049353 CNR1* gene polymorphisms associated with several psychiatric disorders. Heterozygote advantage exists for the *CNR1 rs6454674* polymorphism in manic episode frequency in Turkish bipolar disorder patients.	Alpak et al., 2014
Intervention	Depression	Stable dose/24	Moderate exercise in women with multidepression disorder produced significant elevations in AEA but not in 2-AG, although both eCBs presented significant moderate negative associations between serum changes and mood states	Meyer et al., 2019
Longitudinal study	Depression	20–40 mg daily/154	*CNR1 rs1049353-GG* genotype conferred a better response to citalopram treatment in the subgroup of male patients and G allele carriers (*CNR2 rs2501431*) presented higher Hamilton Depression Rating Score scales in SSRI treatment	Mitjans et al., 2012

a higher expression of CB_1 and CB_2 in senile plaques, with a decreased number of CB_1 in the microglia. In patients with late-onset AD, elevated expression of *FAAH* and a decline in the FAAH promoter gene by decreased DNA methylation in peripheral blood mononuclear cells has been reported. However, FAAH levels in patients with AD are reduced in membranes of the brain cortex (Pascual et al., 2014). Increased FAAH levels in Aβ-enriched neuritic plaques have been found in the surroundings of astrocytes in the brains of patients with AD (Benito et al., 2003).

When studied in patients with AD characterized by late-stage dementia, it has been found that dronabinol reduced nocturnal motor function drastically (Walther et al., 2006). The levels of CB_2 have been found to be increased in patients with AD and associated with molecular markers of AD such as Aβ42 levels. However, there was no correlation of senile score to cognitive status (Solas et al., 2013).

Increased connectivity and stress responsiveness in the frontal amygdala in mice and the FAAH C385A polymorphism in carriers of the human A allele were also enhanced in fear extinction. HD patients have been shown to inhibit FAAH function in the metabolism of AEA. Patients with HD have been reported to possess sixfold increased levels of AEA versus controls (Battista et al., 2007).

Patients with psychosis in the prodromal stage have shown predominantly correlated levels of anandamide in serum and CSF (Potvin et al., 2020). The early phase of psychosis has been associated with cognitive dysfunction, with augmented degradation of NAPE, as well as DAGL expression, along with controlled eCB synthesis in patients. The ECS system has been found to be dysregulated with the impairment of cognitive functions during the first episode of psychosis in schizophrenic patients. An increased risk of schizophrenia associated with defective functioning of CB_2 due to *CNR2 Q63R* gene polymorphism has been evidenced by the postmortem brain tissues of humans and in cultured cells (Ishiguro et al., 2010).

In schizophrenic patients exaggerated binding of CB_1 at the dorsolateral prefrontal cortex with modulated anandamide levels in the brain has been shown (Navarrete et al., 2020). Enhanced clinical efficacy of CBDs in psychotic patients studied in randomized clinical trials has been witnessed. The schizophrenic brain has been studied for endogenous analogues such as AEA, oleoylethanolamide, and palmitoylethanolamide. Lower levels of all endogenous analogs indicated participation of the ECS in the schizophrenia pathophysiology. CBDs act as antipsychotics in patients via 5-HT_{1A} or TRPV1 receptors. This might be the reason for its partial agonistic activity on D_2 receptors (Davies and Bhattacharyya, 2019).

CB_1 positron emission tomography has shown augmented CB_1 expression in the brains of patients with PTSD when compared to a control group. The density and functioning ability of CB_1 and its mRNA were found to be enhanced at the dorsolateral prefrontal cortex area in patients with depression (Neumeister et al., 2014).

An increased risk of schizophrenia was observed with the lower functioning of CB_2 associated with *CNR2* polymorphism. Major depression and bipolar disorder are associated with polymorphism of *CNR1* and *FAAH*, and contribute to enhanced activity in depressive patients (Huang et al., 2016). During chronic treatment with monoamine oxidase inhibitors, increased CB_1 density in the hippocampus of patients was observed. However, treatment with serotonin selective reuptake inhibitors showed reduced levels of CB_1 and their immunoreactivity at the anterior cingulate cortex (Lecue et al., 2018). Female patients with major depressive disorder and a comorbidity of anxiety under antidepressant drug therapy showed resistance to treatment due to polymorphism of the G allele of *rs1049353*. eCB polymorphs of Q63R are abundantly present during eating disorders (both bulimia and anorexia nervosa) in schizophrenic patients. Other serotonin homozygous ('GG') polymorphisms (*rs2180619*) in *CNR1* and *SLC6A4* (5-HTTLPR) have been observed in high-risk patients with anxiety. There was no noteworthy association among overall CB_2 mRNA expression in postmortem dorsolateral prefrontal cortex (BA9) in patients who had suffered from schizophrenia. It is reported that BIA 10-2474 (a FAAH-1 inhibitor) exhibited unanticipated adverse events and toxic reactions such as cerebral syndrome during phase I clinical trials (Kerbrat et al., 2016).

10.5 FUTURE TRENDS AND CHALLENGES

The most current developments in the investigation of the biologic distribution, as well as the functional participation, of the ECS has permitted the progress of several research opportunities directed toward the medicinal exhaustion of eCB signaling, along with the advancement of novel drugs with pharmacologic benefits (Khan et al., 2016). Nearly two decades after the identification of AEA, the entire ECS unit is now depicted as captivating object for the progress of specific entities capable of modifying human pathophysiologic conditions (Maccarrone et al., 2010). A more detailed study of proteins contributing to eCB metabolism, along with signaling pathways, could assist in the design of compounds that may increase the activity of eCBs in a time- and site-reliant manner, eliminating unwanted psychotropic adverse effects. Transgenic animals should be developed to provide a platform for various ECS elements to either be ruled out or ruled in appropriately, thereby permitting ground-breaking therapeutic approaches in a huge array of pathologies (Mai et al., 2002).

10.6 CONCLUSION

The outcomes of the studies referred to in this chapter highlight the influence of CBDs on several phases of memory development. Various preclinical evaluations have revealed multifaceted results of compounds that immediately impact the working of the ECS by means of CBD receptor ligands. Compounds that alter this functioning in a non-specific way such as those that cause degradation of eCBs in brain regions also fall under this category.

The inflection observed by the impact of the CBD receptor ligands on the several stages of memory has been extensively assessed in behavioral studies. Even though CB_1 and CB_2 receptor ligands are capable of both recovering and impairing memory, each effect is brought about in a very dissimilar manner. This issue is still to be explored; therefore, additional evaluation, such as molecular assessment of CBD receptor concentration in different regions of the brain areas, is required. Biochemical assessment in studies involving the impact of CBD receptor ligands on metalloproteinase concentration, along with neurotransmitters on oxidative stress biomarkers, or in the brain, will provide a major breakthrough in the field of neurocognitive research. Hence, it can be concluded that the results obtained from these integrative assessments may offer innovative evidence for the therapeutic benefits of the ECS in the brain.

REFERENCES

Adams IB, Martin BR. Cannabis: pharmacology and toxicology in animals and humans. Addiction. 1996;91:1585–614.

Allen KL, Waldvogel HJ, Glass M, Faull RL. Cannabinoid (CB(1)), GABA(A) and GABA(B) receptor subunit changes in the globus pallidus in Huntington's disease. J Chem Neuroanat. 2009;37:266–81.

Alpak G, Copoglu S, Geyik E, Unal A, Igci M, Igci Y, et al. Rs6454674, Rs806368 and Rs1049353 CNR1 gene polymorphisms in Turkish bipolar disorder patients: A preliminary study. Dis Mol Med. 2014;2:4.

Aso E, Andres-Benito P, Carmona M, Maldonado R, Ferrer I. Cannabinoid receptor 2 participates in amyloid-beta processing in a mouse model of Alzheimer's disease but plays a minor role in the therapeutic properties of a cannabis-based medicine. J Alzheimer's Dis. 2016;51:489–500.

Baquero M, Martin N. Depressive symptoms in neurodegenerative diseases. World J Clin Cases. 2015;3:682–93.

Barzegar S, Komaki A, Shahidi S, Sarihi A, Mirazi N, Salehi I. Effects of cannabinoid and glutamate receptor antagonists and their interactions on learning and memory in male rats. Pharmacol Biochem Behav. 2015;131:87–90.

Basavarajappa BS. Neuropharmacology of the endocannabinoid signaling system-molecular mechanisms, biological actions and synaptic plasticity. Curr Neuropharmacol. 2007;5:81–97.

Batkai S, Rajesh M, Mukhopadhyay , Hasko G, Liaudet L, Cravatt BF. Decreased age-related cardiac dysfunction, myocardial nitrative stress, inflammatory gene expression, and apoptosis in mice lacking fatty acid amide hydrolase. Am J Physiol Heart Circ Physiol. 2007;293:H909–18.

Battista N, Bari M, Tarditi A, Mariotti C, Bachoud-Levi AC, Zuccato C, et al. Severe deficiency of the fatty acid amide hydrolase (FAAH) activity segregates with the Huntington's disease mutation in peripheral lymphocytes. Neurobiol Dis. 2007;27:108–16.

Behl T, Kaur G, Bungau S, Jhanji R, Kumar A, Mehta V. Distinctive evidence involved in the role of endocannabinoid signalling in Parkinson's disease: A perspective on associated therapeutic interventions. Int J Mol Sci. 2020;21:1–27.

Ben Amar M. Cannabinoids in medicine: A review of their therapeutic potential. J Ethnopharmacol. 2006;105:1–25.

Benard G, Massa F, Puente N, J Lourenco J, Bellocchio L, Gomez ES, et al. Mitochondrial CB1 receptors regulate neuronal energy metabolism. Nature Neuroscience. 2012;14(4):558–564.

Benarroch E. ECBs in basal ganglia circuits: implications for Parkinson disease. Neurology. 2007;69:306–9.

Benito C, Nunez E, Tolon RM, Carrier EJ, Rabano A, Hillard CJ. Cannabinoid CB_2 receptors and fatty acid amide hydrolase are selectively overexpressed in neuritic plaque-associated glia in Alzheimer's disease brains. J Neurosci. 2003;23:11136–41.

Bilkei-Gorzo A. The endocannabinoid system in normal and pathological brain ageing. Philos Trans R Soc Lond B Biol Sci. 2012;367:3326–41.

Bioque M, Bueno BG, Macdowell KS, Meseguer A, Saiz PA, Parellada M. Peripheral endocannabinoid system dysregulation in first-episode psychosis. Neuropsychopharmacology. 2013;38:2568–77.

Bolla KI, Brown K, Eldreth D, Tate K, Cadet JL. Dose-related neurocognitive effects of marijuana use. Neurology. 2002;59:1337–43.

Borgan F, Laurikainen H, Veronese M, Marques TR, Haaparanta-Solin M, Solin O, et al. In vivo availability of cannabinoid 1 receptor levels in patients with first-episode psychosis. JAMA Psychiatry. 2019;76:1074–84.

Bortolato M, Campolongo P, Mangieri RA, Scattoni ML, Frau R, Trezza V, et al. Anxiolytic-like properties of the anandamide transport inhibitor AM404. Neuropsychopharmacology. 2006;31:2652–9.

Boundy VA, Gold SJ, Messer CJ, Chen J, Son JH, Joh TH, et al. Regulation of tyrosine hydroxylase promoter activity by chronic morphine in TH9.0-LacZ transgenic mice. J Neurosci. 1998;18:9989–95.

Brotchie JM. CB_1 cannabinoid receptor signalling in Parkinson's disease. Curr Opin Pharmacol. 2003;3:54–61.

Campolongo P, Ratano P, Manduca A, Scattoni ML, Palmery M, Trezza V, et al. The endocannabinoid transport inhibitor AM404 differentially modulates recognition memory in rats depending on environmental aversiveness. Front Behav Neurosci. 2012;6:10.

Campos AC, Guimaraes FS. Involvement of 5HT1A receptors in the anxiolytic-like effects of cannabidiol injected into the dorsolateral periaqueductal gray of rats. Psychopharmacology (Berl). 2008;199:223–30

Ceccarini J, De Hert M, Van Winkel R, Peuskens J, Bormans G, Kranaster L, et al. Increased ventral striatal CB_1 receptor binding is related to negative symptoms in drug-free patients with schizophrenia. Neuroimage. 2013;79:304–12.

Chan PC, Liao MT, Hsieh PS. The dualistic effect of COX-2-mediated signaling in obesity and insulin resistance. Int J Mol Sci. 2019;20:1–13.

Chandrsekar SB, Divakar K, Pawar AT, Goli D, Harisha R, Bhanumathy M. Involvement of opioid, adenosine and 5HT3 receptors in antinociceptive effects of an Ayurvedic polyherbal formulation. Med Princ Pract. 2011;20:66–70.

Chen JY, Wang EA, Cepada C, Levine MS. Dopamine imbalance in Huntington's disease: a mechanism for the lack of behavioral flexibility. Front Neurosci. 2013;7:1–14.

Concannon R, Finn DP, Dowd E. Cannabinoids in Parkinson's disease. In: L. Fattore, editor. Cannabinoids in Neurologic and Mental Disease. Academic Press; 2015.

Currais A, Quehenberger O, M Armando A, Daugherty D, Maher P, Schubert D. Amyloid proteotoxicity initiates an inflammatory response blocked by cannabinoids. NPJ aging and mechanisms of disease. 2016;2(1):1–8.

Davies C, Bhattacharyya S. Cannabidiol as a potential treatment for psychosis. Ther Adv Psychopharmacol. 2019;9:1–16.

Demuth DG, Molleman A. Cannabinoid signalling. Life Sci. 2006;78:549–63.

Desfosses J, Stip E, Bentaleb LA, Potvin S. Endocannabinoids and schizophrenia. Pharmaceuticals (Basel). 2010;3: 3101–26.

Devane WA, Dysarz FA, Johnson MR, Melvin LS, Howlett AC. Determination and characterization of a cannabinoid receptor in rat brain. Mol Pharmacol. 1988,34.605–13.

Devane WA, Hanus L, Breuer A, Pertwee RG, Stevenson LA, Griffin G. Isolation and structure of a brain constituent that binds to the cannabinoid receptor. Science. 1992; 258:1946–9.

Eldik LJV, Carrilo MC, Cole PE, Feuerbach D, Greenbreg BD, Hendrix JA. The roles of inflammation and immune mechanisms in Alzheimer's disease. Alzheimers Dement. 2016;2:99–109.

Esposito G, De Filippis D, Maiuri MC, Stefano DD, Carnuccio R, Iuvone T. Cannabidiol inhibits inducible nitric oxide synthase protein expression and nitric oxide production in beta-amyloid stimulated PC12 neurons through p38 MAP kinase and NF-kappaB involvement. Neurosci Lett. 2006;399:91–5.

Fernandez-Ruiz J, Gonzales S, Romero J, Ramos J. Cannabinoids in neurodegeneration and neuroprotection. In: R.Mechoulam, editor. Cannabinoids as Therapeutics. Birkhaeuser, Basel; 2005.

Gajardo-Gomez R, Labra VC, Maturana CJ, Shoji KF, Santibanez CA, Saez JC, et al. Cannabinoids prevent the amyloid beta-induced activation of astroglial hemichannels: a neuroprotective mechanism. Glia. 2017;65:122–37.

Gerdeman G, Lovinger DM. CB1 cannabinoid receptor inhibits synaptic release of glutamate in rat dorsolateral striatum. J Neurophysiol. 2001;85:468–71.

Gerdeman GL, Lovinger DM. Emerging roles for eCBs in long term synaptic plasticity. Br J Pharmacol. 2003;140:781–9.

Gerdeman GL, Ronesi J, Lovinger DM. Postsynaptic endocannabinoid release is critical to long-term depression in the striatum. Nature Neuroscience. 2002;5:446–51.

Goncalves J, Rosado T, Soares S, Simao AY, Caramelo D, Luis A, et al. Cannabis and its secondary metabolites: Their use as therapeutic drugs, toxicological aspects, and analytical determination. Medicines (Basel). 2019;6:1–47.

Goodman J, Packard MG. The influence of cannabinoids on learning and memory processes of the dorsal striatum. Neurobiol Learn Mem. 2015;125:1–14.

Gorzo AB. The endocannabinoid system in normal and pathological brain ageing. Phil Trans R Soc Biol Sci. 2012;367:3326–41.

Graham WV, Bonito-Oliva A, Sakmar TP. Update on Alzheimer's disease therapy and prevention strategies. Annu Rev Med. 2017;68:413–30.

Grotenhermen F. Mini-review cannabinoids and the endocannabinoid system. Cannabinoids. 2006;1:10–14.

Grotenhermen F, Vahl KRM. The therapeutic potential of cannabis and cannabinoids. Deutsches Arzteblatt International. 2012;109:495–501.

Habtemariam S. Natural products in Alzheimer's disease therapy: would old therapeutic approaches fix the broken promise of modern medicines? Molecules. 2019;24:1–17.

Hariri AR, Gorka A, Hyde LW, Kimak M, Halder I, Ducci F. Divergent effects of genetic variation in endocannabinoid signaling on human threat- and reward-related brain function. Biol Psychiatry. 2009;66:9–16.

Heifets BD and Castillo PE. Endocannabinoid signaling and long-term synaptic plasticity. Annu Rev Physiol. 2009;71:283–306.

Hermanson DJ, Gamble-George JC, Marnett LJ, Patel S. Substrate-selective COX-2 inhibition as a novel strategy for therapeutic endocannabinoid augmentation. Trends Pharmacol Sci. 2014;35:358–67.

Hillard CJ, Huang H, Vogt CD, Rodrigues BE, Neumann TS, Sem DS. Endocannabinoid transport proteins: Discovery of tools to study sterol carrier protein-2. Methods Enzymol. 2017;593:99–121.

Hollister LE. Health aspects of cannabis. Pharmacol Rev. 1986;38:1–20.

Howlett AC. Cannabinoid receptor signalling. Handb Exp Pharmacol. 2005;168:53–79.

Howlett AC, Barth F, Bonner TI, Cabral G, Casellas P, Devane WA, et al. International Union of Pharmacology. XXVII. Classification of cannabinoid receptors. Pharmacol Rev. 2002;54:161–202.

Howlett AC, Reggio PH, Childers SR, Hampson RE, Ulloa NM, Deutsch DG. Endocannabinoid tone versus constitutive activity of cannabinoid receptors. Br J Pharmacol. 2011;163:1329–43.

Huang WJ, Chen WW, Zhang X. Endocannabinoid system: role in depression, reward and pain control (Review). Mol Med Rep. 2016;14:2899–903.

Icick R, Peoc'h K, Karsinti E, Ksouda K, Hajj A, Bloch V, et al. A cannabinoid receptor 1 polymorphism is protective against major depressive disorder in methadone-maintained outpatients. Am J Addict. 2015;24:613–20.

Ishiguro H, Horiuchi Y, Ishikawa M, Koga M, Imai K, Suzuki Y, et al. Brain cannabinoid CB_2 receptor in schizophrenia. Biol Psychiatry. 2010;67:974–82.

Jenko KJ, Hirvonen J, Henter ID, Anderson KB, Zoghbi SS, Hyde TM, et al. Binding of a tritiated inverse agonist to cannabinoid CB_1 receptors is increased in patients with schizophrenia. Schizophr Res. 2012;141:185–8.

Jin W. Regulation of BDNF-TrkB signaling and potential therapeutic strategies for Parkinson's Disease. J Clin Med. 2020;9:1–19.

Johnston LD, O'Malley PM, Bachman JG, Schulenberg JE. Monitoring the future national results on adolescent drug use: 2012 overview of key findings. Available at: https://files.eric.ed.gov/fulltext/ED529133.pdf.

Jung KM, Astarita G, Yasar S, Vasilevko G, Cribbs DH, Head E, et al. An amyloid β42-dependent deficit in anandamide mobilization is associated with cognitive dysfunction in Alzheimer's disease. Neurobiol Aging. 2012;33:1522–32.

Kendall DA, Yudowski GA. Cannabinoid receptors in the central nervous system: Their signaling and roles in disease. Front Cell Neurosci. 2017;10:1–10.

Kerbrat A, Ferre JC, Fillatre P, Ronziere T, Vannier S, Carsin-Nicol B, et al. Acute neurologic disorder from an inhibitor of fatty acid amide hydrolase. N Engl J Med. 2016;375:1717–25.

Khan MI, Sobocinska AA, Czarnecka AM, Krol M, Botta B, Szczylik C. The therapeutic aspects of the endocannabinoid system (ECS) for cancer and their development: from nature to laboratory. Curr Pharm Des. 2016;22:1756–66.

Kim J, Li Y. Chronic activation of CB_2 cannabinoid receptors in the hippocampus increases excitatory synaptic transmission. J Physiol. 2015;593:871–86.

Klein TW. Cannabinoid-based drugs as anti-inflammatory therapeutics. Nat Rev Immunol. 2005;5:400–10.

Kmietowicz Z. Cannabis based drug is licensed for spasticity in patients with MS. BMJ. 2010;340:c3363.

Kofalvi A, Lemos C, Martin-Moreno AM, Pinheiro BS, Garcia LG, Poza MA, et al. Stimulation of brain glucose uptake by cannabinoid CB_2 receptors and its therapeutic potential in Alzheimer's disease. Neuropharmacology. 2016;110:519–29.

Kong X, Miao Q, Lu X, Zhang Z, Chen M, Zhang J, et al. The association of endocannabinoid receptor genes (CNR1 and CNR2) polymorphisms with depression. Medicine (Baltimore). 2019;98:1–8.

Kranaster L, Hoyer C, Aksay SS, Bumb JM, Leweke FM, Janke C, et al. Electroconvulsive therapy enhances endocannabinoids in the cerebrospinal fluid of patients with major depression: a preliminary prospective study. Eur Arch Psychiatry Clin Neurosci. 2017;267:781–6.

Kreitzer AC, Malenka RC. Endocannabinoid-mediated rescue of striatal LTD and motor deficits in Parkinson's disease models. Nature. 2007;445:643–47.

Kunos G, Tam J. The case for peripheral CB_1 receptor blockade in the treatment of visceral obesity and its cardiometabolic complications. Br J Pharmacol. 2011;163:1423–31.

Lastres-Becker I, Cebeira M, de Ceballos ML, Zeng BY, Jenner P, Ramos JA, et al. Increased cannabinoid CB_1 receptor binding and activation of GTP-binding proteins in the basal ganglia of patients with Parkinson's syndrome and of MPTP-treated marmosets. Eur J Neurosci. 2001;14:1827–32.

Lecue II, Cuellar FP, Muguruza, Zanuy EF. The endocannabinoid system in mental disorders: Evidence from human brain studies. Biochem Pharmacol. 2018;1:1–10.

Leung D, Saghatelian A, Simon GM, Cravatt BF. Inactivation of *N*-Acyl phosphatidylethanolamine phospholipase D reveals multiple mechanisms for the biosynthesis of endocannabinoids. Biochemistry. 2006;45:4720–26.

Leweke FM, Giuffrida A, Koethe D, Schreiber D, Nolden BM, Kranaster L. Anandamide levels in cerebrospinal fluid of first-episode schizophrenic patients: Impact of cannabis use. Schizophr Res. 2007;94:29–36.

Lichtman AH, Varvel SA, Martin BR. ECBs in cognition and dependence. Prostaglandins Leukot Essent Fatty Acid. 2002;66:269–85.

Lin XH, Yuece B, Li YY, Feng YJ, Feng JY, Yu LY, et al., A novel CB receptor GPR55 and its ligands are involved in regulation of gut movement in rodents. Neurogastroenterology and Motility. 2011;23(9):862–e342.

Luk T, Jin W, Zvonok A, Lu D, Lin XZ, Chavkin C, et al. Identification of a potent and highly efficacious, yet slowly desensitizing CB_1 cannabinoid receptor agonist. Br J Pharmacol. 2004;142:495–500.

Maccarrone M. Metabolism of the endocannabinoid anandamide: Open questions after 25 years. Front Mol Neurosci. 2017;10:166.

Maccarrone M, Bab I, Biro T, Cabral GA, Dey SK, Marzo VD, et al. Endocannabinoid signalling at the periphery: 50 years after THC. Trends Pharmacol Sci. 2015;36:277–96.

Maccarrone M, Gasperi V, Catani MV, Ai Diep T, Dainese E, Hansen HS, et al. The endocannabinoid system and its relevance for nutrition. Annu Rev Nutr. 2010;30:423–40.

Mackie K, Devane WA, Hille B. Anandamide, an endogenous cannabinoid, inhibits calcium currents as a partial agonist in N18 neuroblastoma cells. Mol Pharmacol. 1993;44:498–503.

Mackie K, Lu HC. An introduction to the endogenous cannabinoid system. Biol Psychiatry. 2016;79:516–25.

McLaughlin R, Hill MN, Gorzalka BB. Monoaminergic neurotransmission contributes to cannabinoid-induced activation of the hypothalamic–pituitary–adrenal axis. Eur J Pharmacol 2009;624:71–6.

Mai B, Miles S, Breeden LL. Characterization of the ECB binding complex responsible for the M/G1-specific transcription of CLN3 and SWI4. Mol Cell Biol. 2002;22:430–41.

Manzanares J, Julian MD, Carrascosa A. Role of the cannabinoid system in pain control and therapeutic implications for the management of acute and chronic pain episodes. Curr Neuropharmacol. 2006;4:239–57.

Marsicano G, Wotjak CT, Azad ZC, Bisogno T, Rammes G, Cascio MG, et al. The endogenous cannabinoid system controls extinction of aversive memories. Nature. 2002;418:530–534.

Mata I, Iglesias RP, Santianez RR, Gutierrez DT, Pazos A, Gutierrez A, et al. Gyrification brain abnormalities associated with adolescence and early-adulthood cannabis use. Brain Res. 2010;1317:297–304.

Matsuda LA, Lolait SJ, Brownstein MJ, Young AC, Bonner TI. Structure of a cannabinoid receptor and functional expression of the cloned cDNA. Nature. 1990;346:561–4.

Meccariello R, Santoro A, Angelo SD, Morrone R, Fasano S, Viggiano A. The epigenetics of the endocannabinoid system. Int J Mol Sci. 2020;21:1–25.

Mechoulam R. Cannabis as Therapeutic Agent. CRC Press: Boca Raton, FL; 1986.

Meschler JP, Howlett AC. Signal transduction interactions between CB_1 cannabinoid and dopamine receptors in the rat and monkey striatum. Neuropharmacol. 2001;40:918–26.

Mesnage V, Houeto JL, Bonnet AM, Clavier I, Arnulf I, Cattelin F, et al. Neurokinin B, neurotensin, and cannabinoid receptor antagonists and Parkinson disease. Clin Neuropharmacol. 2004;27:108–10.

Meyer JD, Crombie KM, Cook DB, Hillard CJ, Koltyn KF. Serum endocannabinoid and mood changes after exercise in major depressive disorder. Med Sci Sports Exerc. 2019;51:1909–17.

Micale V, Cristino L, Tamburella A, Petrosino S, Leggio GM, Di Marzo V. Enhanced cognitive performance of dopamine D3 receptor "knock-out" mice in the step-through passive-avoidance test: assessing the role of the endocannabinoid/endovanilloid systems. Pharmacol Res. 2010;61:531–6.

Milad MR, Orr SP, Lasko NB, Chang Y, Rauch SL, Pitman RK. Presence and acquired origin of reduced recall for fear extinction in PTSD: Results of a twin study. J Psychiatr Res. 2008;42:515–20.

Mitjans M, Gasto C, Catalan R, Fananas L, Arias B. Genetic variability in the endocannabinoid system and 12-week clinical response to citalopram treatment: The role of the CNR1, CNR2 and FAAH genes. J Psychopharmacol. 2012;26:1391–8.

Molina-Holgado E, Molina-Holgado F. Mending the broken brain: Neuroimmune interactions in neurogenesis. J Neurochem. 2010;114:1277–90.

Monory K, Massa F, Egertova M, Eder M, Blaudzun H, Westenbroek R, et al. The endocannabinoid system controls key epileptogenic circuits in the hippocampus. Neuron. 2006;51:455–66.

Monsonego A, Weiner HL. Immunotherapeutic approaches to Alzheimer's disease. Science. 2003;302:834–38.

Moore NL. Role of cannabinoid receptor type 1 desensitization in greater tetrahydrocannabinol impairment of memory in adolescent rats. J Pharmacol Exp Ther. 2010;335:294–301.

Moreira FA, Lutz B. The endocannabinoid system: Emotion, learning and addiction. Addict Biol. 2008;13:196–212.

Muccioli GG. Endocannabinoid biosynthesis and inactivation, from simple to complex. Drug Discov Today. 2010;15:474–83.

Munoz MR, Blazquez PS, Callado LF, Meana JJ, Nino JG. Schizophrenia and depression, two poles of endocannabinoid system deregulation. Transl Psychiatry. 2017;7:1–12.

Munro S, Thomas KL, Abu-Shaar M. Molecular characterization of a peripheral receptor for cannabinoids. Nature. 1993;365:61–5.

Navarrete F, Gutierrez MSG, Barba RJ, Rubio G, Gasparyan A, Olivares AA, et al. Endocannabinoid system components as potential biomarkers in psychiatry. Front Psychiatry. 2020;11:1–30.

Neumeister A, Normaiden MD, Pietrzak RH, Piomelli D, Zheng MQ, Anton AG. Elevated brain cannabinoid CB_1 receptor availability in posttraumatic stress disorder: A positron emission tomography study. Mol Psychiatry. 2014;18:1034–40.

Nie J, Lewis DL. The proximal and distal C-terminal tail domains of the CB_1 cannabinoid receptor mediate G protein coupling. Neuroscience. 2001;107:161–7.

Nuss P. Anxiety disorders and GABA neurotransmission: A disturbance of modulation. Neuropsychiatr Dis Treat. 2015;11:165–75.

Onaivi ES, Ishiguro H, Gong JP, Patel S, Meozzi PA, Myers L, et al. Brain neuronal CB2 cannobinoid receptors in drug abuse and depression: From mice to human subjects. PLoS One. 2008;3:1–10.

Pamplona FA, Takahashi RN. WIN55, 212-2 impairs contextual fear conditioning through the activation of CB_1 cannabinoid receptors. Neurosci Lett. 2006;397:88–92.

Pamplona FA, Takahashi R. Psychopharmacology of the endocannabinoids: Far beyond anandamide. J Psychopharmacol. 2011;26:6–22.

Pape HC and Pare D. Plastic synaptic networks of the amygdala for the acquisition, expression, and extinction of conditioned fear. Physiol Rev. 2010;90:419–63.

Pascual A C, Martin-Moreno AM, Giusto NM, de Ceballos ML, Pasquare SJ. Normal aging in rats and pathological aging in human Alzheimer's disease decrease FAAH activity: Modulation by cannabinoid agonists. Exp Gerontol. 2014;60:92–9.

Pazos MR, Sagredo O, Fernández-Ruiz J. The endocannabinoid system in Huntington's disease. Curr Pharm Des. 2008;14:2317–25.

Pertwee RG. Cannabinoid pharmacology: the first 66 years. Br J Pharmacol. 2006;147:S153–71.

Pertwee RG, Ross A. Cannabinoid receptors and their ligands. Prostaglandins Leukot Essent Fatty Acids. 2002;66:101–21.

Phillips RG, LeDoux JE. Differential contribution of amygdala and hippocampus to cued and contextual fear conditioning. Behav Neurosci. 1992;106:274–85.

Picconi B, Centonze D, Hakansson K, Bernardi G, Greengard P, Fisone G, et al. Loss of bidirectional striatal synaptic plasticity in LDOPA- induced dyskinesia. Nat Neurosci. 2003;6:501–6.

Pisani A, Fezza F, Galati S, Battista N, Napolitano S, Finazzi-Agro A, et al. High endogenous cannabinoid levels in the cerebrospinal fluid of untreated Parkinson's disease patients. Ann Neurol. 2005;57:777–9.

Potvin S, Mahrouche L, Assaf R, Chicoine M, Giguere CE, Furtos A. Peripheral endogenous cannabinoid levels are increased in schizophrenia patients evaluated in a psychiatric emergency setting. Front Psychiatry. 2020;11:1–10.

Rasch B, Born J. About sleeps role in memory. Physiol Rev. 2013;93:681–766.

Ravi J, Sneh A, Shilo K, Nasser MW, Ganju RK. FAAH inhibition enhances anandamide mediated anti-tumorigenic effects in non-small cell lung cancer by downregulating the EGF/EGFR pathway. Oncotarget. 2014;5:2475–86.

Reich CG, Mohammadi MH, Alger BE. Endocannabinoid modulation of fear responses: Learning and state-dependent performance effects. J Psychopharmacol. 2008;22:769–77.

Roberto M, Cruz M, Bajo M, Siggins GR, Parsons LH, Schweitzer P. The endocannabinoid system tonically regulates inhibitory transmission and depresses the effect of ethanol in central amygdala. Neuropsychopharmacology. 2010;35:1962–72.

Rodriguez de Fonseca F, Rocia A Carrera M, Navarro M, Koob GF, Weiss F. Activation of corticotropin-releasing factor in the limbic system during cannabinoid withdrawal. 1997;276(5321):2050–4.

Rojnic Kuzman M, Bosnjak Kuharic D, Ganoci L, Makaric P, Kekin I, Rossini Gajsak L, et al. Association of CNR1 genotypes with changes in neurocognitive performance after eighteen-month treatment in patients with first-episode psychosis. Eur Psychiatry. 2019;61:88–96.

Romero-Sanchiz P, Nogueira-Arjona R, Pastor A, Araos P, Serrano A, Boronat A, et al. Plasma concentrations of oleoylethanolamide in a primary care sample of depressed patients are increased in those treated with selective serotonin reuptake inhibitor-type antidepressants. Neuropharmacology. 2019;149:212–20.

Ruehle S, Rey AA, Remmers F, Lutz B. The endocannabinoid system in anxiety, fear memory and habituation. J Psychopharmacol. 2012;26:23–39.

Russo EB. Taming THC: potential cannabis synergy and phytocannabinoid-terpenoid entourage effects. Br J Pharmacol. 2011;163:1344–64.

Sagredo O, Garcia-Arencibia M, de Lago E, Finetti S, Decio A, Fernandez-Ruiz J. Cannabinoids and neuroprotection in basal ganglia disorders. Mol Neurobiol. 2007;36:82–91.

Samuelsson G, Bohlin L. Drugs of Natural Origin: A Treatise of Pharmacognosy. 6th ed. Swedish Pharmaceutical Press, Stockholm; 2010.

Scherma M, Masia P, Deidda M, Fratta W, Tanda G, Fadda P. New perspectives on the use of cannabis in the treatment of psychiatric disorders. Medicines (Basel). 2018;5:1–17.

Schoeler TJ, Kambeitz I, Murray BR. The effects of cannabis on memory function in users with and without a psychotic disorder: findings from a combined meta-analysis. Psychol Med. 2016; 46:177–88.

Schoenfeld TJ and Cameron HA. Adult neurogenesis and mental illness. Neuropsychopharmacology. 2015;40:113–28.

Solas M, Francis PT, Franco R, Ramirez MJ. CB_2 receptor and amyloid pathology in frontal cortex of Alzheimer's disease patients. Neurobiol Aging. 2013;34:805–8.

Solowij N, Jones KA, Rozman ME, Davis SM, Ciarrochi J, Heaven PCL, et al. Verbal learning and memory in adolescent cannabis users, alcohol users and non-users. Psychopharmacology. 2011a;216:131–44.

Solowij N, Yucel M, Respondek C, Whittle S, Lindsay E, Pantelis C, et al., Cerebellar white-matter changes in cannabis users with and without schizophrenia. Psychol Med. 2011b;41:2349–59.

Straiker A, Mackie K. Depolarization-induced suppression of excitation in murine autaptic hippocampal neurons. J Physiol. 2005;569:501–17.

Straiker A, Mackie K. Metabotropic suppression of excitation in murine autaptic hippocampal neurons. J Physiol. 2007;578:773–85.

Sugiura T, Kobayashi Y, Oka S, Waku K. Biosynthesis and degradation of anandamide and 2-arachidonoylglycerol and their possible physiological significance. Prostaglandins Leukot Essent Fatty Acids. 2002;66:173–92.

Szabo B, Schlicker E. Effects of cannabinoids on neurotransmission. Handb Exp Pharmacol 2005;168:327–65.

Tong D, He S, Wang L, Jin L, Si P, Cheng X. Association of single-nucleotide polymorphisms in the cannabinoid receptor 2 gene with schizophrenia in the Han Chinese population. J Mol Neurosci. 2013;51:454–60.

Toth KF, Adam D, Biro T, Olah A. Cannabinoid signaling in the skin: Therapeutic potential of the "c(ut)annabinoid" system. Molecules. 2019;24:2–56.

Turri M, Teatini F, Donato F, Zanette G, Tugnoli V, Deotto L, et al. Pain modulation after oromucosal cannabinoid spray (SATIVEX®) in patients with multiple sclerosis: a study with quantitative sensory testing and laser-evoked potentials. Medicines (Basel). 2018;5:1–12.

United Nations Office on Drugs and Crimes. World Drug Report 2010. United Nations Publications; 2010.

Van Vliet SA, Vanwersch RA, Jongsma MJ, Olivier B, Philippens IH. Therapeutic effects of Delta(9)-THC and modafinil in a marmoset Parkinson model. Eur Neuropsychopharmacol. 2008;18:383–9.

Vaseghi G, Rabbani M, Hajhashemi V. The effect of AM281, a cannabinoid antagonist, on memory performance during spontaneous morphine withdrawal in mice. Res Pharm Sci. 2013;8:59–64.

Velayudhan L, Diepen EV, Marudkar M, Hands O, Suribhatla S, Prettyman R, et al. Therapeutic potential of cannabinoids in neurodegenerative disorders: A selective review. Curr Pharm Des. 2014;20:2218–30.

Vincenzo DM, Nephi S, Andreas Z. Endocannabinoid signalling and the deteriorating brain. Nat Rev Neurosci. 2015;16:30–42.

Walther S, Mahlberg R, Eichmann U, Kunz D. D-9-tetrahydrocannabinol for nighttime agitation in severe dementia. Psychopharmacology. 2006;185:524–28.

Wang J, Ueda N. Biology of endocannabinoid synthesis system. Prostaglandins Other Lipid Mediat. 2009;89(:112–19.

Waterlow M and Chrisp P. Rimonabant: the evidence for its use in the treatment of obesity and the metabolic syndrome. Core Evid. 2007;2:173–87.

Yan W, Yun Y, Ku T, Li G, Sang N. NO_2 inhalation promotes Alzheimer's disease-like progression: cyclooxygenase-2-derived prostaglandin E2 modulation and monoacylglycerol lipase inhibition-targeted medication. Sci Rep. 2016;6:22429.

Yu W, De Hert M, Moons T, Claes SJ, Correll CU, van Winkel R. CNR1 gene and risk of the metabolic syndrome in patients with schizophrenia. J Clin Psychopharmacol. 2013;33:186–92.

Yuan H, Ma Q, Ye L, Piao G. The traditional medicine and modern medicine from natural products. Molecules. 2016;21:1–18.

Zanettini C, Panlilio LV, Aliczki M, Goldberg SR, Haller J, Yasar S. Effects of endocannabinoid system modulation on cognitive and emotional behavior. Front Behav Neurosci. 2011;5:57.

Zhang J, Hu M, Teng Z, Tang YP, Chen C. Synaptic and cognitive improvements by inhibition of 2-AG metabolism are through upregulation of micro RNA-188-3p in a mouse model of Alzheimer's disease. J Neurosci. 2014;34:14919–33.

Zou S, Kumar U. Cannabinoid receptors and the endocannabinoid system: signaling and function in the central nervous system. Int J Mol Sci. 2018;19:2–23.

11 Role of the Glutaminergic System in Schizophrenia

Renu Kadian
Ram Gopal College of Pharmacy, Sultanpur, Farruknagar, Gurgugram, India

Shrestha Sharma
School of Medical and Allied Sciences, K.R. Mangalam University, Gurugram, India

Faheem Hyder Pottoo
Department of Pharmacology, College of Clinical Pharmacy, Imam Abdulrahman Bin Faisal University, Dammam, Saudi Arabia

Ashwani Arya
Department of Pharmaceutical Education and Research, South Campus, Bhainswal Kalan, BPS Women University, Khanpur Kalan, Haryana, India

Manish Dhall
Swami Dayanand Post Graduate Institute of Pharmaceutical Sciences, Sharma University of Health Sciences, Haryana, India

CONTENTS

DOI: 10.1201/9780429265198-13

11.1 INTRODUCTION

The amino acid glutamate basically acts as an excitatory neurotransmitter. It has an average concentration of about 10–15 mM (Cooper and Jeitner, 2016). Both high and low levels glutamate are damaging to the brain (Zhou and Danbolt, 2014). Glutamate is at the midway point of almost all metabolic pathways, so its presence in perfect concentration is mandatory for the proper functioning of certain pathways (Yelamanchi et al., 2016). Its removal from the site of action at the right time is also necessary (Zhou and Danbolt, 2014). It exhibits exclusion from the synaptic cleft through reuptake or diffusion by glutamate transporters (O'Donovan et al., 2017). A high percentage of glutamate left in the synaptic cleft causes hyperexcitation in nerve cells and, ultimately, death known as 'excitotoxicity', and is the starting point of several other neurological disorders (Puyal et al., 2013). Schizophrenia is a human brain disorder recognized by clear-cut symptoms but improper biological clarification (Tamminga and Medoff, 2000). Recent animal experiments have suggested the role of dopamine to be less critical than previously supposed in the regulation of psychomotor functions (Arya et al., 2018). Of late, more studies have recommended the role of glutamate neurotransmission in schizophrenia (Bondi et al., 2012). There is evidence that glutamatergic transmission expedites glutamate receptor stimulation, improper reuptake, and other synaptic actions (Rubio et al., 2012). Certain symptoms of schizophrenia, negative and cognitive, show an indication for glutamatergic abnormalities and the role of 2-amino-3-3–hydroxy-5-methyl-isoxazol-4-yl propanoic acid (AMPA) and *N*-methyl-D-aspartate (NMDA) receptors in schizophrenia (Menniti et al., 2013). The role of abnormal receptor expression and other reuptake-linked abnormalities related to the glutamate system in schizophrenics has been researched (Rubio et al., 2012). These studies have highlighted specific sites, and explicit improper molecular transcription leading to defective protein formation leading to illness (Rubio et al., 2012). This chapter highlights the role of genetic abnormalities and de novo mutations that could be the focus of new drug development. In light of the glutamatergic system being involved in the pathophysiology of schizophrenia, current drug development attempts have focused on NMDA receptor (NMDAR) systems, but lately additional glutamate receptors and transporters systems have been targeted. Certain glycine transporter inhibitors and allosteric potentiators of metabotropic glutamate receptors are under clinical trial and are thought to have fewer side effects. Efforts are being made to process new drugs related to the modulation of the glutamatergic system. Thus, the authors in this chapter focus on the study of the glutamatergic system and its role in schizophrenia. It holds great potential for the future progress of more successful drugs with new mechanisms based on the glutaminergic system for this serious brain disorder, probably with fewer side effects.

11.2 GLUTAMATE BIOSYNTHESIS

Glutamate is a protein-encoded neurotransmitter distributed throughout the brain (Lajtha et al., 1981). The average concentration of glutamate is higher than that of other amino acids (Cooper and Jeitner, 2016). Glutamate is a non-essential amino acid that cannot penetrate the blood–brain barrier. It is synthesized presynaptically from glucose and glutamine by glutaminase, and is stored in synaptic vesicles (Smith, 2000; Smith and Stoll, 1998). Another method by which glutamate is produced from glucose by a bioprocess called glycolysis, which converts glucose to pyruvate (Hertz and Chen, 2017). The glutamate synthesis pathway is described in Figure 11.1.

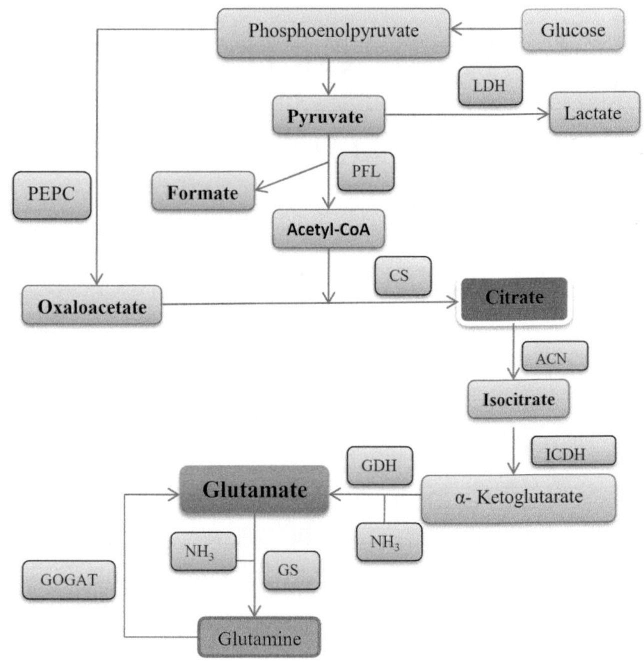

FIGURE 11.1 The glutamate synthesis pathway. PEPC: phosphoenolpyruvate carboxylase; PFL: pyruvate formate lyase; LDH: lactate dehydrogenase; CS: citrate synthase; CAN: aconitase; ICDH: isocitrate dehydrogenase; GDH: glutamate dehydrogenase; GS: glutamine synthetase; GOGAT: glutamine oxoglutarate aminotransferase (adapted with permission from Nishimura J. et al., Biofilm formation by *Streptococcus mutans* and related bacteria. *Advances in Microbiology* 2012; 2(3): 208–215).

Pyruvate enters the tricarboxylic acid cycle (TCA) or Krebs cycle (Hassel et al., 2012). Alpha-ketoglutarate, an intermediate of the citric acid cycle, can produce glutamate. The reaction is catalyzed by glutamate dehydrogenase (Peng et al., 1993). The co-enzyme used in the reaction is nicotinamide adenine dinucleotide (NAD^+) or nicotinamide adenine dinucleotide phosphate ($NADP^+$; Smith et al., 1975). After the generation of an action potential neurotransmitter and its release in the synaptic cleft, it undergoes interaction with its receptors and the action is terminated by reuptake with the help of transporters (Zhou and Danbolt, 2014). It is catabolized mainly by glutamic acid decarboxylase.

11.3 GLUTAMATE RECEPTORS

There are two types of glutamate receptors (ionotropic and metabotropic receptors). Inotropic receptors are ligand-gated ion channels and metabotropic receptors are G protein-coupled receptors (GPCRs) (Kew and Kemp, 2005; Pleasure, 2008). Subtypes of glutamate receptor and its properties mentioned in Figure 11.2.

11.3.1 Inotropic Glutamate Receptors

Widely held excitatory neurotransmission in the brain depends upon ionotropic receptors. The ionotropic type of receptors comprise 4–5 subunits and allow cation influx (Marsman et al., 2013). The NMDA$ is encoded by five genes: *NMDAR1* and *NMDAR2A–NMDAR2D* (Breese et al., 1995; Hollmann et al., 1991; Meador-Woodruff and Healy, 2000). The NMDAR2 subunit contains the binding site for glutamate and competitive antagonists (Anson et al., 1998; Mori and Mishina, 1995). There is a binding site at *NMDAR1* for glycine that must be engaged before glutamate action (Hirai et al., 1996; Kuryatovet al., 1994). Therefore, the pharmacology of binding sites is expressed only after transcription of the NMDAR1 subunit (Hollmann et al., 1991). The ion-permeating channel represents a binding site for antagonists like phencyclidine (PCP), MK-801, and memantine (Zhorov et al., 2004). Receptors associated with NMDAR1, NMDAR2A, or NMDAR2B subunits appear to exhibit greater interaction with MK-801 (Lynch et al., 1994; Rodriguez-Paz et al., 1995). Polyamine modulators are associated with receptors with NMDAR2B subunits (Williams, 1997). There is a voltage-contingent magnesium-binding site that blocks the channel at physiologic concentrations (Mayer et al., 1984). AMPA-type receptor subunits are transcribed from and

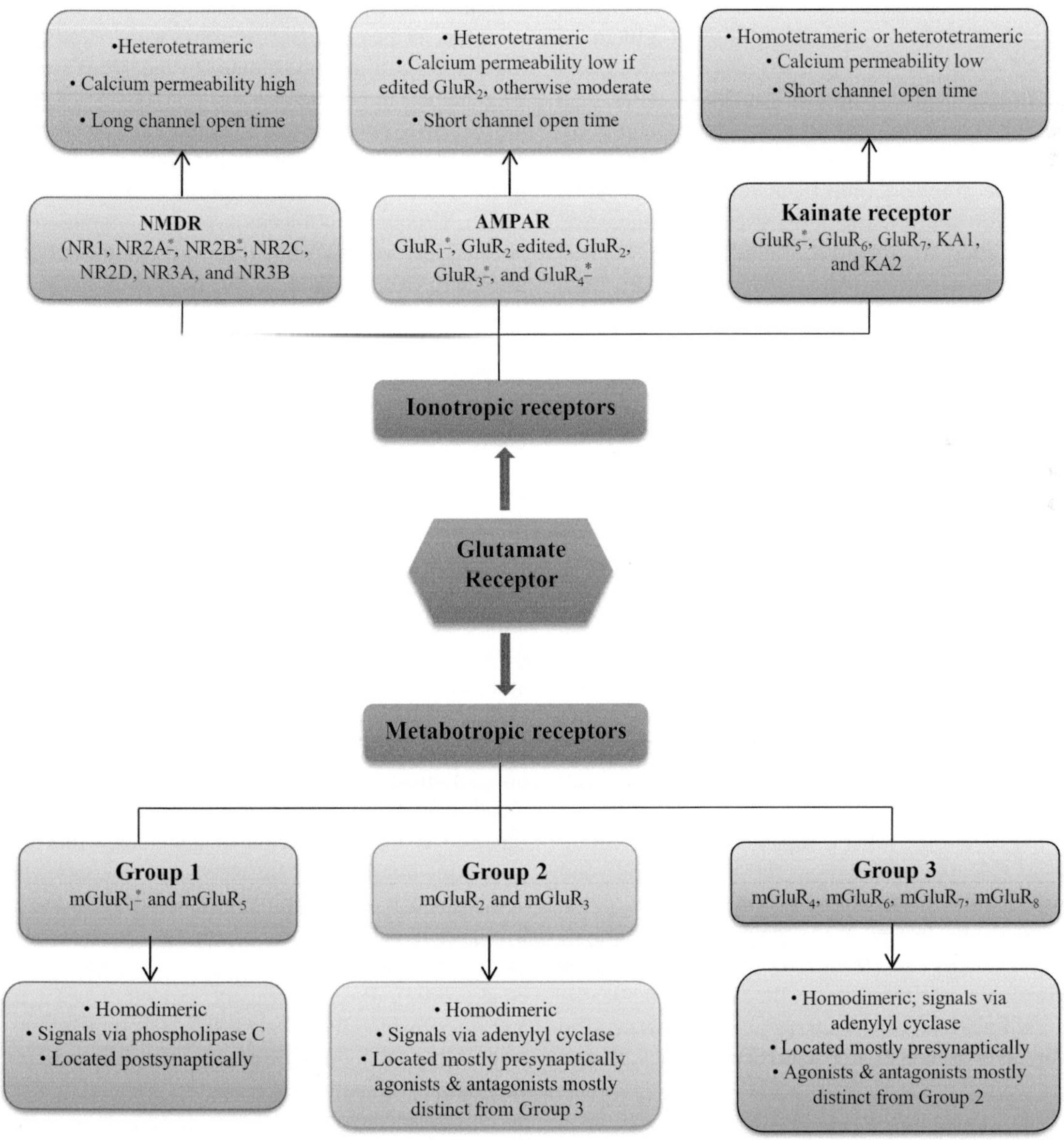

FIGURE 11.2 Subtypes of the glutamate receptor and their properties. NMDR: *N*-methyl-D-aspartate receptors; AMPAR: 2-amino-3-3-hydroxy-5-methyl-isoxazol-4-yl propanoic acid receptor; mGluR: metabotropic glutamate receptors.

expressed by four genes: *GLUR1–GLUR4* (Burnashev et al., 1992; Hollmann et al., 1989). The amassed AMPA receptors consist of three binding sites: one where glutamate binds; a second for binding of competitive antagonists; and a third for desensitization modulators (Hollmann et. al., 1991). Kainate-type receptors comprise subunits derived from the genes *GLUR5–GLUR7* (Kohler et al., 1993). mRNAs encoding the GluR5 and GluR6 subunits are subjected to enzymatic editing by an RNA deaminase (Sommer et al., 1991). The kainite-type receptor is an assembly of five homomer units (Fisher and Fisher, 2014). Proper presynaptic release and postsynaptic depolarization of glutamate are important for smooth brain functioning (Cincotta et al., 1989; Danysz and Parsons, 1998).

11.3.2 Metabotropic Glutamate Receptors

Metabotropic glutamate receptors are GPCRs consisting of seven transmembrane domains. These are mainly responsible for the synaptic modulation and excitability of glutamate (Niswender and Conn, 2010). Eight metabotropic receptors have been cloned, named mGluR1–mGluR8 (Pin and Duvoisin, 1995). The genes for these receptors appear to encode seven membrane-spanning proteins. The receptors are split into three groups: Group I (mGluR1) and Group II (mGluR2, mGluR3 and mGluR5) act via phospholipase C activation; and Group III (mGluR4, mGluR6-mGluR8) acts by adenylyl cyclase inhibition (Alexander et al., 2013). Each group is related to a definite function.

11.4 GLUTAMATE RECEPTOR FUNCTION AND DISTRIBUTION

NMDA and AMPA receptors are postsynaptically co-localized, and may overlap (Rao and Craig, 1997). AMPA receptors are located in the hippocampus, amygdala, lower thalamus ventral striatum, and cortex (Eva and Michael, 2002). Kainate receptors are presynaptically located in infragranular layers of the cortex and the hippocampal CA3 subfield that controls glutamate release (Negrete-Diaz et al., 2018). Together with AMPA these are collectively termed non-NMDA receptors. Both act by conducting Na^+ and K^+ through channels and exhibit rapid conductance, compared to NMDARs, which are slow conductors (Eva and Michael, 2002). Metabotropic receptor families have a totally different distribution and functions, when compared internally. Group I is mainly postsynaptic in location: the hippocampus and thalamus are thought to be the main site of mGluR1 receptors. mGluR5 is abundantly located in the thalamus hippocampus and amygdala (Tamaru et al., 2001). Group II and III receptors modulate glutamate release and are found presynaptically. mGluR2 is confined to the dentate gyrus and cortex. The mGluR3 subtype is located in glia cells (Senter et al., 2016).

11.5 GLUTAMATE RELEASE

Glutamate is constantly discharged to the extracellular fluid. If uptake is inhibited it causes an extracellular build-up of glutamate within an instant (Jabaudon et al., 1999). Its normal released concentration is also far in excess of other neurotransmitters of its kind (Schousboe, 1981). Glutamate is stored on synaptic vesicles in presynaptic neurons and released by exocytosis on arrival of a Ca^{2+}-dependent action potential (Lodish et al., 2000). Its release in the hippocampus is both K^+- and Ca^{2+} -dependent (Fuente-Sandoval et al., 2011). If the release of the substance is Ca^{2+}-dependent it is an important criterion for any substance to be considered as a neurotransmitter (Regehr et al., 2009). However, there are other mechanisms by which glutamate is released (Danbolt, 2001). There are certain non-vesicular mechanisms (e.g., through anion channels; Wang et al., 2013), or by the inverse action of a transporting protein located at the plasma membrane (Grewer et al., 2008). A third mechanism is via cystine transporters. The fourth mechanism is via mature brain astrocytes, although this is intensely doubtful (Bridges et al., 2012). Glutamate released from presynaptic neurons and 20% of glutamate is taken up by glutamate receptors in postsynaptic neurons, which leads to excitatory impulses. The remaining glutamate undergoes astrocytic uptake by glutamate uptake transporters (EAAT-1 and EAAT-2), which are expressed on surface of the astrocytic presynaptic processor. The glutamate is oxidized to alpha-ketoglutarate and is used in the TCA cycle for the synthesis of adenosine triphosphate (ATP). Glutamate is metabolized to glutamine by glutamate synthetase and this glutamine reconverted to glutamate in astrocytes (Kojima et al., 1999; Bergles and Jahr, 1998; Grant and Dowling, 1995; Otis et al., 1997). Glutamate uptake and metabolism in are shown in Figure 11.3.

Thus, there are many opinions regarding the release mechanisms of glutamate, including the gliotransmitter concept. Therefore, a vast literature study is required.

11.6 GLUTAMATE TRANSPORTERS

Five types of glutamate transporter have been replicated for amino acids as a whole, EAAT1–EAAT5, known as excitatory amino acid transporters and previously also recognized as glutamate transporters as they are excitatory neurotransmitters and amino acids (Eulenburg and Gomeza, 2010). However, it has been reported that glutamate transport is mainly related to EAAT2; this is the only transporter expressed on glutamate-rich brain parts (Su et al., 2003). The main location of these transporters is astroglial cells, and their main function is to clear excessive glutamate from the synaptic cleft after interaction with receptors, to control any toxicity and convert it to glutamine. Glutamine from the astrocyte is transported to the neurons, where it is metabolized and packed in vesicles and again released in synaptic cleft (Rothman et al., 2003). Postsynaptically located EAAT3 transporters and presynaptically located EAAT2 transporters are also responsible for about 10% of glutamate transport (O'Donovan et al., 2017). Glutamate transporters associated with glial cells are also associated with glutamate release in certain adverse conditions such as brain damage and stroke (Malarkey and Parpura, 2008). The main mechanism involved is the reversal of transporters (Grewer et al., 2008).

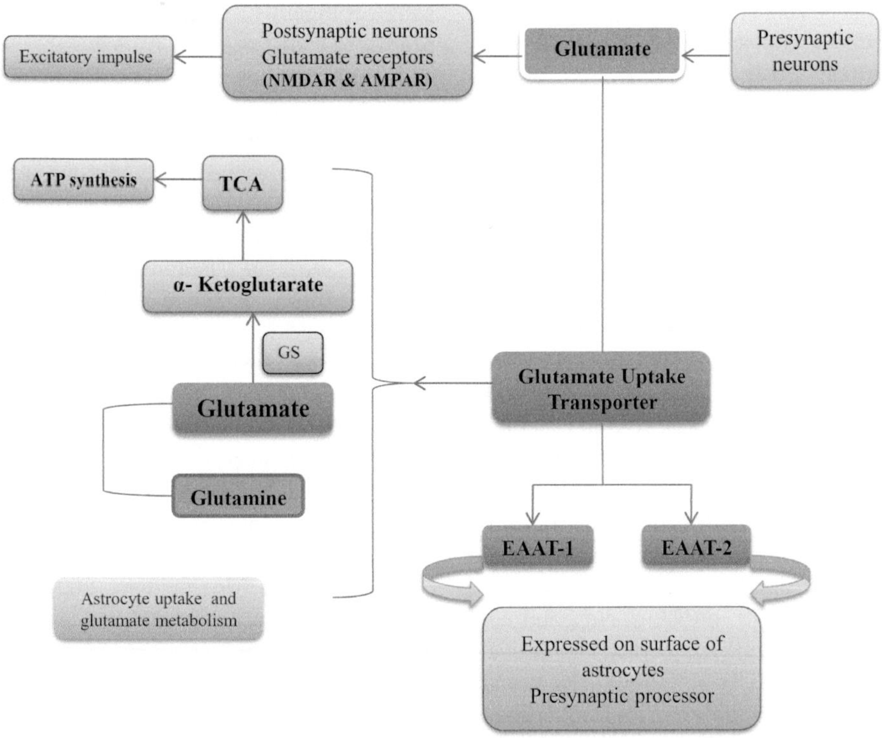

FIGURE 11.3 Glutamate uptake and its metabolism. NMDR: *N*-methyl-D-aspartate receptors; AMPAR: 2-amino-3-3-hydroxy-5-methyl-isoxazol-4-yl propanoic acid receptor; GS: glutamine synthetase; ATP: adenosine triphosphate; TCA: tricyclic acid; EAAT: excitatory amino acid transporter.

11.7 GLUTAMATE IN THE HEALTHY BRAIN

Glutamate is the foremost excitatory signal mediator in the central nervous system (CNS) responsible for plasticity (Johnson and Koerner, 1988). However, it only works properly when it is released in the right amount and at the exact location. The sensitivity of the cell where it is released should be proper and there should not be any abnormality in any type; further, the cell should be energetic enough to show a normal sensitivity level. As with its release, the removal of glutamate should also be at an appropriate rate after the action (Parsons et al., 2007). Excessive or not enough release or less release, and too much or not enough removal, are harmful. Abnormal removal and excessive if left, when act on receptors excite nerve cells to such an extent that it causes cell death due to excessive excitation and toxicity (Mallick, 2007).

11.7.1 Normal Functions of the Glutamate Neurotransmitter in the Brain

Glutamate is a major neurotransmitter that functions at the crossroads between multiple neurotransmission pathways (Zhou and Danbolt, 2014). Glutamate – an amino acid neurotransmitter – is spread throughout the entire nervous system, including the spinal cord, and controls almost all other neurotransmitter pathways. It is responsible for a large number of normal physiological functions. Abnormalities in glutamate result in several diseases and brain abnormalities. Glutamate concentration requires strict control; to do this, several mechanisms are active in the brain. Any perturbation in this control system can lead to deleterious consequences such as hyperexcitability, cytotoxicity, and cell death (Doble, 1999; Choi, 1994). Hyperactivity of glutamate in the hippocampus leads to halted regeneration of neurons and branching of dendrites, resulting in impeded spatial learning (Cortese and Luan Phan, 2005). Retarded reuptake of glutamate from the synaptic cleft is connected to a reduction in reward sensitivity, considered to be a negative symptom of schizophrenia (Bechtholt-Gompf et al., 2010). Long-term potentiation (LTP) is the mainstay of memory and learning, and glutamate, being the central element, along with its receptors, plays a very important role in this process (Lujan et al., 2005). The hippocampus is a broadly studied area of the brain with regard to LTP, although the neocortex, the amygdala, and the striatum are also considered (Morris et al., 2003). The stimulation of NMDARs, along with activated non-NMDARs, plays a prominent role in LTP; it is linked with lifelong changes in synaptic plasticity (Watkins and Jane, 2006). Glutamate, acting as a neurotransmitter, is a chief conciliator of emotions, sensitivity, perception, and motor coordination. Although different areas of the brain follow different molecular mechanisms, glutamate signaling plays the substantial role in them all (Lujan et al., 2005). However, what is most important in glutamate neurotransmission is that not even a minute disturbance of its concentration during release and reuptake is permitted. Its concentration, location, and duration of stay must be kept steady for consistent results; any deviations can result in poor communication and major damage (Zhou and Danbolt, 2014).

11.7.2 Glutamate in the Schizophrenic Brain

Psychotic drug-free individuals were compared to normal controls using positron emission tomography. The areas found to be functionally abnormal with respect to receptor concentration and related to positive symptoms were hippocampal formation, the medial surface of the frontal lobe, and the anterior cingulate cortex (Tamminga, 1998). There is a probability that the abnormality in NMDAR-facilitated neurotransmission is further involved in functional hypoactivity of the frontal lobes (Akbarian et al., 1996). Many reports have also reported a decline in the concentration of receptors other than NMDAR in the hippocampus, resulting in the loss of encoding of the mRNA receptor (Eastwood et al., 1995). In schizophrenics, it has been seen that there is are alterations in the expression of mRNA of *NR2* and *NR1* in the prefrontal and superior temporal cortex, respectively (Beneyto et al., 2007; Humphries et al., 1996). The disease accompanied complex alteration at different molecular levels. Basically, NMDARs are broadly studied in the glutamatergic system at different sites of the brain.

11.8 SCHIZOPHRENIA AND THE GLUTAMATE HYPOTHESIS

The pathophysiology of schizophrenia is relatively complex, continuously altered, and updated by new concepts and approaches in neuroscience (Boteva and Lieberman 2003; Ghaemi, 2006). The origin of schizophrenia is related to multiple theories and hypotheses (Fatemi and Folsom, 2009). The most enduring and earliest theory is that related to the involvement of the dopamine neurotransmitter (Howes and Kapur, 2009). The focus of this hypothesis is on prefrontal and subcortical hypo- and hyper-dopaminergia, respectively (Pycock et al., 1980), and is one of the leading theories of schizophrenia; however, it has certain limitations. There is a substantial number of individuals in whom the pathophysiological origin of their indications could not be linked to dopaminergic dysfunction (Demjaha et al., 2012). Despite the high number of D2 receptors occupied, many schizophrenic individuals do not react to typical antipsychotics (Mortimer et al., 2010). It has been observed that dopamine antagonists and partial agonists are not effective on cognitive mutilation and negative indications (Murphy et al., 2006), and may even worsen the condition (Kim et al., 2013), indicating that those other pathways, namely those concerning glutamate, may be the contributor to cognitive dysfunction.

11.8.1 Nmdar Antagonists and Schizophrenia

The involvement of glutamate in schizophrenia by various mechanisms was first proposed almost 20 years ago (Javitt, 2007). Psychotomimetic drugs of certain categories that act as noncompetitive NMDAR/glutamate receptor antagonists prompt a wide variety of schizophrenia-like symptoms, and thus support the hypoglutamatergic hypothesis of schizophrenia (Coyle, 1996; Javitt, 1987; Javitt and Zukin, 1991; Jentsch and Roth, 1999; Olney and Farber, 1995). Consistently blocked NMDARs, as reported in preclinical studies, recreates the crucial pathological characteristics of schizophrenia, namely downregulation of parvalbumin-positive cortical gamma-aminobutyric acid (GABA) ergic neurons, lessened spine density, and pyramidal neuron dendritic dysgenesis (Coyle, 2012). This is further supported by the fact that chemicals which act through glycine-modulatory sites and ultimately affect the functions of NMDARs, both of which are involved in the functional output of receptors, improves the negative indications of schizophrenia (Coyle, 2012). Recently, it has been proposed that NMDA receptor blockage by antagonists at cortical sites produces a reduction in GABAergic interneuron functioning (Homayoun and Moghaddam, 2007; Rotaru et al., 2012). The cause behind this reduced interneuron function may be the production of superoxides (Behrens et al., 2007). A reversal of psychosis is seen if the formation of these reactive oxygen species is reduced, as seen in preclinical models (Levkovitz et al., 2007; Monte et al., 2013; Source et al., 2010). Ketamine causes the activation of the immune enzyme NADPH-oxidase by the production of interleukin (IL)-6, leading to interneuron dysfunction. If the production of superoxide is prevented, the production of IL-6 can be reversed and thus interneurons can be protected (Behrens et al., 2008). Antioxidants can thus play a role in the defense against schizophrenia.

11.8.1.1 Reduced NMDAR and Hyperglutamate Basis of Schizophrenia

A reduced concentration of NMDARs or their reduced function without any change in concentration may be due to the presence of an antagonist or any type of toxicity that interferes with their control over inhibitory neurons (Moghaddam and Javitt, 2012). Thus, in the absence of any inhibitory control, the activity of glutamatergic neurons increases in the synaptic cleft on AMPA receptors, adjacent to NMDA on postsynaptic membranes, but they have high kinetic energy (Moghaddam, 2003). Thus, it has been confirmed that hyperfunctioning rather than hypofunctioning of glutamate is involved in schizophrenia (Moghaddam and Javitt, 2012). However, one hypothesis states that NMDARs need synchronized membrane depolarization and glutamate levels for their functioning, which – being ionic channels – act via the influx of Ca^{2+} ions and the further activation of a secondary messenger system. Thus, this short-term synaptic activity results in long-term changes in synaptic efficacy (Nakanishi, 1998). It is therefore hypothesized that, overall, glutamatergic tone is reduced rather than enhanced (Kim et al., 1980). Over the last two decades, it has gradually been confirmed that hyperglutamatergic function on AMPA, mediated by hypo-NMDAR activation, may be perilous in schizophrenia, and that an ideal treatment should target a reduction in presynaptic glutamate levels (Moghaddam, 2003). This theory has been backed up by findings that, when injected with NMDAR antagonists, normal control animals showed enhanced glutamate efflux in the prefrontal cortex, which impaired cognitive function and lead to stereotypy-behavior similar to the positive symptoms shown in humans (Lorrain et al., 2003; Moghaddam et al., 1997). NMDAR antagonists cause the accumulation of glutamate in synapses, thus enhancing transmission at AMPA receptors (Moghaddam and Adams, 1998). Thus, the behavioral results of NMDAR underactivity are due to a shift of this activity to adjacent non-NMDARs in their inability to act properly at the same glutamate concentration. Therefore, it can be considered that it involves the improper regulation of glutamate neurotransmission rather than hypofunction. This finding is also supported by evidence from functional magnetic resonance imaging studies that neural activity is enhanced in the human prefrontal cortex by NMDAR antagonists (Breier et al., 1997; Vollenweider et al., 1997). It has also been shown that, when activated by an agonist, extrasynaptically located mGlu2/3 receptors reduce the release of glutamate (Schoepp et al., 1997; Battaglia et al., 1997). Thus, this action reduces the behavioral and cellular results of NMDAR antagonists of activated release of glutamate by mGlu2/3 receptor agonist, such as cognitive impairments in healthy volunteers (Homayoun et al., 2005; Krystal et al., 2005). Plunging and controlling the release of glutamate from storage sites by presynaptic autoreceptors, mainly mGlu2, can be used in schizophrenia treatment; another source of new drug development for schizophrenia related to this mechanism is the enzyme glutaminase (Gaisler-Salomon et al., 2009).

11.8.1.2 Role of NMDAR Antagonists in the Firing of Cortical Neurons

NMDARs are basically responsible for the executive functioning of the prefrontal cortex, which includes set, shifting, and functioning memory (Stefani and Moghaddam, 2005). GABA interneurons control the action of neocortical pyramidal and hippocampal neurons (Buzsaki et al., 2004). The impact of use-dependent inhibition of pyramidal neurons and GABA interneurons in the prefrontal cortex by NMDARs has been examined. It has been found the NMDARs, if inhibited or hypoactive in the early stages of disease, accelerate the firing proportion of major pyramidal neurons, resulting in enhanced activity; this also happens in later stages, although at delayed rate, resulting in depressed symptoms (Homayoun and Moghaddam, 2007). Thus, what is hypothesized and implicated in the hypofunctioning of NMDARs is that they diminish the inhibitory control of the prefrontal cortex on output neurons.

11.9 ABNORMALITY OF THE GLUTAMATE TRANSPORTER IN SCHIZOPHRENIA

The EAAT2 transporters are involved in the clearance of 90% of glutamate from the site of action and are disturbed in patients with schizophrenia. An enzyme called hexokinase1, related to mitochondria, is found to detached from EAAT2, thus altering metabolism. There is also decreased production of ATP, which further fulfils Na^+/K^+ ATPase efficiency and ultimately glutamate reuptake (O'Donovan et al., 2017). The NMDAR antagonist PCP damages prepulse inhibition, which is steadily reduced in diseased persons (Braff et al., 1978). Some studies have reported a reduction in EAAT2, while others have described an increased or unchanged amount of EAAT2, in schizophrenia (Hu et al., 2015). Abnormality of the transporter is highly speculated. The mRNA expression of EAAT2 is enhanced in provoked relay neurons in schizophrenics (McCullumsmith et al., 2016). Mice lacking EAAT1 showed higher locomotor activity, which was reversed by either the antipsychotic drug haloperidol or the mGlu2/3 agonist LY379268 (Karlsson et al., 2008). A reduction in EAAT function in the hippocampus enhances the amount of NMDAR intervened at excitatory postsynaptic potentials (EPSCs) and delays kinetic mechanisms, but AMPA-mediated EPSCs remain unaffected (Arnth-Jensen et al., 2002). EAATs are responsible for limiting glutamate spillover into nearby synapses in the absence synaptic plasticity; memory and learning are affected (Tzingounis and Wadiche, 2007).

11.10 GENETIC ABNORMALITIES OF RECEPTORS IN SCHIZOPHRENIA

A complete functional NMDAR is a hetero-oligomer comprised of several subunits: NRA–D, NR1, and NR3A–B (Kristiansen et al., 2007). In schizophrenia, there is a genetic component involving polymorphisms in *NR1* and *NR2B* (Allen et al., 2008; Begni et al., 2003; Qin et al., 2005). Thus, it can be concluded that the abnormal expression of NMDAR and thus function may be the basis of schizophrenia. Along with NMDARs, many other modulatory sites, including disturbances in the expression and functioning of glycine, serine, and glutathione, are also is becoming targets for understanding the etiology of and therapeutic interventions in schizophrenia (Gysin et al., 2007; Javitt, 2007). However, few reports provide evidence for disturbed glycine metabolism in schizophrenia (Javitt, 2007); however, abnormality in serine at the synthetic and degradation levels is reported. Serine is produced and broken down in the brain by oxidase and serine racemase, respectively. These enzymes are modulated by G72 protein. The involvement of both enzymes has been noted in schizophrenia. Further, it has also been supported by the low amount of serine in the cerebrospinal fluid and plasma of schizophrenic individuals (Labrie and Roder, 2010). Reduced glutathione levels and numerous glutathione synthetic enzymes have also been reported in schizophrenia (Berk et al., 2008; Dodd et al. 2008; Gysin et al., 2007; Rodriguez-Santiago et al., 2010).

11.10.1 Chromosomal Abnormalities in Schizophrenia

There is strong evidence for the involvement of genetic abnormalities in schizophrenia, but no specific location has been identified (Bassett et al., 2000). There are only a few reports of the involvement of chromosomal abnormalities in schizophrenia (Craddock and Owen, 1994). Microdeletion of chromosome 22q11.2 has the most substantial evidence for the involvement of a genetic abnormality in schizophrenia (Demczuk and Aurias, 1995). The detection of microdeletion syndrome was done using a commercially available 22q11.2 probe (D22S75) and fluorescence in situ hybridization (Turk and Hill, 1995). The 22qDS phenotype involves impaired learning ability, mental obstruction, typical physiognomy, and the subsequent onset of psychiatric illnesses such as schizophrenia (Bassett et al., 1998). Mood disorder, anxiety, attention deficit, autism, and substance use disorders are also reported (Bassett et al., 1998). A study has reported that 13 of 50 adults (26%) with 22q11.2 microdeletions had schizophrenia (Murphy et al., 1999). In one study, 22q11.2 deletions were seen in 9% of 22 patients with schizophrenia (Murphy et al., 1998), and in 32% of 28 patients with proposed clinical screening criteria for schizophrenia (Bassett and Chow, 1999). Certain surveys reporting genetic anomalies in schizophrenic patients have also shown the involvement of sex chromosomes (DeLisi et al. 1994) but without consistent results. Therefore, a systemic study is needed to prove the involvement of sex chromosome abnormality in schizophrenia. Beyond this, certain newly introduced mutations also contribute to the risk of schizophrenia. There is a higher risk of these mutations when the parents are of older age at the time of conception (McGrath et al., 2014). Early evidence linking a de novo mutation to schizophrenia arose from studies of copy number variations (CNVs; Kirov et al., 2012). The rate of CNV de novo mutation was found to be substantially higher in schizophrenic patients. These mutations are largely associated with genes encoding members of the NMDARs and neuronal cytoskeleton-associated protein, both of which are involved in regulation of synaptic plasticity (Kirov et al., 2012).

11.11 ANIMAL MODELS OF SCHIZOPHRENIA BASED ON GLUTAMATE

The improper regulation of glutamate neurotransmission has been the main focus in schizophrenia, and the main concern is the NMDA subtype of glutamate receptors, which, when blocked by antagonists, aggravate pre-existing indications for schizophrenia in individuals, and result in behavioral abnormalities in healthy individuals similar to those seen in schizophrenia (Bondi et al., 2012). Ketamine and PCP impair spatial alternation strategy in rodents (Wiley et al., 2003), and cognition in monkeys (Taffe et al., 2002). These drugs also induce stereotypy, aggression, and abnormal sensory gating (Moghaddam and Jackson, 2003). Chronic administration of these drugs gives better results and impairment in working memory is seen (Jentsch and Roth, 1999). The symptoms produced by these antagonists are reversible by atypical antipsychotic drugs (Swerdlow and Geyer, 1998). These antagonists produce glutamate efflux in the prefrontal cortex (Moghaddam et al., 1997). Some postulate that enhanced glutaminergic tone is only compensatory (Gluck et al., 2002). Behavioral changes due to NMDAR antagonists have also been considered because of the action of excessive glutamate on AMPA receptors due to blockage of NMDARs (Takahata and Moghaddam, 2003). Thus, if the release of the glutamate is decreased, behavioral changes can be reversed (Moghaddam and Adams, 1998). Excessive glutamate causes excitotoxicity; thus, treatment with various antioxidants such as *N*-acetyl cystine – a precursor of glutathione – has been revealed to recover schizophrenic indications in patients (Berk et al., 2008). These are also seen to improve NMDAR function (Kohr et al., 1994). Chronic administration of amphetamine induces hallucinatory–paranoid psychosis, which very closely mimics schizophrenic symptoms (Tomiyama, 1990). These symptoms last for months, even after the discontinuation of drugs (Machiyama et al., 1970). Chronic amphetamine administration in monkeys reduces social interaction and enhances aggression (Machiyama et al., 1970), and produces a repetitive head movement and stereotypy in rats. The acute action of amphetamine is considered to be related to increased dopamine release, but the long-term effects are caused by defects in glutamate (Wolf and Jeziorski, 1993). Amphetamine-induced intoxication is reported as a result of overstimulation of AMPA receptors and NMDARs (Cador et al., 1999). Repeated dosing of amphetamine also results in an increased level of glutamate in the prefrontal cortex (Stephans and Yamamoto, 1995). D-Lysergic acid diethylamine (LSD) acts primarily by stimulating 5-hydroxytryptamine 2A (5-HT_{2A}) receptors (Maurel-Remy et al., 1995). These compounds ultimately enhance glutamatergic neurotransmission in the stimulation of NMDARs in the cerebral cortex (Aghajanian and Marek, 2000). Animal models based on glutamate mechanisms have improved our understanding of schizophrenia, and have furthered new drug research (Table 11.1).

TABLE 11.1
Animal Models Based on Glutaminergic Antagonism

Animal Model	Principal	Reference
Antagonism of MK-801-induced locomotion and falling in mice	MK-801, a non-competitive NMDA antagonist induces a characteristic stereotype in mice marked by locomotion and falling behavior	(Verma and Kulkarni, 1992)
Phencyclidine (PCP)-induced social withdrawal	PCP is a NMDA receptor antagonist and induces a characteristic stereotype behavior, hyperactivity, and social withdrawal in rats	(Sams-Dodd, 1996)
Ketamine-induced stereotypy in mice	Ketamine is classified as a NMDA receptor antagonist and induces a characteristic stereotype behavior	(Parle et. al., 2013)

11.12 INTERACTION BETWEEN GLUTAMATE AND DOPAMINE IN SCHIZOPHRENIA

There is a vast literature describing the involvement of dopamine in schizophrenia and several drugs are available on the market that act on this mechanism to control the disease. As dopamine has been found to be not solely responsible for schizophrenia, as drugs do not reverse all the symptoms, the involvement of glutaminergic system is undergoing research (Fuente-Sandoval et al., 2011). Furthermore, interactions between the two neurotransmitter systems is also being studied as glutamate is widely distributed and is related to almost all synaptic pathways (David et al., 2005; Fuente-Sandoval et al., 2011; Kulagina et al., 2001; West et al., 2003). The main interactions being studied are in the frontal cortical tracts, which are involved in motor and cognitive functions. It has been shown that disturbed cortical glutamatergic afferents affect the tonic, with a consequent disturbance of phasic dopamine secretion and atypical reactions to trivial stimuli (Grace, 2016). There is significant evidence for interactions of both systems within the dorsal striatum and nucleus accumbens. These are the highly interactive brain regions and are maximally involved in channeling the information flow (West et al., 2003). Alteration in the glutaminergic system leads to dysregulation of dopaminergic system in a final common pathway (Bartolomeis et al., 2005) that is accountable for the positive indications of schizophrenia.

11.13 POSSIBLE GLUTAMATE-BASED TREATMENT

Dopamine D2 receptor blockers have been used as treatment tools for schizophrenia in the last half century, highlighting the hyperdopamine basis for the disease (Seeman, 2009). However, unbearable side effects lead to discontinuation of treatment. The most refractory symptoms to the current treatments are negative and cognitive. None of the current therapies is effective against these symptoms. Atypical antipsychotics have fewer side effects and similar efficacy to conventional antipsychotic agents. However, there are still concerns over several side effects, namely weight gain, increases in glucose measurements, and lipid metabolism (Lieberman et al., 2005). D2 receptor occupancy may be essential but not adequate for the management of psychosis. This fact is further supported by studies which have shown that non-responders to treatment with these medications show high D2 blockage but no improvement in condition (Pilowsky et al., 1993). Therefore, one can conclude that another mechanism is involved in the disease and drugs acting by other mechanisms are required. The search for new and improved antipsychotic agents that act by alternate mechanisms are under research (Carlsson et al., 1999). PCP is a psychogenic substance that acts by blocking NMDARs and is the basis for studies on the involvement of glutamate transmission in several neuropsychiatric disorders, including schizophrenia (Goff and Coyle, 2001; Javitt and Zukin, 1991; Jentsch and Roth, 1999; Moghaddam and Javitt, 2012). Research carried out over the last two decades has uncovered significant objectives for novel drug advancement on the basis of presynaptic, postsynaptic, and glial mechanisms (Javitt, 2007; Moghaddam and Javitt, 2012). Ketamine-induced NMDAR blockage is basically responsible for the development of negative symptoms and has no relation to positive symptoms (Stone et al., 2008). Thus, it can be concluded that glutamate and dopamine are related to different symptoms. Cognitive decline is one of the symptoms that usually appears in early phase of schizophrenia (Bilder et al., 2006); the reason for this decline is abnormality in either NMDAR or glutamatergic transmission. When injected in rats, an NMDAR antagonist resulted in neurotoxic changes in the cortical regions of the brain, which can be related to a decrease in gray matter in the brains of schizophrenic patients (Olney and Farber, 1995). This toxicity was blocked by AMPA receptor antagonists, and it was hypothesized that this toxicity was due to excess cortical glutamate release (Olney and Farber, 1995).

11.14 RECENT DEVELOPMENTS AND CHALLENGES

Although existing pharmacologic agents have a remarkable effect on the management of schizophrenia, immense challenges still exist (Kane and Malhotra, 2003). Second-generation antipsychotic medications present many important advantages over conventional antipsychotics. Negative symptoms and cognitive dysfunction, and decrements in psychosocial and vocational functioning often continue to endure in spite of the best-available treatment (Geddes et al., 2000; Leucht et al., 1999). Clozapine is the only drug with proven efficacy in patients who are poor or partial responders (Chakos et al., 2001; Kane et al., 1988; Wahlbeck et al., 2000). It also has the advantages of a reduction in suicidal behavior in schizophrenic patients – another important reason for more extensive use; however, patients require regular blood monitoring, as its overuse can cause problems (Meltzer et al., 2003). New-generation antipsychotics have proven to be more effective on negative and cognitive symptoms, but the results are inconsistent (Kane and Malhotra, 2003; Keefe et al., 1999), leaving substantial room for further improvement. Traditionally, dopamine antagonists are used for the treatment of schizophrenia, but there has been considerable interest in the use of antagonists and agonists to modify dopaminergic function (Lawler et al., 1999). Compounds have also been developed which act at receptors other than dopamine D2. However, none of these compounds has been found to be clinically effective as an antipsychotic agent, but issues of appropriate dose finding remain potential concerns (Bristow et al., 1997). In addition, these antipsychotics have their own substantial side effects, specifically weight gain and metabolic syndrome. NMDAR antagonists produce schizophrenia-like symptoms, leading to development of animal models and testing of certain agents in humans (Javitt and Zukin, 1991). Therefore, the hypothesis is that drugs which facilitate or enhance NMDA function might have some therapeutic potential in schizophrenia. Glycine – an agonist – and

D-cycloserine – a partial agonist at the glycine regulatory site – have been employed in clinical trials with some success, particularly on negative symptoms (Javitt et al., 1994; Javitt and Zukin, 1991).

The summary of schizophrenia research provided above generates a number of concerns about the status of future drug development for schizophrenia. The following points may lead to more rapid identification and development of new therapeutic agents for schizophrenia.

- The proper identification of molecular targets may help in drug development, with high predictive validity for therapeutic benefit.
- The creation and validation of a proper model system for specific pathologic domains is needed.
- NMDA channel opening requires both glutamate and glycine binding, and is modulated by numerous substances. Therefore, allosteric sites on the NMDAR complex, particularly the glycine binding site, have been targeted in the development of pharmacotherapy in schizophrenia (Javitt, 2004).
- Metabotropic glutamate receptors (mGluR), currently in preclinical development, are being explored as potential treatments in schizophrenia (Moghaddam, 2004).
- Another glutamatergic approach to drug development in schizophrenia is the development of compounds that stimulate AMPA and kainate glutamate receptors (Black, 2005).

11.14.1 Drugs Targeting Glutamate Under Development in Schizophrenia

Glycine transporter inhibitors are drugs under development; several companies have reported data from clinical trials. GlyT1 receptor inhibitor RG1678, as reported by Roche, was successful in handling negative symptoms in a phase II clinical trial, but they have not reported further data on this compound (Pinard et al., 2010). The GlyT1 inhibitor R231857 showed reversal of scopolamine-induced cognitive impairments in healthy volunteers, as reported by Johnson and Johnson (Liem-Moolenaar et al., 2010). However, respiratory depression related to glycine transporter inhibitors was reported, but it is not clear whether this is a side effect of all compounds of this class or not (Pinard et al., 2010). Another major issue linked to the glycine transporter is that they show effects only in a limited dose range, enhancing NMDAR current and LTP within this dose range only, and causing a reduction in current at higher doses (Martina et al., 2004). Beyond this, allosteric potentiators of metabotropic glutamate receptors are also under development by Lilly and Merck (Johnson et al., 2004). Allosteric potentiators have been reported to have several theoretical advantages. For their action, they require endogenous ligands and thus they have a lower tendency for side effects. LY379268 is drug under development by Lilly that comes under this category and acts on the mGlu2 receptor by blocking ketamine-induced glutamate, dopamine, and histamine release in the prefrontal cortex, and norepinephrine in the hippocampus (Fell et al., 2010). It has also been reported to be neuroprotective in early psychosis by inhibiting MK-801-induced retro splenial cortex damage (Carter et al., 2004). Merck introduced 3-Cyano-N-(1,3-diphenyl-1H-pyrazol-5-yl)benzamide (CDPPB), a compound in this category that acts by increasing the affinity of glutamate for mGlu5 receptors, resulting in augmentation of NMDAR activity (Lindsley *et al.*, 2004). The drug, which acts on excess glutamate release, is of benefit in the early phase of illness, as an elevated level of glutamate in the cortex seems to be more prominent during this phase (Théberge et al., 2002). The administration of drugs such as mGlu2/3 agonists that act by reducing glutamate release in the cortex, or AMPA antagonists that act by inhibiting its effects, may useful if given in the early phase of the disease, that is, in treating patients during their first episodes schizophrenia. However, much research is required to explore these hypotheses as trials to date have generally beleaguered individuals with chronic schizophrenia.

11.15 CONCLUSION

The first decade of the twenty-first century remained focused on the dopamine hypothesis of schizophrenia. Most antipsychotic drugs being used today act by blocking D2 and D4 dopaminergic receptors. These drugs are associated with many side effects as D2 receptor blockers are associated with extrapyramidal side effects and D4 receptor blockers result in weight gain and agranulocytosis for which regular blood monitoring is needed. Further, these drugs show no effects on negative and cognitive impairments. Cognitive impairments are recorded in the early phase of the schizophrenia and remain almost unrecorded. These symptoms are mainly due to NMDAR abnormality leading to imperfection in glutamate neurotransmission. Being a major neurotransmitter, glutamate also creates abnormalities in other neurotransmitters, leading to increased severity of the disease. Glutamatergic models of schizophrenia came into the picture much later than the dopamine hypothesis. They are based on the fact that CNS stimulants, which block NMDA-type glutamate receptors, have been shown to induce schizophrenic symptoms in animals, as well humans. NMDAR antagonists are being used extensively as a tool for testing upcoming preclinical treatment strategies. Many drugs which act to normalize the glutamate level in the brain are in clinical trials. More recent studies have been advanced to genetic and postmortem approaches, which have been accompanied by the conclusion that proteins other than NMDARs are involved in the pathophysiology of schizophrenia, for example AMPA receptors and glycine transporter inhibitors. Thus, with increased biological understanding of the disease, these findings enhance the targets and scope for novel drug development. However, clinical drug development is still slow and must be advanced to a stage at which estimated treatment based on the glutamate hypothesis can be tested. In this chapter, the authors have discussed the scope for the development of drugs based on the glutamate hypothesis.

REFERENCES

Aghajanian GK, Marek GJ. Serotonin model of schizophrenia: emerging role of glutamate mechanisms. Brain Res Brain Res Rev. 2000; 31:302–11.

Akbarian S, Sucher NJ, Bradley D, Tafazzoli A, Trinh D, Hetrick WP, et al. Selective alterations in gene expression for NMDA receptor subunits in prefrontal cortex of schizophrenics. J Neurosci. 1996; 16:19–30.

Alexander SP, Benson HE, Faccenda E, Pawson AJ, Sharman JL, Spedding M, et al. The concise guide to pharmacology 2013/14: G protein-coupled receptors. Br J Pharmacol. 2013;170:1459–581.

Allen NC, Bagade S, McQueen MB, Ioannidis JP, Kavvoura FK, Khoury MJ, et al. Systematic meta-analyses and field synopsis of genetic association studies in schizophrenia: the SzGene database. Nat Genet. 2008; 40:827–34.

Anderson MC, Bunce JG, Barbas H. Prefrontal-hippocampal pathways underlying inhibitory control over memory. Neurobiol Learn Mem. 2016; 134:145–61.

Anson LC, Chen PE, Wyllie DJ, Colquhoun D, Schoepfer R. Identification of amino acid residues of the NR2A subunit that control glutamate potency in recombinant NR1/NR2A NMDA receptors. J Neurosci. 1998;18:581–9.

Arnth-Jensen N, Jabaudon D, Scanziani M. Cooperation between independent hippocampal synapses is controlled by glutamate uptake. Nat Neurosci. 2002; 5:325–31.

Arya A, Sindhwani G, Kadian R. Neurotransmitter and brain parts involved in schizophrenia. Asian J Pharm Clin Res. 2018; 11:4–11.

Bartolomeis A, Fiore G, Iasevoli F. Dopamine-glutamate interaction and antipsychotics mechanism of action: implication for new pharmacological strategies in psychosis. Curr Pharm Des. 2005; 11:3561–94.

Bassett AS, Chow EWC. 22q11 deletion syndrome: a genetic subtype of schizophrenia. Biol Psychiatry. 1999, 46:882–91.

Bassett AS, Chow EW, Weksberg R. Chromosomal abnormalities and schizophrenia. Am J Med Genet. 2000; 97:45–51.

Bassett AS, Hodgkinson K, Chow EWC, Correia S, Scutt LE, Weksberg R. 22q11 deletion syndrome in adults with schizophrenia. Am J Med Genet (Neuropsychiatr Genet). 1998; 81:328–37.

Battaglia G, Monn JA, Schoepp DD. In vivo inhibition of veratridine-evoked release of striatal excitatory amino acids by the group II metabotropic glutamate receptor agonist LY354740 in rats. Neurosci Lett. 1997; 229:161–4.

Bechtholt-Gompf AJ, Walther HV, Adams MA, Carlezon WA, Ongur D, Cohen BM. Blockade of astrocytic glutamate uptake in rats induces signs of anhedonia and impaired spatial memory. Neuropsychopharmacol. 2010; 35:2049–59.

Begni S, Moraschi S, Bignotti S, Fumagalli F, Rillosi L, Perez J, Gennarelli M. Association between the G1001C polymorphism in the *GRIN1* gene promoter region and schizophrenia. Biol Psychiatry. 2003; 53:617–19.

Behrens M, Ali S, Dao D, Lucero J, Shekhtman G, Quick K, et al. Ketamine-induced loss of phenotype of fast-spiking interneurons is mediated by NADPH-oxidase. Science. 2007; 318:1645–7.

Behrens MM, Ali SS, Dugan LL. Interleukin-6 mediates the increase in NADPH-oxidase in the ketamine model of schizophrenia. J Neurosci. 2008; 28:13957–66.

Beneyto M, Kristiansen LV, Oni-Orisan A, McCullumsmith RE, Meador-Woodruff JH. Abnormal glutamate receptor expression in the medial temporal lobe in schizophrenia and mood disorders. Neuropsychopharmacol. 2007; 32:1888–902.

Bergles DE, Jahr CE. Glial contribution to glutamate uptake at Schaffer collateral-commissural synapses in the hippocampus. J Neurosci. 1998;18:7709–16.

Berk M, Copolov D, Dean O, Lu K, Jeavons S, Schapkaitz I, et al. *N*-acetyl cysteine as a glutathione precursor for schizophrenia – a double-blind, randomized, placebo-controlled trial. Biol Psychiatry. 2008; 64:361–8.

Bilder RM, Reiter G, Bates J, Lencz T, Szeszko P, Goldman RS, et al. Cognitive development in schizophrenia: follow-back from the first episode. J Clin Exp Neuropsychol. 2006; 28:270–82.

Black MD. Therapeutic potential of positive AMPA modulators and their relationship to AMPA receptor subunits. A review of preclinical data. Psychopharmacology (Berl) 2005; 179:154–63.

Bondi C, Matthews M, Moghaddam B. Glutamatergic animal models of schizophrenia. Curr Pharm Des. 2012; 18:1593–604.

Boteva K, Lieberman J. Reconsidering the classification of schizophrenia and manic depressive illness – a critical analysis and new conceptual model. World J Biol Psychiatry.2003; 4:81–92.

Braff D, Stone C, Callaway E, Geyer M, Glick I, Bali L. Prestimulus effects on human startle reflex in normals and schizophrenics. Psychophysiology. 1978; 15:339–43.

Breese CR, Freedman R, Leonard SS. Glutamate receptor subtype expression in human postmortem brain tissue from schizophrenics and alcohol abusers. Brain Res. 1995; 674:82–90.

Breier A, Malhotra AK, Pinals DA, Weisenfeld NI, Pickar D. Association of ketamine-induced psychosis with focal activation of the prefrontal cortex in healthy volunteers. Am J Psychiatry. 1997; 154:805–11.

Bridges RJ, Natale NR, Patel SA. System xc- cystine/glutamate antiporter: An update on molecular pharmacology and roles within the CNS. Br J Pharmacol. 2012; 165:20–34.

Bristow LJ, Kramer MS, Kulagowski J, Patel S; Ragan CI, Scabrook GR. Schizophrenia and L-745,870, a novel dopamine-D4 receptor antagonist. Trends Pharmacol Sci. 1997; 18:186–8.

Burnashev N, Monyer H, Seeburg PH, Sakmann B. Divalent ion permeability of AMPA receptor channels is dominated by the edited form of a single subunit. Neuron. 1992; 8:189–98.

Busato GF, Pilowsky LS, Kerwin RW, Costa DC, Ell PJ, Verhoeff NPLG. Dopamine D_2 receptor blockade *in vivo* with the novel antipsychotics risperidone and remoxipride- an^{123}I-IBZM single photon emission tomography (SPET) study. Psychopharmacology. 1995; 117:55–61.

Buzsaki G, Geisler C, Henze DA, Wang XJ. Interneuron diversity series: circuit complexity and axon wiring economy of cortical interneurons. Trends Neurosci. 2004; 27:186–93.

Cador M, Bjijou Y, Cailhol S, Stinus L. D-amphetamine-induced behavioral sensitization: implication of a glutamatergic medial prefrontal cortex–ventral tegmental area innervation. Neuroscience. 1999; 94:705–21.

Carlsson A, Waters N, Carlsson ML. Neurotransmitter interactions in schizophrenia – therapeutic implications. Biol Psychiatry. 1999; 46:1388–95.

Carter K, Dickerson J, Schoepp DD, Reilly M, Herring N, Williams J, et al. The mGlu2/3 receptor agonist LY379268 injected into cortex or thalamus decreases neuronal injury in retrosplenial cortex produced by NMDA receptor antagonist MK-801: possible implications for psychosis. Neuropharmacol. 2004; 47:1135–45.

Chakos M, Lieberman J, Hoffman E, Bradford D, Sheitman B. Effectiveness of second- generation antipsychotics in patients with treatment-resistant schizophrenia: a review and meta-analysis of randomized trials. Am J Psychiatry. 2001; 158:518–26.

Choi DW. Glutamate receptors and the induction of excitotoxic neuronal death. Prog Brain Res. 1994; 100:47–51.

Cincotta M, Beart PM, Summers RJ, Lodge D. Bidirectional transport of NMDAR and ionophore in the vagus nerve. Eur J Pharmacol. 1989;160:167–71.

Cooper AJL, Jeitner TM. Central role of glutamate metabolism in the maintenance of nitrogen homeostasis in normal and hyperammonemic brain. Biomolecules. 2016; 6:16.

Cortese BM, Luan Phan K. The role of glutamate in anxiety and related disorders. CNS Spectrums. 2005;10:820–30.

Coyle JT. The glutamatergic dysfunction hypothesis for schizophrenia. Harv Rev Psychiatry. 1996; 3:241–53.

Coyle JT. NMDA receptor and schizophrenia: a brief history. Schizophr Bull. 2012; 38:920–6.

Craddock N, Owen M. Chromosomal aberrations and bipolar affective disorder. Br J Psychiatry. 1994;164:507–12.

Danbolt NC. Glutamate uptake. Prog Neurobiol. 2001; 65:1–105.

Danysz W, Parsons AC. Glycine and *N*-methyl-D-aspartate receptors: physiological significance and possible therapeutic applications. Pharmacol Rev. 1998; 50:597–664.

David HN, Ansseau M, Abraini JH. Dopamine-glutamate reciprocal modulation of release and motor responses in the rat caudate-putamen and nucleus accumbens of "intact" animals. Brain Res Brain Res Rev. 2005; 50:336–60.

DeLisi LE, Friedrich U, Wahlstrom J, Boccio-Smith A, Forsman A, Eklund K, Crow TJ. Schizophrenia and sex chromosome anomalies. Schizophr Bull. 1994; 20:495–505.

Demczuk S, Aurias A. DiGeorge syndrome and related syndromes associated with 22q11.2 deletions: a review. Ann Genet. 1995; 38:59–76.

Demjaha A, Murray RM, McGuire PK, Kapur S, Howes OD. Dopamine synthesis capacity in patients with treatment-resistant schizophrenia. Am J Psychiatry. 2012; 169:1203–10.

Doble A. The role of excitotoxicity in neurodegenerative disease: implications for therapy. Pharmacol Ther. 1999; 81:163–221.

Dodd S, Dean O, Copolov DL, Malhi GS, Berk M. *N*-acetylcysteine for antioxidant therapy: pharmacology and clinical utility. Expert Opin Biol Ther. 2008; 8:1955–62.

Eastwood SL, McDonald B, Burnet PW, Beckwith JP, Kerwin RW, Harrison PJ. Decreased expression of mRNAs encoding non-NMDA glutamate receptors GluR1 and GluR2 in medial temporal lobe neurons in schizophrenia. Brain Res Mol Brain Res. 1995; 29:211–23.

Eulenburg V, Gomeza J. Neurotransmitter transporters expressed in glial cells as regulators of synapse function. Brain Res Rev. 2010; 63:103–11.

Eva MT, Michael JT. Glutamate and psychiatric disorders. Adv Psychiatr Treat. 2002; 8:189–97.

Fatemi SH, Folsom TD. The neurodevelopmental hypothesis of schizophrenia, revisited. Schizophr Bull. 2009; 35:528–48.

Fell MJ, Katner JS, Johnson BG, Khilevich A, Schkeryantz JM, Perry KW, et al. Activation of metabotropic glutamate (mGlu)2 receptors suppresses histamine release in limbic brain regions following acute ketamine challenge. Neuropharmacol. 2010; 58:632–9.

Fisher MT, Fisher JL. Contributions of different kainate receptor subunits to the properties of recombinant homomeric and heteromeric receptors. Neuroscience. 2014; 278:70–80.

Fuente-Sandoval C, Leon-Ortiz P, Favila R, Stephano S, Mamo D, Bermudez JR, et al. Higher levels of glutamate in the associative-striatum of subjects with prodromal symptoms of schizophrenia and patients with first-episode psychosis. Neuropsychopharmacol. 2011; 36:1781–91.

Gaisler-Salomon I, Miller GM, Chuhma N, Lee S, Zhang H, Ghoddoussi F, et al. Glutaminase-deficient mice display hippocampal hypoactivity, insensitivity to pro-psychotic drugs and potentiated latent inhibition: relevance to schizophrenia. Neuropsychopharmacol. 2009; 34:2305–22.

Geddes J, Freemantle N. Harrison P, Bebbington P. Atypical antipsychotics in the treatment of schizophrenia: systematic overview and metaregression analysis. BMJ. 2000; 321:1371–6.

Ghaemi SN. Paradigms of psychiatry: eclecticism and its discontents Curr Opin Psychiatry. 2006; 19:619–24.

Gluck MR, Thomas RG, Davis KL, Haroutunian V. Implications for altered glutamate and GABA metabolism in the dorsolateral prefrontal cortex of aged schizophrenic patients. Am J Psychiatry. 2002; 159:1165–73.

Goff DC, Coyle JT. The emerging role of glutamate in the pathophysiology and treatment of schizophrenia. Am J Psychiatry. 2001; 158:1367–77.

Grace AA. Dysregulation of the dopamine system in the pathophysiology of schizophrenia and depression. Nat Rev Neurosci. 2016; 17:524–32.

Grant GB, Dowling JE. A glutamate-activated chloride current in cone-driven ON bipolar cells of the white perch retina. J Neurosci. 1995; 15:3852–62.

Grewer C, Gameiro A, Zhang Z, Tao Z, Braams S, Rauen T. Glutamate forward and reverse transport: from molecular mechanism to transporter-mediated release after ischemia. IUBMB Life. 2008; 60:609–19.

Gysin R, Kraftsik R, Sandell J, Bovet P, Chappuis C, Conus P, et al. Impaired glutathione synthesis in schizophrenia: convergent genetic and functional evidence. Proc Natl Acad Sci U S A. 2007; 104:16621–6.

Hassel B, Dingledine R. Glutamate and GABA receptors. In: Brady ST, Siegel GJ, Albers RW, Price DL, editors. Basic Neurochemistry. 8th ed. Amsterdam: Academic Press; 2011.

Hertz L, Chen Y. Integration between glycolysis and glutamate-glutamine cycle flux may explain preferential glycolytic increase during brain activation, requiring glutamate. Front Integr Neurosci. 2017; 11:18.

Hirai H, Kirsch J, Laube B, Betz H, Kuhse J. The glycine binding site of the *N*-methyl-D-aspartate receptor subunit NR1: identification of novel determinants of co-agonist potentiation in the extracellular M3–4 loop region. Proc Natl Acad Sci U S A. 1996; 93:6031–6.

Hollmann M, Hartley M, Heinemann S. Ca^{2+} permeability of KA-AMPA-gated glutamate receptor channels depends on subunit composition. Science. 1991; 252:851–3.

Hollmann M, O'Shea-Greenfield A, Rogers SW, Heinemann S. Cloning by functional expression of a member of the glutamate receptor family. Nature. 1989; 342:643–8.

Homayoun H, Jackson ME, Moghaddam B. Activation of metabotropic glutamate 2/3 receptors reverses the effects of NMDA receptor hypofunction on prefrontal cortex unit activity in awake rats. J Neurophysiol. 2005; 93:1989–2001.

Homayoun H, Moghaddam B. NMDA receptor hypofunction produces opposite effects on prefrontal cortex interneurons and pyramidal neurons. J Neurosci. 2007; 27:11496–500.

Howes OD, Kapur S. The dopamine hypothesis of schizophrenia: version III – the final common pathway. Schizophr Bull. 2009; 35:549–62.

Hu W, MacDonald ML, Elswick DE, Sweet RA. The glutamate hypothesis of schizophrenia: evidence from human brain tissue studies. Ann N Y Acad Sci. 2015; 1338:38–57.

Humphries C, Mortimer A, Hirsch S, de Belleroche J. NMDA receptor mRNA correlation with antemortem cognitive impairment in schizophrenia. Neuroreport. 1996; 7:2051–5.

Jabaudon D, Shimamoto K, Yasuda-Kamatani Y, Scanziani M, Gähwiler BH, Gerber U. Inhibition of uptake unmasks rapid extracellular turnover of glutamate of nonvesicular origin. Proc Natl Acad Sci U S A. 1999; 96:8733–8.

Javitt, DC. Negative schizophrenic symptomatology and the PCP (phencyclidine) model of schizophrenia Hillside. J Clin Psychiatry. 1987; 9:12–35.

Javitt DC. Glutamate as a therapeutic target in psychiatric disorders. Mol Psychiatry. 2004; 9:984–97.

Javitt DC. Glutamate and schizophrenia: phencyclidine, *N*-methyl-D-aspartate receptors, and dopamine-glutamate interactions. Int Rev Neurobiol. 2007; 78:69–108.

Javitt DC, Zukin SR. Recent advances in the phencyclidine model of schizophrenia. Am J Psychiatry. 1991; 148:1301–8.

Javitt DC, Zylberman I, Zukin SR, Heresco-Levy U, Lindenmayer JP. Amelioration of negative symptoms in schizophrenia by glycine. Am J Psychiatry. 1994; 151:234–6.

Jentsch JD, Roth HR. The neuropsychopharmacology of phencyclidine: From NMDA receptor hypofunction to the dopamine hypothesis of schizophrenia. Neuropsychopharmacol. 1999; 20:201–25.

Johnson MP, Nisenbaum ES, Large TH, Emkey R, Baez M, Kingston AE. Allosteric modulators of metabotropic glutamate receptors: Lessons learnt from mGlu1, mGlu2 and mGlu5 potentiators and antagonists. Biochem Soc Trans. 2004; 32:881–7.

Johnson RL, Koerner JF. Excitatory amino acid neurotransmission. J Med Chem. 1988; 31:2057–66.

Kane JM, Honigfeld G, Singer J, Meltzer H. Clozapine for the treatment resistant schizophrenic: a double-blind comparison versus chlorpromazine/ benztropine. Arch Gen Psychiatry. 1988; 45:789–96.

Kane JM, Malhotra A. The future of pharmacotherapy for schizophrenia. World Psychiatry. 2003;2:81–6.

Karlsson RM, Tanaka K, Heilig M, Holmes A. Loss of glial glutamate and aspartate transporter (excitatory amino acid transporter 1) causes locomotor hyperactivity and exaggerated responses to psychotomimetics: rescue by haloperidol and metabotropic glutamate 2/3 agonist. Biol Psychiatry. 2008; 64:810–14.

Keefe RS, Silva SG, Perkins DO, Lieberman JA. The effects of atypical antipsychotic drugs on neurocognitive impairment in schizophrenia: a review and meta-analysis. Schizophr Bull. 1999; 25:201–22.

Kew JN, Kemp JA. Ionotropic and metabotropic glutamate receptor structure and pharmacology. Psychopharmacology (Berl). 2005; 179:4–29.

Kim E, Howes OD, Turkheimer FE, Kim BH, Jeong JM, Kim JW, et al. The relationship between antipsychotic D2 occupancy and change in frontal metabolism and working memory: a dual [(11)C] raclopride and [(18) F]FDG imaging study with aripiprazole. Psychopharmacology (Berl) 2013; 227:221–9.

Kim JS, Kornhuber HH, Schmid-Burgk W, Holzmuller B. Low cerebrospinal fluid glutamate in schizophrenic patients and a new hypothesis on schizophrenia. Neurosci Lett. 1980; 20:379–82.

Kirov G, Pocklington AJ, Holmans P, Ivanov D, Ikeda M, Ruderfer D, et al. De novo CNV analysis implicates specific abnormalities of postsynaptic signalling complexes in the pathogenesis of schizophrenia. Mol Psychiatry. 2012; 17:142–53.

Kohler M, Burnashev N, Sakmann B, Seeburg PH. Determinants of Ca^{2+} permeability in both TM1 and TM2 of high affinity kainate receptor channels: diversity by RNA editing. Neuron. 1993;10:491–500.

Kohr G, Eckardt S, Luddens H, Monyer H, Seeburg PH. NMDA receptor channels: subunit-specific potentiation by reducing agents. Neuron 1994; 12:1031–40.

Kojima S, Nakamura T, Nidaira, T, Nakamura, K, Ooashi, N, Ito E, et al. Optical detection of synaptically induced glutamate transport in hippocampal slices. J. Neurosci. 1999; 19:2580–8.

Kristiansen LV, Huerta I, Beneyto M, Meador-Woodruff JH. NMDA receptors and schizophrenia. Curr Opin Pharmacol. 2007; 7:48–55.

Krystal JH, Abi-Saab W, Perry E, D'Souza DC, Liu N, Gueorguieva R, et al. Preliminary evidence of attenuation of the disruptive effects of the NMDA glutamate receptor antagonist, ketamine, on working memory by pretreatment with the group II metabotropic glutamate receptor agonist, LY354740, in healthy human subjects Psychopharmacol (Berl). 2005; 179:303–9.

Kulagina NV, Zigmond MJ, Michael AC. Glutamate regulates the spontaneous and evoked release of dopamine in the rat striatum. Neuroscience. 2001; 102:121–8.

Kuryatov A, Laube B, Betz H and Kuhse J. Mutational analysis of the glycine-binding site of the NMDA receptor: structural similarity with bacterial amino acid-binding proteins. Neuron 1994; 12:1291–300.

Labrie V, Roder JC. The involvement of the NMDA receptor D-serine/glycine site in the pathophysiology and treatment of schizophrenia. Neurosci Biobehav Rev. 2010; 34:351–72.

Lajtha AB, Maker HS, Clarke DD. Metabolism and transport of carbohydrates and amino acids. In: Siegel DJ, Albers RW, Agranoff BW, Katzman R, editors. Basic Neurochemistry. 3rd ed. Boston, MA: Little, Brown and Company; 1981.

Lawler CP, Prioleau C, Lewis MM, Mak C, Jiang D, Schetz JA, et al. Interactions of the novel antipsychotic aripiprazole (OPC-14597) with dopamine and serotonin receptor subtypes. Neuropsychopharmacology 1999; 20:612–27.

Leucht S. Pitschel-Walz G. Abraham D, Kissling W. Efficacy and extrapyramidal side-effects of the new antipsychotics olanzapine, quetiapine, risperidone, and sertindole compared to conventional antipsychotics and placebo. A meta-analysis of randomized controlled trials. Schizophr Res. 1999; 35:51–68.

Levkovitz Y, Levi U, Braw Y, Cohen H. Minocycline, a second-generation tetracycline, as a neuroprotective agent in an animal model of schizophrenia. Brain Res. 2007; 1154:154–62.

Lieberman JA, Stroup TS, McEvoy JP, Swartz MS, Rosenheck RA, Perkins DO, et al. Effectiveness of antipsychotic drugs in patients with chronic schizophrenia. N Engl J Med. 2005; 353:1209–23.

Liem-Moolenaar M, Zoethout RW, de Boer P, Schmidt M, de Kam ML, Cohen AF, et al. The effects of a glycine reuptake inhibitor R231857 on the central nervous system and on scopolamine-induced impairments in cognitive and psychomotor function in healthy subjects. J Psychopharmacol. 2010; 24:1681–7.

Lodish H, Berk A, Zipursky SL, Matsudaira P, Baltimore D, Darnell J. Neurotransmitters, synapses, and impulse transmission. In: Molecular Cell Biology. 4th ed. New York: WH Freeman; 2000.

Lorrain DS, Baccei CS, Bristow LJ, Anderson JJ, Varney MA. Effects of ketamine and *N*-methyl-D-aspartate on glutamate and dopamine release in the rat prefrontal cortex: modulation by a group II selective metabotropic glutamate receptor agonist LY379268. Neuroscience. 2003; 117:697–706.

Luby ED, Gottlieb JS, Cohen BD, Rosenbaum G, Domino EF. Model psychoses and schizophrenia Am J Psychiatry 1962; 119:61–7.

Lujan R, Shigemoto R, Lopez-Bendito G. Glutamate and GABA receptor signalling in the developing brain. Neuroscience. 2005; 130:567–80.

Lynch DR, Anegawa NJ, Verdoorn T, Pritchett DB. *N*-methyl-D-aspartate receptors: different subunit requirements for binding of glutamate antagonists, glycine antagonists, and channel-blocking agents. Mol Pharmacol. 1994; 45:540–5.

McCullumsmith RE, O'Donovan SM, Drummond JB, Benesh FS, Simmons M, Roberts R, et al. Cell-specific abnormalities of glutamate transporters in schizophrenia: sick astrocytes and compensating relay neurons. Mol Psychiatry. 2016; 21:823–30.

McGrath JJ, Petersen L, Agerbo E, Mors O, Mortensen PB, Pedersen CB. A comprehensive assessment of parental age and psychiatric disorders. JAMA psychiatry. 2014; 71:301–9.

Machiyama Y, Utena H, Kikuchi M. Behavioral disorders in Japanese monkeys produced by the long-term administration of methamphetamine. Proc Japan Acad. 1970; 46:738–43.

Malarkey EB, Parpura V. Mechanisms of glutamate release from astrocytes. Neurochem Int. 2008; 52:142–54.

Mallick HN. Understanding safety of glutamate in food and brain. Indian J Physiol Pharmacol. 2007; 51 :216.

Marsman A, van den Heuvel MP, Klomp DW, Kahn RS, Luijten PR, Hulshoff Pol HE. Glutamate in schizophrenia: a focused review and meta-analysis of 1H-MRS Studies. Schizophr Bull. 2013; 39:120–9.

Martina M, Gorfinkel Y, Halman S, Lowe JA, Periyalwar P, Schmidt CJ, et al. Glycine transporter type 1 blockade changes NMDA receptor-mediated responses and LTP in hippocampal CA1 pyramidal cells by altering extracellular glycine levels. J Physiol. 2004; 557:489–500.

Maurel-Remy S, Bervoets K, Millan MJ. Blockade of phencyclidine-induced hyperlocomotion by clozapine and MDL 100, 907 in rats reflects antagonism of 5-HT2A receptors. Eur J Pharmacol. 1995; 280:R9–11.

Mayer ML, Westbrook GL, Guthrie PB. Voltage-dependent block by Mg of NMDA responses in spinal cord neurons. Nature. 1984; 309:261–3.

Meador-Woodruff JH, Healy DJ. Glutamate receptor expression in schizophrenic brain. Brain Res Brain Res Rev. 2000; 31:288–94.

Meltzer HY. Alphs L. Green, AI, Altamura AC, Anand R, Bertoldi A, et al. Clozapine treatment for suicidality in schizophrenia: International Suicide Prevention Trial (InterSePt) Arch Gen Psychiatry. 2003;60:82–91.

Menniti FS, Lindsley CW, Conn PJ, Pandit J, Zagouras P, Volkmann RA. Allosteric modulators for the treatment of schizophrenia: targeting glutamatergic networks. Curr Top Med Chem. 2013;13:26–54.

Milind P, Kadian R, Kaura S. Non-behavioural model of psychosis. Int Res J Pharm. 2013; 4:89–95.

Moghaddam B. Bringing order to the glutamate chaos in schizophrenia. Neuron. 2003; 40:881–4.

Moghaddam B. Targeting metabotropic glutamate receptors for treatment of the cognitive symptoms of schizophrenia. Psychopharmacology (Berl). 2004; 174:39–44.

Moghaddam B, Adams B. Reversal of phencyclidine effects by a group II metabotropic glutamate receptor agonist in rats. Science 1998; 281:1349–52.

Moghaddam B, Adams B, Verma A, Daly D. Activation of glutamatergic neurotransmission by ketamine: a novel step in the pathway from NMDA receptor blockade to dopaminergic and cognitive disruptions associated with the prefrontal cortex. J Neurosci. 1997; 17:2921–7.

Moghaddam B, Jackson ME. Glutamatergic animal models of schizophrenia. Ann N Y Acad Sci. 2003; 1003:131–7.

Moghaddam B, Javitt D. From revolution to evolution: the glutamate hypothesis of schizophrenia and its implication for treatment. Neuropsychopharmacology. 2012; 37:4–15.

Monte A, de Souza G, McIntyre R, Soczynska J, dos Santos J, Cordeiro R, et al. Prevention and reversal of ketamine-induced schizophrenia related behavior by minocycline in mice: possible involvement of antioxidant and nitrergic pathways. J Psychopharmacol. 2013; 27:1032–43.

Mori H, Mishina M. Structure and function of the NMDA receptor channel. Neuropharmacol. 1995; 34:1219–37.

Morris RGM, Moser EI, Riedel G, Martin SJ, Sandin J, Day M, et al. Elements of a neurobiological theory of the hippocampus: the role of activity-dependent synaptic plasticity in memory. Philos Trans R Soc Lond B Biol Sci. 2003; 358:773–86.

Mortimer A, Singh P, Shepherd C, Puthiryackal J. Clozapine for treatment-resistant schizophrenia: National Institute of Clinical Excellence (NICE) Guidance in the Real World. Clin Schizophr Relat Psychoses. 2010; 4:49–55.

Murphy BP, Chung Y-C, Park T-W, McGorry PD. Pharmacological treatment of primary negative symptoms in schizophrenia: a systematic review. Schizophr Res. 2006; 88:5–25.

Murphy KC, Jones RG, Griffiths E, Thompson PW, Owen MJ. Chromosome 22q11 deletions. An under-recognized cause of idiopathic learning disability. Br J Psychiatry. 1998; 172:180–3.

Murphy KC, Jones AL, Owen MJ. High rates of schizophrenia in velo-cardio-facial syndrome. Arch Gen Psychiatry. 1999; 56:940–5.

Nakanishi S, Nakajima Y, Masu M, Ueda Y, Nakahara K, Watanabe D, et al. Glutamate receptors: brain function and signal transduction. Brain Res Rev. 1998; 26:230–5.

Negrete-Diaz JV, Sihra TS, Flores G, Rodriguez-Moreno A. Non-canonical mechanisms of presynaptic kainate receptors controlling glutamate release. Front Mol Neurosci. 2018; 11:128.

Niciu MJ, Kelmendi B, Sanacora G. Overview of glutamatergic neurotransmission in the nervous system. Pharmacol Biochem Behav. 2012; 100:656–64.

Niswender CM, Conn PJ. Metabotropic glutamate receptors: physiology, pharmacology, and disease. Annu Rev Pharmacol Toxicol. 2010; 50:295–322.

O'Donovan SM, Sullivan CR, McCullumsmith RE. The role of glutamate transporters in the pathophysiology of neuropsychiatric disorders. NPJ Schizophr. 2017; 3:32.
Olney JW, Farber NB. Glutamate receptor dysfunction and schizophrenia. Arch Gen Psychiatry. 1995; 52:998–1007.
Otis TS, Kavanaugh, MP, Jahr, CE. Postsynaptic glutamate transport at the climbing fiber-Purkinje cell synapse. Science 1997; 277:1515–18.
Parle M, Kadian R, Kaura S. Non-behavioural model of psychosis. Int Res J Pharm. 2013; 4:89–95.
Parsons CG, Stoffler A, Danysz W. Memantine: a NMDA receptor antagonist that improves memory by restoration of homeostasis in the glutamatergic system-too little activation is bad, too much is even worse. Neuropharmacol. 2007; 53:699–723.
Peng L, Hertz L, Huang R, Sonnewald U, Petersen SB, Westergaard N, et al. Utilization of glutamine and of TCA cycle constituents as precursors for transmitter glutamate and GABA. Dev Neurosci. 1993; 15:367–77.
Pilowsky LS, Costa DC, Ell PJ, Murray RM, Verhoeff NP, Kerwin RW. Antipsychotic medication, D2 dopamine receptor blockade and clinical response: a 123IIBZMSPET (single photon emission tomography) study. Psychol Med. 1993; 23:791–7.
Pin JP, Duvoisin R. The metabotropic glutamate receptors: structure and functions, Neuropharmacology. 1995; 34:1–26.
Pinard E, Alanine A, Alberati D, Bender M, Borroni E, Bourdeaux P, et al. Selective GlyT1 inhibitors: discovery of [4-(3-fluoro-5-trifluoromethylpyridin-2-yl)piperazin-1-yl][5-methanesulfonyl-2-((S)-2,2,2-trifluoro-1-methylethoxy)phenyl]methanone (RG1678), a promising novel medicine to treat schizophrenia. J Medicinal Chem. 2010; 53:4603–14.
Pleasure D. Diagnostic and pathogenic significance of glutamate receptor autoantibodies. Arch Neurol. 2008; 65:589–92
Puyal J, Ginet V, Clarke PG. Multiple interacting cell death mechanisms in the mediation of excitotoxicity and ischemic brain damage: a challenge for neuroprotection. Prog Neurobiol. 2013; 105:24–48
Pycock CJ, Kerwin RW, Carter CJ. Effect of lesion of cortical dopamine terminals on subcortical dopamine receptors in rats. Nature. 1980; 286:74–6.
Qin S, Zhao X, Pan Y, Liu J, Feng G, Fu J, et al. An association study of the *N*-methyl-D-aspartate receptor NR1 subunit gene (*GRIN1*) and NR2B subunit gene (*GRIN2B*) in schizophrenia with universal DNA microarray. Eur J Hum Genet. 2005; 13:807–14.
Rao A, Craig AM. Activity regulates the synaptic localization of the NMDA receptor in hippocampal neurons. Neuron. 1997;19:801–11.
Regehr WG, Carey MR, and Best AR. Activity-dependent regulation of synapses by retrograde messengers. Neuron. 2009; 63:154–70.
Rodriguez-Paz JM, Anantharam Y, Treistman SN. Block of the *N*-methyl-D-aspartate receptor by phencyclidine-like drugs is influenced by alternative splicing. Neurosci Lett. 1995; 190:147–50.
Rodriguez-Santiago B, Brunet A, Sobrino B, Serra-Juhe C, Flores R, Armengol L, et al. Association of common copy number variants at the glutathione S-transferase genes and rare novel genomic changes with schizophrenia. Mol Psychiatry. 2010; 15:1023–33.
Rotaru D, Lewis D, Gonzalez-Burgos G. The role of glutamatergic inputs onto parvalbumin-positive interneurons: relevance for schizophrenia. Rev Neurosci. 2012; 23:97–109.
Rothman DL, Behar KL, Hyder F, Shulman RG. In vivo NMR studies of the glutamate neurotransmitter flux and neuroenergetics: implications for brain function. Annu Rev Physiol. 2003; 65:401–27.
Rubio M, Drummond JB, Meador-Woodruff JH. Glutamate receptor abnormalities in schizophrenia: implications for innovative treatments. Biomol Ther (Seoul). 2012; 20:1–18.
Sams-Dodd F. Pencyclidine-induced stereotyped behavior and social isolation in rats: a possible animal model of schizophrenia. Behav Pharmacol. 1996; 7:3–23.
Schoepp DD, Johnson BG, Wright RA, Salhoff CR, Mayne NG, Wu S, et al. LY354740 is a potent and highly selective group II metabotropic glutamate receptor agonist in cells expressing human glutamate receptors. Neuropharmacology. 1997; 36:1–11.
Schousboe A. Transport and metabolism of glutamate and GABA in neurons are glial cells. Int Rev Neurobiol. 1981; 22:1–45.
Seeman P. Glutamate and dopamine components in schizophrenia. J Psychiatry Neurosci. 2009; 34:143–9.
Senter RK, Ghoshal A, Walker AG, Xiang Z, Niswender CM, Conn PJ. The role of mGlureceptors in hippocampal plasticity deficits in neurological and psychiatric disorders: implications for allosteric modulators as novel therapeutic strategies. Curr Neuropharmacol. 2016; 14:455–473.
Smith QR. Transport of glutamate and other amino acids at the blood-brain barrier. J Nutr. 2000; 130(4S):1016S–22S.
Smith QR, Stoll J. Blood-brain barrier amino acid transport. In: Pardridge WM, editor, 1st ed. Introduction to the blood-brain barrier: methodology, biology, and pathology. Cambridge: Cambridge University Press; 1998.
Sommer B, Kohler M, Sprengel R, Seeburg PH. RNA editing in brain controls a determinant of ion flow in glutamate-gated channels. Cell. 1991; 67:11–19.
Sorce S, Schiavone S, Tucci P, Colaianna M, Jaquet V, Cuomo V, et al. The NADPH oxidase NOX2 controls glutamate release: a novel mechanism involved in psychosis-like ketamine responses. J Neurosci. 2010; 30:11317–25.
Stefani MR, Moghaddam B. Systemic and prefrontal cortical NMDA receptor blockade differentially affect discrimination learning and set-shift ability in rats. Behav Neurosci. 2005; 119:420–8.
Stephans SE, Yamamoto BY. Effect of repeated methamphetamine administrations on dopamine and glutamate efflux in rat prefrontal cortex. Brain Res. 1995; 700:99–106.
Stone JM. Glutamatergic antipsychotic drugs: a new dawn in the treatment of schizophrenia? Ther Adv Psychopharmacol. 2011; 1:5–18.
Stone JM, Erlandsson K, Arstad E, Squassante L, Teneggi V, Rodrigo A, et al. Relationship between ketamine-induced psychotic symptoms and NMDA receptor occupancy: a [(123)I]CNS-1261 SPET study. Psychopharmacol (Berl). 2008; 197:401–8.
Su ZZ, Leszczyniecka M, Kang DC, Sarkar D, Chao W, Volsky DJ, et al. Insights into glutamate transport regulation in human astrocytes: cloning of the promoter for excitatory amino acid transporter 2 (EAAT2). Proc Natl Acad Sci. 2003; 100(4): 1955–60.
Swerdlow NR, Geyer MA. Using an animal model of deficient sensorimotor gating to study the pathophysiology and new treatments of schizophrenia. Schizophr Bull. 1998; 24:285–301.
Taffe MA, Davis SA, Gutierrez T, Gold LH. Ketamine impairs multiple cognitive domains in rhesus monkeys. Drug Alcohol Depend. 2002; 68:175–87.

Takahata R, Moghaddam B. Activation of glutamate neurotransmission in the prefrontal cortex sustains the motoric and dopaminergic effects of phencyclidine. Neuropsychopharmacol. 2003; 28:1117–24.

Tamaru Y, Nomura S, Mizuno N, Shigemoto R. Distribution of metabotropic glutamate receptor mGluR3 in the mouse CNS: differential location relative to pre- and postsynaptic sites. Neuroscience. 2001; 106:481–503.

Tamminga CA. Schizophrenia and glutamate. Crit Rev Neurobiol. 1998; 12:21–36.

Tamminga CA, Medoff DR. The biology of schizophrenia. Dialogues Clin Neurosci. 2000; 2:339–48.

Théberge J, Bartha R, Drost DJ, Menon RS, Malla A, Takhar J, et al. Glutamate and glutamine measured with 4.0 T proton MRS in never-treated patients with schizophrenia and healthy volunteers. Am J Psychiatry. 2002;159:1944–6.

Tomiyama G. Chronic schizophrenia-like states in methamphetamine psychosis. Jpn J Psychiatry Neurol. 1990; 44:531–9.

Turk J, Hill P. Behavioral phenotypes in dysmorphic syndromes. Clin Dysmorphol. 1995;4:105–15.

Tzingounis AV, Wadiche JI. Glutamate transporters: confining runaway excitation by shaping synaptic transmission. Nat Rev Neurosci. 2007; 8:935–47.

Verma A, Kulkarni SK. Modulation of MK-801 response by dopaminergic agents in mice. Psychopharmacol. 1992; 107:431–6.

Vollenweider FX, Leenders KL, Oye I, Hell D, Angst J. Differential psychopathology and patterns of cerebral glucose utilisation produced by (S)- and (R)-ketamine in healthy volunteers using positron emission tomography (PET) Eur Neuropsychopharmacol. 1997; 7:25–38.

Wahlbeck K, Cheine M, Essali MA. Clozapine versus typical neuroleptic medication for schizophrenia. Cochrane Database Syst Rev. 2000: CD000059.

Wang F, Smith NA, Xu Q, Goldman S, Peng W, Huang JH, et al. Photolysis of caged Ca^{2+} but not receptor-mediated Ca^{2+} signaling triggers astrocytic glutamate release. J Neurosci. 2013; 33:17404–11.

Watkins JC, Jane DE. The glutamate story. Br J Pharmacol. 2006;147:S100–8.

West AR, Floresco SB, Charara A, Rosenkranz JA, Grace AA. Electrophysiological interactions between striatal glutamatergic and dopaminergic systems. Ann NY Acad Sci. 2003; 1003:53–74

Wiley JL, Buhler KG, Lavecchia KL, Johnson KM. Pharmacological challenge reveals long term effects of perinatal phencyclidine on delayed spatial alternation in rats. Prog Neuropsychopharmacol Biol Psychiatry. 2003; 27:867–73.

Williams K. Interactions of polyamines with ion channels. Biochem J. 1997; 325:289.

Wolf ME, Jeziorski M. Co-administration of MK-801 with amphetamine, cocaine or morphine prevents rather than transiently masks the development of behavioral sensitization. Brain Res. 1993; 613:291–4.

Yelamanchi SD, Jayaram S, Thomas JK, Gundimeda S, Khan AA, Singhal A, et al. A pathway map of glutamate metabolism. J Cell Commun Signal. 2016; 10:69–75.

Zhorov BS, Tikhonov DB. Potassium, sodium, calcium and glutamate-gated channels: Pore architecture and ligand action. J Neurochem. 2004;88:782–99.

Zhou Y, Danbolt NC. Glutamate as a neurotransmitter in the healthy brain. J Neural Transm. (Vienna) 2014; 121:799–817.

12 Myasthenia Gravis

Molecular Pathogenesis and Therapeutic Advances

Faisal Ashraf Bhat and Ashif Iqubal
Department of Pharmacology, School of Pharmaceutical Education and Research, Jamia Hamdard, New Delhi, India

Sadat Shafi
Pharmaceutical Medicine, Department of Pharmacology, School of Pharmaceutical Education and Research, Jamia Hamdard, New Delhi, India

Bakr Ahmed Hameed
Department of Pharmaceutics, School of Pharmaceutical Education and Research, Jamia Hamdard, New Delhi, India

Faizana Fayaz
Department of Pharmaceutical Chemistry, Delhi Institute of Pharmaceutical Sciences and Research, Department of Pharmaceutical Chemistry, New Delhi, India

Ozaifa Kareem
Department of Pharmaceutical Sciences, University of Kashmir, Hazratbal, Srinagar, India

Faheem Hyder Pottoo
Department of Pharmacology, College of Clinical Pharmacy, Imam Abdulrahman Bin Faisal University, Dammam, Saudi Arabia.

CONTENTS

DOI: 10.1201/9780429265198-14

12.1 INTRODUCTION

Myasthenia gravis (MG), a rare autoimmune disorder that tends to affect the neuromuscular junction (NMJ), leads to an antibody-mediated blockade of neurotransmitters and thus causes weakness and fatigue in the skeletal muscles (Mi et al., 2019). It is believed that the causative factors of MG are the autoantibodies that eventually bind at the NMJ of postsynaptic neurons (Guastafierro et al., 2020). These autoantibodies are formed in opposition to the different types of receptors such as muscle-specific kinase (MuSK), acetylcholine receptor (AChR), and lipoprotein related protein 4 (LRP4), which are recognized as particular and sensitive factors, as well as diagnostic markers, as shown in Figure 12.1. However, in 80% of patients, autoantibodies are formed against AChR, while in the remainder, these autoantibodies are formed against MuSK and LRP4. Findings have also shown that in 10–15% of patients, MuSK and LPR4 autoantibodies against MuSK or LPR4 are not detected because of the lower sensitivity of the available bioassay (Gilhus and Verschuuren, 2015a). These autoantibodies alter end-plate potential, resulting in obstruction of effective neuromuscular transmission (Lindstrom et al., 1976). The global prevalence of MG is estimated to be 150–200 per million population and the annual incidence of MG is reported to be 8–10 cases/million/person-years (Carr et al., 2010). Considering the risk factors, both, genetic and environmental attributes play a prominent role in the pathogenesis of MG. Alteration in the genes such as *HLA*, *CTLA4*, *TNF*, *CD86*, *TNFSF138*, *AKAP12*, and *IFNG* is a major contributor to the development, as well as progression, of MG (Hong et al., 2018). Apart from dysfunctional genes, alterations in microRNAs (miRNA) such as miR-150-5p and 21-5p are also reported to be contributor allies in disease progression (Hong et al., 2018). However, at present, the exact role of environmental factors is unclear, but studies have reported that environmental pollutant-induced damage of the thymus could be a possible mechanism (Cavalcante et al., 2010). Treatment of MG mostly depends on pathogenic antibodies involved like acetylcholinesterase inhibitors, immunosuppressant drugs, corticosteroids, intravenous immunoglobulins (IVIg), plasma therapy, and thymectomy, as shown in Figure 12.1 (Gilhus et al., 2019). However, with advances in the understanding of the pathogenesis of MG, immunotherapy plays a significant role in the management and treatment of this autoimmune disease (Gilhus et al., 2019). In 2017, eculizumab – a monoclonal antibody – was approved by the U.S. Food and Drugs Administration (FDA) for the treatment of generalized MG in patients diagnosed with anti-AChR antibody-positive MG, and many such drugs are in various stages of clinical trials (Gilhus and Verschuuren, 2015a). Therefore, in this chapter, we focus on the molecular pathogenesis of MG, and discuss in detail the various available therapeutic modalities, recent advancements in treatment, challenges, and investigational drugs in different stages of clinical trials.

12.2 CLASSIFICATION OF MYASTHENIA GRAVIS

12.2.1 Osserman's Classification, Based on Disease Severity

12.2.1.1 Generalized Myasthenia Gravis

MG involving several body muscles is termed generalized MG. Generalization hinders neck, bulbar, and proximal limb movements (Utsugisawa et al., 2019; Gilhus and Verschuuren,

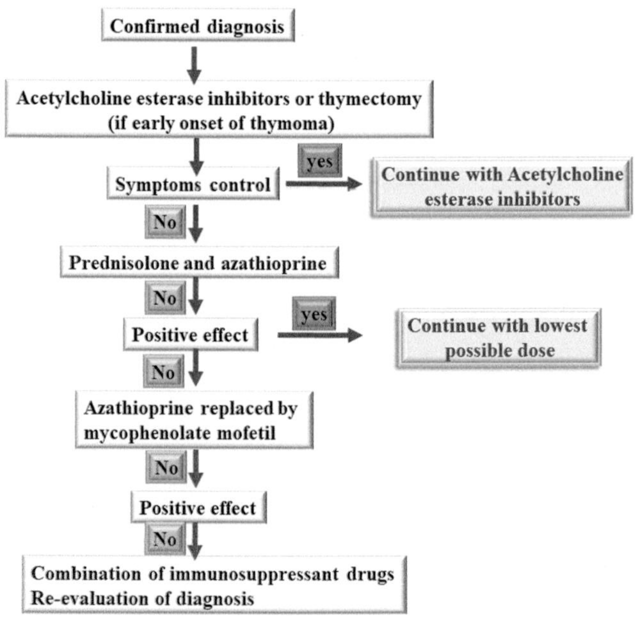

FIGURE 12.1 The treatment algorithm for chronic myasthenia gravis. Acetylcholinesterase inhibitors are used as the first-line drug; if there is no clinical benefit, prednisolone or azathioprine is used. If a benefit is seen with acetylcholinesterase inhibitors, it is continued; if the benefit is seen with prednisolone or azathioprine, it is continued at the lowest possible dose. However, if no benefit is observed then therapy is either switched to other immunosuppressant drugs alone or in combination, otherwise the diagnosis is repeated.

2015b; Akaishi et al., 2014). There is also difficulty in chewing, swallowing, and in neck extensor movements (Utsugisawa et al., 2019). A serious weakness in respiratory muscles is considered to be the most severe aspect of MG (Vinge et al., 2019), and when breathing is affected so severely that mechanical ventilation is needed, the situation is considered to be a 'myasthenic crisis' (Oh et al., 2019).

12.2.1.2 Ocular Myasthenia

Ocular MG (OMG) is a one type of MG where the weakness is limited to the ocular muscles (Bever Jr et al., 1983). The initial signs of OMG are diplopia and ptosis, which are reported in more than 50% of patients with MG patients, 50–80% of whom may develop generalized MG (Bever Jr et al., 1983). Patients with OMG have noticeable AChR antibodies, although MuSK antibodies are rarely found (Bever Jr et al., 1983).

12.2.1.3 Paraneoplastic Myasthenia Gravis

Paraneoplastic MG is also termed thymoma-associated MG. In this disease, healthy cells are attacked by immune cells and is supplemented by neoplasm, mostly gynecological and lung tumors (Kirzinger et al., 2017). Paraneoplastic MG produces a characteristic non-limb symptom at its onset, featured by symptoms related to the eyes, neck, bulbar, and respiratory system (Simonsen et al., 2018; Skeie and Romi, 2008). Autoantibodies against AChR, titin, and ryanodine receptors (RyRs) have been reported to be correlated with disease progression (Marx et al., 2013; Kirzinger et al., 2017).

12.2.1.4 Neonatal Myasthenia Gravis

Neonatal or transient MG occurs due to vertical transmittance of pathogenic autoantibodies against AChR from mother to fetus, and is reported in 15–20% of babies born to mothers with MG (Papazian, 1992). The disease signs and symptoms commonly present a few hours after birth, but may also be hindered for 24 hours or longer, and needs continual checking by the physician (Vinge et al., 2019). Facial weakness and pathetic cry are characteristic markers in the majority of affected babies, followed by difficulty in feeding (Lee et al., 2017). Additionally, respiratory weakness may lead to a blockage of the airways and causes cyanosis (Thanvi and Lo, 2004).

12.2.2 Osserman's Classification, Based on Causative Agents

12.2.2.1 Muscle-Specific Kinase Antibody Myasthenia Gravis

MuSK, a protein functionally linked to AChR and regulator of AChR functions is located in the postsynaptic membrane. Serum MuSK antibodies are detected in approximately 1–4% of patients with MG (Borges and Richman, 2020). However, owing to the presence of new sensitive diagnostic tests, more cases will be documented. MuSK-associated MG predominantly affects bulbar and cranial muscles, with few cases presented with diplopia and ptosis (Guptill et al., 2011). In patients with MuSK MG, characteristic symptoms include bulbar weakness, and facial, tongue, and pharyngeal weakness, frequently linked to neck and respiratory involvement (Guptill et al., 2011).

12.2.2.2 Lipoprotein-Related Protein 4-Associated Myasthenia Gravis

LRP4 is primarily expressed in the postsynaptic membrane (Zhang et al., 2012). It functions as an activator of MuSK and receptor for nerve-derived agrin (Zhang et al., 2012). It is also essential in upholding AChR function (Higuchi et al., 2011). In patients suffering from LRP4 MG, the presence of atrophic thymus and hyperplasia has been described (Zisimopoulou et al., 2013).

12.2.2.3 Drug-Induced Myasthenia Gravis

Several therapeutic agents have been known to alter neuromuscular transmission, including antibiotics (macrolides and aminoglycosides), cardiovascular agents, and antiepileptics, (Shelton et al., 2000). Some patients exposed to penicillamine show minor signs of MG with increased AChR antibody level and electrophysiological irregularities of MG (Russell and Lindstrom, 1978). Interferon-α2b, which is mostly used as a medication for cancer and chronic hepatitis, might also contribute to the induction of severe MG (Batocchi et al. 1995). However, the underlying mechanism is unknown

TABLE 12.1
Clinical Classification of MG According to the Myasthenia Gravis Foundation of America (Gravis, 2000)

Class	Characteristics
I	Weakness in ocular muscles
	Weakness in closing the eyes
	The rest of the muscular strength is normal
II	Weakness in other muscles, apart from eyes
IIa	Weakness in the limb or axial muscles
	Oropharyngeal muscles are least involved
IIb	Weakness in oropharyngeal and respiratory muscles
	Limb and axial muscles are least involved
III	Mild weakness in muscles other than eyes
	Ocular weakness of variable severity
IIIa	Limb and axial muscles are predominantly involved
	Oropharyngeal muscles are least involved
IIIb	Oropharyngeal and respiratory muscles are mostly affected
	Limb and axial muscles are moderately involved
IV	Except for ocular muscles, the rest of the muscles are severely affected
	Muscles of eyes may have a variable degree of weakness
IVa	Weakness in the limb, as well as in axial muscles
	Oropharyngeal muscles least involved
IVb	Respiratory and oropharyngeal muscles are extensively involved
	Limb and axial muscles are least involved
V	Mechanical ventilation is needed

(Batocchi et al., 1995). Calcium channel blockers such as nifedipine and felodipine have been reported to induce MG symptoms, including generalized weakness, dysphagia, and ptosis, in elderly patients (Erm et al., 2001). Beta blockers have also been documented to induce the symptoms of MG in patients with unknown neuromuscular transmission defects (Jonkers et al., 1996).

12.2.3 Clinical Classification of Myasthenia Gravis

The clinical classification of MG is based on the Myasthenia Gravis Foundation of America (MGFA) system that divides MG into five primary classes numbered class I to class IV, and these classes are further categorized into subclasses, as shown in Table 12.1 (Jaretzki et al., 2000). The MGFA system of classification is considered as the ideal approach in recognizing subcategories of patients who possess MG with different clinical manifestations or different disease severities. However, this system of classification is not used to estimate or evaluate the outcome of treatment, but it is used to evaluate the therapy response (Jayam Trouth et al., 2012).

12.3 MOLECULAR MECHANISMS/PATHOPHYSIOLOGY

The chemical communication between a muscle and nerve fiber occurs at the NMJ, where the transmittance of motor nerve impulses to muscle cells occurs (Plomp et al., 1995). Transmission in the neuromuscular system is started by an action potential that results in the acetylcholine (ACh) release at the NMJ, which diffuses through the synapse and binds to AChR on striated muscle causing depolarization of the postsynaptic membrane, leading to muscle contraction. However, in the case of MG, NMJ abnormalities occur as a result of the antibody-mediated process (Lewis et al., 1995). These pathogenic autoantibodies are divided into two categories, one against extracellular or transmembrane autoantigens, and the other against intracellular autoantigens (Figure 12.2).

12.3.1 Neurochemical Aspects of Myasthenia Gravis

It is well established that ACh is released from the NMJ in the form of quanta, which further diffuses across the synaptic clefts, and ultimately binds with AChR in the end-plate membrane of muscle and regulates various physiological functions (Desmedt, 1966). To be more specific, in response to action potential and depolarization, voltage-gated calcium channel opens (located on presynaptic membranes) and stimulates the release of ACh, which further diffuses and reaches postsynaptic neurons, where it binds with the AChR on the end-plate membrane (Zhong et al., 2019). However, in MG, this membrane becomes distorted and the concentration of AChR is reduced in response to the autoantibodies produced. This causes muscle weakness (bulbar, ocular, and proximal limb skeletal muscle), locomotor dysfunction, and fatigue (Zhong

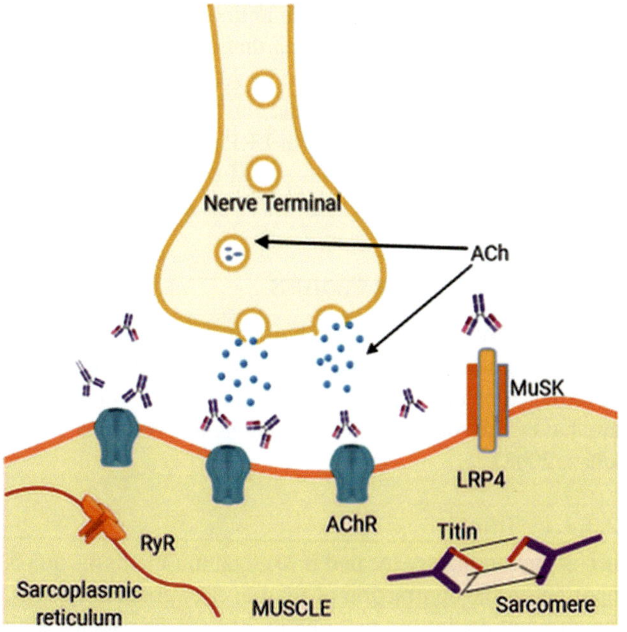

FIGURE 12.2 The involvement of various neurochemicals in the pathogenesis of myasthenia gravis (MG). MG is caused by antibodies produced against the acetylcholine receptor (AChR), muscle-specific kinase (MuSK), lipoprotein related protein 4 (LRP4), or agrin, and all this takes place in the postsynaptic membrane located at the neuromuscular junction. All these attributes cause abnormal neurotransmission, muscle weakness, and fatigue. ACh, acetylcholine; RyR, ryanodine receptor.

et al., 2019). Thus, to restore normal muscle physiology, the first-line treatment is acetylcholinesterase inhibitors, which can prevent the excess breakdown of ACh and thus more ACh will be available for binding at the AChR (Farmakidis et al., 2018). Genetic factors, environmental pollutants, sex hormones, and viral infections are responsible for the pathogenesis of MG (Gilhus et al., 2019). Genetic variation found in the promoter region of *CHRNA1* is responsible for encoding the α-subunit of AChR that stimulates the etiology of MG (Giraud et al., 2007). It is not only genes, but also dysfunctional miRNA (miR-150-5p and miR-21-5p) that have been reported to aggravate the production of autoantibodies against AChR (Punga and Punga, 2018). Aside from these two factors, thymitis, thymoma, or altered thymus physiology is a major contributor to the production of autoantibodies against AChR and worsens MG (Cron et al., 2019). Considering the role of sex hormones, it has been reported that the prevalence of MG is three times higher in females than in males (Marx et al., 2013). The reason for this is hypothesized to be the role of estrogen, as estrogen is responsible for thymus hyperplasia (Gilhus and Verschuuren, 2015a). Further, it has been reported that estrogen modulates the expression of autoimmune regulator (a transcription factor of the thymus) and aggravates the production of autoantibodies for AchR and MG (Dragin et al., 2016).

12.3.2 Immune Pathogenesis of Myasthenia Gravis (Role of the Thymus in Autoantibodies Against Acetylcholine Receptor)

In the previous section, we discussed the role of environmental factors in the pathogenesis of MG. In this section we will discuss the etiology of MG in the context of an alteration in thymus physiology (Singman et al., 2011). A number of studies have shown the involvement of thymic abnormality in MG. Approximately 75% of patients suffer from a dysfunctional thymus, whereas 85% of patients suffer from thymic hyperplasia (presence of germinal centers; Singman et al., 2011). In normal physiological conditions, germinal centers are absent and the thymus is responsible for the development, maturation, and differentiation of T cells (Marx et al., 2013). These T cells help in the differentiation of B cells into antibody-secreting cells that together maintain the homeostatic immune system of the body (Phillips and Vincent, 2016). However, in MG, T helper cells (CD^+ cells) are activated and expended via major histocompatibility complex class II and stimulate B cells to undergo proliferation and the secretion of AChR autoantibodies (Thiruppathi et al., 2012). Thus, it seems persuasive that CD^+ cells are a critical immune player in the pathogenesis of MG (Thiruppathi et al., 2012). Further, studies have also shown the direct association of thymoma with MG. Approximately 10–15% of patients with MG suffer from thymoma (Gilhus et al., 2011). In thymoma, an excess of CD^+ thymocytes is produced from the bone marrow progenitor cells and hence these CD^+ cells stimulate the production of AChR autoantibodies (Vernino and Lennon, 2004).

12.3.3 Extracellular or Transmembrane Proteins

Antibodies targeting transmembrane or extracellular proteins are pathogenic for MG and can alter AChR function at the NMJ, resulting in ionic transport impairment through the muscle membrane, thus decreasing muscle contraction, as shown in Figure 12.2 (Gilhus and Verschuuren, 2015a), by acting directly (anti-AChR antibodies) or indirectly (anti-LRP4 and anti-MuSK antibodies; Gilhus and Verschuuren, 2015a).

12.3.3.1 AChR

Muscle nicotinic AChR is one of the commonly reported autoantigens present in MG and is located at postsynaptic membrane tips (Bindu et al., 2008). Anti-AChR antibodies are reported in about 80% of patients with MG (Luo et al., 2009). These antibodies impair signal transduction by binding to extracellular AChR domains (Gilhus and Verschuuren, 2015a). These antibodies exert a pathogenic effect by two mechanisms:

1. Anti-AChR antibodies exert a pathogenetic effect through complement activation. These antibodies

activate the complement cascade upon binding to antigen, leading to the formation of membrane attack complex and damage to the postsynaptic membrane (Tüzün and Christadoss, 2013).

2. Another important pathogenetic mechanism of these antibodies is the modulation of antigens, internalization of AChR, and annihilation arbitrated by the cross-linking of AChRs via bivalent antibodies (Kordas et al., 2014). This cross-linking of antigens leads to AChR loss at the postsynaptic membrane (Kordas et al., 2014). This loss of AChR is not completely reimbursed by the augmented AChR synthesis that takes place as a result of the degradation of AChR caused by the elevated autoantibody-induced mechanism (Meriggioli and Sanders, 2012). Additionally, some anti-AChR antibodies can directly block the attachment of ACh to AChR and thus inhibit the downstream signaling cascade (Zisimopoulou et al., 2013; Gilhus et al., 2016; Hara et al., 1993).

12.3.3.2 MuSK

MuSK is a single-subunit transmembrane protein managing AChR clustering at the NMJ and maintenance of the postsynaptic membrane (Messéant et al., 2015). MuSK is activated by LRP4–agrin complex-induced phosphorylation, followed by AChR clustering (Rivner et al., 2018). The AChR clustering process involves rapsyn, a protein responsible for the linking of AChR with the cytoskeleton (Li et al., 2016). Anti-MuSK antibodies are only reported in 1–4% of patients with MG (Li et al., 2016). Their pathogenic mechanism is different from AChR antibodies as they are unable to induce complement cascade stimulation and antigenic modulation since they are monovalent antibodies (Koneczny et al., 2017). Anti-MuSK antibodies function by masking MuSK binding sites that permit connections with various binding proteins such as LRP4 and collagen Q; therefore, these activate MuSK (Huijbers et al., 2013). The inactivation of MuSK results in reduced AChR postsynaptic density and impaired alignment in the postsynaptic membrane (Koneczny et al., 2017). Among the patients of MG, the level of MuSK antibody changes as the disease progresses. (Evoli et al., 2018).

12.3.3.3 LRP4

LRP4 is also a transmembrane protein (single-subunit) with a huge extracellular domain containing numerous units of low-density lipoproteins (Shen et al., 2015). LRP4 is mostly concentrated within the NMJ, where it normally binds to agrin released by nerves. As discussed above, this LRP4 agrin complex initiates MuSK activation (Kim et al., 2008). Anti-LRP4 antibodies are reported in 8% of patients with MG diagnosed as positive for anti-AChR antibodies, in 15% of patients with anti-MuSK antibodies, and in 4% of patients with other different neuroimmune diseases (Hong et al., 2017). However, thee pathogenic mechanism of MG needs to be fully established (Hong et al., 2017). The pathogenetic mechanism of anti-LRP4 antibodies was evaluated in LRP4-immunized mice (Ulusoy et al., 2017). These antibodies are reported to disrupt the LRP4–agrin interaction that cause suppression of neuromuscular transmission aided by AChR via inhibition of agrin-induced MuSK activation and clustering of AChR (Shen et al., 2013). The disease form in LRP4-immunized mice is similar to that reported in AChR- and MuSK-immunized mice (Shen et al., 2013).

12.3.4 Intracellular Proteins

Antibodies that work against intracellular proteins are clinically significant as characteristic MG markers, including the presence of thymoma, disease severity, and so on, rather than having a role in disease pathogenicity (Inohara and Nuñez, 2003).

12.3.4.1 Titin

Titin is predominantly located in skeletal muscle tissue and is important in the contraction of muscle (Powers et al., 2014). Anti-titin antibodies are reported in 20–30% of patients with MG who have anti-AChR antibodies, mainly in MG that is of relatively late onset or thymoma (Romi et al., 2017). The existence of such antibodies specifies a more-severe type of MG accompanied with a not-so-severe myopathy (Romi et al., 2017).

12.3.4.2 Ryanodine Receptor

RyRs are preset on the surface of the endoplasmic reticulum and regulate the entry and exit of calcium, thus modulating muscle contraction (Iqubal et al., 2019). Antibodies against RyR are identified in 14% of cases with the onset of MG and 70% of patients with thymoma MG (Mygland et al., 1992). The presence of anti-RyR antibodies is a diagnostic marker of thymoma, indicating severe MG (Mygland et al., 1992).

12.4 ONGOING THERAPEUTIC MANAGEMENT OF MYASTHENIA GRAVIS

Considering the current developments in the field of immunopharmacotherapeutics, the ongoing treatment of MG is mostly based on general immunosuppression, with the lack of a targeted mechanism of action (Farmakidis et al., 2018). The management outlook for MG generally entails the recognition of the disease in its early stage, starting the treatment instantly, and seeking support for bulbar/respiratory function; the patient should be treated according to the symptoms using the pharmacologic enhancement of neuromuscular transmission, and immunotherapy (Farmakidis et al., 2018). Immunomodulation that is either fast- or short-lasting, and immunosuppression that is long-term or thymectomy are general therapeutic management options (Guptill et al., 2016).

12.4.1 Pyridostigmine

The primary action of autoantibodies is to decrease the accumulation of AChR, which is necessary for the efficiency of the synapse is and controlled by impeccable communication between cardinal proteins such as MuSK and LPR4 (Evoli, 2017). Thus, immunotherapy is considered the basic treatment of MG, irrespective of its type, and aims at declining the adverse outcomes of autoantibodies. Pyridostigmine is an inhibitor of acetylcholinesterase and is also antibody-positive. This drug is considered to be the first-line treatment of AChR based on symptoms. It is started at a dose of 60 mg and administered four times daily. If needed, the dose can be increased from 60 mg to 120 mg and administered at intervals of 3 hours (Farmakidis et al., 2018). The onset of action occurs after 30 minutes of administration and the defect lasts for 4 hours (Farmakidis et al., 2018). If patients suffer from sleep disturbance and weakness, delayed-release pyridostigmine at a dose of 180 mg is recommended (Farmakidis et al., 2018). However, the inhibitors of cholinesterase do not offer everlasting relief from symptoms and therefore the majority of the patients are unable to attain proper or long-lasting relief of MG symptoms and escalation of the disease remains unaffected (Singer et al., 2006). Patients suffering from non-progressive, mild, or ocular disease may find this treatment to be sufficient. It is also possible that people suffering from MuSK MG may not find this treatment effective at all. Additionally, when pyridostigmine is used at a high dose it can cause muscle weakness, which happens as a result of AChRs becoming unsensitized (Punga et al., 2006). Additionally, the use of pyridostigmine is associated with abdominal cramps, diarrhea, cramps, and perspiration (Osserman et al., 1958). Further acetylcholinesterase inhibitors are not recommended in myasthenia crisis, as these drugs may cause respiratory depression or a 'cholinergic crisis' (Farmakidis et al., 2018). However, to manage such muscarinic adverse effects, hyoscyamine 0.125 mg, glycopyrrolate 1 mg, or loperamide 2 mg are used orally (Farmakidis et al., 2018).

12.4.2 Prednisone

In accordance with the treatments employed in MG, the use of corticosteroids as a potential treatment is only supported by retrospective series (Benatar et al., 2016). Based on the available evidence, it can be said that among all the patients who were administered prednisone, 70–75% either experienced considerable improvement or there was complete remission (Evoli, 2017; Guptill et al., 2016; Sghirlanzoni et al., 1984). If there comes a stage when cholinesterase inhibitors become unable to efficiently control the symptoms of MG, only then is prednisone usually introduced (Bruce and Kupersmith, 2012; Wolfe et al., 2019). In case of severe and moderate MG, the treatment is generally started by administering high doses (i.e., 0.75–1.0 mg per kg daily for a few months), after which the dose is reduced and that low dose (often taken on alternate days) is then continued for several years (Bruce and Kupersmith, 2012). However, if the patient suffers from milder or predominantly ocular MG, high doses at the initial stage of treatment are not required. It may take 2–3 weeks for a clinical response to be noticeable (Bruce and Kupersmith, 2012). Without treatment, complete remission of MG in the long run is highly uncommon, even though the mortality due to MG has been considerably lowered with the use of oral corticosteroids at increased doses with acceleration and de-acceleration (Nacu et al., 2015). The MGFA post-intervention status of complete and stable remission (CSR) is only achieved by >10% of patients with MG (Nagane et al., 2011; Masuda et al., 2012; Sanders et al., 2016). Therefore, the probability of prolonged treatment should be considered by the treatment strategies, and these should aim at maintaining the patients' quality of life (QoL), which is related to both physical and mental health (Masuda et al., 2012). The international standard for the management of MG puts forward minimal manifestation (MM) status or better as the target for treating MG (Sanders et al., 2016). The Japanese clinical guidelines propose that, in comparison to CSR, 'MM status or better with that of prednisolone [PSL] at the dose of 5 mg per day or lower', is an achievable treatment target and this target manages to produce patient satisfaction equal to that of CSR based on the patients' QoL (Masuda et al., 2012). According to a Japanese a cross-sectional study it has been put forward that a low-dose PSL regimen combined with some other treatment can ensure early accomplishment of the targeted treatment (i.e., 'MM or better 5 mg' in the case of generalized MG) (Imai, 2019).

A new strategy has been suggested that would lead to an increased success rate of minimal manifestation status or better clinical outcome and includes an early fast-acting treatment with a low dose of corticosteroids that are given orally in combination with calcineurin inhibitors. Then, to rectify the remaining symptoms as quickly as possible, plasma exchange (PE), IVIg, plasmapheresis (PP), and methylprednisolone pulse therapy are given intravenously (Menon et al., 2020). However, certain Japanese studies have also pointed out that an ample proportion of patients (40–50%) suffering from MG faced a number of difficulties while bringing the disease under control, even after employing the above-mentioned new treatment (Imai et al. 2019). Such patients who do not acknowledge the sufficient dose and adequate duration of conventional immunotherapies require quick-acting treatment therapy like PE, PP, and IVIg excessively (Sanders et al., 2016).

12.4.3 Azathioprine

Several immunosuppressant drugs are competent in treating MG (Sathasivam, 2008). Azathioprine (AZA) is a purine antimetabolite drug and an immunosuppressant that has the best track record when it comes to the treatment of MG. It works by interfering with the proliferation of B and T cells. The benefit of AZA may be evident after 6–12 months, and it plays a great role in improving weakness in the majority of patients. A prospective randomized study pointed out that when AZA is added to PSL, it creates the possibility of the reducing the dose of PSL required for the maintenance of remission. The study also showed that this strategy may result in fewer treatment failures (Palace et al., 1998). Some other immunosuppressant drugs include ciclosporin, tacrolimus, mycophenolate mofetil (MMF), and methotrexate, and are used as steroid-sparing

TABLE 12.2
Pharmacotherapeutics Currently Used in Myasthenia Gravis (MG): Dosing, Onset of Action, Monitoring, and Use (Guptill et al., 2016)

Treatment	Initial Dosing/ Frequency	Onset of Action	Comments	Side Effects
Pyridostigmine	30–60 mg at 4 hour intervals	15–30 min after oral or intramuscular administration 2–5 min after intravenous administration	First-line treatment of AChR	Abdominal pain nausea, excess salivation, sweating, cough, muscle cramps, and blurred vision
Prednisone	a. 0.75–1.00 mg/kg/day b. 20–40 mg/kg/day	3–6 weeks	First-line treatment for short-term use	Hypertension, insomnia, obesity, diabetes, osteoporosis, visibility issue, and increased appetite
AZA	2–3 mg/kg/day	6–12 months	First-line drug (steroid-sparing)	Pyrexia, muscle cramps, hepatotoxicity, and hypotension
PE	4–6 mg/kg on alternate days	1–5 days	Use in case of exacerbating MG	Well tolerated in patients with MG
IVIg	1–2 g/kg at an interval of 2–5 days	3–10 days	Use in case of exacerbating MG	It is as equally tolerated as PE

AChR = acetylcholine receptor; AZA = azathioprine; PE = plasma exchange; IVIg = intravenous immunoglobulin.

drugs. Generally, such drugs are combined with corticosteroids and then administered to the patient to achieve better control of the disease to reduce or discontinue steroids (Lorenzoni et al., 2020). Due to the favorable side effect profile of MMF, it may be preferred over AZA; however, the other treatment options are usually employed when AZA is not effective enough or is not tolerated by the patient (Niakan et al., 1986). Efficacy and resistance fluctuate, and patient conformity may be substandard. Immunosuppressants can prove to be effective; however, a long-term commitment is required as not many patients can maintain improvement without undergoing the treatment therapy at adequate doses (Niakan et al., 1986). Certain reports show that in patients with MG who are under immunosuppression treatment, there is an increased chance of developing malignancy; however, it has not been proven if there is any long-term risk of malignancy (Jones and Loftus Jr, 2007). Therefore, the lowest dose of immunosuppressant medications essential in keeping the disease under control should be used.

12.4.4 Intravenous Immununoglobulin and Plasma Exchange

The employment of IVIg and PE is done in case of severe aggravation in MG; it also provides the patient with adequate strength to cope with surgery (Gamez et al., 2019). It is also used if there is a need for quick, but short-term, immunotherapy (Yanagihashi et al., 2020). PE primarily equips patients with MG with adequate strength (Gajdos et al., 1997, Batocchi et al., 1995). On the contrary, IVIg is generally employed by patients who are suffering from acute MG. Certain controlled randomized trials have bolstered the use of IVIg in the clinic based on its efficacy being equal to PE (Gajdos et al., 1997), and a placebo-controlled, double-blinded study was conducted in patients with MG with severe weakness (Zinman et al., 2007). Based on randomized controlled trials that have compared IVIg and PP, it has been reported that that there is no noticeable difference between the two treatment options while treating an exacerbation of acute MG (Barth et al., 2011). However, PE may act quickly and is thus the preferred treatment in severe MG. Further details of various pharmacotherapeutics used in MG are shown in Table 12.2 (Zinman et al., 2007).

12.5 THERAPEUTIC ADVANCEMENTS AND FUTURE PERSPECTIVES IN MYASTHENIA GRAVIS

The current standards of therapeutic management in MG primarily comprise general immunosuppression which lacks antigen-specificity and is usually not selective when it comes to immune cellular targets (López-Cano et al., 2003). If the currently used drugs are employed for a lengthy period, their side effects may reduce the patients' QoL; therefore, novel therapies that are targeted in nature (based on the immunopathogenesis of MG) should be developed that can help in improving patients' QoL (Melzer et al., 2016). Keeping in mind the disease heterogeneity and variance in MG, it is necessary that future therapeutic should preferably be customized according to MG subtype (Ahmed et al., 2005). Considering the prevalence and impact of of MG on quality of life of patients, various changes have been made in its management. Recently thymectomy was found to be beneficial among the AChR positive patients and among the refractory MG patients, U.S. FDA approves the use of various pharmacotherapy. (Evoli, 2017). Various rational therapeutic agents are in different phases of clinical trials, and many other options are being explored (Gilhus et al., 2019).

12.5.1 Thymectomy

A surgical approach has been developed to safeguard patients with thymomatous MG against neighboring intrusion and the cancerous potential of the tumor (thymoma). Thymoma is considered to be an absolute indication for thymectomy, but removing the tumor sometimes does not result in improvement or remission. Such patients with thymomatous MG are at higher risk of severe disease (Romi et al., 2006), and there is a requirement for a powerful immunotherapeutic approach that ought to be cautiously reduced after thymectomy (Kas et al., 2001). There is also a role for thymectomy in non-thymomatous MG, and it is considered to be a treatment option in patients suffering from anti-AChR-positive generalized MG (Gamez and Ponseti, 2017). However, the condition is that disease onset should be before the age of 50 years, and some would also contemplate it as a therapeutic advantage in patients lacking anti-AChR antibodies. To completely avert or lower the load of immunosuppressive medications in patients with non-thymomatous MG, or to combat a failed response to these medications, or due to the side effects of medications being intolerable, thymectomy is employed (Sanders et al., 2016, 2018).The very first trial, which was randomized, compared thymectomy with medical management in non-thymomatous patients (Wolfe et al., 2016). The study population included patients with MGFA class II–IV MG, along with patients with AChR-positive MG; patient age ranged from 18 to 65 years, and length of time with the disease was 3–5 years (Wolfe et al., 2016). To further add to its role in non-thymomatous MG, an international randomized trial is currently being conducted (NCT00294658), and it is hoped the results with clear up the problem.

The first randomized trial to compare thymectomy with medical management in patients with non-thymomatous MG was published in 2016 (Wolfe et al., 2016). Before the results of the trial were published, it had already been determined that in all patients with AChR-positive MG – ocular or general – and has an established thymoma, thymectomy is the benchmark of care. It is worth noting that patients suffering from MuSK or LRP4 antibodies did not fulfil the study inclusion criteria. Trial patients underwent thymectomy along with prednisone or only prednisone on a random basis (Wolfe et al., 2016). The quantitative MG (QMG) score and the appropriate prednisone dose over 3 years were set as the primary study endpoints. The results showed that patients randomly placed in the thymectomy group had an increase of about 3 points (2.85) in QMG score (Gajdos et al., 1997). A lower dose of prednisone over 3 years was required by patients in this cohort. No treatment-related complications were observed in either group (Wolfe et al., 2016). The study indicated that thymectomy is a potential therapeutic (surgical) option for patients suffering from AChR-positive generalized MG, irrespective of thymoma status on imaging (Hara et al., 1993).

12.5.2 Robotic-Assisted Thoracoscopic Surgery

Thymectomy has been used as an additional treatment option for patients with MG since Blalock and colleagues reported that a case was handled with trans-sternal thymectomy (TST) in 1939 (Oosterhuis, 1981; Fulghum, 1950). Since then, many other cases have been reported in which thymectomy has been used as an approach, and may be categorized into open surgery that includes sternotomy, thoracotomy, and nominal invasive surgery that includes video-assisted thoracoscopic surgery, and cervicectomy. Although TST is considered to be the best option for thymectomy, nominal invasive surgery for thymectomy has managed to achieve global attention in the last few years (Jaretzki, 2003; Sanders et al., 2018). Available data suggest that robotic-assisted thorascopic surgery (RATS) thymectomy is much safer and more convenient, and the clinical outcomes associated with this technique are promising compared to other surgical procedures (Sanders et al., 2016; Kawaguchi et al., 2018).

The first surgical experience of RATS in MG was described in 2003 by Ashton and colleagues; since then, a series of RATS thymectomy approaches have been described from one-sided to two-sided to subxiphoid (Ashton Jr et al., 2003; Rückert et al., 2011; Kawaguchi et al., 2018; Suda et al., 2015). RATS thymectomy may be considered as a better surgical option in MG only if a proper enucleation with complete thymectomy can be achieved and well executed by skilled surgeons (Liu et al., 2014). It is not a routine recommended approach in patients with thymoma (Liu et al., 2014; Pennathur et al., 2011; Ye et al., 2014). Studies have tended to suggest that RATS thymectomy may be a safe and effective approach in early-stage thymoma, as well as in developed thymoma (large thymoma) (Liu et al., 2014; Pennathur et al., 2011; Ye et al., 2014; Kiessling et al., 2017). To reduce the excessive load of immunosuppressive therapy and the drastic side effects associated with these medications, it can also be an option in non-thymomatous MG (Sanders et al., 2016, 2018). Various approaches to RATS surgical procedures in MG have been employed and reported over time. A huge discussion has been held over decades about the comparative benefits of left-sided surgery versus a right-sided surgical approach (Niakan et al., 1986; Sanders et al., 2018). To date, researchers have proposed left-sided three-trocar RATS thymectomy as a better treatment option in patients with MG, yet comparative studies on the long-term clinical outlook of patients with MG after different RATS approaches to thymectomy are yet to be carried out (Sanders et al., 2018). Future studies should address the comparability of various surgical methods, along with continuing follow-up.

12.5.3 Eculizumab

Eculizumab is a C5 monoclonal antibody whose target is C5 complement protein; it does not allow terminal complement complex C5-9 to form, and has been employed as a treatment in paroxysmal nocturnal hemoglobinuria and atypical hemolytic uremic syndromes (Dubois et al., 2011). Clinical and preclinical studies of this drug evaluated complement-aided harm at the endplate of the NMJ in the neuromuscular system in MG, and proved the rationale for MG treatment (Engel and Arahata, 1987; Morgan et al., 2006). A phase 3 trial (REGAIN; NCT01997229) was commenced in 2017. Based on this trial, eculizumab was approved for MG (Howard et al., 2019; Jones and Loftus Jr, 2007). In this double-blinded, placebo-controlled trial, patients aged 18 years or older suffering from refractory generalized anti-AChR antibody-positive MG were included. Patients with abnormal activities of daily living (ADL) are described according to MG-ADL score of 6 or more, and class II–IV disease as per the MGFA classification (Jaretzki, 2003). The basic definition of refractory MG in the study was MG with no curb on symptoms for 1 year prior to treatment with at least two immunosuppressive treatments, or, at a minimum, one immunosuppressant drug with IVIg, or PE administered quarterly (Mantegazza and Antozzi, 2018). A noticeable difference was seen from baseline was seen in patients treated with eculizumab versus placebo-treated groups at week 26, according to patient-administered QoL, along with ADL surveys, as well as physician-administered scoring systems. The set parameters in the study, such as Myasthenia Gravis Quality of Life 15-item Scale (MG-QoL15) and QMG scores, markedly differed between the groups. Disease aggravation, medication requirements, and hospitalization were reduced in patients who received eculizumab, in comparison to placebo (Dhillon, 2018). Commonly seen adverse effects were severe upper respiratory tract infections, headache, and nasopharyngitis (Dhillon, 2018). The risk of meningococcal diseases with eculizumab is increased by 1000–2000-fold, so there is a need vaccination against *Neisseria meningitidis* before the initiation of any medicine.

12.5.4 Ravulizumab (ALXN1210)

Ravulizumab was developed by involving the substitutions of four novel amino acid residues into the eculizumab heavy chain (McKeage, 2019). Due to such replacements, ravulizumab produces a serum terminal elimination half-life that is much longer and has a many-fold increased time period of pharmacological activity compared to eculizumab (Kulasekararaj et al., 2019). A phase III study of ravulizumab in treating generalized AChR-MG is in progress.

12.5.5 Neonatal Fc Receptor Antibodies

The neonatal Fc receptor (FcRn) plays a significant role in protecting circulating serum IgGs from any lysosomal degradation and directing them into recycling processes (Pyzik et al., 2019). In MG, there is an increased level of IgG isotype autoantibodies that are highly pathogenic; therefore, an agent that reduces these levels would be a very good therapeutic option. That agent is the FcRn monoclonal antibody, which constrains the role of FcRn (Gable and Guptill, 2020). Many FcRn antibodies, like efgartigimod (ARGX-113) (Howard et al., 2019; Huijbers et al., 2013) and rozanolixizumab (UCB7665; Kiessling et al., 2017), are, at present, undergoing various clinical trial stages.

12.5.5.1 Efgartigimod (ARGX-113)

A phase II trial of the FcRn monoclonal antibody efgartigimod was done to evaluate how safe and efficacious this drug is if used as a treatment option in generalized MG, and also to check its pharmacokinetics and pharmacodynamics (Howard et al., 2019). The phase II trial was conducted for 6 weeks. Of all the subjects, efgartigimod was used as a treatment option in 75%, and they showed development in MG-ADL scores and experienced lowered IgG levels and could tolerate the drug well versus the other 25%, who were administered placebo (Imai, 2019). No serious adverse events were reported during the study period (Gilhus et al., 2011; Imai, 2019). The main target of the current phase III trial is to affirm how safe, efficacious, and tolerable efgartigimod is in the treatment of patients with general MG classified as MGFA class II–IVb, and therefore confirming that the lowering of serum IgG levels can be used as a treatment option in MG (Imai, 2019).

12.5.5.2 Rozanolixizumab (UCB7665)

Just as efgartigimod, rozanolixizumab is also a FcRn monoclonal antibody aimed at lowering pathogenic IgGs levels in serum (Gable and Guptill, 2020). A phase I clinical trial (NCT02220153) was conducted in which 41 healthy adult volunteers were enrolled. How safe, efficacious, and tolerable rozanolixizumab is was studied, as were the pharmacokinetics and pharmacodynamics of the drug (Kiessling et al., 2017). A multicentric, random, placebo-controlled double-blind, phase II study has now begun to evaluate how effective, safe, and tolerable rozanolixizumab is patients with MG ranging in intensity from moderate to serious (NCT03052751).

12.5.6 Rituximab

Many prospective trials, retrospective studies, and case series have indicated that MG responds well to rituximab, a $CD20^+$ B cell that depletes monoclonal antibody (Pierpont et al., 2018). However, researchers did not include a large number of patients and their design was neither randomized nor blinded. Additionally, these surveys also had different criteria for including volunteers, their outcome measures were different, and the treatment regimens also differed, and that is why these cannot be compared. In 2017, a re-examination of how efficacious and safe rituximab is in 169 patients suffering from MG based on case reports and case series was done. It was concluded that, among all the patients studied, AChR antibodies were present in 59%, and 34% of patients showed the presence of MuSK antibodies. In the responder subset of rituximab, 26% who were suffering from AChR-MG and 82% suffering from MuSK-MG, there was a lowered antibody load after the treatment. A meta-analysis established MuSK-MG as a moderate disorder and also pointed out that treatment at a younger age can predict response to rituximab. It further proposed that patients were generally able to tolerate rituximab very well (Tandan et al., 2017). Another large-scale, multicenter, blinded review was published in 2017 and compared 24 patients with MuSK-MG receiving rituximab ('rituximab-treated') with 31 patients suffering from MuSK-MG who were receiving some other immunosuppressive medications (control group; Hehir et al., 2017). With Myasthenia Gravis Status and Treatment Intensity (MGSTI) being the fundamental endpoint, the rituximab-treated group managed to achieve a MGSTI score of 2 or higher, which was statistically important compared with the control group. However, the retrospective data collection is a limitation of this review. The primary statistical results of a phase II randomized, double-blinded, placebo-controlled clinical trial (BeatMG; NCT02110706) for AChR-MG were presented at the American Academy of Neurology in 2018. To assess the safety and versatility of rituximab, 52 volunteers that had AChR antibodies were selected on a random basis to either rituximab or placebo (in addition to their baseline immunotherapy regimens). Along with the safety profile of the drug, the fundamental outcome measures were steroid requirement over the 52-week study. Primary endpoints were kept almost the same in this research. The final findings have yet to be published.

12.5.7 Monarsen (EN101)

Monarsen is an antisense oligonucleotide that brings about a reduction in generating acetylcholinesterase and works by intermixing with mRNA that codes for the enzyme (Angelini et al., 2013). In 2007, it was established in an open-label, non-placebo-controlled, phase Ib study that monarsen potentially improved the QMG score in 87% of participants and no significant side effects were seen (Argov et al., 2007). Following this study, a phase II trial was conducted which established that there was an improved QMG score in 31 patients administered the drug (Sussman et al., 2012). However, during the clinical trial, deaths attributed to cerebral hemorrhage and other such events occurred, yet the survey investigators stated that there was no relationship between the drug and these adverse events (Sussman et al., 2012). At this time, further studies are not being planned. The scope of using monarsen in the future is presently unknown.

12.5.8 Autologous Hemopoietic Stem Cell Transplantation

It has been reported that autologous hemopoietic stem cell transplant has great efficacy in severe refractory MG (Melo et al., 2019). One study found that seven non-responders to conventional treatments achieved remission without any drug after an autologous hemopoietic stem cell transplant (Bryant et al., 2016). An instance where a patient suffering from refractory MG, who had failed immunosuppression with rituximab, bortezomib, and sirolimus, exhibited momentous development after an autologous hemopoietic stem cell transplant, without any supplementary treatment (Håkansson et al., 2017).

12.6 RECENT DEVELOPMENTS AND CHALLENGES

As we know, MG is primarily characterized by persistent muscle weakness and fatigue, but it is sometimes accompanied by various comorbid conditions (Gilhus et al., 2015). Hence, it is challenging to differentiate and diagnose the symptoms as due to MG or to coexisting comorbidities. It was recently reported that in China that vaccination against Japanese encephalitis is associated with childhood MG, and findings have been confirmed in preclinical studies (He et al., 2018). However, newer techniques, where single cells are deeply sequenced, can provide detailed information on the molecular and cellular pathology of MG (Cron et al., 2018). Additionally, study of thymic miRNA provides detailed mechanistic insight into the association of thymic abnormality and MG (Cron et al., 2018; Sabre et al., 2018). Challenges not only exist in the diagnosis, but also in the management and treatment of MG. Currently, various treatment paradigms are available for the treatment and management of MG, but none of them offers a complete cure. Thus, a major challenge is to look for therapeutic options that can be curative rather than prophylactic. Additionally, most of the available drugs are immunosuppressants and thus offer serious systemic side effects such as secondary infection, leukopenia, hepatorenal toxicity, response to host infection, thrombocytopenia, and overall worsening of patient QoL (Pedersen et al., 2014). Steroids are also routinely used and their use is reported to cause damage to bones, reducing bone mineral density (Grennan and Wang, 2019). When patients are on steroidal therapy, supplementation with calcium or vitamins should be provided and, whenever needed, bisphosphonates should be used (Allen et al., 2016). Additionally, when patients are on glucocorticoid drugs, increased ocular pressure and tests for cataracts should be done periodically (Fini et al., 2017). However, with the recent advances in targeted-based therapy with various monoclonal antibodies, such problems have resolved significantly.

However, to date, a major challenge that still exists is finding the exact cause of MG so that antigen-specific therapeutics can be used to restore the normal muscular physiology. Newer and advanced diagnostic tools, as well as a biomarkers, would be beneficial in the diagnosis of MG. Currently, electrophysiological and antibody tests are used for diagnosis, but sometimes the results of these tests are false-positive or false-negative; hence, the focus has shifted toward the development and validation of a cell-based antibody assay, as this method offers higher specificity and sensitivity (Jing et al., 2015; Cruz et al., 2015). Apart from a cell-based assay, the development of non-radioactive tests such as an immunostick enzyme-linked immunosorbent assay for the detection of the anti-AchR antibody could be a better approach (Trakas and Tzartos, 2018). Further, advances in imaging techniques could provide an alternate option in identifying structural damage in ocular muscles or the NMJ (Oishi et al., 2007). Currently, the diagnosis of thymoma and thymic hyperplasia in AchR-MG is not available; hence, magnetic resonance imaging (MRI) techniques are being developed where MRI is combined with antibody markers to obtain an improved and precise diagnosis (Priola et al., 2016). Further, MG is a complex disease where the immune system, thymus, and bone marrow play a pivotal role in the etiology; hence, the drug used must be multifactorial in origin with an established safety profile.

It has been reported that more than 10–30% of patients with MG are refractory to conventional pharmacotherapeutics; hence, rituximab is currently considered the drug of choice (Tandan et al., 2017). Rituximab has a better safety profile and is well tolerated by patients who were previously on immunosuppressant therapy (Kanth et al., 2016). Additionally, the result of a phase II clinical trial (NCT62110706) showed the safety and efficacy of this drug in refractory MG. Unlike rituximab, eculizumab did not show significant improvement in the symptoms of MG in a phase III clinical trial (Howard Jr et al., 2017). However, eculizumab is currently approved by the European Medicines Agency and U.S. FDA for refractory MG; however, being expensive, it is difficult for average middle-class patients to afford it (Dhillon, 2018). Additionally, based on recent research, high-dose cyclophosphamide, hematopoietic stem cell transplant, plasma therapy, and IVIg have been found to be effective in such a situation (Gilhus et al., 2019; Sargiannidou et al., 2015). However, it is important to understand that IVIg and PE exhibit a rapid onset of action and hence have a very short half-life (Molina et al., 2006). Therefore, in such patients immunotherapy is used as a maintenance therapeutic regimen, but the cost of treatment subsequently increases (Molina et al., 2006). However, it is also important to understand that plasma therapy has many limitations: it causes hypotension, catheter-related issues, the chances of infection, fluid replacement problems, and allergies (Gilhus et al., 2019). Therefore, because of these aforementioned issues, newer drugs are being investigated that may offer additional benefits over existing therapy. Fc receptor antagonists (efgartigimod) have shown promising effects in the initial study and, in the future, it may be used as an add-on to existing therapy (Gilhus et al., 2019). Seldegs (engineered antibody-based reagents), imlifidase (IgG-degrading enzyme), and tirasemtiv (activator of the fast skeletal muscle troponin complex) are being investigated as future therapeutics in MG (Devanaboyina et al., 2017; Lonze et al., 2018).

12.7 CONCLUSION

MG is an autoimmune disease characterized by muscle weakness and fatigue due to autoantibodies produced against AchR at the NMJ of postsynaptic neurons. For the last 30 years, our understanding of its molecular and immunological pathogenesis has increased significantly. Based on these findings, newer diagnostic tests have been developed that eventually led to timely treatment and improved patient QoL (Yan et al., 2018). The current treatment paradigm for MG is to weaken the autoimmune response and improve neuromuscular transmission. Various newer therapies are being developed, such as the use of subcutaneous immunoglobulins, the depletion of specific MG antibodies, and the repurposing of amifampridine. Salbutamol, as well as ephedrine, have shown positive outcomes on MG associated with cognitive dysfunction (Lazaridis et al., 2017; Sala et al., 2018; Lee et al., 2018). Recently, Fc receptor antagonists (efgartigimod) have been found to be effective in the treatment of MG (Ulrichts et al., 2018). Therefore, it can be concluded that that because of the plethora of various drugs being developed for MG, and the cost of treatment, critical analysis is needed to prioritize the rational use of drugs in MG.

REFERENCES

Ahmed, S., Kirmani, J. F., Janjua, N., Alkawi, A., Khatri, I., Yahia, A. M., et al. 2005. An update on myasthenic crisis. *Current Treatment Options in Neurology*, 7, 129–141.

Akaishi, T., Yamaguchi, T., Suzuki, Y., Nagane, Y., Suzuki, S., Murai, H., et al. 2014. Insights into the classification of myasthenia gravis. *PLoS ONE*, 9, e106757.

Allen, C. S., Yeung, J. H., Vandermeer, B. & Homik, J. 2016. Bisphosphonates for steroid-induced osteoporosis. *Cochrane Database of Systematic Reviews*, 10, CD001347.

Angelini, C., Martignago, S. & Bisciglia, M. 2013. New treatments for myasthenia: a focus on antisense oligonucleotides. *Drug Design, Development and Therapy*, 7, 13.

Argov, Z., McKee, D., Agus, S., Brawer, S., Shlomowitz, N., Yoseph, O. B., et al. 2007. Treatment of human myasthenia gravis with oral antisense suppression of acetylcholinesterase. *Neurology*, 69, 699–700.

Ashton Jr, R. C., McGinnis, K. M., Connery, C. P., Swistel, D. G., Ewing, D. R. & Derose Jr, J. J. 2003. Totally endoscopic robotic thymectomy for myasthenia gravis. *The Annals of Thoracic Surgery*, 75, 569–571.

Barth, D., Nouri, M. N., Ng, E., Nwe, P. & Bril, V. 2011. Comparison of IVIg and PLEX in patients with myasthenia gravis. *Neurology*, 76, 2017–2023.

Batocchi, A. P., Evoli, A., Servidei, S., Palmisani, M., Apollo, F. & Tonali, P. 1995. Myasthenia gravis during interferon alfa therapy. *Neurology*, 45, 382–383.

Benatar, M., McDermott, M. P., Sanders, D. B., Wolfe, G. I., Barohn, R. J., Nowak, R. J., et al. 2016. Efficacy of prednisone for the treatment of ocular myasthenia (EPITOME): a randomized, controlled trial. *Muscle & Nerve*, 53, 363–369.

Bever Jr, C. T., Aquino, A. V., Penn, A. S., Lovelace, R. E. & Rowland, L. P. 1983. Prognosis of ocular myasthenia. *Annals of Neurology*, 14, 516–519.

Bindu, P., Nirmala, M., Patil, S. & Taly, A. 2008. Myasthenia gravis and acetylcholine receptor antibodies: a clinico immunological correlative study on South Indian patients. *Annals of Indian Academy of Neurology*, 11, 242.

Borges, L. S. & Richman, D. P. 2020. Muscle-specific kinase myasthenia gravis. *Frontiers in Immunology*, 11, 707.

Bruce, B. B. & Kupersmith, M. J. 2012. Safety of prednisone for ocular myasthenia gravis. *Journal of Neuro-Ophthalmology*, 32, 212.

Bryant, A., Atkins, H., Pringle, C. E., Allan, D., Anstee, G., Bence-Bruckler, I., et al. 2016. Myasthenia gravis treated with autologous hematopoietic stem cell transplantation. *JAMA Neurology*, 73, 652–658.

Carr, A. S., Cardwell, C. R., McCarron, P. O. & McConville, J. 2010. A systematic review of population based epidemiological studies in myasthenia gravis. *BMC Neurology*, 10, 46.

Cavalcante, P., Serafini, B., Rosicarelli, B., Maggi, L., Barberis, M., Antozzi, C., et al. 2010. Epstein–Barr virus persistence and reactivation in myasthenia gravis thymus. *Annals of Neurology*, 67, 726–738.

Cron, M. A., Maillard, S., Delisle, F., Samson, N., Truffault, F., Foti, M., et al. 2018. Analysis of microRNA expression in the thymus of myasthenia gravis patients opens new research avenues. *Autoimmunity Reviews*, 17, 588–600.

Cron, M. A., Maillard, S., Truffault, F., Gualeni, A. V., Gloghini, A., Fadel, E., et al. 2019. Causes and consequences of miR-150-5p dysregulation in myasthenia gravis. *Frontiers in Immunology*, 10, 539.

Cruz, P. R., Huda, S., López-Ruiz, P. & Vincent, A. 2015. Use of cell-based assays in myasthenia gravis and other antibody-mediated diseases. *Experimental Neurology*, 270, 66–71.

Desmedt, J. E. 1966. Presynaptic mechanism in myasthenia gravis. *Annals of the New York Academy of sciences,* 135, 209–246.

Devanaboyina, S. C., Khare, P., Challa, D. K., Ober, R. J. & Ward, E. S. 2017. Engineered clearing agents for the selective depletion of antigen-specific antibodies. *Nature Communications*, 8, 1–6.

Dhillon, S. 2018. Eculizumab: a review in generalized myasthenia gravis. *Drugs*, 78, 367–376.

Dragin, N., Bismuth, J., Cizeron-Clairac, G., Biferi, M. G., Berthault, C., Serraf, A., et al. 2016. Estrogen-mediated downregulation of AIRE influences sexual dimorphism in autoimmune diseases. *The Journal of Clinical Investigation*, 126, 1525–1537.

Dubois, E. A., Rissmann, R. & Cohen, A. F. 2011. New drug mechanisms. *British Journal of Clinical Pharmacology*, 71, 804–806.

Engel, A. G. & Arahata, K. 1987. The membrane attack complex of complement at the endplate in myasthenia gravis. *Annals of the New York Academy of Sciences*, 505, 326–332.

Erm, A., Arst, H., Trei, T., Reinart, A. & Hussainov, M. 2001. Optical and biological properties of Lake Ülemiste, a water reservoir of the city of Tallinn I: Water transparency and optically active substances in the water. *Lakes & Reservoirs: Research & Management*, 6, 63–74.

Evoli, A. 2017. Myasthenia gravis: new developments in research and treatment. *Current Opinion in Neurology*, 30, 464–470.

Evoli, A., Alboini, P. E., Damato, V., Iorio, R., Provenzano, C., Bartoccioni, E. & Marino, M. 2018. Myasthenia gravis with antibodies to MuSK: an update. *Annals of the New York Academy of Sciences*, 1412, 82–89.

Farmakidis, C., Pasnoor, M., Dimachkie, M. M. & Barohn, R. J. 2018. Treatment of myasthenia gravis. *Neurologic Clinics*, 36, 311–337.

Fini, M. E., Schwartz, S. G., Gao, X., Jeong, S., Patel, N., Itakura, T., et al. 2017. Steroid-induced ocular hypertension/glaucoma: Focus on pharmacogenomics and implications for precision medicine. *Progress in Retinal and Eye Research*, 56, 58–83.

Fulghum, C. B. 1950. Cystic tumor of the thymic region and myasthenia gravis: report of a case in which the tumor was removed. *Annals of Internal Medicine*, 32, 327–335.

Gable, K. L. & Guptill, J. T. 2020. Antagonism of the neonatal Fc receptor as an emerging treatment for myasthenia gravis. *Frontiers in Immunology*, 10, 3052.

Gajdos, P., Chevret, S., Clair, B., Tranchant, C., Chastang, C. & Group, M. G. C. S. 1997. Clinical trial of plasma exchange and high-dose intravenous immunoglobulin in myasthenia gravis. *Annals of Neurology*, 41, 789–796.

Gamez, J. & Ponseti, J. M. 2017. Thymectomy for non-thymomatous myasthenia gravis: the end of controversy, albeit fifty years later. *Journal of Xianga Medicine*, 2, 4.

Gamez, J., Salvadó, M., Carmona, F., De Nadal, M., Romero, L., Ruiz, D., et al. 2019. Intravenous immunoglobulin to prevent myasthenic crisis after thymectomy and other procedures can be omitted in patients with well-controlled myasthenia gravis. *Therapeutic Advances in Neurological Disorders*, 12, 1756286419864497.

Gilhus, N. E. & Verschuuren, J. J. 2015a. Myasthenia gravis: subgroup classification and therapeutic strategies. *The Lancet Neurology*, 14, 1023–1036.

Gilhus, N. E. & Verschuuren, J. J. 2015b. Myasthenia gravis: subgroup classification and therapeutic strategies. *Lancet Neurol*, 14, 1023–1036.

Gilhus, N., Nacu, A., Andersen, J. & Owe, J. 2015. Myasthenia gravis and risks for comorbidity. *European Journal of Neurology*, 22, 17–23.

Gilhus, N. E., Owe, J. F., Hoff, J. M., Romi, F., Skeie, G. O. & Aarli, J. A. 2011. Myasthenia gravis: a review of available treatment approaches. *Autoimmune Diseases*, 2011, 847393.

Gilhus, N. E., Skeie, G. O., Romi, F., Lazaridis, K., Zisimopoulou, P. & Tzartos, S. 2016. Myasthenia gravis—autoantibody characteristics and their implications for therapy. *Nature Reviews Neurology*, 12, 259–268.

Gilhus, N. E., Socrates, T., Amelia, E., Jacqueline, P. & Burns, T. M. 2019. Myasthenia gravis. *Nature Reviews: Disease Primers*, 5, 30.

Giraud, M., Taubert, R., Vandiedonck, C., Ke, X., Lévi-Strauss, M., Pagani, F., et al. 2007. An IRF8-binding promoter variant and AIRE control CHRNA1 promiscuous expression in thymus. *Nature*, 448, 934–937.

Grennan, D. & Wang, S. 2019. Steroid side effects. *JAMA*, 322, 282–282.

Guastafierro, E., Tramacere, I., Toppo, C., Leonardi, M., Mantegazza, R., Bonanno, S., et al. 2020. Employment in myasthenia gravis: a systematic literature review and meta-analysis. *Neuroepidemiology*, 54, 304–312.

Guptill, J. T., Sanders, D. B. & Evoli, A. 2011. Anti-MuSK antibody myasthenia gravis: clinical findings and response to treatment in two large cohorts. *Muscle & Nerve*, 44, 36–40.

Guptill, J. T., Soni, M. & Meriggioli, M. N. 2016. Current treatment, emerging translational therapies, and new therapeutic targets for autoimmune myasthenia gravis. *Neurotherapeutics*, 13, 118–131.

Håkansson, I., Sandstedt, A., Lundin, F., Askmark, H., Pirskanen, R., Carlson, K., et al. 2017. Successful autologous haematopoietic stem cell transplantation for refractory myasthenia gravis–a case report. *Neuromuscular Disorders*, 27, 90–93.

Hara, H., Hayashi, K., Ohta, K., Itoh, N., Nishitani, H. & Ohta, M. 1993. Detection and characterization of blocking-type anti-acetylcholine receptor antibodies in sera from patients with myasthenia gravis. *Clinical Chemistry*, 39, 2053–2057.

He, D., Zhang, H., Xiao, J., Zhang, X., Xie, M., Pan, D., et al. 2018. Molecular and clinical relationship between live-attenuated Japanese encephalitis vaccination and childhood onset myasthenia gravis. *Annals of Neurology*, 84, 386–400.

Hehir, M. K., Hobson-Webb, L. D., Benatar, M., Barnett, C., Silvestri, N. J., Howard, J. F., et al. 2017. Rituximab as treatment for anti-MuSK myasthenia gravis: multicenter blinded prospective review. *Neurology*, 89, 1069–1077.

Higuchi, O., Hamuro, J., Motomura, M. & Yamanashi, Y. 2011. Autoantibodies to low-density lipoprotein receptor-related protein 4 in myasthenia gravis. *Annals of Neurology*, 69, 418–422.

Hong, Y., Li, H. F., Romi, F., Skeie, G. & Gilhus, N. 2018. HLA and Mu SK-positive myasthenia gravis: a systemic review and meta-analysis. *Acta Neurologica Scandinavica*, 138, 219–226.

Hong, Y., Zisimopoulou, P., Trakas, N., Karagiorgou, K., Stergiou, C., Skeie, G., et al. 2017. Multiple antibody detection in 'seronegative'myasthenia gravis patients. *European Journal of Neurology*, 24, 844–850.

Howard, J. F., Bril, V., Burns, T. M., Mantegazza, R., Bilinska, M., Szczudlik, A., et al. 2019. Randomized phase 2 study of FcRn antagonist efgartigimod in generalized myasthenia gravis. *Neurology*, 92, e2661-e2673.

Howard Jr, J. F., Utsugisawa, K., Benatar, M., Murai, H., Barohn, R. J., Illa, I., et al. 2017. Safety and efficacy of eculizumab in anti-acetylcholine receptor antibody-positive refractory generalised myasthenia gravis (REGAIN): a phase 3, randomised, double-blind, placebo-controlled, multicentre study. *The Lancet Neurology*, 16, 976–986.

Huijbers, M. G., Zhang, W., Klooster, R., Niks, E. H., Friese, M. B., Straasheijm, K. R., et al. 2013. MuSK IgG4 autoantibodies cause myasthenia gravis by inhibiting binding between MuSK and Lrp4. *Proceedings of the National Academy of Sciences*, 110, 20783–20788.

Imai, T. 2019. Why is development of new treatments necessary for myasthenia gravis? Recent advances in clinical trials. *Neurology and Clinical Neuroscience*, 7, 161–165.

Inohara, N. & Nuñez, G. J. N. R. I. 2003. NODs: intracellular proteins involved in inflammation and apoptosis. *Nature Review Immunology*, 3, 371–382.

Iqubal, A., Iqubal, M. K., Sharma, S., Ansari, M. A., Najmi, A. K., Ali, S. M., et al. 2019. Molecular mechanism involved in cyclophosphamide-induced cardiotoxicity: old drug with a new vision. *Life Sciences*, 218, 112–131.

Jaretzki, A., III. 2003. Thymectomy for myasthenia gravis: analysis of controversies—patient management. *The Neurologist*, 9, 77–92.

Jaretzki, A., 3rd, Barohn, R. J., Ernstoff, R. M., Kaminski, H. J., Keesey, J. C., Penn, A.. & Sanders, D. B. 2000. Recommendations of clinical research standards. *Neurology*, 55, 16–23.

Jayam Trouth, A., Dabi, A., Solieman, N., Kurukumbi, M. & Kalyanam, J. 2012. Myasthenia gravis: a review. *Autoimmune Diseases*, 2012, 874680.

Jing, F., Cui, F., Chen, Z., Yang, F., Ling, L. & Huang, X. 2015. Clinical and electrophysiological markers in myasthenia gravis patients. *European Neurology*, 74, 22–27.

Jones, J. L. & Loftus Jr, E. V. 2007. Lymphoma risk in inflammatory bowel disease: is it the disease or its treatment? *Inflammatory Bowel Diseases*, 13, 1299–1307.

Jonkers, I., Swerup, C., Pirskanen, R., Bjelak, S. & Matell, G. 1996. Acute effects of intravenous injection of beta-adrenoreceptor- and calcium-channel antagonists and agonists in myasthenia gravis. *Muscle & Nerve*, 19, 959–965.

Kanth, K. M., Solorzano, G. E. & Goldman, M. D. 2016. PML in a patient with myasthenia gravis treated with multiple immunosuppressing agents. *Neurology: Clinical Practice*, 6, e17–e19.

Kas, J., Kiss, D., Simon, V., Svastics, E., Major, L. & Szobor, A. 2001. Decade-long experience with surgical therapy of myasthenia gravis: early complications of 324 transsternal thymectomies. *The Annals of Thoracic Surgery*, 72, 1691–1697.

Kawaguchi, K., Fukui, T., Nakamura, S., Taniguchi, T. & Yokoi, K. 2018. A bilateral approach to extended thymectomy using the da Vinci Surgical System for patients with myasthenia gravis. *Surgery Today*, 48, 195–199.

Kiessling, P., Lledo-Garcia, R., Watanabe, S., Langdon, G., Tran, D., Bari, M., et al. 2017. The FcRn inhibitor rozanolixizumab reduces human serum IgG concentration: A randomized phase 1 study. *Science Translational Medicine*, 9, eaan1208.

Kim, N., Stiegler, A. L., Cameron, T. O., Hallock, P. T., Gomez, A. M., Huang, J. H., et al. 2008. Lrp4 is a receptor for Agrin and forms a complex with MuSK. *Cell*, 135, 334–342.

Kirzinger, L., Schneider, R., Mayer, C., Schötz, S., Backhaus, R., Gerken, M., et al. 2017. The influence of thymoma-surgery on paraneoplastic myasthenia gravis. *Journal of the Neurological Sciences*, 381, 474.

Koneczny, I., Stevens, J. A., De Rosa, A., Huda, S., Huijbers, M. G., Saxena, A., et al. 2017. IgG4 autoantibodies against muscle-specific kinase undergo Fab-arm exchange in myasthenia gravis patients. *Journal of Autoimmunity*, 77, 104–115.

Kordas, G., Lagoumintzis, G., Sideris, S., Poulas, K. & Tzartos, S. J. 2014. Direct proof of the in vivo pathogenic role of the AChR autoantibodies from myasthenia gravis patients. *PLoS One*, 9, e108327.

Kulasekararaj, A. G., Hill, A., Rottinghaus, S. T., Langemeijer, S., Wells, R., Gonzalez-Fernandez, F. A., et al. 2019. Ravulizumab (ALXN1210) vs eculizumab in C5-inhibitor-experienced adult patients with PNH: the 302 study. *Blood*, 133, 540–549.

Lazaridis, K., Dalianoudis, I., Baltatzidi, V. & Tzartos, S. J. 2017. Specific removal of autoantibodies by extracorporeal immunoadsorption ameliorates experimental autoimmune myasthenia gravis. *Journal of Neuroimmunology*, 312, 24–30.

Lee, J.-Y., Min, J.-H., Han, S.-H. & Han, J. 2017. Transient neonatal myasthenia gravis due to a mother with ocular onset of anti-muscle specific kinase myasthenia gravis. *Neuromuscular Disorders*, 27, 655–657.

Lee, M., Beeson, D. & Palace, J. 2018. Therapeutic strategies for congenital myasthenic syndromes. *Annals of the New York Academy of Sciences*, 1412, 129–136.

Lewis, R. A., Selwa, J. F. & Lisak, R. P. 1995. Myasthenia gravis: immunological mechanisms and immunotherapy. *Annals of Neurology*, 37, 51–62.

Li, L., Cao, Y., Wu, H., Ye, X., Zhu, Z., Xing, G., et al. 2016. Enzymatic activity of the scaffold protein rapsyn for synapse formation. *Neuron*, 92, 1007–1019.

Lindstrom, J., Seybold, M., Lennon, V., Whittingham, S. & Duane, D. 1976. Acetylcholine receptor antibody in myasthenia gravis: incidence, clinical correlates and usefulness as a diagnostic test. *Neurology (NY)*, 26, 1054.

Liu, T. J., Lin, M.-W., Hsieh, M.-S., Kao, M.-W., Chen, K.-C., Chang, C.-C., et al. 2012. Video-assisted thoracoscopic surgical thymectomy to treat early thymoma: a comparison with the conventional transsternal approach. *Annals of Surgical Oncology*, 21, 322–328.

Lonze, B. E., Tatapudi, V. S., Weldon, E. P., Min, E. S., Ali, N. M., Deterville, C. L., et al. 2018. IdeS (Imlifidase): a novel agent that cleaves human IgG and permits successful kidney transplantation across high-strength donor-specific antibody. *Annals of Surgery*, 268, 488–496.

López-Cano, M., Ponseti-Bosch, J. M., Espin-Basany, E., Sánchez-García, J. L. & Armengol-Carrasco, M. 2003. Clinical and pathologic predictors of outcome in thymoma-associated myasthenia gravis. *The Annals of Thoracic Surgery*, 76, 1643–1649.

Lorenzoni, P. J., Kay, C. S. K., Zanlorenzi, M. F., Ducci, R. D.-P., Werneck, L. C. & Scola, R. H. 2020. Myasthenia gravis and azathioprine treatment: Adverse events related to thiopurine S-methyl-transferase (TPMT) polymorphisms. *Journal of Neurological Sciences*, 412, 116734.

Luo, S., Ainslie, G., Giragosian, L. & Monterosso, J. R. 2009. Behavioral and neural evidence of incentive bias for immediate rewards relative to preference-matched delayed rewards. *Journal of Neuroscience*, 29, 14820–14827.

McKeage, K. 2019. Ravulizumab: first global approval. *Drugs*, 79, 347–352.

Mantegazza, R. & Antozzi, C. 2018. When myasthenia gravis is deemed refractory: clinical signposts and treatment strategies. *Therapeutic Advances in Neurological Disorders*, 11, 1756285617749134.

Marx, A., Pfister, F., Schalke, B., Saruhan-Direskeneli, G., Melms, A. & Ströbel, P. 2013. The different roles of the thymus in the pathogenesis of the various myasthenia gravis subtypes. *Autoimmunity Reviews*, 12, 875–884.

Masuda, M., Utsugisawa, K., Suzuki, S., Nagane, Y., Kabasawa, C., Suzuki, Y., et al. 2012. The MG-QOL15 Japanese version: validation and associations with clinical factors. *Muscle & Nerve*, 46, 166–173.

Melo, C. L. S., Peña, A. M., Salazar, L. A., Jiménez, S. I., Gómez, E. D., Chalela, C. M., et al. 2019. Autologous hematopoietic stem cell transplantation in a patient with refractory seropositive myasthenia gravis: a case report. *Neuromuscular Disorders*, 29, 142–145.

Melzer, N., Ruck, T., Fuhr, P., Gold, R., Hohlfeld, R., Marx, A., et al. 2016. Clinical features, pathogenesis, and treatment of myasthenia gravis: a supplement to the Guidelines of the German Neurological Society. *Journal of Neurology*, 263, 1473–1494.

Menon, D., Barnett, C. & Bril, V. 2020. Novel treatments in myasthenia gravis. *Frontiers in Neurology*, 11, 538.

Meriggioli, M. N. & Sanders, D. B. 2012. Muscle autoantibodies in myasthenia gravis: beyond diagnosis? *Expert Review of Clinical Immunology*, 8, 427–438.

Messéant, J., Dobbertin, A., Girard, E., Delers, P., Manuel, M., Mangione, F., et al. 2015. MuSK frizzled-like domain is critical for mammalian neuromuscular junction formation and maintenance. *Journal of Neuroscience*, 35, 4926–4941.

Mi, W., Silvestri, N. J. & Wolfe, G. I. 2019. A neurologist's perspective on thymectomy for myasthenia gravis: current perspective and future trials. *Thoracic Surgery Clinics*, 29, 143–150.

Molina, V., Haj-Yahia, S., Solodeev, I., Levy, Y., Blank, M. & Shoenfeld, Y. 2006. Immunomodulation of experimental pulmonary fibrosis by intravenous immunoglobulin (IVIG). *Autoimmunity*, 39, 711–717.

Morgan, B., Chamberlain-Banoub, J., Neal, J., Song, W., Mizuno, M. & Harris, C. 2006. The membrane attack pathway of complement drives pathology in passively induced experimental autoimmune myasthenia gravis in mice. *Clinical & Experimental Immunology*, 146, 294–302.

Mygland, Å., Tysnes, O. B., Aarli, J. A., Gilhus, N. E., Matre, R. & Volpe, P. 1992. Ryanodine receptor autoantibodies in myasthenia gravis patients with a thymoma. *Annals of Neurology*, 32, 589–591.

Nacu, A., Andersen, J. B., Lisnic, V., Owe, J. F. & Gilhus, N. E. 2015. Complicating autoimmune diseases in myasthenia gravis: a review. *Autoimmunity*, 48, 362–368.

Nagane, Y., Suzuki, S., Suzuki, N. & Utsugisawa, K. 2011. Early aggressive treatment strategy against myasthenia gravis. *European Neurology*, 65, 16–22.

Niakan, E., Harati, Y. & Rolak, L. A. 1986. Immunosuppressive drug therapy in myasthenia gravis. *Archives of Neurology*, 43, 155–156.

Oh, S. J., Jeong, D., Lee, I. & Alsharabati, M. 2019. Repetitive nerve stimulation test in myasthenic crisis. *Muscle & Nerve*, 59, 544–548.

Oishi, A., Miyamoto, K., Kashii, S. & Yoshimura, N. 2007. Abducens palsy and Sjögren's syndrome induced by pegylated interferon therapy. *British Journal of Ophthalmology*, 91, 843–844.

Oosterhuis, H. J. 1981. Observations of the natural history of myasthenia gravis and the effect of thymectomy. *Annals of the New York Academy of Sciences*, 377, 678–690.

Osserman, K. E., Kornfeld, P., Cohen, E., Genkins, G., Mendelow, H., Goldberg, H., et al. 1958. Studies in myasthenia gravis: review of two hundred eighty-two cases at the Mount Sinai Hospital, New York City. *AMA Archives of Internal Medicine*, 102, 72–81.

Palace, J., Newsom-Davis, J., Lecky, B. & Group, M. G. S. 1998. A randomized double-blind trial of prednisolone alone or with azathioprine in myasthenia gravis. *Neurology*, 50, 1778–1783.

Papazian, O. 1992. Topical review article: transient neonatal myasthenia gravis. *Journal of Child Neurology*, 7, 135–141.

Pedersen, E., Pottegård, A., Hallas, J., Friis, S., Hansen, K., Jensen, P. & Gaist, D. 2012. Risk of non-melanoma skin cancer in myasthenia patients treated with azathioprine. *European Journal of Neurology*, 21, 454–458.

Pennathur, A., Qureshi, I., Schuchert, M. J., Dhupar, R., Ferson, P. F., Gooding, W. E., et al. 2011. Comparison of surgical techniques for early-stage thymoma: feasibility of minimally invasive thymectomy and comparison with open resection. *The Journal of Thoracic and Cardiovascular Surgery*, 141, 694–701.

Phillips, W. D. & Vincent, A. 2016. Pathogenesis of myasthenia gravis: update on disease types, models, and mechanisms. *F1000Research*, 5, F1000.

Pierpont, T. M., Limper, C. B. & Richards, K. L. 2018. Past, present, and future of rituximab—the world's first oncology monoclonal antibody therapy. *Frontiers in Oncology*, 8, 163.

Plomp, J. J., Van Kempen, G. T. H., De Baets, M. B., Graus, Y. M., Kuks, J. B. & Molenaar, P. C. 1995. Acetylcholine release in myasthenia gravis: regulation at single end-plate level. *Annals of Neurology*, 37, 627–636.

Powers, K., Schappacher-Tilp, G., Jinha, A., Leonard, T., Nishikawa, K. & Herzog, W. 2012. Titin force is enhanced in actively stretched skeletal muscle. *Journal of Experimental Biology*, 217, 3629–3636.

Priola, A., Priola, S., Gned, D., Giraudo, M. T., Fornari, A. & Veltri, A. 2016. Comparison of CT and chemical-shift MRI for differentiating thymoma from non-thymomatous conditions in myasthenia gravis: value of qualitative and quantitative assessment. *Clinical Radiology*, 71, e157–e169.

Punga, A. R. & Punga, T. 2018. Circulating microRNAs as potential biomarkers in myasthenia gravis patients. *Annals of the New York Academy of Sciences*, 1412, 33–40.

Punga, A. R., Flink, R., Askmark, H. & Stålberg, E. V. 2006. Cholinergic neuromuscular hyperactivity in patients with myasthenia gravis seropositive for MuSK antibody. *Muscle & Nerve*, 34, 111–115.

Pyzik, M., Sand, K. M., Hubbard, J. J., Andersen, J. T., Sandlie, I. & Blumberg, R. S. 2019. The neonatal Fc receptor (FcRn): a misnomer? *Frontiers in Immunology*, 10, 1540.

Rivner, M. H., Pasnoor, M., Dimachkie, M. M., Barohn, R. J. & Mei, L. 2018. Muscle-specific tyrosine kinase and myasthenia gravis owing to other antibodies. *Neurologic Clinics*, 36, 293–310.

Romi, F., Gilhus, N. & Aarli, J. 2006. Myasthenia gravis: disease severity and prognosis. *Acta Neurologica Scandinavica*, 113, 24–25.

Romi, F., Hong, Y. & Gilhus, N. E. 2017. Pathophysiology and immunological profile of myasthenia gravis and its subgroups. *Current Opinion in Immunology*, 49, 9–13.

Rückert, J. C., Swierzy, M. & Ismail, M. 2011. Comparison of robotic and nonrobotic thoracoscopic thymectomy: a cohort study. *The Journal of Thoracic and Cardiovascular Surgery*, 141, 673–677.

Russell, A. & Lindstrom, J. J. N. 1978. Penicillamine-induced myasthenia gravis associated with antibodies to acetylcholine receptor. *Neurology*, 28, 847–849.

Sabre, L., Maddison, P., Sadalage, G., Ambrose, P. A. & Punga, A. R. 2018. Circulating microRNA miR-21-5p, miR-150-5p and miR-30e-5p correlate with clinical status in late onset myasthenia gravis. *Journal of Neuroimmunology*, 321, 164–170.

Sala, T. P., Crave, J.-C., Duracinsky, M., Bompeka, F. L., Tadmouri, A., Chassany, O. & Cherin, P. 2018. Efficacy and patient satisfaction in the use of subcutaneous immunoglobulin immunotherapy for the treatment of auto-immune neuromuscular diseases. *Autoimmunity Reviews*, 17, 873–881.

Sanders, D. B., Wolfe, G. I., Benatar, M., Evoli, A., Gilhus, N. E., Illa, I., et al. 2016. International consensus guidance for management of myasthenia gravis: executive summary. *Neurology*, 87, 419–425.

Sanders, D. B., Wolfe, G. I., Narayanaswami, P. & MGFA Task Force on MG Treatment Guidance. 2018. Developing treatment guidelines for myasthenia gravis. *Annals of the New York Academy of Sciences*, 1412, 95–101.

Sargiannidou, I., Kagiava, A., Bashiardes, S., Richter, J., Christodoulou, C., Scherer, S. S. & Kleopa, K. A. 2015. Intraneural GJB 1 gene delivery improves nerve pathology in a model of X-linked Charcot–Marie–Tooth disease. *Annals of Neurology*, 78, 303–316.

Sathasivam, S. 2008. Steroids and immunosuppressant drugs in myasthenia gravis. *Nature Clinical Practice Neurology*, 4, 317–327.

Sghirlanzoni, A., Peluchetti, D., Mantegazza, R., Fiacchino, F. & Cornelio, F. 1984. Myasthenia gravis: prolonged treatment with steroids. *Neurology*, 34, 170–170.

Shelton, G. D., Ho, M. & Kass, P. H. 2000. Risk factors for acquired myasthenia gravis in cats: 105 cases (1986–1998). *Journal of the American Veterinary Medical Association*, 216, 55–57.

Shen, C., Lu, Y., Zhang, B., Figueiredo, D., Bean, J., Jung, J., et al. 2013. Antibodies against low-density lipoprotein receptor–related protein 4 induce myasthenia gravis. *The Journal of Clinical Investigation*, 123, 5190–5202.

Shen, C., Xiong, W.-C. & Mei, L. 2015. LRP4 in neuromuscular junction and bone development and diseases. *Bone*, 80, 101–108.

Simonsen, M., Miyabe, M. M., Ouki, H. T., Galvão, A. C. R., Leite, D., Murayama, B. A. R., et al. 2018. Myasthenia as a paraneoplastic manifestation of ovarian Cancer. *Gynecologic Oncology Reports*, 25, 35–36.

Singer, W., Sandroni, P., Opfer-Gehrking, T. L., Suarez, G. A., Klein, C. M., Hines, S., et al. 2006. Pyridostigmine treatment trial in neurogenic orthostatic hypotension. *Archives of Neurology*, 63, 513–518.

Singman, E. L., Matta, N. S. & Silbert, D. I. 2011. Use of the Cogan lid twitch to identify myasthenia gravis. *Journal of Neuro-Ophthalmology*, 31, 239–240.

Skeie, G. & Romi, F. J. E. J. O. N. 2008. Paraneoplastic myasthenia gravis: immunological and clinical aspects. *European Journal of Neurology*, 15, 1029–1033.

Suda, T., Tochii, D., Tochii, S. & Takagi, Y. 2015. Trans-subxiphoid robotic thymectomy. *Interactive CardioVascular and Thoracic Surgery*, 20, 669–671.

Sussman, J., Argov, Z., Wirguin, Y., Apolski, S., Milic-Rasic, V. & Soreq, H. 2012. Further developments with antisense treatment for myasthenia gravis. *Annals of the New York Academy of Sciences*, 1275, 13–16.

Tandan, R., Hehir, M. K., Waheed, W. & Howard, D. B. 2017. Rituximab treatment of myasthenia gravis: a systematic review. *Muscle & Nerve*, 56, 185–196.

Thanvi, B. & Lo, T. 2004. Update on myasthenia gravis. *Postgraduate Medical Journal*, 80, 690–700.

Thiruppathi, M., Rowin, J., Jiang, Q. L., Sheng, J. R., Prabhakar, B. S. & Meriggioli, M. N. 2012. Functional defect in regulatory T cells in myasthenia gravis. *Annals of the New York Academy of Sciences*, 1274, 68.

Trakas, N. & Tzartos, S. J. 2018. Immunostick ELISA for rapid and easy diagnosis of myasthenia gravis. *Journal of Immunological Methods*, 460, 107–112.

Tüzün, E. & Christadoss, P. 2013. Complement associated pathogenic mechanisms in myasthenia gravis. *Autoimmunity Reviews*, 12, 904–911.

Ulrichts, P., Guglietta, A., Dreier, T., Van Bragt, T., Hanssens, V., Hofman, E., et al. 2018. Neonatal Fc receptor antagonist efgartigimod safely and sustainably reduces IgGs in humans. *The Journal of Clinical Investigation*, 128, 4372–4386.

Ulusoy, C., Çavuş, F., Yilmaz, V. & Tüzün, E. 2017. Immunization with recombinantly expressed LRP4 induces experimental autoimmune myasthenia gravis in C57BL/6 mice. *Immunological Investigations*, 46, 490–499.

Utsugisawa, K., Benatar, M., Murai, H., Barohn, R., Illa, I., Jacob, S., et al. 2019. Long-term safety and efficacy of eculizumab in generalized myasthenia gravis. *Muscle & Nerve*, 60, 14–24.

Vernino, S. & Lennon, V. A. 2004. Autoantibody profiles and neurological correlations of thymoma. *Clinical Cancer Research*, 10, 7270–7275.

Vinge, L., Jakobsen, J. & Andersen, H. 2019. Muscle weakness and functional disability in patients with myasthenia gravis. *Muscle & Nerve*, 59, 218–223.

Wolfe, G. I., Kaminski, H. J., Aban, I. B., Minisman, G., Kuo, H.-C., Marx, A., et al. 2016. Randomized trial of thymectomy in myasthenia gravis. *New England Journal of Medicine*, 375, 511–522.

Wolfe, G. I., Kaminski, H. J., Aban, I. B., Minisman, G., Kuo, H.-C., Marx, A., et al. 2019. Long-term effect of thymectomy plus prednisone versus prednisone alone in patients with non-thymomatous myasthenia gravis: 2-year extension of the MGTX randomised trial. *The Lancet Neurology*, 18, 259–268.

Yan, M., Liu, Z., Fei, E., Chen, W., Lai, X., Luo, B., et al. 2018. Induction of anti-agrin antibodies causes myasthenia gravis in mice. *Neuroscience*, 373, 113–121.

Yanagihashi, M., Okamoto, R., Morioka, H., Sawada, M., Matsumoto, S., Ikeda, T. & Kano, O. J. B. N. 2020. Coronary spastic angina after the administration of intravenous immunoglobulin in myasthenia gravis: a case report. *BMC Neurology*, 20, 1–4.

Ye, B., Tantai, J.-C., Ge, X.-X., Li, W., Feng, J., Cheng, M., et al. 2012. Surgical techniques for early-stage thymoma: video-assisted thoracoscopic thymectomy versus transsternal thymectomy. *The Journal of Thoracic and Cardiovascular Surgery*, 147, 1599–1603.

Zhang, B., Tzartos, J. S., Belimezi, M., Ragheb, S., Bealmear, B., Lewis, R. A., et al. 2012. Autoantibodies to lipoprotein-related protein 4 in patients with double-seronegative myasthenia gravis. *Archives of Neurology*, 69, 445–451.

Zhong, H., Zhao, C. & Luo, S. 2019. HLA in myasthenia gravis: From superficial correlation to underlying mechanism. *Autoimmunity Reviews*, 18, 102349.

Zinman, L., Ng, E. & Bril, V. 2007. IV immunoglobulin in patients with myasthenia gravis: a randomized controlled trial. *Neurology*, 68, 837–841.

Zisimopoulou, P., Brenner, T., Trakas, N. & Tzartos, S. J. 2013. Serological diagnostics in myasthenia gravis based on novel assays and recently identified antigens. *Autoimmunity Reviews*, 12, 924–930.

Section III

Typical Neurochemical Processes

13 Neurogenesis in the Embryonic and Adult Brain

Newman Osafo, Joseph Owusu-Sarfo, and Oduro Kofi Yeboah
Kwame Nkrumah University of Science and Technology, Kumasi, Ghana

Nana Akua Acheampong
Kennesaw State University, Atlanta, GA, USA

CONTENTS

13.1 INTRODUCTION

Neurogenesis is defined as the process by which neurons are made from neural progenitor cells. Previously, it was believed that neural stem cells were depleted and that the process of neurogenesis ceases immediately before and after birth. Studies by Altman and Das identified newly generated dentate granule cells (Altman and Das, 1965). Years later, Reynolds and Weiss (1992) also detected multipotent neural progenitors from the adult mammalian brain. Several other works have established that neural stem cells persist throughout the adult life of mammals, including humans (Bergmann et al., 2015; Lindsey and Tropepe, 2006).

The process of neurogenesis encompasses proliferation, differentiation, and migration of neural precursor cells, in a coordinated fashion (Eriksson et al., 1998). Through the aging process, brain injury, and genetic mutations, the continuous loss of neural progenitors may contribute to the development of neurodegenerative diseases (NDs).

In this chapter, we discuss neurogenesis in the adult and embryonic brain, as well as other key players and regulators in the neurogenesis process. We also touch on neurogenesis and some human pathologies and the association between neurogenesis and the molecular mechanism of memory.

13.2 NEUROGENESIS IN EMBRYONIC BRAIN

The bulk of nervous system development occurs during the embryonic stage, commencing with the appearance of the primordial neural plate approximately 19 days after conception.

DOI: 10.1201/9780429265198-16

Before the emergence of the neural plate, the blastocyst, which is a cluster of cells that forms after fertilization, flattens and give rise to the embryonic disc in the second week of embryonic development. Next, the embryonic disc reorganizes into an outer ectodermal layer, a middle mesodermal layer, and an inner endodermal layer. Mesodermal cells give rise to the notochord, which is a flexible, rod-like structure that later becomes part of the vertebral column (Matthews and Lentz, 2020; De Vellis and Carpenter, 1999). The notochord and mesoderm induce part of the ectoderm, via chemical signaling, to form the neural plate along the dorsal side of the embryo around the day 19 after conception. The neural plate then lengthens and its edges fuse at the midline of the embryo, creating the neural tube by week 4 of development. The neural tube eventually differentiates into the central nervous system (CNS). After neural tube formation, the anterior region of the neural tube differentiates into three primary anatomic brain regions: the prosencephalon, the mesencephalon, and the rhombencephalon. Lateral aspects of the forebrain form the optic vesicle, which develops into the optic nerve, retina, and iris. The posterior region differentiates into the spinal cord. Multipotent neural crest cells, which temporarily arise from the ectoderm, line the walls of the neural tube, proliferate, and differentiate into neurons and glia that make up the peripheral nervous system (Matthews and Lentz, 2020; De Vellis and Carpenter, 1999)

The neuronal make-up of the CNS emerges from neural precursor cells known as neuroepithelial cells that line the center of the neural tube. Neuroepithelial cells give rise to neuroblasts, which differentiate into neurons. Neuroepithelial cells also differentiate into glioblasts, which are precursors to neuroglia. Newly formed neurons drift to different parts of the developing brain and self-assemble into unique brain structures (De Vellis and Carpenter, 1999). Once neurons settle into respective brain regions, they form axons and dendrites to enable synaptic communication with other neurons. Finally, they establish functional neuronal circuits to execute biological processes (Matthews et al., 2020).

13.3 NEUROGENESIS IN SOMATOSENSORY PLASTICITY

The somatosensory system of humans comprises spinal cord circuits, thalamic nuclei, stem nuclei, peripheral nerve afferents, and a group of cortical areas that are interconnected in a complex fashion to form a processing system that allows information from cutaneous and other peripheral receptors to guide behavior (Hollyday, 2001). When there is an injury to any region of the somatosensory system, people tend to experience brief perceptual and sensory impairments, which is followed by a partial or complex recovery of lost abilities and may take place over weeks to months (Eriksson et al., 1998; Cooper et al., 1993; Qi et al., 2011). The recovery can be simply learning to use what is left such that new ways of performing tasks are obtained, probably by using the intact portions of somatosensory system. However, much of the recovery actually depends on the reorganization of the somatosensory system such that the deactivated neurons regain responsiveness to other inputs and hence contribute perceptual and sensory functions.

Neuroplasticity is defined as the ability of the brain to reorganize itself by forming new functional synaptic connections throughout life (Quadrato et al., 2014). As part of adaptation to new situations, there is a continuous remodeling of cortical maps and neuronal connections in response to experiences (Taupin, 2008). Also, the reorganization is important in allowing healthy neurons to compensate for injured neurons and this is very important in maintaining sensory and perceptual functions in the somatosensory system (Sbordone et al., 1995).

Neurogenesis in somatosensory plasticity is not a well-explored area. Whether newborn neurons play a role in compensatory neuroplasticity following brain injury is yet to be answered. Nonetheless, what we currently know is that because of the unique properties possessed by newborn neurons, it is believed that they contribute significantly to the reorganization of brain circuitry (Quadrato et al., 2014). It has also been established that by virtue of their reduced threshold for the generation of long-term potentiation (Schmidt-Hieber et al., 2004), newborn neurons have a great capacity to facilitate synaptic plasticity, making them ideal to adapt to the reorganization of brain networks.

13.4 NEUROGENESIS IN ADULT BRAIN

Contrary to the current school of thought, it was widely believed that *de novo* formation of neurons did not occur in the adult CNS in a belief known as the 'central dogma of neurobiology'. Twentieth-century modern neuroscience pioneer Santiago Ramón y Cajal categorically debunked adult neurogenesis. The Spanish neuroscientist believed that once neurons fully develop, their growth and regeneration is halted irreversibly. By extrapolation, Cajal believed that once neurons die and synaptic formation is disrupted, regeneration from progenitor cells was implausible. Despite evidence of the existence of stem cells throughout the stages of life, this reigning dogma against the regeneration of neural cells in adulthood remained virtually unchallenged until the 1960s when evidence emerged supporting neurogenesis in the postnatal rat hippocampus (Ming and Song, 2011). Even then, the lack of robust experimental evidence prevented the scientific community from shifting its stance. It was not until the late twentieth century that evidence for the nature and extent of neurogenesis in the mammalian brain began to gain widespread acceptance (Akins and Garcia, 2016).

From a mechanistic perspective, neurogenesis was considered unlikely due to the high level of neuronal complexity. The pattern of dendritic arborization coupled with polysynaptic axonal combinations in most neurons suggested that they had permanently left the cell cycle and could no longer proliferate. Another reason for the dismissal of adult neurogenesis was purely conceptual, questioning how new neurons, with their dendrites, axons, and synapse, would incorporate into the existing adult brain circuitry

(Akins and Garcia, 2016). The advent of bromodeoxyuridine (BrdU), a synthetic nucleoside that detects proliferating cells and tissues, allowed *in vivo* monitoring of adult neurogenesis in the mammalian brain. In 1998, scientists discovered an indication of hippocampal neurogenesis in cadavers of cancer patients who had received BrdU injections to halt cancer cell proliferation (Akins and Garcia, 2016). In addition to BrdU, the discovery of neural stem cells that produce neurons, astrocytes, and oligodendrocytes in the adult brain demonstrated that neurons are not refractory to replication.

Scientists have identified two main neurogenic niches that contain neural stem cells for adult neurogenesis. The sub-granular zone (SGZ) of the hippocampal dentate gyrus (DG), which regulates learning and memory, contains neural progenitor cells that differentiate into neurons before they migrate into the granule cell layer, which lies adjacent to the SGZ (Greenberg, 2017). The subventricular zone (SVZ), which resides along the lateral ventricle, is another prominent site of neurogenesis in the adult mammalian brain. In rodents, neuroblasts in the SVZ proliferate and differentiate into interneurons as they migrate along the rostral migratory stream to the olfactory bulb. The human analog of this pathway becomes less prominent after infancy and replenishes both the olfactory bulb and prefrontal cortex (Greenberg, 2017). Besides the two prominent neurogenic niches, neurogenesis may also occur in the cortex, amygdala, substantia nigra, and hypothalamus, although evidence for this is rare and inconclusive (Thompson et al., 2008).

13.5 NEUROCHEMISTRY OF ADULT NEUROGENESIS

Constitutive neurogenesis that takes place in the SVZ and SGZ is functionally important in processes such as mood regulation, memory processing, and olfaction (Enwere et al., 2004; Kempermann et al., 2004). In part, the regulation of the generation of SVZ- and SGZ-born neuroblasts and their migration, differentiation, integration, and survival is dependent on diffusible neurochemicals such as neurotransmitters (Hagg, 2005; Lledo and Saghatelyan, 2005). Thus, it is hardly surprising that changes in brain neurotransmitter levels, albeit pathological or physiological, would affect adult neurogenesis. The following subsections will discuss the role of neurotransmitters such as dopamine (DA), glutamate, and gamma-aminobutyric acid (GABA) in adult neurogenesis.

13.5.1 Dopamine

Dopaminergic afferents originate from the substantia nigra pars compacta and project to the SVZ in rodents and primates, where they are believed to affect neuronal progenitor cell lineage (Höglinger et al., 2004; Freundlieb et al., 2006). The proposition that DA regulates neurogenesis in the SVZ stems from the immunohistochemical and electron microscopy findings that precursor cells in this region, including neuroblasts and transit-amplifying cells (C-cells), express DA receptors (Diaz et al., 1997; Höglinger et al., 2004). This is supported by evidence from preclinical studies that showed that ablation of DA neurons in the midbrain of rodents, by injection of selective neurotoxins, reduced the proliferation of SVZ progenitor cells (Winner et al., 2009; L'Episcopo et al., 2012). Predominantly, D_2-like DA receptors are expressed on transit-amplifying cells, whereas (PSA-NCAM+)-restricted neural precursors (A-cells) express both D_1- and D_2-like receptors (Höglinger et al., 2004).

This reduction in the proliferation of progenitor cells can be partly averted by enhancing dopaminergic activity in the affected brain, by increasing DA receptor signaling and by administering DA receptor agonists (Yang et al., 2008; Winner et al., 2009). Interestingly, in a DA D_2 receptor-dependent manner, administration of the DA receptor antagonist haloperidol resulted in increased proliferation of label-retaining cells (stem-like cells) in a study by Kippin et al. (2005). On the face of it, it may appear that these findings are contradictory and irreconcilable. However, based on the premise that one neurotransmitter may exert contrasting effects on neural stem cells and amplifying cells, it is believed that DA signaling might inhibit the proliferation of neural stem cells but promote proliferation of the transient-amplifying cells (Berg et al., 2013).

D_2 receptor-induced stimulation of neurogenesis is dependent on ciliary neurotrophic factor, which have been shown to promote proliferation in the SVZ (Emsley and Hagg, 2003; Yang et al., 2008). This proliferation is proposed to be mediated via Akt and extracellular signal-regulated kinase 1/2 signaling (Reynolds and Weiss, 1996; O'Keeffe et al., 2009). However, it has also been suggested that DA stimulates the release of epidermal growth factor (EGF), a known promoter of neural stem cell proliferation (Lao et al., 2013).

In summary, diverse studies have established that DA endogenously enhances adult neurogenesis in the SVZ via activation of D_2-like receptors expressed on transit-amplifying progenitor cells. In the adult mammalian brain, precursor cells are pharmacologically reachable and can be targeted by systemic administration of dopaminergic drugs to stimulate endogenous neurogenesis. This may therefore serve as a potential strategy for regenerative therapy in ND in which SVZ-derived cells potentially account for the repair progression (Curtis et al., 2003; Jin et al., 2004).

13.5.2 Glutamate

The role of glutamatergic transmission has been studied mostly in the hippocampus and is identified as an important player in setting the timing for survival, proliferation, migration, synapse formation, and integration of newly formed neurons in established synaptic networks (Vicini, 2008). There are three main sources of glutamatergic input into the DG: the dentate granule cells; contralateral hilar mossy cells; and neurons in layer II of the entorhinal cortex (Witter, 2007; Kumamoto et al., 2012).

In the injured brain, acutely giving group II mGluR agonists reduces injury-induced proliferation, whereas chronic activation of ionotropic glutamate receptors increases proliferation in the DG, especially during an ischemic injury (Feng et al., 2011). These findings indicate that the outcome on hippocampal progenitor varies depending on the predominant receptor subtypes involved, and whether the brain is injured. The exact downstream mechanisms involved in glutamate-mediated regulation of progenitor cell proliferation in the adult brain is yet to be fully understood. Nonetheless, further studies are needed to fully elucidate the mechanism(s) and illuminate the function of glutamatergic signaling on cell proliferation and subsequent neurogenesis (Berg et al., 2013).

13.5.3 GABA

Growing evidence suggests that the establishment of synaptic networks from the integration of newly formed neurons in the adult vertebrate brain is significantly altered by GABAergic neurotransmission (Berg et al., 2013). GABA, the main inhibitory neurotransmitter in the brain, is released mainly by interneurons but can also be released by astrocytes (Vicini, 2008). Its action is dependent on the unique presence and tonic activation of non-synaptic $GABA_A$ receptors which, in the SVZ, functions as a feedback regulator on the production and migration of neurons (Bordey, 2007). Additionally, GABAergic neurotransmission regulates differentiation and the timing of synaptic integration in the SGZ (Overstreet et al., 2004; Ge et al., 2007). The depolarizing outcome of GABA on immature neurons and neural progenitor cells, during developmental and adult neurogenesis provides a mechanistic basis for *in vivo* 'excitation–neurogenesis coupling' (Dammerman et al., 2000; Ge et al., 2006).

Activation of the $GABA_A$ receptor in the adult SVZ induces phosphorylation of histone H2AX, which then mediates the inhibition of cell cycle by GABA in this region, and also embryonic stem cell proliferation (Fernando et al., 2011; Andäng et al., 2008). To counteract the outcome of tonic activation of $GABA_A$ on neurogenesis, diazepam-binding inhibitor expressed by SVZ progenitor cells binds to a subunit of $GABA_A$ (Alfonso et al., 2012).

The role of $GABA_B$ in neurogenesis is largely unreported. However, it has been shown that $GABA_B$ receptor-mediated potassium responses are absent in newborn DG granules. Nevertheless, the reasons for the absence of $GABA_B$ receptor-mediated potassium responses is not completely understood, whether it is due to lack of functional $GABA_B$ receptors, G protein-coupled inward-rectifying potassium channels, or coupling between the two is unknown (Markwardt and Overstreet-Wadiche, 2008). As such, GABAergic signaling in newly developed DG is optimized for depolarization-mediated functions.

13.5.4 Other Neurochemicals

Aside the above-mentioned neurotransmitters, other forms of neurotransmission such as transmission through noradrenergic, cholinergic, and serotonergic neurons also affect proliferation, migration, differentiation, and integration of newly formed neurons. Other neurochemicals such as neuropeptide Y and the atypical neurotransmitter, nitric oxide, also play important roles in neurogenesis. Additionally, neurogenesis may be altered indirectly upon release of these neurotransmitters, via the action of neurotrophic factors or through upregulation of second messenger or downstream signaling systems without upregulating ligand expression (Berg et al., 2013).

Cholinergic input into the DG originates from the medial septum, and has been shown to alter proliferation (Swanson and Cowan, 1979; Dougherty and Milner, 1999). As proof of concept, long-term treatment with nicotine has been established to curtail proliferation in the DG (Abrous et al., 2002; Jang et al., 2002). In this part of the brain, neurons expressing choline acetyltransferase have been observed near progenitor cells. However, this observation has not been made in the SVZ (Kaneko et al., 2006). In the SGZ, polysialic acid-neural cell adhesion molecule-positive (PSA-NCAM+) and radial glial-like cells express muscarinic acetylcholine receptors, and ablation of these neurons reduces proliferation and impairs spatial memory (Van Kampen and Eckman, 2010; Itou et al., 2011). As opposed to nicotinic acetylcholine receptor activity, activation of muscarinic M_1 receptors in the SGZ results in enhanced proliferation in the SGZ (Van Kampen and Eckman, 2010). Moreover, acutely administering physostigmine, an indirect cholinomimetic, results in the inhibition of cell proliferation in the DG, whereas prolonged administration of donepezil, another indirect cholinomimetic, does not alter the quantity of cells that express proliferating cell nuclear antigen (Mohapel et al., 2005; Kotani et al., 2006).

Serotonergic projections found in the DG originate from the raphe nucleus, and destruction of these neurons results in a decrease in proliferation within the DG (Mongeau et al., 1997; Brezun and Daszuta, 1999). The resultant effect of this destruction has been shown to be salvaged by the grafting of fetal raphe neurons, implying that serotonin (hydroxytryptamine) stimulates neurogenesis in the DG (Brezun and Daszuta, 2000).

13.6 ROLE OF MICRORNAS IN NEUROGENESIS

MicroRNAs (miRNAs) represent short non-coding RNAs that are involved in the post-transcriptional level regulation of gene expression. They do this via mediation of translational inhibition or degradation of messenger RNA (mRNA; Lang and Shi, 2012). These types of RNA are involved in several biological processes, including neurogenesis. Specifically, miRNAs bind to the 3′-untranslated regions of mRNAs to regulate gene expression. The base pairing of a miRNA with this region of the mRNA can result in either translational inhibition or degradation of the mRNA. It is established that the percentage of the human genome that is under the control of miRNA regulation exceeds 60% (Friedman et al., 2009).

miRNAs are involved in several steps of the neurogenesis pathway.

A number of miRNAs are regulated during the process of neural development. The crucial roles that different miRNAs play in neurogenesis within the embryonic cortex have been demonstrated by a number of scientific studies.

13.6.1 MicroRNAs and Embryonic Neurogenesis

miR-124 is one well-elucidated miRNA specific to the brain. This one deserves a mention because it is heavily expressed during periods of brain development and has been proposed to facilitate differentiation of neurons in diverse ways (Ji et al., 2013). For instance, miR-124 is able to induce neurogenesis via the suppression of small C-terminal domain phosphatase 1 (SCP1) expression during development of the CNS (Visvanathan et al., 2007).

One study also reported that miR-124 facilitates the differentiation of neurons via repression of the neural progenitor-specific BAF complex 53 subunit (BAF53a), which is replaced by BAF53b during neuronal differentiation in the mouse embryo (Yoo et al., 2009). Another studied miRNA is miR-9, which represents another fascinating miRNA in the brain. The expression of miR-9 is activated during the formation of neurosphere and it is upregulated before terminal differentiation (Ji et al., 2013). Delaloy et al. (2010) reported that loss of the function of miR-9 leads to the suppression of neurogenic processes in human embryonic neural progenitor cells, by targeting stathmin, which, in turn, increases the instability of microtubules in migrating neuroblasts.

miR-137 is another miRNA that plays a key role in controlling the fate of neural stem cells. The overexpression of this important miRNA has been shown to inhibit the proliferation of cells and promote the neural differentiation of embryonic neural stem cells.

miR-128 is also upregulated during the period of brain development. When miR-128 is overexpressed, it promotes neural differentiation and also increases the mean dendritic spine length of neural stem cells (Ji et al., 2013). The targets of this miRNA play important roles in the nonsense-mediated decay (NMD) pathway. The NMD is involved in the degradation of abnormal mRNAs that harbor premature termination codons (Cheng et al., 2009).

13.6.2 MicroRNAs and Adult Neurogenesis

In the differentiation and proliferation of adult neural progenitors, the roles of miRNAs have been extensively studied. There is convincing evidence that miRNAs play a role in adult neurogenesis by fine-tuning the whole process. Zhao et al. (2009) reported that miR-9 regulates neural stem cell proliferation and differentiation by binding to the 3′-untranslated regions of TLX mRNA. During the process of neural differentiation, miR-9 is upregulated and TLX is suppressed. This leads to a reduction in neural progenitor cell proliferation and the promotion of neural differentiation.

The results of a study by Cheng et al. (2009) showed that miR-124 is upregulated during the transition from the transit-amplifying cell stage to the neuroblast stage in the SVZ. This upregulation of miR-124 promotes neural maturation. The observation means that miR-124 is therefore relevant in neuronal differentiation. Targets of miR-124 have been identified, including laminin $\gamma 1$ and integrin $\beta 1$. These are downregulated by the miRNA during neuronal differentiation (Cao et al., 2007).

One controller of adult neurogenesis known as the SRY-box transcription factor SOX9 is another target inhibited by miR-124 (Cheng et al., 2009). The significance of miR-124 has further been highlighted by the introduction of miR-124 (Ambasudhan et al., 2011; Yoo et al., 2011).

The roles of miR-137 have been studied in adult neural stem cell differentiation and proliferation. By targeting the histone H3 lysine 27 methyltransferase EZHZ, miR-137 is able to modulate the proliferation and differentiation of adult progenitor cells (Szulwach et al., 2010). Smrt et al. (2010) also reported that miR-137 regulates neuronal maturation by binding to Mind bomb 1 and inhibiting dendrite formation. Several other miRNAs have been studied extensively, together with their target in neurogenesis of the adult brain. Targets and functions of some miRNAs are summarized in Table 13.1.

13.7 HIPPOCAMPAL NEUROGENESIS

13.7.1 The Role of Omega-3 Fatty Acids

The hippocampal formation, composed of the hippocampus proper, DG, and subicular cortex, has been identified as a major neurogenic niche in the adult mammalian brain. Omega-3 polyunsaturated fatty acids (PUFAs) contribute to the function and development of the CNS and enhance the function of the hippocampus. In embryonic and postnatal rodents, a deficiency in PUFAs downregulates neurogenesis and inhibits normal development (Dyall, 2011). Several rodent models have demonstrated that PUFAs boost hippocampal neurogenesis in the adult brain and, consequently, may enhance hippocampal function. Kawakita et al. (2006) conducted a study in which they fed adult rats with docosahexaenoic acid (DHA), a type of omega-3 PUFA. The results revealed an increase in newborn neurons in the DG, demonstrating that DHA enhances neuronal proliferation.

A transgenic mouse line named fat-1 expresses the *Fat1* gene, which enables them to convert omega-6 PUFAs to omega-3 PUFAs. These transgenic mice exhibit increased brain DHA versus wild-type mice and, more importantly, display enhanced hippocampal neurogenesis. Additionally, the mice perform better in spatial learning tasks, suggesting that DHA has positive effects on neurogenesis (Dyall, 2011). The lobster brain – a non-mammalian model of adult neurogenesis – has also exhibited enhanced neurogenesis in the presence of

TABLE 13.1
MicroRNA (miRNA) Targets and Their Functions

miRNA	miRNA Target	Function of miRNA	Reference
let-7b	Hmga2	Promotion of neuronal differentiation	(Nishino et al., 2008)
	TLX, cyclin D1	Promotion of cell-cycle exit and neuronal differentiation	(Zhao et al., 2010)
miR-124	BAF53a	Promotion of neuronal differentiation	(Yoo et al., 2009)
	SOX9	Promotion of neuronal differentiation	(Cheng et al., 2009)
	PTBP1	Promotion of neuronal differentiation	(Makeyev et al., 2007)
	Ephrin-B1	Promotion of neuronal differentiation	(Arvanitis et al., 2010)
	SCP1	Promotion neuronal differentiation	(Visvanathan et al., 2007)
	CREB1	Synaptic activity inhibition	(Rajasethupathy et al., 2009)
miR-184	NUMBL1	Promotion of neural stem proliferation and inhibition of differentiation	(Liu et al., 2010)
miR-137	LSD1	Inhibition of neural stem cell proliferation	(Sun et al., 2010)
	MIB1	Inhibition of dendritic development	(Smrt et al., 2010)
	EZH2	Promotion of neural stem proliferation	(Szulwach et al., 2010)
miR-134	LIMK1	Inhibition of dendritic development	(Schratt et al., 2006)
MiR-128	UPF1	Promotion of neuronal maturation	(Bruno et al., 2011)
miR-9	TLX	Promotion of neural differentiation	(Zhao et al., 2009)
	STATHMIN	Promotion of progenitor proliferation	(Delaloy et al., 2010)
	HAIRY1	Inhibition of progenitor proliferation, promotion of neuronal fate, enhancement progenitor survival	(Bonev et al., 2011)
	SENSELESS	Promotion of neuronal differentiation	(Li et al., 2008)
	FOXG1	Promotion of neuronal differentiation	(Shibata et al., 2008)
	SIRT1	Promotion of neuronal differentiation	(Saunders et al., 2010)
miR-138	APTI	Inhibition of dendritic development	(Siegel et al., 2009)

APT1, acyl protein thioesterase 1; CREB1, cAMP response element-binding protein 1; EZH2, enhancer of zeste homolog 2; FOXG1, forkhead box protein G1; Hmga2, high mobility group AT-hook 2; Let-7, lethal-7; LIMK1, LIM domain kinase 1; MIB1, mind bomb homolog 1; NUMBL1, numblike 1; PTBP1, polypyrimidine tract binding protein 1; SCP1, small CTD; Phosphatases 1; SIRT1, sirtuin 1; SOX9, SRY-box containing gene 9.

omega-3 PUFAs. These collective observations demonstrate that omega-3 PUFAs upregulate hippocampal neurogenesis.

13.8 NEUROGENESIS IN OLFACTORY AND VOMERONASAL SENSORY EPITHELIA

Neurogenesis is reported to take place in three primary regions of the nervous system: the SGZ, the SVZ, and the olfactory neuroepithelia. The SVG supplies new granule cells to the DG, whereas the SVZ – located outside the walls of the lateral ventricles – supplies new relay neurons to the olfactory bulb. However, the olfactory neuroepithelia generate new sensory neurons that innervate the olfactory bulb (Brann and Firestein, 2014). The primary peripheral olfactory epithelia consist of the vomeronasal organ and the main olfactory epithelial (Figure 13.1 A). Neurogenesis in those regions is still under investigation, even though significant progress has been made.

The olfactory epithelial and the vomeronasal organ share similar functional and anatomic characteristics (Figure 13.1). Both are pseudostratified columnar epithelia comprising basal cells, immature and mature sensory neurons, sustentacular cells, and Bowman's gland cells (Brann and Firestein, 2014). The vomeronasal organs seem to be restricted to the detection of non-volatile odorants (Garrosa et al., 1998). The detection of odor takes place through a family of metabotropic receptors known as the vomeronasal receptors. It is reported that stem cells in the olfactory epithelia have the capacity to generate neurons (Leung et al., 2007). Furthermore, the developmental process of aging regulates olfactory neurogenesis (Brann and Firestein, 2014). Neurogenesis in these areas is thus complex, governed by several regulatory mechanisms.

Over the past five decades, knowledge on neurogenesis in the olfactory epithelial in vertebrates has advanced. Basal cells in the olfactory and vomeronasal epithelial retain the ability to generate new neurons throughout life (Brann and Firestein, 2014). It has been shown that the molecular and genetic determinants of neurogenesis in the olfactory epithelia are conserved between embryonic and postnatal stages. Some studies have demonstrated that neurogenesis in the olfactory epithelia is a result of the turnover of immature neurons rather than replenishment of mature sensory neurons (Hinds et al., 1984; Mackay-Sim and Kittel, 1991). This evidence thus implies that the newborn neurons do not attain maturity. However, a study has shown that newborn neurons actually attain maturity (Brann and Firestein, 2010).

In the vomeronasal organ, basal progenitor cells have the ability to undergo extensive neurogenesis, despite the fact that a large number of newborn cells die before becoming functional (Martinez-Marcos et al., 2005). Using lesion paradigms, the regenerative capacity in the olfactory and vomeronasal

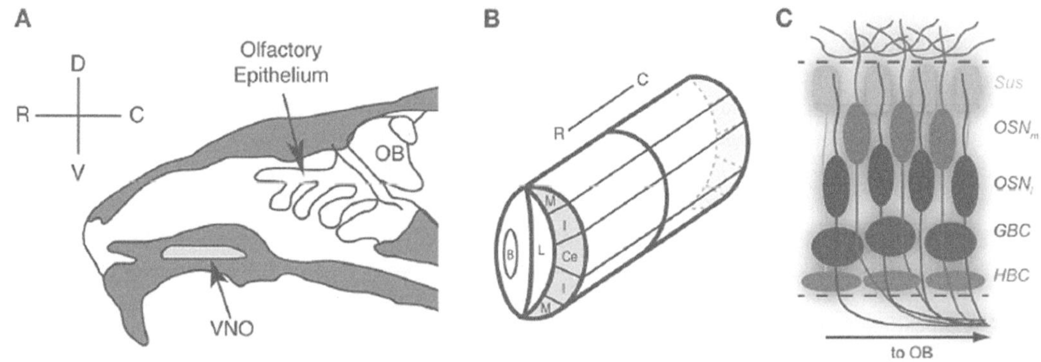

FIGURE 13.1 Organization and zones of the mouse olfactory epithelium. (A) Sagittal schematic of the rodent nose depicting the locations of the olfactory epithelium and the vomeronasal organ (VNO). (B) One half of a coronal plane of VNO, which is a bilaterally symmetrical tubular structure. (C) The five primary cell types of the olfactory epithelia: horizontal basal cell (HBC); globose basal cell (GBC); immature olfactory sensory neuron (OSN_i); mature olfactory sensory neuron (OSN_m); and sustentacular cell (Sus). OB, olfactory bulb; OE, olfactory epithelium; B, blood vessel; C, caudal; L, lumen; R, rostral. (Adapted from Brann and Firestein, 2014.)

epithelial has been studied extensively. A number of these methods target the mature neurons, such as the removal of the olfactory bulb or severing of the olfactory nerve. With these techniques, the result is the initial degeneration of sensory neurons, which is followed by a tremendous upregulation of proliferation of progenitor cells. Methimazole and methylbromide gas are types of chemical ablation that can also be used; these methods cause the most severe forms of lesions (Chen et al., 2005). The available literature on recovery following a lesion in the vomeronasal organ is scant. However, the regenerative capacity of the olfactory epithelia has been investigated more thoroughly. Studies have shown that recovery is evident within 45 days (Blanco-Hernández et al., 2012) and odorant receptor expression patterns are re-established within 90 days (Iwema et al., 2004). Studies using chemical ablation methods have shown that olfactory bulb removal is rapidly followed by the proliferation of basal cells (Matulionis, 1975; Hurtt et al., 1988), producing more that 8–10 million new neurons in total (Carter et al., 2004; Suárez et al., 2012). These studies show that remarkable neurogenesis occur in these epithelia.

13.9 NEUROGENESIS AND HUMAN PATHOLOGIES

13.9.1 Epilepsy

Epilepsies are a group of neurological disorders that cause hyperactivity in neuronal cells and are associated with hippocampal neuronal cell death, cognitive impairment, depression, and higher rates of NDs. Some epilepsies are idiopathic and may have an underlying genetic basis, while other forms are secondary to pathologies such as tumors, stroke, and infection. The hippocampal DG regulates the flow of excitatory input into the hippocampus, a function that is impaired in epilepsy (Peng and Bonaguidi, 2018). Epilepsy causes restructuring of the dentate network circuitry in both humans and animals in a manner that may increase excitatory input. In adult rodent studies, epileptic seizures initially induce neuronal cell death and subsequently stimulate neurogenesis in the DG (Thompson et al., 2008). However, long-term findings demonstrate an overall decrease in neurogenesis in epileptic patients. This may be due to the formation of abnormal synaptic contacts between newborn neurons and existing neurons (Peng and Bonaguidi, 2018). Furthermore, newly generated neurons may quickly adapt to the hyperexcitable, epileptic environment and become epileptogenic (Peng and Bonaguidi, 2018).

Mesial temporal lobe epilepsy (mTLE) is the most common form of intractable epilepsy in adults and possesses a distinct hippocampal pathology. Rodent models of mTLE have provided insight into the etiology of mTLE. mTLE involves spontaneous recurrent seizures that arise from mesial structures and involve the hippocampus within the temporal lobe. It is characterized by abnormal maturation of neurons in hippocampal neurogenesis, including neuronal hypertrophy, mossy fiber sprouting, abnormal hilar basal dendrites, and accelerated maturation. These abnormal characteristics of adult-generated neurons suggest that epilepsy-induced neurogenesis may not be reparative, but rather contribute to the persistence of epileptic seizures in rodent models. Although the exact role of epilepsy-induced neurogenesis remains unclear, the phenomenon represents a possible target for the development of drug therapies.

13.9.2 Cerebrovascular Disease

Cerebrovascular diseases are ubiquitous and are the leading cause of death and disability worldwide. Effective therapies for cerebrovascular diseases remain elusive, despite decades of research. There is evidence to support the upregulation of neurogenesis and the partial recovery of the adult brain after ischemic injury. Research has shown that cerebrovascular diseases like ischemic stroke and intracerebral hemorrhage can enhance proliferation of neuroblasts in the SVZ in rodents (Thompson et al., 2008). In humans, stroke-induced neurogenesis has been observed in affected brain regions after post-mortem brain tissue examinations (Thompson et al., 2008). The occurrence of neurogenesis in response to cerebrovascular damage or disease may help provide therapeutic approaches that stimulate neurogenesis to replace dead or damaged neurons.

Investigations into stroke-induced neurogenesis have identified growth factors, neurotransmitters, and hormones as modulators of cell proliferation, survival, and migration after stroke occurrence. Basic fibroblast growth factor (FGF-2) – a growth factor – has been shown to have a prominent function in neurogenesis, although there is conflicting evidence on its specific role (Thompson et al., 2008). Brain-derived neurotrophic factor (BDNF), is another important growth factor that, at best, functions as a neuroprotective agent but may increase neurogenesis in the hippocampus and SVZ and enhance cell migration.

13.10 NEUROGENESIS AND THE MOLECULAR MECHANISM OF MEMORY

The hippocampus of the brain plays an important role in memory function. Two structures, namely the DG and the Ammon's horn, make up the hippocampus. The hippocampus is involved in consolidation of information from short-term to long-term memory. It is also involved in spatial memory. In dementia, the hippocampus is one of the regions of the brain that is greatly damaged.

Several studies have reported the role of hippocampus in the formation of episodic memory during the mnemonic process (Morris et al., 2003; Neves et al., 2008). The identification of newly generated cells in the hippocampus opened the door to extensive studies on the functions of hippocampus-dependent learning and formation of memory (Altman, 1963) There is also compelling evidence that hippocampal neurogenesis is associated with some forms of hippocampus-dependent learning and memory formation (Leuner et al., 2006; Koehl and Abrous, 2011). It is reported that increased adult neurogenesis in the DG is associated with improvement of memory formation, maintenance, and acquisition of memory (Dupret et al., 2008; Saxe et al., 2006; Shors et al., 2002; Winocur et al., 2006; Imayoshi et al., 2008; Snyder et al., 2005)

Cognitive impairment that occurs in a number of NDs has been associated with dysfunction of the hippocampus, which, in turn, is partially linked to neurogenesis dysregulation in the adult (Yau et al., 2014). As such, a good understanding of the mechanisms involving neurogenesis in learning and memory is of great significance in the development of new cognitive enhancers.

13.10.1 Neurogenesis in the Adult Hippocampus

In adult mammals, the neurogenesis process takes place in the SGZ of the DG (Ming and Song, 2011; Cameron and McKay, 2001). Approximately 1400 newborn neurons have been reported to be added daily to the two hippocampi of adult humans. This accounts for 1.8% of the total renewable neuronal population (Spalding et al., 2005).

Neurogenesis in the adult hippocampus begins with the proliferation of progenitor cells in the SGZ (Zhao et al., 2006). Only a few of the progenitor cells differentiate into glial cells, with the majority differentiating into the granular cells in the DG (Aimone et al., 2010). Studies of morphological and physiological maturation patterns of newborn DG cells show that the granular layer of the DG starts in the first week of life (Aimone et al., 2010), with GABA playing a crucial role in their maturation and survival (Pontes et al., 2013). Other important factors like FGF-2, BDNF, and epidermal growth factor have also been shown to be involved in neuronal generation and development (Waterhouse et al., 2012; Mudò et al., 2009).

There is enough evidence that neurogenesis in the hippocampus plays an immense role in learning and memory. The actual mechanistic process for this contribution remains to be explained. The results of several computer-stimulated and experimental studies (using manipulations of transgenic animals or stimulation or inhibition of neurogenesis) reveal that the DG (the site of adult neurogenesis) is in charge of learning of new information, the formation of distinct memory, and consolidation of pattern separation of similar information (Kaptan and Üzüm, 2016). Pattern separation – an important characteristic of episodic memory – is thought to be a result of neurogenesis in the adult hippocampus (Deng et al., 2010).

As mentioned earlier, the exact role of adult hippocampal neurogenesis in the formation of memory remains to be fully understood. However, some evidence shows that it takes about 4 weeks for newly generated neurons to become fully integrated into hippocampal neural circuits. This slow process cannot explain the mechanism of plasticity alone. Besides their morphological changes, newly generated neurons have electrophysiological characteristics that are different from mature neurons. With a low threshold for excitability, these newborn neurons are able to induce long-term potentiation at a lower threshold than mature neurons (Schmidt-Hieber et al., 2004; Wang et al., 2000). These finding thus suggest that newly generated adult hippocampal neurons can enhance synaptic plasticity via their unique physiologic characteristics.

Despite the immense progress made in understanding the mechanisms of neurogenesis and its role in memory formation and consolidation, there is still not enough evidence to explain the functional importance of newborn neurons in the adult brain and the exact role played in memory formation and learning.

13.11 RECENT DEVELOPMENTS AND FUTURE RESEARCH DIRECTIONS

Some recent developments have occurred in the area of neurogenesis. It is without doubt that neurogenesis is a key component in the development of the CNS. One remarkable discovery is the association between zinc and neurogenesis.

Zinc plays a catalytic role in several biological process and is a cofactor for enzymes such as DNA ligases (Cotner-Gohara et al., 2010) and histone deacetylases (Marks, 2010). Zinc deficiency has been shown to reduce stem-cell proliferation in mice during brain development (Levenson and Morris, 2011). Zinc deficiency has also been reported to reduce the levels of hippocampal double cortin, which is a marker of neuronal differentiation (Suh et al., 2009; Gao et al., 2009). These studies thus reveal the implications of zinc deficiency on brain development and the critical role of zinc in neurogenesis, though several questions remain to be answered.

Other studies have looked at the role of neurogenesis as a potential therapeutic target in NDs like Parkinson's disease and Alzheimer's disease (Radad et al., 2017). Due to the chronic loss of neuronal cells in PD, stimulation of neurogenesis in the substantia nigra of these patients represents an important therapeutic interest. However, the development of the adult, mammalian substantia nigra dopaminergic neuron from precursor cells is yet to be unequivocally demonstrated *in vivo*. Further studies are thus needed to make useful this therapeutic approach.

To further deepen our understanding of the role of neurochemicals in the organization and differentiation of progenitor cells in different parts of the brain, the design of future studies should include a combination of cell-tracking studies with experimental manipulation of the release of neurotransmitters. The use of advanced imaging strategies will help provide additional insights into the development of adult neural stem cells from quiescence and proliferation. This will also contribute to further understanding the molecular mechanisms that underlie the proliferation, differentiation, and function of adult-born neurons.

13.12 CONCLUSION

Over the last few years, significant findings on neurogenesis in the adult mammalian brain have been reported. Increasingly, the role of neurotransmitters and neurotrophic factors in the proliferation, differentiation, migration, and integration are becoming apparent and rooted. Neurotransmission through dopaminergic, glutamatergic, GABAergic, adrenergic, cholinergic, and serotonergic neurons have been demonstrated to alter and regulate the processes of neurogenesis, dependent on the region of the brain involved. Thus, neurochemicals in the brain represent potential targets for current and novel therapies in the treatment of NDs, and management of the spectrum of symptoms associated with them. However, several questions remain unanswered, and this calls for more research work to be done. Furthermore, a better understanding of the primary regulatory mechanisms and the functions of neurogenesis in the adult brain in both pathological and physiological states will help hasten the development of new methods for investigating neurological diseases in humans. The discovery of miRNAs has also created a new direction in understanding the mechanisms that lead to neurogenesis at the post-transcriptional level by the repression of specific target genes. This advance may provide an approach to understanding or unraveling miRNA-mediated neurodevelopment and its role in neuropathology.

REFERENCES

Abrous DN, Adriani W, Montaron MF, Aurousseau C, Rougon G, Le Moal M, Piazza PV. Nicotine self-administration impairs hippocampal plasticity. J Neurosci. 2002;22:3656–62.

Aimone JB, Deng W, Gage FH. Adult neurogenesis: integrating theories and separating functions. Trends Cogn Sci. 2010;14 :325–37.

Akins MR, Garcia ADR. Neurogenesis in the adult brain. Encyclopedia of Cell Biology. 2016;4:134–40.

Alfonso J, Le Magueresse C, Zuccotti A, Khodosevich K, Monyer H. Diazepam binding inhibitor promotes progenitor proliferation in the postnatal SVZ by reducing GABA signaling. Cell Stem Cell. 2012;10:76–87.

Altman J. Autoradiographic investigation of cell proliferation in the brains of rats and cats. Anat Rec. 1963;145:573–91.

Altman J, Das GD. Autoradiographic and histological evidence of postnatal hippocampal neurogenesis in rats. J Comp Neurol. 1965;124:319–35.

Ambasudhan R, Talantova M, Coleman R, Yuan X, Zhu S, Lipton SA, Ding S. Direct reprogramming of adult human fibroblasts to functional neurons under defined conditions. Cell Stem Cell. 2011; 9:113–18.

Andäng M, Hjerling-Leffler J, Moliner A, Lundgren TK, Castelo-Branco G, Nanou E, et al. Histone H2AX-dependent GABA(A) receptor regulation of stem cell proliferation. Nature. 2008;451:460–4.

Arvanitis DN, Jungas T, Behar A. Ephrin-B1 reverse signaling controls a posttranscriptional feedback mechanism via miR-124. Mol Cell Biol. 2010;30 :2508–17.

Berg DA, Belnoue L, Song H, Simon A. Neurotransmitter-mediated control of neurogenesis in the adult vertebrate brain. Development. 2013;140:2548–61.

Bergmann O, Spalding KL, Frisén J. Adult neurogenesis in humans. Cold Spring Harb Perspect Biol. 2015;7:a018994.

Blanco-Hernández E, Valle-Leija P, Zomosa-Signoret V, Drucker-Colín R, Vidaltamayo R. Odor memory stability after reinnervation of the olfactory bulb. PLoS One. 2012;7:e46338.

Bonev B, Pisco A, Papalopulu N. MicroRNA-9 reveals regional diversity of neural progenitors along the anterior-posterior axis. Dev Cell. 2011;20:19–32.

Bordey A. Enigmatic GABAergic networks in adult neurogenic zones. Brain Res Rev. 2007;53:124–34.

Brann JH, Firestein S. Regeneration of new neurons is preserved in aged vomeronasal epithelia. J Neurosci. 2010;30: 15686–94.

Brann JH, Firestein SJ. A lifetime of neurogenesis in the olfactory system. Front Neurosci. 2014;8:182.

Brezun JM, Daszuta A. Depletion in serotonin decreases neurogenesis in the dentate gyrus and the subventricular zone of adult rats. Neuroscience. 1999;89:999–1002.

Brezun JM, Daszuta A. Serotonin may stimulate granule cell proliferation in the adult hippocampus, as observed in rats grafted with foetal raphe neurons. Eur J Neurosci. 2000;12:391–6.

Bruno, IG, Karam R, Huang L, Bhardwaj A, Lou CH, Shum EY, et al. Identification of a microRNA that activates gene expression by repressing nonsense-mediated RNA decay. Mol Cell. 2011;42:500–10.

Cameron HA, McKay RDG. Adult neurogenesis produces a large pool of new granule cells in the dentate gyrus. J Comp Neurol. 2001;435:406–17.

Cao X, Pfaff SL, Gage FH. A functional study of miR-124 in the developing neural tube. Genes Dev. 2007;21:531–6.

Carter LA, Macdonald JL, Roskams AJ. Olfactory horizontal basal cells demonstrate a conserved multipotent progenitor phenotype. J Neurosci. 2004;24:5670.

Chen H, Kohno K, Gong Q. Conditional ablation of mature olfactory sensory neurons mediated by diphtheria toxin receptor. J Neurocytol. 2005;34:37–47.

Cheng L-C, Pastrana E, Tavazoie M, Doetsch F. miR-124 regulates adult neurogenesis in the subventricular zone stem cell niche. Nat Neurosci. 2009;12:399.

Cooper BY, Glendinning DS, Vierck CJ. Finger movement deficits in the stumptail macaque following lesions of the fasciculus cuneatus. Somatosens Mot Res. 1993;10:17–29.

Cotner-Gohara E, Kim I-K, Hammel M, Tainer JA, Tomkinson AE, Ellenberger TJB. Human DNA ligase III recognizes DNA ends by dynamic switching between two DNA-bound states. Biochemistry. 2010;49:6165–76.

Curtis MA, Penney EB, Pearson AG, van Roon-Mom WM, Butterworth NJ, Dragunow M, et al. Increased cell proliferation and neurogenesis in the adult human Huntington's disease brain. Proc Natl Acad Sci U S A. 2003;100:9023–7.

Dammerman RS, Flint AC, Noctor S, Kriegstein AR. An excitatory GABAergic plexus in developing neocortical layer 1. J Neurophysiol. 2000;84:428–34.

De Vellis J, Carpenter E. General Development of the Nervous System. In: Siegel GJ, Agranoff BW, Albers RW, et al., editors. Basic Neurochemistry: Molecular, Cellular and Medical Aspects. 6th edition. Philadelphia: Lippincott-Raven. Available from: www.ncbi.nlm.nih.gov/books/NBK28253/. 1999.

Delaloy C, Liu L, Lee J-A, Su H, Shen F, Yang GY, et al. MicroRNA-9 coordinates proliferation and migration of human embryonic stem cell-derived neural progenitors. Cell Stem Cell. 2010;6:323–35.

Deng W, Aimone JB, Gage FH. New neurons and new memories: how does adult hippocampal neurogenesis affect learning and memory? Nat Rev Neurosci. 2010;11:339–50.

Diaz J, Ridray S, Mignon V, Griffon N, Schwartz JC, Sokoloff P. Selective expression of dopamine D3 receptor mRNA in proliferative zones during embryonic development of the rat brain. J Neurosci. 1997;17:4282–92.

Dougherty KD, Milner TA. Cholinergic septal afferent terminals preferentially contact neuropeptide Y-containing interneurons compared to parvalbumin-containing interneurons in the rat dentate gyrus. J Neurosci. 1999;19:10140–52.

Dupret D, Revest J-M, Koehl M, Ichas F, De Giorgi F, Costet P, et al. Spatial relational memory requires hippocampal adult neurogenesis. PLoS One. 2008;3:e1959.

Dyall SC. The role of omega-3 fatty acids in adult hippocampal neurogenesis. OCL. 2011;18:242–5.

Eichenbaum H. A cortical-hippocampal system for declarative memory. Nat Rev Neurosci. 2000;1:41–50.

Emsley JG, Hagg T. Endogenous and exogenous ciliary neurotrophic factor enhances forebrain neurogenesis in adult mice. Exp Neurol. 2003;183:298–310.

Enwere E, Shingo T, Gregg C, Fujikawa H, Ohta S, Weiss S. Aging results in reduced epidermal growth factor receptor signaling, diminished olfactory neurogenesis, and deficits in fine olfactory discrimination. J Neurosci. 2004;24:8354–65.

Eriksson PS, Perfilieva E, Björk-Eriksson T, Alborn AM, Nordborg C, Peterson DA, Gage FH. Neurogenesis in the adult human hippocampus. Nat Med. 1998;4:1313–17.

Feng YB, Yao H, Man X, Chi LY, Chi ZF. Effects of the group II mGlu receptor agonist 2R,4R-APDC on dentate gyrus cell proliferation in the adult rat brain after diffuse brain injury. Neurol Res. 2011;33:381–8.

Fernando RN, Eleuteri B, Abdelhady S, Nussenzweig A, Andäng M, Ernfors P. Cell cycle restriction by histone H2AX limits proliferation of adult neural stem cells. Proc Natl Acad Sci U S A. 2011;108:5837–42.

Freundlieb N, François C, Tandé D, Oertel WH, Hirsch EC, Höglinger GU. Dopaminergic substantia nigra neurons project topographically organized to the subventricular zone and stimulate precursor cell proliferation in aged primates. J Neurosci. 2006;26:2321–5.

Friedman RC, Farh KK-H, Burge CB, Bartel DP. Most mammalian mRNAs are conserved targets of microRNAs. Genome Res. 2009;19:92–105.

Gao H-L, Zheng W, Xin N, Chi ZH, Wang ZY, Chen J, Wang ZY. Zinc deficiency reduces neurogenesis accompanied by neuronal apoptosis through caspase-dependent and-independent signaling pathways. Neurotox Res. 2009;16:416.

Garrosa M, Gayoso MJ, Esteban FJ. Prenatal development of the mammalian vomeronasal organ. Microsc Res Tech. 1998;41:456–70.

Ge S, Goh EL, Sailor KA, Kitabatake Y, Ming GL, Song H. GABA regulates synaptic integration of newly generated neurons in the adult brain. Nature. 2006;439:589–93.

Ge S, Pradhan DA, Ming GL, Song H. GABA sets the tempo for activity-dependent adult neurogenesis. Trends Neurosci. 2007;30:1–8.

Greenberg DA. Neurogenesis in Cerebrovascular Disease. In: Primer on Cerebrovascular Diseases: Second Edition, 2017; 89–91.

Hagg T. Molecular regulation of adult CNS neurogenesis: an integrated view. Trends Neurosci. 2005;28:589–95.

Hinds JW, Hinds PL, McNelly NA. An autoradiographic study of the mouse olfactory epithelium: Evidence for long-lived receptors. Anat Rec. 1984;210:375–83.

Höglinger GU, Rizk P, Muriel MP, Duyckaerts C, Oertel WH, Caille I, Hirsch EC. Dopamine depletion impairs precursor cell proliferation in Parkinson disease. Nat Neurosci. 2004;7:726–35.

Hollyday M. Neurogenesis in the vertebrate neural tube. Int J Dev Neurosci. 2001;19:161–73.

Hurtt ME, Thomas DA, Working PK, Monticello TM, Morgan KT. Degeneration and regeneration of the olfactory epithelium following inhalation exposure to methyl bromide: pathology, cell kinetics, and olfactory function. Toxicol Appl Pharmacol. 1988;94:311–28.

Imayoshi I, Sakamoto M, Ohtsuka T, Takao K, Miyakawa T, Yamaguchi M, et al. Roles of continuous neurogenesis in the structural and functional integrity of the adult forebrain. Nat Neurosci. 2008;11:1153–61.

Itou Y, Nochi R, Kuribayashi H, Saito Y, Hisatsune T. Cholinergic activation of hippocampal neural stem cells in aged dentate gyrus. Hippocampus. 2011;21:446–59.

Iwema CL, Fang H, Kurtz DB, Youngentob SL, Schwob JE. Odorant receptor expression patterns are restored in lesion-recovered rat olfactory epithelium. J Neurosci. 2004;24:356.

Jang MH, Shin MC, Jung SB, Lee TH, Bahn GH, Kwon YK, et al. Alcohol and nicotine reduce cell proliferation and enhance apoptosis in dentate gyrus. Neuroreport. 2002;13:1509–13.

Ji F, Lv X, Jiao J. The role of microRNAs in neural stem cells and neurogenesis. Genet Genom. 2013;40:61–6.

Jin K, Galvan V, Xie L, Mao XO, Gorostiza OF, Bredesen DE, Greenberg DA. Enhanced neurogenesis in Alzheimer's disease transgenic (PDGF-APPSw,Ind) mice. Proc Natl Acad Sci U S A. 2004;101:13363–7.

Kaneko N, Okano H, Sawamoto K. Role of the cholinergic system in regulating survival of newborn neurons in the adult mouse dentate gyrus and olfactory bulb. Genes Cells. 2006;11:1145–59.

Kaptan Z, Üzüm G. The role of adult hippocampal neurogenesis in learning and memory function. Turk J Neurol. 2016;22:149–55

Kawakita E, Hashimoto M, Shido O. Docosahexaenoic acid promotes neurogenesis in vitro and in vivo. Neuroscience. 2006;139:991–7.

Kempermann G, Wiskott L, Gage FH. Functional significance of adult neurogenesis. Curr Opin Neurobiol. 2004;14;186–91.

Kippin TE, Kapur S, van der Kooy D. Dopamine specifically inhibits forebrain neural stem cell proliferation, suggesting a novel effect of antipsychotic drugs. J Neurosci. 2005;25:5815–23.

Koehl M, Abrous DN. A new chapter in the field of memory: adult hippocampal neurogenesis. Eur J Neurosci. 2011;33:1101–14.

Kotani S, Yamauchi T, Teramoto T, Ogura H. Pharmacological evidence of cholinergic involvement in adult hippocampal neurogenesis in rats. Neuroscience. 2006;142:505–14.

Kumamoto N, Gu Y, Wang J, Janoschka S, Takemaru K, Levine J, Ge S. A role for primary cilia in glutamatergic synaptic integration of adult-born neurons. Nat Neurosci. 2012;15:399–405.

Lang M-F, Shi YJFIN. Dynamic roles of microRNAs in neurogenesis. Front Neurosci. 2012;6:71.

Lao CL, Lu CS, Chen JC. Dopamine D3 receptor activation promotes neural stem/progenitor cell proliferation through AKT and ERK1/2 pathways and expands type-B and -C cells in adult subventricular zone. Glia. 2013;61:475–89.

L'Episcopo F, Tirolo C, Testa N, Caniglia S, Morale MC, Deleidi M, et al. Plasticity of subventricular zone neuroprogenitors in MPTP (1-methyl-4-phenyl-1,2,3,6-tetrahydropyridine) mouse model of Parkinson's disease involves cross talk between inflammatory and Wnt/β-catenin signaling pathways: functional consequences for neuroprotection and repair. J Neurosci. 2012;32:2062–85.

Leuner B, Gould E, Shors TJ. Is there a link between adult neurogenesis and learning? Hippocampus. 2006;16:216–24.

Leung CT, Coulombe PA, Reed RR. Contribution of olfactory neural stem cells to tissue maintenance and regeneration. Nat Neurosci. 2007;10:720–6.

Levenson CW, Morris DJAIN. Zinc and neurogenesis: making new neurons from development to adulthood. Adv Nutr. 2011;2:96–100.

Li W, Sun G, Yang S, Qu Q, Nakashima K, Shi Y. Nuclear receptor TLX regulates cell cycle progression in neural stem cells of the developing brain. Mol Endocrinol. 2008;22:56–64.

Lindsey BW, Tropepe V. A comparative framework for understanding the biological principles of adult neurogenesis. Prog Neurobiol. 2006;80:281–307.

Liu C, Teng ZQ, Santistevan NJ, Szulwach KE, Guo W, Jin P, Zhao X. Epigenetic regulation of miR-184 by MBD1 governs neural stem cell proliferation and differentiation. Cell Stem Cell. 2010;6:433–44.

Lledo PM, Saghatelyan A. Integrating new neurons into the adult olfactory bulb: joining the network, life-death decisions, and the effects of sensory experience. Trends Neurosci. 2005;28:248–54.

Mackay-Sim A, Kittel P. Cell dynamics in the adult mouse olfactory epithelium: a quantitative autoradiographic study. Eur J Neurosci. 1991;11:979–84.

Makeyev EV, Zhang J, Carrasco MA, Maniatis T. The MicroRNA miR-124 promotes neuronal differentiation by triggering brain-specific alternative pre-mRNA splicing. Mol Cell. 2007;27:435–48.

Marks PA. Histone deacetylase inhibitors: a chemical genetics approach to understanding cellular functions. J Biochim Biophys. 2010;1979:717–25.

Markwardt S, Overstreet-Wadiche L. GABAergic signalling to adult-generated neurons. J Physiol. 2008;586:3745–9.

Martinez-Marcos A, Jia C, Quan W, Halpern M. Neurogenesis, migration, and apoptosis in the vomeronasal epithelium of adult mice. J Neurobiol. 2005;63:173–87.

Matthews PBC, Lentz TL. Morphological development. Encyclopædia Britannica, 2020; www.britannica.com/science/human-nervous-system/Morphological-development.

Matulionis DH. Ultrastructural study of mouse olfactory epithelium following destruction by $ZnSO_4$ and its subsequent regeneration. Am J Anat. 1975;142:67–89.

Ming G-L, Song H. Adult neurogenesis in the mammalian brain: significant answers and significant questions. Neuron. 2011;70:687–702.

Mohapel P, Leanza G, Kokaia M, Lindvall O. Forebrain acetylcholine regulates adult hippocampal neurogenesis and learning. Neurobiol Aging. 2005;26:939–46.

Mongeau R, Blier P, de Montigny C. The serotonergic and noradrenergic systems of the hippocampus: their interactions and the effects of antidepressant treatments. Brain Res Brain Res Rev. 1997;23:145–95.

Moreno-López B, Romero-Grimaldi C, Noval JA, Murillo-Carretero M, Matarredona ER, Estrada C. Nitric oxide is a physiological inhibitor of neurogenesis in the adult mouse subventricular zone and olfactory bulb. J Neurosci. 2004;24:85–95.

Morris RG, Moser EI, Riedel G, Martin SJ, Sandin J, Day M, O'Carroll C. Elements of a neurobiological theory of the hippocampus: the role of activity-dependent synaptic plasticity in memory. Philos Trans R Soc Lond B Biol Sci. 2003;358:773–86.

Mudò, G, Bonomo A, Di Liberto V, Frinchi M, Fuxe K, Belluardo N. The FGF-2/FGFRs neurotrophic system promotes neurogenesis in the adult brain. J Neural Transm (Vienna). 2009;116:995–1005.

Neves G, Cooke SF, Bliss TV. Synaptic plasticity, memory and the hippocampus: a neural network approach to causality. Nat Rev Neurosci. 2008;9:65–75.

Nishino J, Kim I, Chada K, Morrison SJ. Hmga2 promotes neural stem cell self-renewal in young but not old mice by reducing p16Ink4a and p19Arf Expression. J Cell Biol. 2008;135:227–39.

O'Keeffe GC, Tyers P, Aarsland D, Dalley JW, Barker RA, Caldwell MA. Dopamine-induced proliferation of adult neural precursor cells in the mammalian subventricular zone is mediated through EGF. Proc Natl Acad Sci U S A. 2009;106:8754–9.

Overstreet LS, Hentges ST, Bumaschny VF, de Souza FS, Smart JL, Santangelo AM, et al. A transgenic marker for newly born granule cells in dentate gyrus. J Neurosci. 2004;24:3251–59.

Peng L, Bonaguidi MA. Function and dysfunction of adult hippocampal neurogenesis in regeneration and disease. Am J Pathol. 2018;188:23–8.

Pontes A, Zhang Y, Hu W. Novel functions of GABA signaling in adult neurogenesis. Front Biol. 2013;8:10.1007/s11515-013-1270-2.

Qi H-X, Chen LM, Kaas JH. Reorganization of somatosensory cortical areas 3b and 1 after unilateral section of dorsal columns of the spinal cord in squirrel monkeys. J Neurosci. 2011;31:13662–75.

Quadrato G, Elnaggar M, Di Giovanni S. Adult neurogenesis in brain repair: cellular plasticity vs. cellular replacement. Front Neurosci. 2014;8:17.

Radad K, Moldzio R, Al-Shraim M, Kranner B, Krewenka C, Rausch WD. Recent advances on the role of neurogenesis in the adult brain: therapeutic potential in Parkinson's and Alzheimer's diseases. CNS Neurol Disord Drug Targets. 2017;16:740–8.

Rajasethupathy P, Fiumara F, Sheridan R, Betel D, Puthanveettil SV, Russo JJ, et al. Characterization of small RNAs in aplysia reveals a role for miR-124 in constraining synaptic plasticity through CREB. Neuron. 2009;63:803–17.

Reynolds BA, Weiss S. Generation of neurons and astrocytes from isolated cells of the adult mammalian central nervous system. Science. 1992;255:1707–10.

Reynolds BA, Weiss S. Clonal and population analyses demonstrate that an EGF-responsive mammalian embryonic CNS precursor is a stem cell. Dev Biol. 1996;175:1–13.

Saunders LR, Sharma AD, Tawney J, Nakagawa M, Okita K, Yamanaka S, et al. miRNAs regulate SIRT1 expression during mouse embryonic stem cell differentiation and in adult mouse tissues. Aging. 2010;2:413.

Saxe MD, Battaglia F, Wang J-W, Malleret G, David DJ, Monckton JE, et al. Ablation of hippocampal neurogenesis impairs contextual fear conditioning and synaptic plasticity in the dentate gyrus. Proc Natl Acad Sci U S A. 2006;103:17501.

Sbordone RJ, Liter J, Pettler-Jennings P. Recovery of function following severe traumatic brain injury: a retrospective 10-year follow-up. J Brain injury. 1995;9:285–99.

Schmidt-Hieber C, Jonas P, Bischofberger J. Enhanced synaptic plasticity in newly generated granule cells of the adult hippocampus. Nature. 2004;429:184–7.

Schratt GM, Tuebing F, Nigh EA, Kane CG, Sabatini ME, Kiebler M, Greenberg ME. A brain-specific microRNA regulates dendritic spine development. Nature. 2006;439:283–9.

Scoville WB, Milner B. Loss of recent memory after bilateral hippocampal lesions. Neurol Neurosurg Psychiatry. 1957;20:11–21.

Shibata M, Kurokawa D, Nakao H, Ohmura T, Aizawa S. MicroRNA-9 modulates Cajal–Retzius cell differentiation by suppressing Foxg1 expression in mouse medial pallium. J Neurosci. 2008;28:10415–21.

Shors TJ, Townsend DA, Zhao M, Kozorovitskiy Y, Gould E. Neurogenesis may relate to some but not all types of hippocampal-dependent learning. Hippocampus. 2002;12:578–84.

Siegel G, Obernosterer G, Fiore R, Oehmen M, Bicker S, Christensen M, et al. A functional screen implicates microRNA-138-dependent regulation of the depalmitoylation enzyme APT1 in dendritic spine morphogenesis. Nat Cell Biol. 2009;11:705–16.

Smrt RD, Szulwach KE, Pfeiffer RL, Li X, Guo W, Pathania M, et al. MicroRNA miR-137 regulates neuronal maturation by targeting ubiquitin ligase mind bomb-1. Stem cells. 2010;28:1060–70.

Snyder JS, Hong NS, McDonald RJ, Wojtowicz JM. A role for adult neurogenesis in spatial long-term memory. Neuroscience. 2005;130:843–52.

Song J, Zhong C, Bonaguidi MA, Sun GJ, Hsu D, Gu Y, et al. Neuronal circuitry mechanism regulating adult quiescent neural stem-cell fate decision. Nature. 2012;489:150–4.

Spalding KL, Bhardwaj RD, Buchholz BA, Druid H, Frisén J. Retrospective birth dating of cells in humans. Cell. 2005;122:133–43.

Suárez R, García-González D, De Castro F. Mutual influences between the main olfactory and vomeronasal systems in development and evolution. Front Neuroanat. 2012;6:50.

Suh SW, Won SJ, Hamby AM, Yoo BH, Fan Y, Sheline CT, et al. Decreased brain zinc availability reduces hippocampal neurogenesis in mice and rats. J Cereb Blood Flow Metab. 2009;29:1579–88.

Sun G, Alzayady K, Stewart R, Ye P, Yang S, Li W, Shi Y. Histone demethylase LSD1 regulates neural stem cell proliferation. Mol Cell Biol. 2010;30:1997–2005.

Swanson LW, Cowan WM. The connections of the septal region in the rat. J Comp Neurol. 1979;186:621–55.

Szulwach K, Chang S, Jin P. Crosstalk between microRNA and epigenetic regulation in stem cells. macro roles for microRNAs in the life and death of neurons. J Cell Biol. 2010;189:127–41.

Taupin P. Stem Cells and Regenerative Medicine: Volume I: Adult Neurogenesis and Neural Stem Cells. Nova Publishers. 2008.

Thompson BAK, Van Dam A-M, Lucassen PJ. Changes in adult neurogenesis in neurodegenerative diseases: cause or consequence? Genes Brain Behav. 2008;7:28–42.

Van Kampen JM, Eckman CB. Agonist-induced restoration of hippocampal neurogenesis and cognitive improvement in a model of cholinergic denervation. Neuropharmacology. 2010;58:921–9.

Vicini S. The role of GABA and glutamate on adult neurogenesis. J Physiol. 2008;586:3737–8

Visvanathan J, Lee S, Lee B, Lee JW, Lee SK. The microRNA miR-124 antagonizes the anti-neural REST/SCP1 pathway during embryonic CNS development. Genes Dev. 2007;21:744–9.

Wang S, Scott BW, Wojtowicz JM. Heterogenous properties of dentate granule neurons in the adult rat. J Neurobiol. 2000;42:248–57.

Waterhouse EG, An JJ, Orefice LL, Baydyuk M, Liao GY, Zheng K, et al. BDNF promotes differentiation and maturation of adult-born neurons through GABAergic transmission. J Neurosci. 2012;32:14318–30.

Winner B, Desplats P, Hagl C, Klucken J, Aigner R, Ploetz S, et al. Dopamine receptor activation promotes adult neurogenesis in an acute Parkinson model. Exp Neurol. 2009;219:543–52.

Winocur G, Wojtowicz JM, Sekeres M, Snyder JS, Wang S. Inhibition of neurogenesis interferes with hippocampus-dependent memory function. Hippocampus. 2006;16:296–304.

Witter MP. The perforant path: projections from the entorhinal cortex to the dentate gyrus. Prog. Brain Res. 2007;163:43–61.

Yang P, Arnold SA, Habas A, Hetman M, Hagg T. Ciliary neurotrophic factor mediates dopamine D2 receptor-induced CNS neurogenesis in adult mice. J Neurosci. 2008;28:2231–41.

Yau S-Y, Gil-Mohapel J, Christie BR, So K-F. Physical exercise-induced adult neurogenesis: a good strategy to prevent cognitive decline in neurodegenerative diseases? Biomed Res Int. 2014;2014:403120.

Yoo AS, Staahl BT, Chen L, Crabtree GR. MicroRNA-mediated switching of chromatin-remodelling complexes in neural development. Nature. 2009;460:642–6.

Yoo AS, Sun AX, Li L, Shcheglovitov A, Portmann T, Li Y, et al. MicroRNA-mediated conversion of human fibroblasts to neurons. Nature. 2011;476:228–31.

Zhao C, Sun G, Li S, Lang MF, Yang S, Li W, Shi Y. MicroRNA let-7b regulates neural stem cell proliferation and differentiation by targeting nuclear receptor TLX signaling. Proc Natl Acad Sci U S A. 2010;107:1876–81.

Zhao C, Sun G, Li S, Shi Y. A feedback regulatory loop involving microRNA-9 and nuclear receptor TLX in neural stem cell fate determination. Nat Struct Mol Biol. 2009;16:365.

Zhao C, Teng EM, Summers RG, Ming GL, Gage FH. Distinct morphological stages of dentate granule neuron maturation in the adult mouse hippocampus. J Neurosci. 2006;26:3–11.

14 Neuronal Proliferation and Associated Diseases

Vivek Kumar Gupta and Bechan Sharma
Department of Biochemistry, Faculty of Science, University of Allahabad, Allahaabd, India

14.1 INTRODUCTION

The complexity of the brain begins along with its evolution after the crucial external form is established by a series of sequential events. These events involve neuronal proliferation in the brain (Lee, 2019). These proliferating neurons migrate to the specific sites of the brain in the central nervous system (CNS). This series of organizational events results in the complex circuitry characteristic of the brain (i.e., its neuron-specific membrane). These actions span a period from the second month of gestation to adult life, including the perinatal period (Barnette et al., 2009). Abnormalities in the development of human brain may be a significant consequence of genetic alterations or mutations, as well as a variety of prenatal and perinatal circumstances at crucial times in its development (Barnette et al., 2009).

The brain (i.e., CNS) is accountable for the normal functioning of the whole body. The development of the human brain occurs in a wonderful way, due to genes and the environment (Esposito, 2017). Usually, the genes a child inherits from its parents are sufficient for the good development of the brain. The environment comprises the situation within the womb and delivery, in addition to both the physical and emotional conditions and events after birth. A normal, healthy pregnancy provides the best environment in which to support the good development of the brain (Schore, 2003).

It is a very exciting and active process, the understanding of which is still evolving (Figure 14.1). Brain development includes the formation of the neural tube (day 18 – week 7 of gestation), prosencephalon (weeks 2–3 of gestation), neuronal proliferation (months 2–3 of gestation), migration (months 3–5 of gestation), organization (month 5 of gestation to years postnatally), and myelination (birth to years postnatally) (Herculano-Houzel et al., 2016). Among these brain development events, the organization of neurons and their myelination may also occur postnatally. The maturation of the CNS can be divided into six major sequential events: dorsal induction; ventral induction; proliferation; migration; organization; and myelination (Lowery and Sive, 2004). However, in reality, several stages overlap.

Proliferation is the third phase in the formation of the CNS. It is a two-phase process comprising glioblasts and neuroblasts (Mangoura et al., 1997). The development of neural cells (first phase) and the basic structure to support newly formed neural cells (second phase) is formed by neuroblasts and glioblasts in a mature brain (Silbereis et al., 2016). The major events of proliferation occur in the first trimester of gestation and span months 2–4 of gestation (Herculano-Houzel et al., 2016). The major event that occurs during neuronal proliferation is the production of a proliferative unit. The proliferative unit is formed by the symmetrical division of progenitor cells, followed by enlargement of the proliferative unit (Lahti et al., 2011). The enlargement of the proliferative unit occurs by the asymmetric division of progenitor cells before neuronal migration. The second phase of development takes place from months 2 to 4 of gestation, and goes through its most rapid proliferation during postnatal months 5–12 (Herculano-Houzel

DOI: 10.1201/9780429265198-17

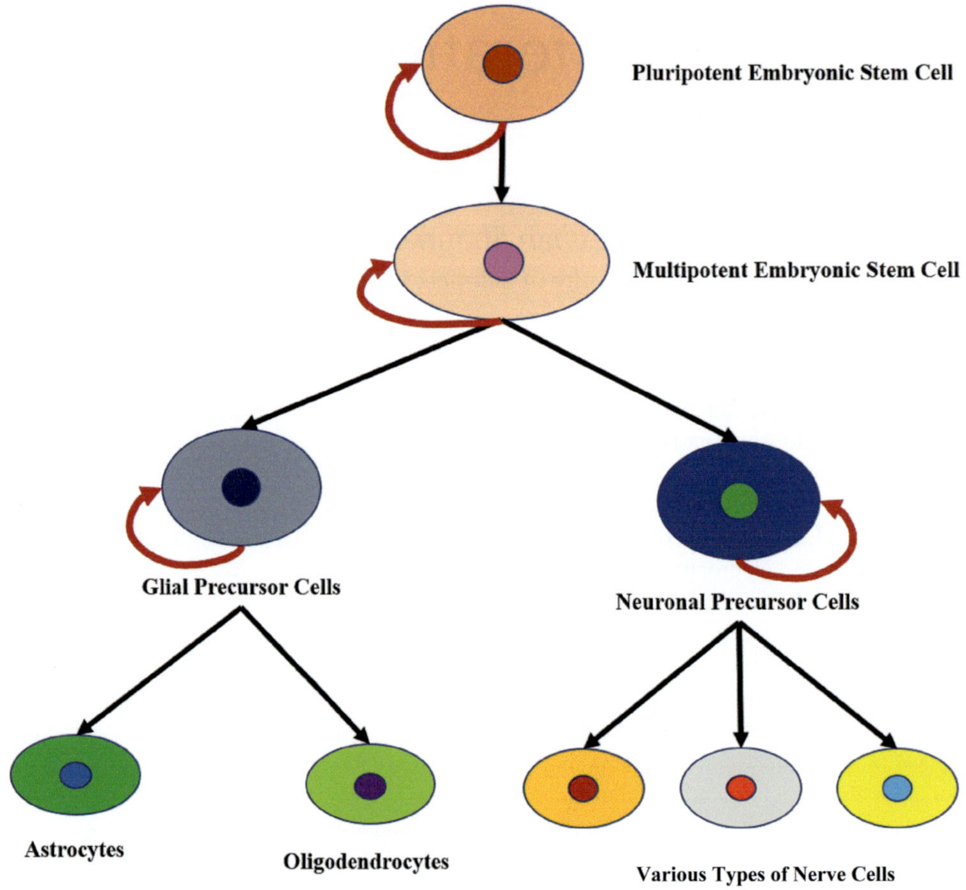

FIGURE 14.1 A diagrammatic representation of the normal events of neuronal proliferation.

et al., 2016). The glioblasts and four neuroblasts begin to grow quickly during the events of neuronal proliferation to form a number of neural cells an individual will have for life (Mashanov et al., 2013).

The neuron (nerve cell) is the basic unit of the nervous system (Branco and Hausser, 2010; Huang et al., 2019). The number of nerve cells formed during neuronal proliferation is approximately 40–50% more than required for any mature functional brain. Once proliferation halts, new neurons cannot be formed (Gross, 2000). However, reorganization may happen in some neurons (Weis et al., 2019). When any stroke causes a neuron to expire, it is possible for other nerve cells to take over the functions of the dead cell and offer some recovery. The function may not be as well coordinated as the original function of the dead neuron. Interruption in the proliferation of neurons in the developing nervous system may result in an abnormal brain size (i.e., microcephaly or megalocephaly) (Rice and Barone 2000; Budday et al., 2015). Microcephaly (Cox et al., 2006), where the brain is too small, is related to mild-to-moderate mental disability; a brain that is too large in size is termed megalocephaly/megalencephaly (De Myer 1986). Neuronal proliferation and migration occurs simultaneously in the embryo.

This chapter explains the aspects of normal neuronal proliferation and associated syndromes that occur when normal brain development is disturbed. This chapter is devoted to the major processes of brain development (i.e., neuronal proliferation events) and what happens when they are disturbed. The information you will find in this chapter will enhance your understanding of how proliferation occurs during brain development. Increased knowledge of the structure and function of the brain can be beneficial in making sense of reports on the brain or help in discussions with clinicians.

14.2 NEURONAL PROLIFERATION UNDER NORMAL CONDITIONS/NORMAL DEVELOPMENT

Major proliferative events initially span months 2 and 4 of gestation, peaking at the end of the first trimester and into the onset of the second trimester (i.e., months 3 and 4; Table 14.1) (Herschkowitz, 1988; Webb et al., 2001; Poduri and Volpe, 2018). Initially, a tangential stream of neuronal migration from the ganglionic eminence results in the establishment of the preplate or marginal zone. All glia and radially migrating neurons are derivatives of the subventricular and ventricular zones, present at the subependymal position in the developing brain and nervous system (Herschkowitz, 1988; Goldman, 2003; Malatesta et al., 2003).

TABLE 14.1
Proliferation of Neurons

Period	Major events
Months 2–4 of gestation	1. Subventricular and ventricular zones are the sites of proliferation
	2. Formation of proliferative units by symmetric division of neural progenitor cells
	3. Enlargement of proliferative units by asymmetrical division of progenitor cells

TABLE 14.2
Functions of Radial Glial Cells

Serial No.	Functions
1	Progenitor cell for cortical neurons
2	Guides the migration of neurons
3	Progenitor cell for astrocytes and radial glia
4	Neural stem cell present in the subventricular zone

Information regarding the proliferation of nerve cells is the result of studies of the deposition of the brain's genetic material and the biochemical correlates of a number of cells (Samuelsen et al., 2003). A leptomeningeal plexus of vessels is seen at the end of the first trimester (i.e., in the third month of gestation). These primary unbranched vessels in the first month of the second trimester (i.e., month 4 of gestation) later progres into a horizontal branching (Norman and O'Kusky, 1986). The fundamental characteristics of nerve cell proliferation in the wall of neural tubes were defined foremost on the basis of morphology (Sauer, 1935; Puelles, 2001), and were further described by several researchers, using autoradiography (Sidman, 1962; Berry et al., 1964; Rakic and Sidman, 1968; Rakic, 1985). These characteristics have also been established by immunocytochemistry, time-lapse multiphoton imaging, computer-assisted serial electron micrographic reconstruction, and various molecular techniques (Noctor et al., 2007; Miller et al., 2014; Hoerder-Suabedissen and Molnar, 2015).

During the development of cortical plate (CP), the cerebral wall differentiates into major zones: the ventricular zone (VZ); subventricular zone (SZ); intermediate zone (IZ); and the marginal zone (Rakic, 1975). Cells at the periphery of the VZ divide and migrate toward its surface (Noctor et al., 2007). The result is that offspring cells migrate back toward the periphery of VZ. In monkey cerebrum, the VZ generates the maximum number of neurons; the SZ is the place of origin for some neurons and glia (Rakic, 1974; Kostovic and Rakic, 1980). When cells stop the mitotic cell cycle and proliferative activity, they migrate into the IZ to make the CP (Cicero and Herrup, 2005; Ueno et al., 2006). Progenitor cells divide into two supplementary progenitor cells (a proliferative unit) in the initial phases of neuronal proliferation (Rakic, 2005). In humans, in the second half of month 2 of gestation, the quantity of these proliferative units becomes constant as the progenitor cells begin to split asymmetrically, where each division results in two dissimilar cells, one of which is a postmitotic neuronal cell and the other a stem cell (Weissman et al., 2003). These divisions regulate the size of the proliferative unit and, as proliferation moves onward, fewer stem cells are formed in comparison to additional postmitotic neurons (Kornack and Rakic, 1998). The neurons of these proliferative units travel collectively to make a neuronal column of the cerebral cortex (Rakic, 2005). Two types of neuronal progenitors exist in the VZ: radial glial cells and neural precursors (Silbereis et al., 2016). Studies have shown that the radial glia may give rise to many neurons in the VZ. Therefore, the term radial glial cell might be substituted by glial progenitor or radial progenitor (Anthony et al., 2004; García and Harwell, 2017).

These actions of proliferation comprise radial glia as progenitor cells, controlled by several signaling pathways, such as ErbB receptor, Notch receptor, β-catenin (Chenn and Walsh, 2002), and fibroblast growth factor receptor (Rakic, 2005) pathways. The generation of neurons is controlled by calcium (Ca) waves propagating over the connexin channels of glial cells (Deitmer et al., 1998; Stout et al., 2002; Perea and Araque, 2005). The entrance of Ca is accountable for the regulation of the cell cycle, radial astrocytes, and the production of glial cells (Stout et al., 2002; Merkle et al., 2004). Radial glial cells give rise to the development of stem cells (capable of producing neurons), which continues in the SZ (Merkel et al., 2004). The brain contains several centers for the proliferation of neurons (Villar-Cervino et al., 2015; Sorrells et al., 2018). The multiple functions of radial glial cells are summarized in Table 14.2.

The median ganglionic eminence (MGE) contains several centers for neuronal proliferation (Villar-Cervino et al., 2015). It is known that the proliferation and migration of neurons present in the VZ and SZ are radial. In addition, there are vital proliferative centers in the MGE that give rise to striatal and cortical interneurons with tangential migration. Several investigations have reported that interneurons ascending from the identical MGE progenitor can uphold some clustering properties (Bolon et al., 2013; Ennis et al., 2015). Another study has suggested that numerous progenitors in the MGE give rise to interneurons that scatter widely in the adult brain. There are several significant interspecies problems to consider as animal models for studying the development of nervous system, particularly proliferation, and our understanding of the development of an adult human brain, principally concerning circuitry and interneuron development, is constantly evolving. Cell lineage investigations in organotypic slice cultures of human embryonic forebrain have indicated two γ-aminobutyric acid (GABA)ergic subpopulations in the human brain (Anwar et al., 2008). The first GABAergic subpopulations ascend from the VZ and SZ in the dorsal telencephalon, and express transcription factors such as Mash1 and Dlx1/2. These may represent approximately two-thirds of human neocortical GABAergic neurons. The second GABAergic subpopulation ascends from the MGE of the ventral telencephalon, which may comprise neurons that are transcriptionally distinguished from the first GABAergic subpopulation by the fact that they are Dlx1/2-positive but Mash1-negative (Petanjek et al., 2008; Heinrich et al., 2010).

14.3 ABERRANT DEVELOPMENT OF THE BRAIN AND RELATED DISORDERS

Aberrant neuron development, including proliferation, leads to the onset of several distorted functions in the CNS, followed by brain diseases (Zlokovic, 2008). Proliferative disorders are often difficult to define by conventional neuropathological examination due to difficulties with large populations of neurons (Hemphill et al., 2015). According to alterations in brain size, brain disorders can be divided into two categories: (1) microcephaly (undersized brain) and (2) macrocephaly (oversized brain), which are briefly summarized in Table 14.3.

14.3.1 Disorders Linked to Aberrant Neuronal Proliferation

14.3.1.1 Macrocephaly

The term macrocephaly is often used to describe a large head or brain. Macrocephaly is an important characteristic of heterogeneous disorders (McCaffery and Deutsch, 2005). Numerous organisms exist in which the developed brain is usually well formed but unusually large in size. The genetic variabilities and expressive developmental program for the proliferation of neurons have been defined (Mirzaa and Poduri, 2014). However, proliferative disorders can only be elucidated by excluding additional rare syndromes of macrocephaly. The anomalous proliferation of neurons may arise during brain development in the embryo. In addition, macrocephaly may also arise as a result of later-occurring flaws in apoptosis of the cells of the nervous system. The clinical symptoms in several macrocephalies range from no neurologic deficit to severe intellectual disability. In some disorders, the extraneural features can influence the clinical presentation. Some of the individual clinical aspects are covered briefly in the following subsections.

14.3.1.1.1 Isolated Macrocephaly

Familial isolated macrocephaly is the most common form of macrocephaly that occurs in the absence of any other extraneural indication (Saudubray et al., 2000). There are two genetic types: autosomal dominant and autosomal recessive. In autosomal dominant macrocephaly, the head is large in size at birth and continues to enlarge postnatally at a comparatively faster rate than normal (Laubscher et al., 1990). The genetic constituent of the syndrome is every so often disregarded until the head perimeter is measured. The analysis of fetal macrocephaly can be made in week 34 of gestation in a woman with benign macrocephaly (De Rosa et al., 1992). Autosomal dominant macrocephaly is categorized by numerous terms: benign subdural effusions of infancy; enlargement of the extracerebral space; and idiopathic external hydrocephalus (Wilms et al., 1993). In a few cases, subarachnoid and subdural fluid can be present (de Vries et al., 1990). The growth of the head in the first year is rapid but stops after 1 year. Over the following years the extracerebral space becomes less. Because of the large size of the brain, the isolated macrocephaly will be detected easily, even if the infant is assessed after the second year. The genetic characteristics are alike among the infants with autosomal dominant isolated macrocephaly, because about 90% of such newborns have a parent with large head size. Patients with isolated macrocephaly due to autosomal recessive inheritance are expected to display epilepsy, definite intellectual disability, and motor deficits (Warkany et al., 1981).

Sporadic isolated macrocephaly is also known as megalencephaly (Olney, 2007; Williams et al., 2008). There is no history of an ancestral disorder, and measurement of the circumference of the parents' head occurs slightly less often than with the autosomal dominant syndrome described earlier (Pettit et al., 1980). Clinical progression is similar to familial isolated macrocephaly.

14.3.1.1.2 Associated Disturbance of Growth

Macrocephaly may be comprehensively associated with growth disorders, such as Beckwith syndrome, achondroplasia, cerebral gigantism or Sotos syndrome, Marshall-Smith syndrome, fragile X syndrome, thanatophoric dysplasia, Cowden syndrome, and Weaver syndrome (Mirzaa and Poduri, 2014). Excluding Beckwith syndrome, the neurologic features in the neonatal period are unusual.

14.3.1.1.3 Neurocutaneous Syndromes

Several neurocutaneous syndromes are related to the excessive proliferation of cells in the brain (Ruggieri and Pratico, 2015). Macrocephaly arises most often in this situation in multiple hemangiomatosis syndromes (Jones, 2006).

TABLE 14.3
Disorders of the Brain Related to Neuron Proliferation

Serial No.	Macrocephaly	Microcephaly
1	**Isolated Macrocephaly** Familial (autosomal recessive and autosomal dominant) Sporadic	Microcephaly vera (autosomal recessive and autosomal dominant)
2	**Growth-Associated Disturbance** Achondroplasia Beckwith syndrome Cerebral gigantism Fragile X syndrome Thanatophoric dysplasia Marshall-Smith syndrome Weaver syndrome	X-linked recessive
3	**Neurocutaneous Syndromes** Multiple hemangiomatoses Lipomas, hemangiomas, lymphangiomas, pseudopapilledema neurocutaneous syndromes (Bannayan–Riley–Ruvalcaba) Asymmetrical hypertrophy, hemangiomata, varicosities neurocutaneous syndromes (Klippel–Trenaunay–Weber) Asymmetrical hypertrophy, telangiectatic lesions, flame nevus of the face, neurocutaneous syndromes (cutis marmorata, telangiectatica congenita) Neurofibromatosis, tuberous sclerosis, Sturge–Weber syndrome, neurocutaneous syndromes Epidermal nevus syndrome	Teratogenic Irradiation Metabolic-toxic Infection (rubella, cytomegalovirus, HIV, Zika virus)
4	**Chromosomal Disorders** Fragile X syndrome (relative macrocephaly) Klinefelter syndrome	Syndromic Chromosomal Familial Sporadic
5	**Unilateral Macrocephaly (Hemimegalencephaly)** Isolated Syndromic: Proteus syndrome and epidermal nevus syndrome	Nonsyndromic sporadic syndrome

Neurofibromatosis is an autosomal dominant disorder in which the major proliferative aberration involves glial cells, mostly astrocytes (Le and Parada, 2007). Thus, the beginning of the proliferative syndrome in this disease happens mainly after the time period of proliferative events of the neurons. Almost 40 percent of newborns exhibit more than five spots (light-brown color) at birth greater than 5 mm (Ward and Gutmann, 2005). Approximately 40–50% of such infants have macrocephaly, generally after the neonatal period (North, 1988). Consistent with the predominant glial rather than neural involvement, the megalencephaly chiefly indicates an increase in the volume of cerebral white matter and space, primarily in parietal and frontal areas (Cutting et al., 2002). Hemimegalencephaly with neonatal seizures and associated flaws in neuronal migration has also been reported (Balestri et al., 2003). Macrocephaly with glial tumors has also been recognized by ultrasound in the prenatal period (Drouin et al., 1997). The gene responsible for this ailment is positioned on chromosome 17. *NF1* has been known to code for a protein involved in negative regulation of the Ras pathway, which communicates mitogenic signals to the nucleus (Ward and Gutmann, 2005). Thus, the absence of neurofibromin (neurofibromatosis protein) results in the enhancement of mitogenic signaling, leading to the progression of proliferative abnormalities (Gutmann et al., 2013).

Sturge–Weber syndrome is a sporadic disorder which affects leptomeningeal blood vessels (Comi, 2003). Thus, the disease probably begins to develop during neuronal proliferation. The vital flaw in this syndrome is failure in the formation of superficial cortical veins, which results in the development of leptomeninges with abnormal vascular channel formation (Griffiths, 1996). Aberrations of fibronectin in the vessels of the cerebrum in the developing brain might play an important role in the origin of vascular abnormalities (Comi et al., 2005). It is very difficult to identify newborns with intracranial involvement. Seizures and calcifications in the cerebrum have occasionally been noted in newborns (Kitahara and Maki, 1978). Cerebral calcifications most commonly appear after the sixth postnatal month, and often later (Griffiths, 1996). Newborns with Sturge–Weber syndrome who have bilateral cerebral disease have inferior outcomes, compared to those with unilateral cerebral disease (Comi, 2015; Torre et al., 2018). Reports have suggested that Sturge–Weber disorder is lined to postzygotic or somatic mutations in *GNAQ* (Shirley et al., 2013).

Tuberous sclerosis complex (TSC) is one of the main proliferative aberrations that affects cooperation in neurons and glial cells (Curatolo et al., 2015; Martin et al., 2017). The molecular and neuropathological aspects of TSC reflect abnormal differentiation and migration (Narayanan, 2003). Seizures may arise in the neonatal period (Curatolo et al., 2002). Cardiac tumors have been recognized to affect the fetus and provide the first evidence for diagnosis in the prenatal period (Datta et al., 2008). The diagnosis of TSC is recognized in 80–95% of fetuses with cardiac tumors (Fesslova et al., 2004). Subependymal nodules in the common cerebral lesion can be seen in the 21-week-old fetus (Levine et al., 2000). There is a useful group of features for a TSC diagnosis in the neonatal period involving cardiac tumors, depigmented macule, subependymal nodule, and cortical tuber.

The TSC phenotype is related to dysfunction of genes on either chromosome 9 or chromosome 16 (Crino and Henske, 1999). The disorder is autosomal dominant and has a high rate of spontaneous mutation. *TSC1* and *TSC2*, present on chromosome 9 and chromosome 16 respectively, code for hamartin and tuberin respectively (Rate et al., 2009). Both proteins (hamartin and tuberin) interact to create a complex associated with the molecular signaling responsible for cell proliferation, growth, and adhesion/migration (Henske, et al. 2016). Trouble in these processes triggers proliferative aberrations (subependymal nodules; astrocytomas; and renal, cardiac, and other tumors), cell size abnormalities (large, bizarre cells in nodules and tubers), and disorders in radial white matter (apparent disorders in radial migration and the release of migrating cells). Mutations in *TSC2* result in the most severe phenotypes and account for most neonatal cases (McEneaney and Tee 2019; Symonds et al., 2019).

14.3.1.1.4 Chromosomal Disorders

Fragile X disorder might comprise a proliferative abnormality. It is elevated by the outcome of relative macrocephaly in infants (approximately 40%) (Simko et al., 1989). However, the disturbance of cerebral organizational appears to be the dominant neural abnormality in fragile X syndrome. Klinefelter syndrome and partial trisomy of chromosome 7 are rare chromosomal ailments with macrocephaly (Drigo et al., 1996).

14.3.1.2 Hemimegalencephaly

Hemimegalencephaly is a disorder of neuronal proliferation in the nervous system (Bosman et al., 1996). In this syndrome, the excessive proliferation of astrocytes and neurons occurs, but defects in their subsequent migration and organization are also noted (Barth, 1987). The expansion of one hemisphere or a portion of the hemisphere occurs, which is often followed by the formation of a disordered and unusually thick cortex (abnormal cortical gyration), larger neurons, heterotopic neurons in subcortical white matter, and an increased number and size of astrocytes (D'Agostino et al., 2004; Di Rocco et al., 2006). The clinical condition comprises severe seizures from early infancy, with ensuing disturbed neurological development (D'Agostino et al., 2004). Computed tomography can establish a grade of increase in the size of the hemisphere, which may be indicative of a congenital neoplasm. Magnetic resonance imaging (MRI) is favored and is the best imaging technique with which to assess neuronal heterotopias and abnormal gyration of the cerebral cortex (Manoranjan and Provias, 2010). Electroencephalography in newborn babies displaying seizures might establish a distinguishing pattern of larger-amplitude triphasic complexes (Sutter and Kaplan, 2012). Hemispherectomy, in the first months of development, can improve the results (Flores-Sarnat et al., 2003). Prognosis based on hemispherectomy depends mainly on the status of the hemisphere. The various syndromes of hemimegalencephaly are multiple, and include Proteus syndrome, hypomelanosis of Ito, and epidermal nevus syndrome (Flores-Sarnat et al., 2003), with neurocutaneous disorder in half of cases (Flores-Sarnat, 2002). Several infants display facial hemihyperplasia (Zarate et al., 2009).

14.3.1.3 Microcephaly

In microencephaly (small brain), the size of the head in living patients is used as an estimate of brain size. Excluding severe cranial defects, premature skull closure is generally considered to be the reason for a small head size (Mochida et al., 2009). The main microcephalies, including autosomal recessive syndromes, have been shown to be related to impaired neuron proliferation (Naveed et al., 2018). Microcephaly vera denotes a group of heterogeneous ailments which appear to have a common denominator and small brain size. Thus, there is no sign of intrauterine destructive illness and the anomalous brain size can be observed as early as the third trimester of gestation (Ganeshwaran and Christopher, 2001; Faheem et al., 2015).

The name microcephaly vera refers to an autosomal recessive disorder (Naveed et al., 2018). Outstanding insights into the genetic and molecular bases of this disorder have been gained. Anatomic investigations provide an understanding of the essential alterations in brain development in microcephaly vera (Evrard et al., 1992). The simplification of gyral shape/formation occurs with no other external anomalies and no evidence of any aberrant process. The number of neuronal columns formed in the brain cortex appears to be normal; however, the neuronal complement of each column is distinctly reduced, particularly the superficial layers of the cortex. An additional sign of alterations in proliferative actions is the absence of the residual germinal matrix in the 26-week-old brain of a fetus (Evrard et al., 1992). The timing of microcephaly syndromes involves the period of development at 6–18 weeks in the human fetus due to later proliferative events through asymmetrical divisions in the neuronal progenitors. The severely underdeveloped brains are likely to have the initial onset but a marked shortage of neurons in the cortical column.

The clinical presentation of newborns with autosomal recessive microcephaly vera is interesting, with the seizures accompanying recessive mutations in *PNKP*

(Shen et al., 2010; Naveed et al., 2018). Magnetic resonance imaging (MRI) is vital in the assessment of microcephaly, especially in assessing the growth of the gyral region in the brain and the presence of migration-associated anomalies (Barkovich, 1998). Rare cases are correlated with severe disturbances in the migration of neurons, and include lissencephaly, periventricular heterotopia, and posterior fossa deficits, especially cerebellar hypoplasia (Sztriha et al., 2005).

Radial microbrain is a syndrome of specific interest since it seems to offer an initial example of significant alterations in the quantity of proliferative units of neurons, resulting in a reduction in the size of the brain (Poduri and Volpe, 2018). The very small brain has no noticeable anomaly in the gyral region. This is an indication of a damaging process and disturbance in cortical lamination (Giustina, 2020). The disturbance includes the early phase of neuronal proliferative events which is based on the finding of a clear decrease in the number of neuronal columns in the cortical region of the brain (Evrard et al., 1992; Kuzniecky and Barkovich, 2001). The assumed timing of radial microbrain syndrome is not later than the initial phase of neuronal proliferative events in the second month of gestation. The important anomaly involves symmetrical divisions of neuronal progenitors in the column to form extra progenitors, thereby affecting the normal number of neuronal proliferative units (Kuzniecky and Barkovich, 2001; Poduri and Volpe, 2018; Giustina, 2020). The clinical significance is still not entirely clear, since this disorder is rare. Reported cases have involved full-term newborns who died in their first month of life. The syndrome is remarkably familial and may be inherited in an autosomally recessive manner.

14.3.1.3.1 Autosomal Recessive Microcephaly

Multiple genes responsible for autosomal recessive microcephaly have been recognized (Abdel-Hamid et al., 2016). These genes can play an important function in mitosis (Kaindl et al., 2010). Microcephalin is key in the control of the cell cycle, chromosomal condensation, and DNA repair (Table 14.4). A centrosomic protein (CDK5RAP2) involved in microtubular function is required for the construction of the mitotic spindle (Issa et al., 2013). Others, such as abnormal spindle in microcephaly (ASPM), are essential for the function of microtubules, and centromere protein J (CENPJ) is likewise involved in the formation of the spindle fibers in mitotic cells. Of these, the ASPM is the most common (Zhong et al., 2005; Tang, 2006; Kaindl et al., 2010; Mirzaa et al., 2014). The MRI findings of autosomal recessive microcephaly show microcephaly with cerebral hemorrhage and calcification, as well as congenital cataracts, associated with recessive mutations in *JAM3* (Mochida et al., 2010). It is the primary microcephaly correlated with dysfunction in vascular endothelial cadherin (Aran et al., 2016).

Syndromic cases might be linked to chromosomal syndromes/monogenic flaws. They may also arise sporadically. The nature of neuronal proliferative syndrome in this group is unknown. However, the clinical facts are available (Jones, 2006).

Sporadic nonsyndromic cases are the most common form of microcephaly vera. They are identified in individuals with no family history, recognizable teratogens, and/or recognizable syndromes (Krauss et al., 2003). The nature of such neuronal proliferative distress is unknown. Any autosomal dominant/X-linked conditions may also arise sporadically by mutation during meiosis, during gametogenesis.

14.4 GENETICS OF MEGALENCEPHALY AND RELATED DISORDERS

The relation between hemimegalencephaly and Proteus syndrome offers major insight into the crucial role of the protein kinase B (AKT)/mammalian target of rapamycin (mTOR) pathway in regulating the proliferation of a nerve cell (Xu et al., 2020). Mutations in *AKT1* have been recognized as resulting in Proteus syndrome (Lindhurst et al., 2011). However, alterations in *AKT3* (Lee et al., 2012) and related genes in the Akt/mTOR pathway (Riviere et al., 2012) have been recognized as resulting in hemimegalencephaly (Figure 14.2). These findings have been extended to the

TABLE 14.4
Molecular Genetics of Autosomal Recessive Primary Microcephaly (Microcephaly Vera)

Serial No.	Gene	Locus	Affected Function
1	*MCPH1* (microcephalin 1)	MCPH1	Cell-cycle control
2	*WDR62* (WD repeat domain 62)	MCPH2	Mitosis (centrosome)
3	*CDK5RAP2* (CDK5 regulatory subunit associated protein 2)	MCPH3	Mitotic spindle formation
4	*CASC5* (cancer-associated candidate 5)	MCPH4	Mitotic spindle formation
5	*ASPM* (assembly factor for spindle microtubules)	MCPH5	Mitotic spindle formation
6	*CENPJ* (centromere protein J)	MCPH6	Mitotic spindle formation
7	*SCL*/TAL1 interrupting locus (Stil)	MCPH7	Mitotic spindle formation
8	*CEP135* (centrosomal protein 135 kD)	MCPH8	Mitosis (centrosome)
9	*CEP152* (centrosomal protein 152 kD)	MCPH9	Mitosis (centrosome)
10	*ZNF335* /zinc finger protein 335	MCPH10	Regulation of neurogenesis

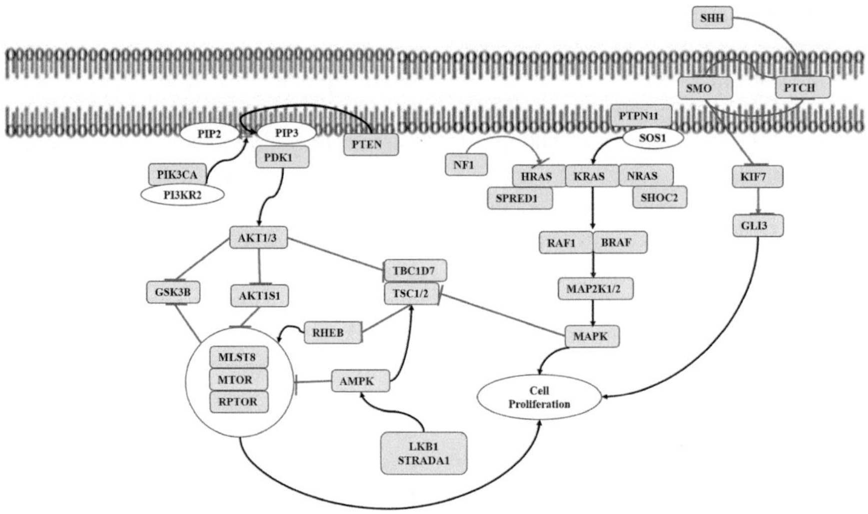

FIGURE 14.2 Mammalian target of rapamycin (mTOR)/AKT pathway, which is involved in cases of hemimegalencephaly, megalencephaly, and focal cortical dysplasia. Red lines indicate the inhibition of genes and green arrows indicate gene activation.

TABLE 14.5
Mutations of the PI3K/AKT/mTOR Pathway Related to Brain Malformations

Serial No.	Brain Malformation	Inheritance	Genes
1	Hemimegalencephaly/Proteus syndrome	De novo, postzygotic	*AKT1*
			AKT3
2	Hemimegalencephaly/FCD	De novo, postzygotic	*CCND2*
3	Polymicrogyria/MPPH syndrome	De novo/dominant	*DEPDC5*
			MTOR
4	Hemimegalencephaly/FCD (also nonlesional focal epilepsy)	De novo/dominant ± postzygotic	*PIK3CA*
	Hemimegalencephaly/FCD	De novo, postzygotic	
	Hemimegalencephaly/MCAP syndrome	De novo, postzygotic	
5	MPPH syndrome	De novo/dominant	*PIK3R2*
6	Hemimegalencephaly/FCD, megalencephaly/autism, Cowden syndrome, Bannayan–Riley–Ruvalcaba syndrome	De novo/dominant	*PTEN*
7	Hemimegalencephaly/FCD	De novo/dominant ± postzygotic	*TSC1/TSC2*
		De novo/dominant	

FCD, focal cortical disorder; MPPH, megalencephaly–polymicrogyria–polydactyly–hydrocephalus; MCAP, megalencephaly–capillary malformation syndrome.

inheritance of focal cortical dysplasia and megalencephaly (D'Gama et al., 2015). Any damage to the regulation of the Akt/mTOR pathway in early neuronal progenitor cells may lead to focal brain lesions or widespread overgrowth of the brain (Gomez et al., 2011). The level of these lesions in the brain is facilitated by the location, timing (i.e., how early or late), and the nature of these alterations in gene function (Jansen et al., 2015). Malformations of the brain may be due to alterations in the phosphatidylinositol 3-kinae (PI3K)/Akt/mTOR pathway (Table 14.5).

A report on the genetics of hemimegalencephaly demonstrated that the lesion in the developing brain could be connected to a mutation in *AKT3* (Lee et al., 2012), causing duplication of *AKT3* or to a point mutation in AKT3 protein (Poduri et al., 2012). The presence of mutations in the majority of cells indicates that the mutation starts in the postzygotic period in progenitor cells, depending on the timing of the mutational event (Poduri et al., 2013). Hemimegalencephaly can arise from such mutations not only at the particular locus of *AKT3*, but also at other *AKT3* loci and other genes in the PI3K/AKT/mTOR pathway (D'Gama et al., 2015). A single-cell study of hemimegalencephaly due to a postzygotic increase in the copy number of the chromosomes 1q region containing *AKT3* revealed the mutation was a tetrasomy (Cai et al., 2014).

Focal cortical disorder (FCD) shares neuropathological characteristics with hemimegalencephaly and TSC (Poduri et al., 2013). The inheritance of FCD has been explained in a very small percentage of cases, owing to technical limitations which restrict the detection of postzygotic mutations.

Recessive FCD is defined by intellectual disability and associated epilepsy (Strauss et al., 2006), which is related to significant mutations in *CNTNAP2*. Another hereditary cause of FCD is de novo, including all genes involved in the PI3K/AKT/mTOR pathway. FCD can be associated with the mosaic upsurge in AKT3 copy number (Conti et al., 2015). FCD has also been associated with point mutations in *PTEN*, *PIK3CA*, *AKT3*, *MTOR*, and *DEPDC5* (Baulac et al., 2015; Qin et al., 2010; Strauss et al., 2016). Thus, the understanding of proliferative ailments related to neurons in the brain, particularly hemimegalencephaly, megalencephaly, and FCD, is concerned with the signaling pathway of genes associated with the regulation and synthesis of proteins involved in the growth of the brain (Lindhout, 2008).

14.5 FUTURE PROSPECTS

In the early development of neurons, neural cells migrate from the subependymal layer to form the caudate nucleus. It would not be irrational to propose that these juvenile neurons migrate toward the caudate nucleus in response to the loss of neurons taking place in the brain. In one study, it was reported that the increase in the quantity of active progenitor cells in the subependymal layer of brain with evolving neuropathological grades indicates that the proliferation of neurons is in response to more collapsing neurons in the caudate nucleus (Thomas, 1999). However, it is still not clear which factors and/or signals regulate cellular proliferation and migration. A number of pharmaceutical agents and many growth factors have been found to induce neurogenesis and cellular proliferation. Endogenous neural replacement potential can be enhanced pharmacologically with the use of exogenous pharmaceuticals and/or growth factor that are reported to increase the rate of formation of neural progenitors, as well as the migration and maturation of the newly formed neurons. Hence, the rate of loss of neural cells can be reduced with clinical improvements. That the diseased brain might be capable of regenerated neurons in response to neuronal loss is of major relevance for the expansion of therapeutic approaches. The molecular mechanisms or genetics of several diseases involving the abnormal development, growth, and proliferation of nerve cells are still unclear. Studies should be conducted on the inheritance of diseases associated with aberrant neuronal proliferation. The molecular mechanisms of these neuronal proliferative events should also be included in future studies.

14.6 CONCLUSION

Neurodevelopmental syndromes are complex pathologic conditions with enhanced genomic heterogeneity. Research on understanding disorders associated with neuronal development have focused on the molecular mechanisms of disease. Neural stem cells divide to form a neuron, which is the basic convergence point of diseases related to neuronal development. Alterations in the function of genes that play a key role in the developmental process of neurons are implicated in several neurodevelopmental syndromes. Genes (i.e., *MBD5*, *CDKL5*, and *MECP2*) with truncating mutations are involved in disorders of neuronal development and clustered to alter neural stem cell proliferation and differentiation. Genes related to neurodevelopmental ailments have an important role in the proliferation and differentiation of cells. The reciprocal duplication or deletion of copy number variants (such as 1q21.1 and 16p11.2 associated with neurodevelopmental syndromes) may cause the reverse effects on macrocephaly or microcephaly, which is an assessable yield of the cellular proliferation discrepancies. Neuronal progenitor cells may also be used to quantity the level of differentiation discrepancies or the degree of impairment in neurons. Moreover, alterations in cell signaling pathways (mTOR, mitogen-activated protein kinase, and Ras) also play a crucial role in disorders associated with neuronal developmental. These discrepancies are quantifiable and may be an important diagnostic tool in clinical intervention and molecular therapies.

REFERENCES

Abdel-Hamid MS, Ismail MF, Darwish HA, et al. Molecular and phenotypic spectrum of ASPM-related primary microcephaly: identification of eight novel mutations. Am J Med Genet A. 2016;170:2133–2140.

Anthony TE, Klein C, Fishell G, et al. Radial glia serve as neuronal progenitors in all regions of the central nervous system. Neuron. 2004;41:881–890.

Anwar MA, Andreasen CM, Lippert SK, et al. Dopaminergic differentiation of human neural stem cells mediated by co-cultured rat striatal brain slices. J Neurochem. 2008;105:460–470.

Aran A, Rosenfeld N, Jaron R, et al. Loss of function of PCDH12 underlies recessive microcephaly mimicking intrauterine infection. Neurology. 2016;86:2016–2024.

Au KS, Northrup H. Genetics of neurocutaneous disorders. Neurocutaneous Disorders. 2004 8:6.

Balestri P, Vivarelli R, Grosso S, et al. Malformations of cortical development in neurofibromatosis type 1. Neurology. 2003;61:1799–1801.

Barkovich AJ, Ferriero DM, Barr RM, et al. Microlissencephaly: a heterogeneous malformation of cortical development. Neuropediatrics. 1998;29:113–119.

Barnette AR, Neil JJ, Kroenke CD, et al. Characterization of brain development in the ferret via MRI. Pediatr Res. 2009;66:80–84.

Barth P. Disorders of neuronal migration. Can J Neuro Sci. 1987;14:1–16.

Baulac S, Ishida S, Marsan E, et al. Familial focal epilepsy with focal cortical dysplasia due to DEPDC5 mutations. Ann Neurol. 2015;77:675–683.

Berry M, Rogers AW, Eayrs JF. Pattern of cell migration during cortical histogenesis. Nature. 1964;203:591.

Bolon B, Butt MT, Garman RH, et al. Nervous system. In: Haschek WM, Rousseaux CG, Wallig WA (eds.). Haschek and Rousseaux's Handbook of Toxicologic Pathology, 3rd edition, vol. III. 2013;2005–2093.

Bosman C, Boldrini R, Dimitri L, et al. Hemimegalencephaly. Child Nerv Syst. 1996;12:765–775.

Branco T, Hausser M. The single dendritic branch as a fundamental functional unit in the nervous system. Curr Opinion Neurobiol. 2010;20:494–502.

Budday S, Steinmann P, Kuhl E. Physical biology of human brain development. Front Cell Neurosci. 2015;9:257.

Cai X, Evrony GD, Lehmann HS, et al. Single-cell, genome-wide sequencing identifies clonal somatic copy-number variation in the human brain. Cell Rep. 2014;8:1280–1289.

Chenn A, Walsh CA. Regulation of cerebral cortical size by control of cell cycle exit in neural precursors. Science. 2002;297:365–369.

Cicero S, Herrup K. Cyclin-dependent kinase 5 is essential for neuronal cell cycle arrest and differentiation. J Neurosci. 2005;25:9658–9668.

Comi A. Topical review: pathophysiology of Sturge–Weber syndrome. J Child Neurol. 2003;18:509–516.

Comi A. Current therapeutic options in Sturge-Weber syndrome. Seminars Ped Neurol. 2015;22:295–301.

Comi AM, Weisz CJC, Highet BH, et al. Sturge–Weber syndrome: altered blood vessel fibronectin expression and morphology. J Child Neurol. 2005;20:572–577.

Conti V, Pantaleo M, Barba C, et al. Focal dysplasia of the cerebral cortex and infantile spasms associated with somatic 1q21.1-q44 duplication including the *AKT3* gene. Clin Genet. 2015;88:241–247.

Cox J, Jackson AP, Bond J, et al. What primary microcephaly can tell us about brain growth. Trends Mol Med. 2006;12:358–366.

Crino PB, Henske EP. New developments in the neurobiology of the tuberous sclerosis complex. Neurology. 1999;53:1384–1390.

Curatolo P, Verdecchia M, Bombardieri R. Tuberous sclerosis complex: a review of neurological aspects. Eur J Paediatr Neurol. 2002;6:15–23.

Curatolo PP, Moavero R, Vries PPJ. Neurological and neuropsychiatric aspects of tuberous sclerosis complex. Lancet Neurol. 2015;14:733–745.

Cutting LE, Cooper KL, Koth CW, et al. Megalencephaly in NF1. Predominantly white matter contribution and mitigation by ADHD. Neurology. 2002;59:1388–1394.

D'Agostino MD, Bastos A, Piras C, et al. Posterior quadrantic dysplasia or hemi-hemimegalencephaly: a characteristic brain malformation. Neurology. 2004;62:2214–2220.

Datta AN, Hahn CD, Sahin M. Clinical presentation and diagnosis of tuberous sclerosis complex in infancy. J Child Neurol. 2008;23:268–273.

De Myer W. Megalencephaly: types, clinical syndromes, and management. Pediatr Neurol. 1986;2:321–328.

De Rosa MJ, Secor DL, Barsom M, et al. Neuropathologic findings in surgically treated hemimegalencephaly—immunohistochemical, morphometric, and ultrastructural study. Acta Neuropathol. 1992;84:250–260.

de Vries LS, Smet M, Ceulemans B, et al. The role of high resolution ultrasound and MRI in the investigation of infants with macrocephaly. Neuropediatrics. 1990;21:72–75.

Deitmer JW, Verkhratsky AJ, Lohr C. Calcium signalling in glial cells. Cell Calcium. 1998;24:405–416.

D'Gama AM, Geng Y, Couto JA, et al. Mammalian target of rapamycin pathway mutations cause hemimegalencephaly and focal cortical dysplasia. Ann Neurol. 2015;77:720–725.

Di Rocco C, Battaglia D, Pietrini D, et al. Hemimegalencephaly: clinical implications and surgical treatment. Childs Nerv Syst. 2006;22:852–866.

Drigo P, Carra S, Laverda AM, et al. Macrocephaly and chromosome disorders: a case report. Brain Dev. 1996;18:312–315.

Drouin V, Marret S, Petitcolas J, et al. Prenatal ultrasound abnormalities in a patient with generalized neurofibromatosis type 1. Neuropediatrics. 1997;28:120–121.

Ennis M, Puche AC, Holy T, et al. The olfactory system. In: Paxinos G, ed.The Rat Nervous System, 4th edition. St Louis: Elsevier, 2015;761–803.

Esposito G. The development of attachment: integrating genes, brain, behavior, and environment. Behav Brain Res. 2017;325:87–89.

Evrard P, Miladi N, Bonnier C, et al. Normal and abnormal development of the brain. In: Rapin I, Segalowitz SJ, eds. Handbook of Neuropsychology, Vol. 6: Child Neuropsychology. Amsterdam: Elsevier Science; 1992:11–44.

Faheem M, Naseer I, Rasool M, et al. Molecular genetics of human primary microcephaly: an overview. BMC Med Genomics. 2015;8:S4.

Fernandes J, Saudubray JM, Van den Berghe G, Walter JH. Clinical approach to inherited metabolic diseases. Inborn metabolic diseases. 4th ed. Heidelberg: Springer Medizin Verlag;2006:7–13.

Fesslova V, Villa L, Rizzuti T, et al. Natural history and long-term outcome of cardiac rhabdomyomas detected prenatally. Prenat Diagn. 2004;24:241–248.

Flores-Sarnat L. Hemimegalencephaly: part 1. Genetic, clinical, and imaging aspects. J Child Neurol. 2002;16:373–384.

Flores-Sarnat L, Sarnat HB, Davila-Gutierrez G, et al. Hemimegalencephaly: part 2. Neuropathology suggests a disorder of cellular lineage. J Child Neurol. 2003;18:776–785.

Ganeshwaran MH, Christopher WA. Molecular genetics of human microcephaly. Curr Opin Neurol. 2001;14:151–156.

García MT, Harwell CC. Radial glia in the ventral telencephalon. FEBS Lett. 2017;591:3942–3959.

Giustina ED. Neurologic development and objective neurological examination of the infant and the child. In: C. Di Rocco et al. (eds.) Textbook of Pediatric Neurosurgery. New York: Springer;2020;79–103.

Goldman S. Glia as neural progenitor cells. Trends Neurosci. 2003;26:590–596.

Gomez O, Sanchez-Rodriguez A, Le MQU, et al. Cannabinoid receptor agonists modulate oligodendrocyte differentiation by activating PI3K/Akt and the mammalian target of rapamycin (mTOR) pathways. Br J Pharmacol. 2011;163:1520–1532.

Griffiths PD. Sturge Weber syndrome revisited: the role of neuroradiology. Neuropediatrics. 1996;27:284–294.

Gross C. Neurogenesis in the adult brain: death of a dogma. Nat Rev Neurosci. 2000;1:67–73.

Gutmann DH, McLellan MD, Hussain I, et al. Somatic neurofibromatosis type 1 (NF1) inactivation characterizes NF1-associated pilocytic astrocytoma. Genome Res. 2013;23:431–439.

Heinrich C, Blum R, Gascon S, et al. Directing astroglia from the cerebral cortex into subtype specific functional neurons. PLoS Biol. 2010;8:e1000373.

Hemphill MA, Dauth S, Yu CJ, et al. Traumatic brain injury and the neuronal microenvironment: a potential role for neuropathological mechanotransduction. Neuron. 2015;85:1177–1192.

Henske EP, Jozwiak S, Kingswood JC, et al. Tuberous sclerosis complex. Nat Rev Dis Primers. 2016;2:16035.

Herculano-Houzel S, Kaas JH, de Oliveira-Souza R. Corticalization of motor control in humans is a consequence of brain scaling in primate evolution. J Comp Neurol. 2016;524:448–455.

Herschkowitz N. Brain development in the fetus, neonate and infant. Biol Neonate. 1988;54:1–19.

Hoerder-Suabedissen A, Molnar Z. Development, evolution and pathology of neocortical subplate neurons. Nat Rev Neurosci. 2015;16:133–146.

Huang L, Nakamura Y, Lo EH, et al. Astrocyte signaling in the neurovascular unit after central nervous system injury. Int J Mol Sci. 2019;20:282.

Issa L, Mueller K, Seufert K, et al. Clinical and cellular features in patients with primary autosomal recessive microcephaly and a novel CDK5RAP2 mutation. Orphanet J Rare Dis 2013;8:59.

Jansen LA, Mirzaa GM, Ishak GE, et al. PI3K/AKT pathway mutations cause a spectrum of brain malformations from megalencephaly to focal cortical dysplasia. Brain. 2015;138:1613–1628.

Jones KL. Smith's Recognizable Patterns of Human Malformation. 6th ed. Philadelphia, PA: Elsevier Saunders;2006.

Kaindl AM, Passemard S, Kumar P, et al. Many roads lead to primary autosomal recessive microcephaly. Progress Neurobiol. 2010;90:363–383.

Kitahara T, Maki Y. A case of Sturge–Weber disease with epilepsy and intracranial calcification at the neonatal period. Eur Neurol. 1978;17:8–12.

Kornack DR, Rakic P. Changes in cell-cycle kinetics during the development and evolution of primate neocortex. Proc Natl Acad Sci U S A. 1998;95:1242–1246.

Kostovic I, Rakic P. Cytology and time of origin of interstitial neurons in the white matter in infant and adult human and monkey telencephalon. J Neurocytol. 1980;9:219–242.

Krauss MJ, Morrissey AE, Winn HN, et al. Microcephaly: an epidemiologic analysis. Am J Obstet Gynecol. 2003;188: 1484–1489.

Kuzniecky RI, Barkovich AJ. Malformations of cortical development and epilepsy. Brain Dev. 2001;23:2–11.

Lahti L, Saarimaki-Vire J, Rita H, et al. FGF signaling gradient maintains symmetrical proliferative divisions of midbrain neuronal progenitors. Dev Biol. 2011;349:270–282.

Laubscher B, Deonna T, Uske A, et al. Primitive megalencephaly in children: natural history, medium term prognosis with special reference to external hydrocephalus. Eur J Pediatr. 1990;149:502–507.

Le L, Parada L. Tumor microenvironment and neurofibromatosis type I: connecting the GAPs. Oncogene. 2007;26: 4609–4616.

Lee J, Huynh M, Silhavy J, et al. De novo somatic mutations in components of the PI3K-AKT3-mTOR pathway cause hemimegalencephaly. Nat Genet. 2012;44:941–945.

Lee SW. A copernican approach to brain advancement: the paradigm of allostatic orchestration. Front Human Neurosci. 2019;13:129.

Levine D, Barnes P, Korf B, et al. Tuberous sclerosis in the fetus: second-trimester diagnosis of subependymal tubers with ultrafast MR imaging. AJR Am J Roentgenol. 2000;175:1067–1069.

Lindhout D. Somatic mosaicism as a basic epileptogenic mechanism? Brain. 2008;131:900–901.

Lindhurst MJ, Sapp JC, Teer JK, et al. A mosaic activating mutation in AKT1 associated with the Proteus syndrome. N Engl J Med. 2011;365:611–619.

Lowery LA, Sive H. Strategies of vertebrate neurulation and a re-evaluation of teleost neural tube formation. Mechan Dev. 2004;121:1189–1197.

McCaffery P, Deutsch CK. Macrocephaly and the control of brain growth in autistic disorders. Prog Neurobiol. 2005;77:38–56.

McEneaney LJ, Tee AR. Finding a cure for tuberous sclerosis complex: From genetics through to targeted drug therapies. Adv Genet. 2019;103:91–118.

Malatesta P, Hack MA, Hartfuss E, et al. Neuronal or glial progeny: regional differences in radial glia fate. Neuron. 2003;37:751–764.

Mangoura DA, Pelletiere C, Wang D, et al. Plasticity in astrocytic phenotypes. A role for protein kinase C, tyrosine kinases, and cytokine signaling. Adv Exp Med Biol. 1997;429:39–53.

Manoranjan B, Provias JP. Hemimegalencephaly: a fetal case with neuropathological confirmation and review of the literature. Acta Neuropathol. 2010;120:117–130.

Martin K, Zhou W, Bowman MJ, et al. The genomic landscape of tuberous sclerosis complex. Nat Comm. 2017;8:15816.

Mashanov VS, Zueva OR, Garcia-Arraras JE. Radial glial cells play a key role in echinoderm neural regeneration. BMC Biol. 2013;11:49.

Merkle FT, Tramontin AD, Garcia-Verdugo JM, et al. Radial glia give rise to adult neural stem cells in the subventricular zone. Proc Natl Acad Sci U S A. 2004;101:17528–17532.

Miller JA, Ding SL, Sunkin SM, et al. Transcriptional landscape of the prenatal human brain. Nature. 2014;508:199–206.

Mirzaa GM, Poduri A. Megalencephaly and hemimegalencephaly: breakthroughs in molecular etiology. Am J Med Genet C Semin Med Genet. 2014;166:156–172.

Mirzaa GM, Vitre B, Abramaowicz CGI, et al. Mutations in CENPE define a novel kinetochore-centromeric mechanism for microcephalic primordial dwarfism. Hum Genet. 2014;133:1023–1039.

Mochida GH, Ganesh VS, Felie JM, et al. A homozygous mutation in the tight-junction protein JAM3 causes hemorrhagic destruction of the brain, subependymal calcification, and congenital cataracts. Am J Hum Genet. 2010;87:882–889.

Mochida GH, Mahajnah M, Hill AD, et al. A truncating mutation of TRAPPC9 is associated with autosomal-recessive intellectual disability and postnatal microcephaly. Am J Hum Genet. 2009;85:897–902.

Narayanan V. Tuberous sclerosis complex: genetics to pathogenesis. Pediatr Res. 2003;29:404–409.

Naveed M, Kazmi S, Amin M, et al. Comprehensive review on the molecular genetics of autosomal recessive primary microcephaly (MCPH). Genetics Res. 2018;100:E7.

Noctor SC, Martinez-Cerdeno V, Kriegstein AR. Contribution of intermediate progenitor cells to cortical histogenesis. Arch Neurol. 2007;64:639–642.

Norman MG, O'Kusky JR. The growth and development of microvasculature in human cerebral cortex. J Neuropathol Exp Neurol. 1986; 45:222–232.

North KN. Neurofibromatosis 1 in childhood. Semin Pediatr Neurol. 1998;5:231–241.

Olney AH. Macrocephaly syndromes. Semin Pediatr Neurol. 2007;14:128–135.

Perea G, Araque A. Glial calcium signaling and neuron-glia communication. Cell Calcium. 2005;38:375–382.

Petanjek Z, Dujmovi A, Kostovi I, et al. Distinct origin of GABA-ergic neurons in forebrain of man, nonhuman primates and lower mammals. Coll Antropol. 2008;32:9–17.

Pettit RE, Kilroy AW, Allen JH. Macrocephaly with head growth parallel to normal growth pattern: neurological, developmental, and computerized tomography findings in full-term infants. Arch Neurol. 1980;37:518–521.

Poduri A, Evrony GD, Cai X, et al. Somatic activation of AKT3 causes hemispheric developmental brain malformations. Neuron. 2012;74:41–48.

Poduri A, Evrony GD, Cai X, et al. Somatic mutation, genomic variation, and neurological disease. Science. 2013;341:1237758.

Poduri A, Volpe JJ. Neuronal proliferation. In: Volpe J, Inder T, Darras B, et al. (eds). Volpe's Neurology of the Newborn, 6th edition. St Louis, MO: Elsevier, 2018:100–119.

Puelles L. Brain segmentation and forebrain development in amniotes. Brain Res Bull. 2001;55:695–710.

Qin W, Chan JA, Vinters HV, et al. Analysis of TSC cortical tubers by deep sequencing of TSC1, TSC2 and KRAS demonstrates that small second-hit mutations in these genes are rare events. Brain Pathol. 2010;20:1096–1105.

Rakic P. Neurons in rhesus monkey visual cortex: systematic relation between time of origin and eventual disposition. Science. 1974;183:425–427.

Rakic, P. Timing of major ontogenetic events in the visual cortex of the rhesus monkey. In: Buchwald NA and Brazier MAB (eds). Brain Mechanisms in Mental Retardation. New York: Academic Press 1975;3–40.

Rakic P. Limits of neurogenesis in primates. Science. 1985;227:1054–1056.

Rakic P. Less is more: progenitor death and cortical size. Nat Neurosci. 2005;8:981–982.

Rakic P, Sidman RL. Supravital DNA synthesis in the developing human and mouse brain. J Neuropathol Exp Neurol. 1968;27:246–276.

Rice D, Barone S Jr. Critical periods of vulnerability for the developing nervous system: evidence from humans and animal models. Environ Health Perspect. 2000;108:511–533.

Riviere JB, Mirzaa GM, O'Roak BJ, et al. De novo germline and postzygotic mutations in AKT3, PIK3R2 and PIK3CA cause a spectrum of related megalencephaly syndromes. Nat Genet. 2012;44:934–940.

Ruggieri M, Pratico AD. Mosaic neurocutaneous disorders and their causes. Sem Ped Neurol. 2015;22:207–233.

Samuelsen GB, Larsen KB, Bogdanovic N, et al. The changing number of cells in the human fetal forebrain and its subdivisions: a stereological analysis. Cereb Cortex. 2003;13:115–122.

Sauer FC. The cellular structure of the neural tube. J Comp Neurol. 1935;63:13.

Schore AN. The effects of early relational trauma on right brain development, affect regulation, and infant mental health. Infant Ment Health J 2003;22:201–269.

Shen J, Gilmore EC, Marshall CA, et al. Mutations in PNKP cause microcephaly, seizures and defects in DNA repair. Nat Genet. 2010;42:245–249.

Shirley MD, Tang H, Gallione CJ, et al. Sturge–Weber syndrome and port-wine stains caused by somatic mutation in GNAQ. N Engl J Med. 2013;368:1971–1979.

Sidman RL, Angevine JB. Autoradiographic analysis of time of origin of nuclear versus cortical components of mouse telencephalon. Anat Rec. 1962;142:326.

Silbereis JC, Pochareddy S, Zhu Y, et al. The cellular and molecular landscapes of the developing human central nervous system. Neuron. 2016; 89:248–268.

Simko A, Hornstein L, Soukup S, et al. Fragile X syndrome: recognition in young children. Pediatrics. 1989;83:547–552.

Sorrells SF, Paredes MF, Cebrian-Silla A, et al. Human hippocampal neurogenesis drops sharply in children to undetectable levels in adults. Nature. 2018;555:377–381.

Stout CE, Costantin JL, Naus CCG, et al. Intercellular calcium signaling in astrocytes via ATP release through connexin hemichannels. J Biol Chem. 2002;277:10482–10488.

Strauss GM, Campbell CD, Solovieff N, et al. Association of MTOR mutations with developmental brain disorders, including megalencephaly, focal cortical dysplasia, and pigmentary mosaicism. JAMA Neurol. 2016;73:836–845.

Strauss KA, Puffenberger EG, Huentelman MJ, et al. Recessive symptomatic focal epilepsy and mutant contactin-associated protein-like 2. N Engl J Med. 2006;354:1370–1377.

Sutter R, Kaplan PW. Electroencephalographic criteria for nonconvulsive status epilepticus: Synopsis and comprehensive survey. Epilepsia. 2012;53:1–51.

Symonds JD, Zuberi SM, Stewart K, et al. Incidence and phenotypes of childhood-onset genetic epilepsies: a prospective population-based national cohort. Brain. 2019;142:2303–2318.

Sztriha L, Johansen JG, Al Gazali LI. Extreme microcephaly with agyria-pachygyria, partial agenesis of the corpus callosum, and pontocerebellar dysplasia. J Child Neurol. 2005;20:170–172.

Tang BL. Molecular genetic determinants of human brain size. Biochem Biophys Res Commun. 2006;345:911–916.

Thomas LB. Neuron production in the adult mammalian subependymal zone. The University of Tennessee Health Science Center; 1999.

Torre AJD, Luat AF, Juhasz C, et al. A multidisciplinary consensus for clinical care and research needs for Sturge–Weber syndrome. Pediatr Neurol. 2018;84:11–20.

Ueno M, Katayama K, Yamauchi H, et al. Cell cycle progression is required for nuclear migration of neural progenitor cells. Brain Res. 2006;1088:57–67.

Villar-Cervino V, Kappeler C, Nobrega-Pereira S, et al. Molecular mechanisms controlling the migration of striatal interneurons. J Neurosci. 2015;35:8718–8729.

Ward BA, Gutmann DH. Neurofibromatosis 1: from lab bench to clinic. Pediatr Neurol. 2005;32:221–228.

Warkany J, Lemire RJ, Cohen MM Jr. Mental Retardation and Congenital Malformations of the Central Nervous System. Chicago: Year Book;1981.

Webb SJ, Monk CS, Nelson CA. Mechanisms of postnatal neurobiological development: implications for human development. Dev Neuropsychol. 2001;19:147–171.

Weis S, Sonnberger M, Dunzinger A, et al. (eds). Histological constituents of the nervous system. In: Imaging Brain Diseases. Vienna: Springer;2019:225–265.

Weissman T, Noctor SC, Clinton BK, et al. Neurogenic radial glial cells in reptile, rodent and human: From mitosis to migration. Cereb Cort. 2003;13:550–559.

Williams CA, Dagli A, Battaglia A. Genetic disorders associated with macrocephaly. Am J Med Genet A. 2008;146:2023–2037.

Wilms G, Vanderschueren G, Demaerel PH, et al. CT and MR in infants with pericerebral collections and macrocephaly. Benign enlargement of the subarachnoid spaces versus subdural collections. AJNR Am J Neuroradiol. 1993;14:855–860.

Xu F, Na L, Li Y, et al. Roles of the PI3K/AKT/mTOR signalling pathways in neurodegenerative diseases and tumours. Cell Biosci. 2020;10:54.

Zarate YA, Mena R, Martin LJ, et al. Experience with hemihyperplasia and Beckwith-Wiedemann syndrome surveillance protocol. Am J Med Genet Part A. 2009;149A:1691–1697.

Zhong X, Liu L, Zhao A, et al. The abnormal spindle-like, microcephaly-associated (*ASPM*) gene encodes a centrosomal protein. Cell Cycle. 2005;4:1227–1229.

Zlokovic BV. The blood–brain barrier in health and chronic neurodegenerative disorders. Neuron. 2008;57:178–201.

15 Neurobiology in Correlation with Romantic Attraction

Swathi Putta and Eswar Kumar Kilari
A.U. College of Pharmaceutical Sciences, Andhra University, Visakhapatnam, India

Hemavathi G, Chandrasekar SB, and Lokesh Prasad MS
Hi-Tech Lab, Drugs Testing Laboratory, Drugs Control Department, Bangalore, India

Anil T Pawar
School of Pharmacy, Dr. Vishwanath Karad MIT World Peace University, Pune, India

Aman Upaganlawar
SNJB's Shriman Suresh Dada Jain College of Pharmacy, Nashik, India

CONTENTS

15.1 INTRODUCTION

Romantic attraction acts as the glue between partners in the forging of a successful, long-term relationship. The characteristic feature of attraction is focused attention on a chosen mate (Fisher et al., 2006). Attraction includes various aspects of love, such as romance, passion, infatuation, closeness, seeing beauty in another, compassion, and similar interests. Social bonds and attachments are often characterized by physical contact or cognitive attraction toward a preferred partner, and in some cases by distress or vocalization in the absence of the partner (Lieberwirth and Wang, 2014). Changes in physiological measures, including autonomic, endocrine, or immune changes, may also be used to indirectly index selective social relationships. Positive attachments have been recognized to have proximity maintenance, and a lack of social attachment can be related to both physical and psychological sickness. In order to recognize the underlying process of building an attachment and its advantages, and other systems of social sustenance, significant insights into processes that are protective for mental and physical health are essential (Carter, 1998).

A difference in the expression of romance is generally noticed between the sexes. However, the intensity of romantic attraction might be the same in both (Hatfield and Rapson, 1996). Individual feelings experienced during attraction

DOI: 10.1201/9780429265198-18

may result in different psychological responses, such as euphoria, lack of sleep and appetite, exhilaration, a raised heartbeat, anxiety, fear, and accelerated breathing. The initial phase of lust is attraction. Feelings of romantic attraction are regulated by the neuronal mechanisms of the human brain complementing the increased emotions (Saad et al., 2019).

This complex mechanism has also been identified in other mammals and this process results not only in lust and reproduction, but also in the bonding of pairs, and social organization and interaction (Seshadri, 2016). Love and romantic attraction are not controlled casually; various biologic consequences are involved to result in one partner being the sole focus (Jankowiak and Fischer, 1992).

Some common methods are used to assess the attachment processes, such as partner preference based on selective proximity, and physical and eye contact; defense with partner aggression; and physiological and biological systems, specifically autonomic, immune, and endocrine systems in either the presence or absence of partner; and behavioral changes, including crying, agitation, and anxiety which are usually involved with partner's presence. Changes observed include significant functional variations in specific regions of the brain, c-Fos expression involving appearance of expression genes related to specific putative peptides, neurotransmitters, receptor-specific gene expression changes, and functional imaging techniques such as blood oxygen level-dependent (BOLD) functional magnetic resonance imaging (fMRI) or diffusion tensor imaging (Carter, 2017).

The goal of this chapter is to present a comprehensive assessment of the neurobiological consequences of romantic attraction, with evidence from preclinical and clinical investigations. The theories involved in romantic attraction play a major part in evoking interest in the biological events involved in love. Experiments based on exploring the facts behind love and feeling need to be conducted. Two decades of neuroscience research have tried to discover the original and consequent events of brain during the phases of love and attraction (Liebowitz, 1983; Marazziti and Cassano, 2003).

15.2 NEUROBIOLOGICAL BASIS OF ROMANTIC ATTRACTION

Attachment is associated with distinct features in the functional region of the brain. Upon release of vasopressin, ventral palladium activation occurs, which results in the, deactivation of right anterior insula, superior frontal gyrus and hypothalamus, which are involved in stress. It is hypothesized that several predisposing factors change the physiology and functionality of the brain, making it susceptible to other stimuli such as hormonal changes and altered life patterns. These events are contribute to the experience of falling in love (Marazziti and Cassano, 2003). Other stimuli that contribute to falling in love are visual, auditory, tactile, gustatory, and olfactory. Among these stimuli, visual stimuli are important in humans. A series of events occur in thalamus and are divided in two: one is the shortest pathway directed to the amygdala; the other is a longer pathway that reaches the amygdala through the thalamus and cortex. After activation of the amygdala, the neuronal response in cerebral and peripheral organs is modulated. These changes are subjected to altered neurovegetative functions and ultimately produce feelings of emotion (Damasio, 2003). After its activation in the amygdala, information is passed to the cortex, where its action is translated into a feeling of fear, joy, or love. The ventral tegmental area of the brain is a reward region of the brain associated with emotional feelings such as pleasure, euphoria, motivation, common arousal, and focused attention. This region is affected by the consumption of cocaine and chocolate. Romantic love also includes the deactivation of frontal, parietal, and mid-temporal portions of the cortex and amygdala. Deactivation of subsequent regions of the brain results in the relaxation of inhibition. Activation of the ventral palladium is associated with stimulation and specific distribution of vasopressin (V1q), which is responsible for pair bonding and attachment (Aron et al., 2005) (Table 15.1).

During romantic attraction, the stimulated neurotransmitters are steroids, norepinephrine (noradrenaline), dopamine, serotonin, oxytocin, and vasopressin. In addition, other factors like major histocompatibility complex (MHC), pheromones, nerve growth factors, and sympathetic and parasympathetic responses (flight and fight/rest and digest) are also stimulated to attract mates and pair bonding. These stimulations activate brain pathways (Roelofs, 2017), and eventually result in the regulation of the lower limb of the brain, motivation, reward sex, and parenting.

It is reported that neutrophin and brain-derived neurotrophic factor (BDNF) levels in plasma affect romantic attraction in both sexes (Marazziti et al., 2009). BDNF plays a role in promoting a social relationship through decreasing the avoidance and fear of strangers. Attraction is also known to show similar features to obsessive compulsive disorders, with intrusive thoughts about the chosen partner (Fisher, 1992).

15.3 NEUROCHEMICAL AND HORMONAL BALANCE IN ROMANTIC ATTRACTION

The neurotransmitters released during the romantic attraction have been found to be involved in hormonal balance. Lust, attraction, and attachment are the stages of love, and are associated with the release of estrogen and testosterone (lust); adrenaline, noradrenaline, dopamine, phenylethylamine (PEA), and serotonin (pleasure, attraction, anxiety, and attraction); and oxytocin, vasopressin, and endorphins (bonding and attachment) (Dfarhud et al., 2014) (Table 15.2).

Monoamine transmitters such as adrenaline, dopamine, and serotonin have been found to be activated in attraction. Romantic attraction depends on disturbance in neurochemical stimulation, as reported in bipolar disorders (Manji et al., 2003). These events trigger the increased functioning of noradrenaline and dopamine. A trace neurotransmitter similar to amphetamine is PEA, and is called the 'love molecule'. It is found to be elevated in romantic attraction. PEA promotes the early phases of love and infatuation. Sexual desire is obtained as conjugated action of PEA, sex hormones, and other neurotransmitters (Leonti and Casu, 2018) (Figure 15.1).

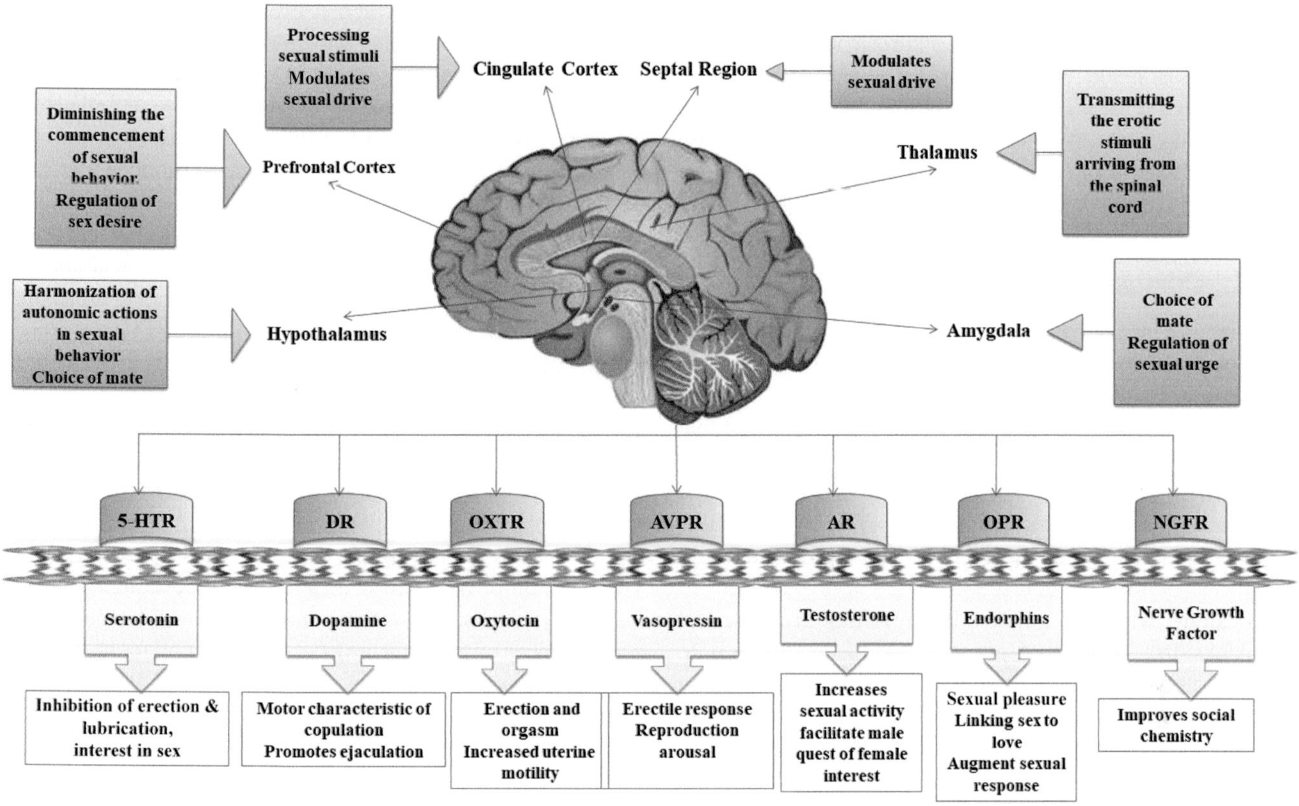

FIGURE 15.1 Role of neurotransmitters in the brain in sexual attraction. 5-HTR: 5-hydroxytryptamine receptor; AR: androgen receptor; AVPR: arginine vasopressin receptor; DR: dopamine receptor; NGFR: nerve growth factor receptor; OPR: opiate receptor; OXTR: oxytocin receptor.

15.3.1 Adrenaline and Noradrenaline

According to the emotion and motivation, monoamine levels may vary in the brain. A complex and dynamic series of events evoke romantic attraction in the affective state. Enhanced noradrenaline levels evoke stimulus-induced memory (Griffin and Taylor, 1995). Increased stress levels in people in troubled marriages are associated with adrenocorticotropic hormone, adrenaline, and noradrenaline. Adrenaline, which is involved in the fight-or-flight response, is characterized by a rise in blood pressure. Heartbeat is found to be raised and intense during focused attention on a lover, and feelings of nervousness and anxiety might be due to the release of adrenaline. Whenever people think about their intimate partners, they have a feeling of 'butterflies' in their stomachs due to adrenaline release. Noradrenaline causes the heart to race, the palms to sweat, and the skin to flush (Steimer, 2002).

Adrenaline is a hormone secreted in response to stress. This hormone plays a role in the progress of attraction along with arousal and leads to a 'love wave' in the heart. Its quick release causes sexual attraction, which makes the partners to approach to a condition of incitement and yearning to "roll mountains". Adrenaline is considered to the most vital hormone secreted by the adrenal medulla. Its release is mainly observed during conditions of fear, such as war or panic. Adrenaline is released mainly into the bloodstream under such situations. The adrenaline release affects the force of contraction, as well as the frequency of beating, of the heart muscles. It also enhances the breathing force and the sequence of metabolic functions (Verberne et al., 2016).

15.3.2 Serotonin

Serotonin (5-hydroxytryptamine) is a monoamine neurotransmitter, and plays a role in building love and rekindling love for a partner. Serotonin simultaneously aims to reduce pain and improve the speed of transmission of pleasant emotions. When we look at the emotional side of tenderness, serotonin levels begin to rise. It has been verified that initial love (the attraction stage) augments romantic relationships. It has also been demonstrated that obsessive compulsive disorder and romantic love have common characteristics of lower serotonin levels. Psychologists believe that such a romantic relationship forces partners to remain together so as to proceed to the next level of love (i.e., attachment). Nevertheless, the chief purpose of serotonin is to amplify emotions that occur in the cerebral cortex. When levels of serotonin are insufficient, it leads to depression. Changes in the serotonergic contribution to numerous regions of the limbic forebrain may reflect changes in serotonergic input into the hypothalamus (Chandrasekar et al., 2019).

Decreased serotonin and increased dopamine concentrations have been found to occur in different patterns of attraction, grooming, and even in compulsions (Marazziti and Cassano, 2003). Fluoxetine is a selective serotonin reuptake inhibitor that results in loss of libido. Hence, it is called an anti-love drug due to its inhibitory activity (Fisher et al., 2005). Reduced serotonin levels have been observed in the attraction phase and cause obsessive thinking about a partner, reflected in a feeling of love for the partner during romantic encounters (Marazziti et al., 1999).

15.3.3 Dopamine

Dopamine is a pleasure chemical. It is found in young couples 'suffering from love'. During romantic attraction, neuronal transmission is triggered via the putamen, right ventral tegmental area, and right caudate nucleus. It is reported that dopamine levels are high in romantic lovers (Harvey, 2012). According to psychiatrists, dopamine triggers the desire and reward acts of sexual attraction, accompanied by boundless pleasure. These functions are regulated centrally, especially through the brain. This course comprises an expenditure of energy along with the gratification resulting from sexual acts (Aleksidze, 2014). The primary role of dopamine is to uplift the nervous system, which is molded in the region of the cerebral cortex. When the levels of this hormone are insufficient, it causes stress and depression.

Dopamine also initiates other neurochemical processes involved feelings of romantic affiliation and addiction. Dopamine helps with intense pleasure and euphoria, similar to the feelings elicited by cocaine. Enhanced dopamine release is linked with greater levels of oxytocin and a lower concentration of serotonin (Earp et al., 2017), and also stimulates the feeling of liking and wanting a romantic partner. Dopamine elicits feelings of desire and reward paths involve pleasure. If a behavior results in gratification with sexual action and in defence, the brain will release the right amount of dopamine in the future, making these expectations equal and activities of this nature more enjoyable (Love, 2014). According to many experts, it is difficult to break free of romantic feelings. When satisfaction and emotional reversal are overturned, the brain continues to use love hormones and improve dopamine synthesis and thus enhances feelings and emotional state. The amount of dopamine released is dependent on particular behavioral or occupational effects, and a sense of satisfaction (Dfarhud et al., 2014).

15.3.4 Oxytocin

Oxytocin is a powerful neurotransmitter in the brain and is popularly called a love hormone. The social interactions and pair-bonding activity of oxytocin are modulated through the mesolimbic dopamine circuits. Oxytocin acts on the adrenal cortex by inhibiting the release of stress hormones involved in uterine contraction and orgasm, the production and ejection of milk, relaxation, and the immune system. It is also involved in the social behavior, care, bonding, memory, support, trust, relaxation, and sexual behavior (Neumann, 2007). The body responds to the actions of oxytocin during physical touch and close contact such as kissing, hugging, cuddling, and other sexual activities, after which it evokes feeling of calmness, security, and contentment. An imbalance in corticosteroids and oxytocin creates difficulties in maintaining a bond, trust, and devotion between pairs, possibly due to an imbalance between stress and relaxation. Oxytocin activities are also coordinated in the presence of dopamine to produce a feeling of pleasure during touch. Oxytocin levels have been found to be increased in the serum of individuals in love, and levels are increased during romantic love. A 'trust elixir' is available on the market; it is an oxytocin-laced scent designed to create the feeling of being in love (Navneet and Kalra, 2011).

Oxytocin is a nanopeptide and is basically a behavioral and trust hormone. Oxytocin is highly commanding and emotionally triggering in nature. It is secreted by both sexes during orgasm. Oxytocin is produced in the inner nucleus region of the hypothalamus. It is then deposited in the neurohypophysis (posterior pituitary), where it is moved by a precise neurophysin protein receptor (Scheele et al., 2012). The function of oxytocin is achieved by redeveloping uterus, mammary glands, and smooth muscle fibers. The biological consequences of oxytocin are established by collaboration with the specific organ's cytoplasmic receptors and the action of fermented adenylate cyclase, which endorses the development of cyclic adenosine monophosphate (cAMP), an intracellular effect related to cAMP-based protein kinase. The secretion of oxytocin enables the uterus to extend in the stage of late pregnancy, along with nipple erection during breastfeeding (Landgraf and Neuman, 2004). Oxytocin also increases the sensation of attachment and brings couples closer following sex. Most captivatingly, before an injection of oxytocin, female mice in one study were unresponsive to the cries of restless mice and displayed violence; however, after an oxytocin injection the activity of the mice was found to be similar to what is usual in maternal mice and they started to tend to the crying mice (Dulac et al., 2014). Oxytocin-sensitive neurons are located in several regions of the brain. When the activities of these neurons is disordered, women lose interest in men. When people have sex more frequently, their love for one another other deepens. Oxytocin also reinforces the attachment between a child and its mother (Cardoso et al., 2013).

Oxytocin has been recognized as a chief supporter of parental attachment, as well as human attraction. In one randomized, placebo-controlled, clinical study, the first behavioral evidence was acquired after oxytocin was administered intranasally in humans. The results showed that oxytocin fortified male-only relationships. Hence, oxytocin can help to endorse faithfulness in homosexual relationships. It is important to note that pheromones are often involved in romantic relationships. Pheromones are chemical substances that behave similarly to hormones, but outside of the body of the secreting individual and are mainly involved in sexual behavior (Carter and Porges, 2013).

15.3.5 Vasopressin

Vasopressin (antidiuretic hormone) has multifunctional roles in the sleep cycle, maintenance of rhythm, immunoprotection, and strengthening social bonds between individuals. It acts on social bonds in association with testosterone and other neurotransmitters involved in romantic attraction. Vasopressin is a traditional neurotransmitter, restricted to local mechanism of action across a synapse. These molecules are present in axons, neuronal soma, and dendrites. They mainly function as neuromodulators in the nervous system. Vasopressin appears to support the effects of autonomy and emotional feelings of romance and orgasm, and coping with early childhood exposure (Kenkel et al., 2013). The early stages of various sexual relationships accompanied by sexual arousal with orgasm can trigger the potency of vasopressin to allow for increased sensitivity without parasympathetic reversal. Vasopressin, along with its receptors, forms a biological and genetic mechanism that regulates attachment and binding (Carter, 2014). It is also capable of altering the autonomic, immune, and metabolic systems that work to coordinate the increase or decrease of responses in these systems when faced with a challenge. It is reported that testosterone also stimulates the production of vasopressin by moderating vasopressin receptors in the hypothalamus, and is associated with aggressiveness (Harvey et al., 2012).

15.3.6 Testosterone

Testosterone levels have been found to be altered depending on the condition and age of the person. There is evidence that testosterone and cortisol levels are linked to the attraction of a romantic partner; testosterone levels are related to a range of sexual behaviors and cortisol is a critical factor in the stress response. Testosterone, a steroid hormone, is involved in facilitating what scientists refer to as 'mating effort', which implies the investment of time and energy into same-sex competition and mate-seeking behavior.

Even when it is not directly associated with enhancing libido, testosterone may serve to facilitate the male quest of female interest (i.e., mating effort). Wild male chimpanzees have augmented testosterone levels when parous females display extremely tumescent sexual swellings, but not in response to maximal swelling in nulliparous females that males do not compete as intensively over. As males copulate with nulliparous and parous females at approximately the same rates, the findings suggest that these changes in testosterone facilitate the male pursuit of females rather than sexual performance per se (Muller and Wrangham, 2004). It was found in one study that a man's testosterone levels increased after he interacted with an attractive woman and chiefly so when the woman assumed he was trying to excite her (Roney et al., 2003).

15.3.7 Endorphins

Endorphins are opioids found naturally in the body and diminish pain, augment attention or gratification, enhance self-esteem, authorize emotions, and increase happiness, and are engaged in the usual reward series. The release of endorphins from the human body results in physical motion, exercise, musical activity, the consumption of certain foods (dark chocolate), sensitivity to ultraviolet light, and, most importantly, childbirth. It is suggested that endorphins control the release of other hormones (prolactin, sex hormones, and growth hormone) involved in sexual activity and love (Hiramoto et al., 2013).

The connection between β-endorphins, sex, and romantic relationships is comparable with endogenous sex steroids that influence the neurobiology of sexual performance via a direct effect on receptors residing at the nuclear and membrane level, or indirectly influencing neurotransmitter neuropeptides. It is known that a minor increase in β-endorphin concentrations generates a feeling of well-being, and that a huge upsurge can result in analgesic, as well as euphoric, conditions (Dunbar et al., 2012). A range of behavioral experiences can activate the release of β-endorphins. For example, exercise triggers the release of corticotropin-releasing hormone, which results in enhanced endorphins and adrenocorticotropic hormone levels that leads to an increase in an individual's sex drive. In addition to aerobic exercise, stopping smoking and illegal drug use, and limited alcohol consumption recover the oxygenation of tissues, encourage metabolism, reduce body mass index, and enhance endorphin release, which can augment sexual response (Frye, 2009).

15.3.8 Nerve Growth Factor

Plasma neurotrophin levels are associated with the initial phase of romantic love. An optimistic relationship between the initial passion of the romantic emotional state and serum concentrations of nerve growth factor (NGF) has been identified (Emanuele, 2011). The first stage of a romantic association can be connected to variations in the levels of circulating neurotrophins (Emanuele et al., 2006). NGF is known to regulate the growth, proliferation, and maintenance of certain nerve cells. However, it also has other functions; for example, it plays an important role in regulating the immune system. NGF also has links with reproduction.

Studies have shown that circulating NGF levels, and not other neurotrophins, are amplified partners who are in love, signifying a noteworthy contribution of this entity to the 'social chemistry' in humans (Emanuele et al., 2006). The explicitness of NGF intensification in the course of the initial stage of love undoubtedly indicates that this neurokine performs a pivotal role in the development of novel bonds; however, it does not appear to play a chief role in their regulation. Even though no physical or mental explanation can be immediately found through study, it is speculated that NGF might exercise its effects by regulating various endocrine functions, together with stimulation of the hypothalamus–pituitary–adrenal axis, and the secretion of vasopressin.

A number of studies have been conducted to understand the influence, mechanisms, and relationships of various neurochemicals involved in love and romantic attraction. Understanding of neurobiological concepts involved in sexual function derives from three major research methodologies: (1) preclinical studies; (2) clinical studies;

and (3) clinical reports of sexual impairment attributed to drug therapy or sickness.

15.4 INVESTIGATIONAL STUDIES

15.4.1 Preclinical Studies

Response to a sexual arousal is an intensely entrenched key biological state that exists in all entire species. The investigation of sexual intensity in animals is a multifaceted subject that might be advanced from various viewpoints, presuming it is based on communication involving genetic impacts, and the nervous and endocrine systems.

Several preclinical models are available to assess the neurobiology associated with romantic attraction. Prairie voles and titi monkeys show partner preference and have been used for adult pair-bonding studies (Bales et al., 2017). Some of the preclinical studies are discussed below.

In one study, when a male prairie vole (*Microtus ochrogaster*) copulated with a female, the female developed a discrete fondness for the male, which was correlated with a 50 percent increase in the dopamine level in the nucleus accumbens (Gingrich et al., 2000). However, the scenario changed when the nucleus accumbens was injected with a dopamine antagonist. The injection resulted in the loss of desire for the male by the female. When the same female was treated with a dopamine agonist, it chose the conspecific animal that existed at the time of injection, even if it had not copulated with that male (Gingrich et al., 2000; Wang et al., 1999).

A study conducted in female sheep revealed an upsurge in central dopamine, which was related to courtship attraction (Fabre-Nys et al., 1997; Fabre-Nys, 1998). Also, in male rats, augmented levels of striatal dopamine release was observed in response to the existence of a receptive female rat (Montague et al., 2004; Robinson et al., 2002).

Studies in animals have revealed that noradrenaline is certainly connected with sexual arousal and motivation (Bockstaele et al., 1989; Clayton et al., 2002; Fraley, 2002; Pfaff, 2005). When a feminine prairie vole had urine droplet from a male applied to its labium superioris, oestrogen was released due to the noradrenaline present in the olfactory bulb. This release of oestrogen was connected to proceptive behavior (Dluzen et al., 1981). The opposite mechanism occurred in a study where oestradiol, along with progesterone, contributed to the release of noradrenaline in the hypothalamus and resulted in lordosis in rats (Etgen et al., 1999). Furthermore, when the ovaries in sexually receptive female rats were removed surgically, and the rats injected with oestrogen followed by copulation, the result was a release of noradrenaline in the region of the ventromedial hypothalamus (Etgen and Morales, 2002). Drug users link the optimistic chemical relation between sex drive and noradrenaline. Precise oral doses of dopamine and noradrenaline agonists result in an augmented sex drive (Buffum et al., 1988).

Oxytocin and vasopressin have been related to partner preference and attachment behavior. Dopaminergic pathways have been shown to be associated a lot more accurately and exceptionally with partner preference only. Therefore, Lim and Young (2004) suggested that when monogamous prairie voles and other pair-bonding animals are involved in copulation, the act triggers the action of vasopressin in the region of the ventral pallidum, along with the effects of oxytocin in the nucleus accumbens. Copulation activities permit the release of dopamine in these reward areas, which inspires both sexes to choose a recent partner and show attachment and pair-bonding behaviors. However, the males of promiscuous species possibly sense attraction but they do not relate this gratifying sensation with its partner and so do not begin an extended term of attachment. In species that do not progress with this type of bonding, the relation with the dopamine reward centers is weak (Kendrick, 2000).

Preclinical studies indicate that the increased activity of dopaminergic pathways can trigger a cascade of events, involving testosterone and oestrogen release (Ferrari and Giuliana, 1995; Hull et al., 1995, 1997, 2002; Kawashima and Takagi, 1994; Wenkstern et al., 1993). Similarly, cumulative levels of testosterone and oestrogen endorse dopamine release (Pfaff, 2005).

In one study, a male rat was positioned in a cage in such a way that it could perceive the odor of a female (in estrous) kept in another cage. This resulted in an increase in central dopamine activity that contributed to sexual stimulation on detecting the female (Hull et al., 1995, 1997, 2002; Wenkstern et al., 1993). When these rats were allowed to mate, dopamine activity in the medial preoptic area continued to increase (Hull et al., 1995). Likewise, dopamine directly injected into particular parts of the brain of the male rat stimulated mating behavior (Ferrari and Giuliani, 1995). Constraining the central dopamine actions in rats reduced several proceptive sexual behaviors (hopping and darting; Herbert, 1996). Lastly, electrochemical studies involving male rats showed augmented release of dopamine in the striatum (dorsal, as well as ventral) due to the presence of a receptive female rat (Montague et al., 2004; Robinson et al., 2002). It is reported that dopamine in male rats increased sexual behavior by increasing sexual arousal, as well as courtship behavior, by potentiating the motor performances of mounting and facilitating responses of the genitalia to the stimulus (Pfaff, 2005).

Rats (*Rattus norvegicus*) housed in a seminatural laboratory setting have been studied to determine copulatory patterns. During copulation, each female (in estrous) mated with each male; similarly, every male mated with each female. Though separate males and females underwent same number of matings, there were a lot of sex variances in order and temporal pattern (Johansen et al., 2008). Males copulated in a pattern of multiple intromissions accompanied by a further ejaculatory sequence when numerous females remained in estrus. On the contrary, females acknowledged intromissions, as well as ejaculations, in a haphazard pattern. Males mated at short-term intermissions as compared to females; a temporal sex variance was observed in the pattern of female implorations and approaches by males. These sex variances are considerate to discuss the contradictory parts of investigation that are appropriate for the sexual behavior of both sexes in rats. Additionally, sex variances in the temporal pattern of coitus gave the impression that group copulation was similar to known sex variances in temporal limitations of neuroendocrine reflexes that help productive reproduction in the native strain (Le Moene and Agmo, 2019).

Several studies have been performed over previous decades to determine the influence of perinatal estradiol and/or testosterone on the development of a fondness for female to choose a male or male to choose a female companion. The preclinical literature on this subject matter has been extensively researched (Adkins-Regan, 1988; Baum, 2006; Wallen and Baum, 2002). Several of the preclinical models employed in these experiments (e.g., ferrets, rats, and zebra finches) do not appear to exhibit passionate similar-sex attraction as part of their characteristic species behavioral selections. As a substitute, same-sex mate preferences are the consequence of exposure to remarkably greater or lower perinatal hormonal concentrations. These exogenic therapies alter both genital and brain sex and therefore are accompanied by similar-sex desirability, along with partner preference. However, noteworthy perceptions perceived from these experiments are confounded by their inability to pinpoint sexual companion preferences from genital structure and further aspects of sexual behavior. In contrast to these models, explicit similar-sex desirability arises spontaneously in domestic populations of rams.

15.4.2 Clinical Studies

Sexual behavior in humans is influenced by cultural situations demanding lively behavioral adaptation. Consequently, multischeme communication is vital to guarantee both the features and the complicated functions distinct to humans. Neural constructions involved in sexual behavior are located throughout the central and peripheral nervous divisions of the human body (Ngun et al., 2011). Recognition of multimodal sexual inducements consist of sensory involvement that triggers autonomic and proper motor activities under intellectual regulation. Although the neural processes fundamental to factors pertaining to attraction, desire, and arousal (besides orgasm) are identical in both sexes, their sexual responses are dissimilar (Balthazart, 2016; Clark and Hatfield, 1989; Petersen and Hyde, 2010).

Several studies have revealed that dopamine is the principal player in stimulating sexual motivation, signifying that increased dopamine levels in regions of the reward system might process behavioral shift in the direction of hypersexuality (Oei et al., 2012). Of note, sexual triggering involves numerous neural substrates, such as those involved in processing reward and multifaceted cognitive mechanisms (e.g., decision-making) (Haber, 2011; Haber and Knutson, 2010).

Many studies reveals that thalamus is the key player in sexual fondness, by releasing neurotransmitters/hormones and communicating them to the temporal lobes, progressive towards behavioral deviations. The thalamus, then, perhaps plays an essential part in companion selection (Spinella, 2004), firming up the evidence that the thalamus is a focus of transmission from the spinal cord to the cortex and vice versa, as well as an amalgamation center which is self-motivated throughout the three segments of desire, arousal, and orgasm (Figure 15.2).

In one fMRI experiment, Brunetti et al. (2008) determined a noteworthy association between bilateral hypothalamus stimulation and deep sexual identity (DSI), which is well-demarcated as a multi-assessed dependability system that merges the different features of sexual identity, including physical, mental, and cultural factors. More precisely, DSI was evaluated subsequent to the re-check of the Franck drawing

TABLE 15.1
Brain Areas and Their Related Human Sexual Behavior

Brain area	Related Human Sexual Behavior
Amygdala	• Provides emotional consequence to arriving erotic trigger • Choice of mate • Regulation of sexual urge
Cingulate cortex	• Sexual trigger processing in situations of conflict • Regulation of sexual desire
Hypothalamus	• Harmonization of autonomic actions in sexual behavior • Choice of mate
Insula	• Conscious tumescence of erectile organs • Regulation of sexual urge
Prefrontal cortex	• Delaying the start of sexual behavior • Regulation of sexual desire
Reward system	• Stimulates sexual motivation • Regulation of sexual desire
Septal region	• Regulation of sexual desire
Thalamus	• Transmitting erotic stimuli arriving from the spinal cord

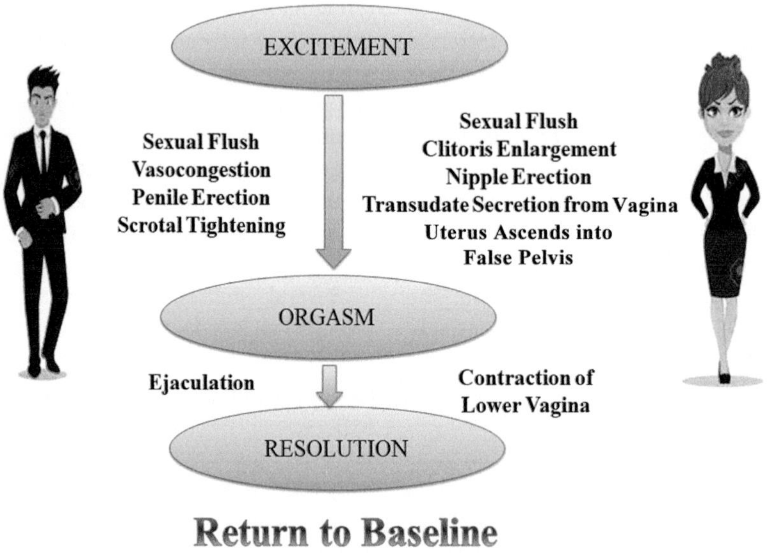

FIGURE 15.2 Human sexual response cycle in men and women.

TABLE 15.2
Neurotransmitters and Hormones Involved in Modulating Human Sexual Behavior

Neurotransmitter/Hormone	Function in Human Sexual Behavior
Acetylcholine	• Penile erection • Reversal of antidepressant-induced ejaculation and erectile complications
Dopamine	• Striatal dopamine for motor characteristic of copulation
Histamine	• Complete or incomplete erection through stimulation of H2 and H3 receptors in the peripheral division • Modulation of sexual behavior and libido at central level
Noradrenaline	• Stimulation of penile erection via activation of the autonomic system • Reversal of sexual inhibition resulting from sexual exhaustion, useful in treating anorgasmia and erectile dysfunction
Opioids	• Interaction between hormones and opioids leads to sexual impairment
Serotonin	• In the periphery, its action on the smooth muscles of the genitals, as well as the vascular system of other sexual organs, resulting in vasoconstriction and vasodilatation • Serotonin causes the inhibition of erection, lubrication, and interest in sex
Sex hormones	• Androgens stimulate and regulate sexual function in men • Estradiol is responsible for behavioral progress in male animals • Prolactin affords the body sexual satisfaction following sexual acts, even though high blood concentrations of prolactin can result in impotence, as well as loss of libido

completion test presented by M. Olivetti Belardinelli in 1994, resulting in a phase of concord between self-informed and mental sexual recognition, when participants were exposed to a sensual provocation (Brunetti et al., 2008). In view of the fact that DSI allows for interindividual variances in emotional attitude to sex, the significant positive relation between BOLD activity of the bilateral hypothalamus and DSI stipulates that the hypothalamus may be a structural substratum for an individual alteration in certain features of sexual behavior.

Several studies have revealed that the removal of exogenously supplied testosterone in men whose testicles have been removed or that are hypogonadal caused a quick and noticeable reduction in sexual attention, and that the action was restored within a few weeks with testosterone replacement therapy (Davidson et al., 1982; Kwan et al., 1983; Skakkebaek et al., 1981). In one study that distinguished between hyper- and hypogonadotropic hypogonadal male mammals, chronic testosterone therapy was determined to be further beneficial in enhancing individual interest in sexual acts/excitement, and the incidence of sexual feelings between males with hypergonadotropic hypogonadism (Gooren, 1988). In adolescent boys, free testosterone concentrations allow for the estimation of sexual feelings (Halpern et al., 1994; Udry et al., 1985), and once-a-month salivary testosterone assays have been wholly related to the commencement and degree of copulation (Halpern et al., 1998). It is probable that, in this latter study, increased incidences of copulation may have resulted in an upsurge in testosterone concentrations. In typical male adults, extensive and distinct variability arises in circulating concentrations of testosterone that are not expected to be connected in any logical way with discrete variances in the stages of sexual drive/behavior (Schiavi and White, 1976).

One study determined that testosterone increased sexual stimulation and gratification in hypogonadal and non-hypogonadal men (Alexander et al., 1997). The study also demonstrated an optimistic effect on mood alone in men with unusually low testosterone levels (Stanworth and Jones, 2008). Other investigational studies in men with hypogonadism have revealed that testosterone does not suggestively effect erection when provoked by erotic stimuli. Testosterone does not vary erectile responses suggestively among normal men and those with hypogonadism (Bancroft and Wu, 1983; Kwan et al., 1983; O'Carroll et al., 1985). Testosterone does not assist with erection in men with normal testosterone levels (O'Carroll and Bancroft, 1984).

Studies performed in premenopausal and healthy women carried out to investigate the relationship of testosterone concentration with sexual desire revealed slightly varying outcomes. Persky et al. (1978b) described a link between the frequency of copulation and mid-cycle testosterone concentration. Bancroft et al. (1983) described a connection between masturbation and the concentration of testosterone but not with the frequency of intercourse, and Udry et al. (1986) described an association between sexual attention and the concentration of testosterone in teenagers, but they revealed that noble relationships were an extra significant factor in sexual behavior. Halpern et al. (1997) also described an important relationship between the commencement of coitus and the concentration of testosterone in adolescent females. Though sexual longing in women is affected by androgen levels, androgens alone are not enough for the involvement of sexual craving. This is obvious from findings that were unsuccessful in elucidating differences in the testosterone levels in women with or without clinically detected hypoactive sexual desire, and from studies that showed that oral contraceptives and androgen inhibitors do not reliably reduce women's overall desire for sexual activity (Schreiner-Engel et al., 1989; Stuart et al., 1987). Testosterone therapy appears to be valuable in increasing sexual desire in a subset of women with hypoactive sexual desire, but safety monitoring is required due to latent lipoprotein fluctuations, cardiovascular illnesses, and androgenic skin variations (e.g., androgenic alopecia acne and hirsutism) (Reis and Abdo, 2014).

Substantial research proposes that estrogens have a slightly influential effect on libido in both sexes. In men, comparatively greater concentration levels of exogenic estrogen have been somewhat beneficial in constricting sexual desire in sex offenders and those with an overpowering libido (Bancroft et al., 1974; Cooper, 1986; Field and Williams, 1970). In females, several early investigational studies suggested that estrogen is noteworthy in the typical desire for sex (Benedek and Rubenstein, 1939); however, many researchers agree that estrogens play a negligible part in the female desire for sex. For instance, Schreiner-Engel et al. (1989) did not find any noteworthy variances in estrogen concentrations in women with and without a clinically detected hypoactive sexual desire; Dennerstein et al. (1994) found that variations in sexual desire during the course of the menstrual cycle are not connected to estrogen concentrations; Abplanalp et al. (1979) found that there was no relationship between estradiol concentration and gratification from heterosexual activity or the frequency of heterosexual activities; and there is a sufficient amount of evidence that there is no alteration in the desire for sex secondary to the administration of exogenous estrogenic substances in women suffering from several gynecologic illnesses (Nathorst-Boos et al., 1993; Burger et al., 1984; Furuhjelm et al., 1984; Salmon and Geist, 1943). Furthermore, during the administration of both sex hormones to normal or surgically induced menopausal females for the restoration of sexual desire, estrogen therapy was not solely revealed to be effective (Haber and Knutson, 2010).

Some orally administered contraceptives that increase the progesterone concentration throughout the menstrual cycle have been related to reduced interest in and desire for sex (Huffer et al., 1970; Kane et al., 1969; Warner and Bancroft, 1988), similar to subfascial implanted progesterone pills that are used in the treatment of several gynecologic ailments (Greenblatt et al., 1942). However, experimentation by Bakke (1965) revealed that estrogen and progesterone therapy improved sexual desire in females whose uterus has been surgically removed to a larger extent compared to estrogen alone. It is commonly agreed that progesterone therapy does not have a substantial effect on sexual desire in pre- or postmenopausal females (Campbell, 1976; Campbell and Whitehead, 1977; Studd et al, 1977; Coope, 1976; Furuhjelm et al., 1984; Leiblum et al., 1983; Persky et al., 1976, 1978a; Schreiner-Engel et al., 1981; Sherwin, 1985; Utian, 1972).

The effect of prolactin hormone on features of sexual behaviour in humans remains ambiguous. Erectile dysfunction is experienced in men with unusually low concentrations of prolactin (Besser and Thorner, 1975; Buvat et al., 1978; Horrobin, 1974). It has also been noted that remarkably low concentrations of prolactin signify a greater inhibition of erection capability (Zeitlin and Rajfer, 2000). In females, an unusually high prolactin concentration been related to amenorrhea, sterility, and reduced sexual activities (Muller et al., 1979). The effect of prolactin on sexual provocation may arise peripherally, centrally, or through dopaminergic regulation (Exton et al., 1999). Several experiments performed in men have revealed that prolactin concentrations either reduce instantly after sexual stimulation or are unaffected after watching an erotic movie, masturbation, or mating (Brown and Heninger, 1975; Carani et al., 1990; Ferla et al., 1978; Lee et al., 1974; Lincoln, 1974; Purvis et al., 1976; Rowland et al., 1987; Stoleru et al., 1993).

Circulating levels of oxytocin increase during sexual arousal and orgasm in both men and women (Blaicher et al., 1999; Carmichael et al., 1987, 1994; Murphy et al., 1987). By means of a continuous blood sampling technique along with anal electromyography, Carmichael et al. (1994) described a positive association between oxytocin levels and the strength, but not period, of orgasmic contractions in males and females.

For multiorgasmic women, the rise in oxytocin level was also positively associated with subjective reports of orgasm intensity. In a few cases, reports have indicated that a synthetic formulation of oxytocin used to enable breastfeeding was associated with increase sexual desire, along with vaginal lubrication (Hunt and Dennerstein, 1994, 1995). A study performed by Turner et al. (1999) determined a positive relationship between plasma oxytocin levels and measures of positive affect.

Studies have shown that the concentration of cortisol in blood does not change much in either sex when blood was drawn at the same time as participants watched a sensual film (Carani et al., 1990; Exton et al., 1999, 2000; Ismail et al., 1972; Kruger et al., 1998). The cortisol concentration was found to be in greater in males with erectile dysfunction that was psychogenic in nature. These subjects had a poor response when a smooth muscle relaxant was introduced by intracavernous injection. These subjects also measured higher on anxiety scores (Granata et al., 1995).

A placebo-controlled, double-blinded clinical trial showed that in males treated with an artificial human male pheromone, an increased frequency of coitus, sleeping with a romantic companion, and caressing and kissing was seen. Nevertheless, no variation in the frequency of masturbation was observed. The researchers concluded that male pheromones enhance the sexual appeal of a male to a female (Verhaeghe et al., 2013).

The initial results from a placebo-controlled, double blinded, two-way crossover clinical trial exhibited a noteworthy increase in the amplitude of the vaginal pulse when the patient was treated with one 50 mg dose of sildenafil citrate. The study was conducted with 12 women who were sexually active. Individual outcomes of sexual stimulation did not suggestively vary with the introduction of sildenafil (Lo Monte et al., 2014). Certain experiments revealed that the treatment of sildenafil counteracted sexual impairment induced by antidepressants in women (Ashton, 1999; Ashton and Bennett, 1999; Nurnberg et al., 1999). One 12-week global trial with 577 premenopausal women with sexual arousal impairment revealed that 30–50 percent of the participants on sildenafil experienced an enhancement in sexual activity versus 43 percent of females who were treated with placebo (Stulberg and Ewigman, 2008). In an open-label, 12-week, non-randomized clinical trial conducted with 33 sexually impaired postmenopausal females, treatment with sildenafil gave rise to a noteworthy satisfactory response in only six patients (Davidson et al., 1982).

Studies performed on masculine rats propose that stimulation of certain subtypes of serotonin receptor enables sexual behavior, whereas stimulation of other receptor subtypes constrains it. In particular, stimulation of the 5-HT_{1A} receptor reduces the urge to ejaculate; inhibition of the 5-HT_2 receptor prevents sexual behaviors; and stimulation of 5-HT_{1B} and 5-HT_{1C} receptors inhibit sexual behaviors (Bitran and Hull, 1987).

Numerous studies have reported that levodopa, used to treat parkinsonism, results in erection (Bowers et al., 1971; Hyyppa et al., 1970; O'Brien et al., 1971). The dopamine (D_1 and D_2) agonist apomorphine enables erection in men accompanied by normal erection capability (Danjou et al., 1988, 1989; Lal et al., 1984, 1987, 1989; O'Sullivan and Hughes, 1998; Segraves et al., 1991). The drug bromocriptine has been found to reduce prolactin concentrations and functions as a dopamine agonist (long-acting), facilitating erection. Antipsychotic medicines, which reduce dopaminergic action, are known to weaken erection (Aizenberg et al., 1995; Marder and Meibach, 1994) and also cause priapism (prolonged erection of the penis) (Chan et al., 1990; Jackson and Walker 1991; Merkin, 1977). In one study, haloperidol antagonized dopamine-induced erection (Gower et al., 1984).

In sexually active men, adrenaline levels in blood plasma show a negligible increase immediately prior to masturbation (Kruger et al., 1998), but the adrenaline concentration in urine remains unaffected subsequent to watching sensual films (Carani et al., 1990). Adrenergic action plays a critical role in regulating the flaccid penis and detumescence.

A number of studies that investigated the concentration of noradrenaline in plasma have reported that the levels of noradrenaline increase during sexual acts. In males, the concentration of noradrenaline in plasma is definitely linked to stimulation and erection throughout masturbation, as well as copulation, and increases by up to 12 times during climax (Kruger et al., 1998; Wiedeking et al., 1977, 1979). Kruger et al. (1998) found that the concentration of noradrenaline reduced to baseline within just 2 minutes of reaching climax. On the contrary, the noradrenaline concentration in urine remained the same in eight men 4 hours before and 4 hours after watching a sensual movie (Carani et al., 1990).

A study investigating the effects of opioids administered intraspinally determined that in the 1-month period following administration, the participants experienced a decline in overall sexual longing, as well as erectile dysfunction (Pake et al., 1994). The removal of opiate addiction was characterized by an increased occurrence of erections in the morning/spontaneous ejaculation and a sluggish reappearance of their sex drive. A number of males even experienced entire loss of erection and orgasm ability (Pfaus and Gorzalka, 1987). Although the process by which opiates disturb sexual operations is indistinct, a few reports indicate that enhanced opioid activity results in a decrease in circulating hormone levels, including testosterone, LH, and so on, leading to sexual dysfunction (Mirin et al., 1980; Pake et al., 1994).

In one in vitro experiment conducted with the corpus cavernosal tissue of males suffering from diabetes mellitus and related erectile impairment proposed that relaxation due to acetylcholine may be diminished (Saenz de Tejada et al., 1989). Bethanechol, a cholinergic drug, has been described to be beneficial in reversing antidepressant-induced erectile dysfunction (Segraves, 1987; Yager, 1986). Though cholinergic fibers are distributed peripherally, reports propose that erectile activity is regulated at the level of brain, as well as spinal cord (Giuliano et al., 1995).

By means of Fos staining performed during copulatory tests, the medial amygdala, as well as the bed nucleus of the stria terminalis, have been recognized as important in the sexual activity of women (Erskine, 1993; Tetel et al., 1993; Turner et al., 1999). The medial amygdala is thought to have a role in the regulation of sex drive in men (Everitt et al., 1989; Meisell and Sachs, 1994). When electrically stimulated, the hippocampus provoked erections (Chen et al., 1992; MacLean, 1975). Stimulation of the septal region has been associated with orgasm.

15.5 RECENT DEVELOPMENTS AND CHALLENGES

Romantic attraction is seen in humans, as well as in the animal and bird kingdoms. These species spend an ample amount of time searching for sex. The act is accomplished by an agreement between two individuals, who may have spent hours, days, or weeks in their quest for a mate. The journey is usually described as an amusing accident and causing brain-dazing stress (LeVay, 1993).

Romantic attraction has been considered since humans came into existence, and it has been an important theme for artists over the centuries. Romantic love also has become a topic of curiosity for researchers (de Boer et al., 2012). Investigations are currently seeking clarification on the brain activities accountable for attraction, love, and feelings. Until recently, the elucidation of romantic attraction has been chiefly occupied by psychology (Kandel, 2018).

Considering the progressional viewpoint, important factors include fulfillment and confining of emotions, required to accustom and reprise. Therefore, these emotions evolve the behavioral strategy of an individual in such a way that promotes an increase in the likelihood of attaining these achievements (Geary, 2005). Then again, do these developmentally assessed behaviors possess a biological substratum and an association with the initiation of precise areas of brain? Fisher (Fisher et al., 2002, 2010; Puigvert et al., 2019) theorized that the choosing of a mate is related to a neuronal mechanism leading to 'display chooser' in order to affirm their breeding energy on particular coupling partners, thereby preserving the time of courtship and energy for metabolism.

Findings in neuroscience have highlighted precise brain areas that facilitate and make the connections at the beginning of romantic attraction. Nevertheless, romantic attraction continues to be a topic of research, principally in the area of psychology, despite the enormous growth in neurobiological research (Barrett, 2009).

In the past few decades, there has been a noteworthy intensification in the quantity of research performed in relation to neuronal love correlates by means of neuroimaging techniques such as fMRI and positron emission tomography. With these techniques, scientists are able to study neurotransmitter functions, as well as the neuroendocrine system in love and romantic attraction (Crosson et al., 2010).

Romantic attraction is a kind of emotion that can be described by assessing brain processes and understanding the neural link of love. Through the study of the neurobiology of romantic attraction, future neurologists will benefit from the worldwide literature on love, as literature provides solid clues as to the make-up of the romantic system in the brain (Zeki, 2007). Hence, there is a wide scope for exploring the independent impacts of the endocrine system, neurotransmitters, and brain organization on romantic attraction.

15.6 CONCLUSIONS

Romantic desirability could be understood as a collection of activities linked to the attainment and retention of feelings that people want to last and reproduce. These emotions alter an individual's behavioral tactics in such a way that will enhance the probability of attaining these goals. Love may be distinct as a developing possession of an early blend of neuropeptides and neurotransmitters. It gives the impression that desire, attachment, and attraction are discrete but entwined progressions in the brain, each mediated by its own neurotransmitters and circuits. These circuits interact with and strengthen each other. Sexual longing is facilitated by sex hormones (testosterone and estrogen) and the amygdala is a significant center. Attraction is arbitrated by stress and reward hormones, including dopamine, noradrenaline, cortisol, and the serotonergic system; the nucleus accumbens and the ventral tegmental area are key mediators.

Human romantic love involves in an extensive range of fluctuating, overlying, and dynamic brain systems, as would be expected of a multifaceted process that has noteworthy social, reproductive, and genetic significance. The chief neural associates connected with intense, early-stage romantic love are probably similar across individuals and cultures – and even between species – as this neural mechanism has progressed to lead to a vital aspect of mammalian reproduction and mate choice.

REFERENCES

Abplanalp JM, Rose RM, Donnelly AF, Vaughan LL. Psycho-endocrinology of the menstrual cycle, II: the relationship between enjoyment of activities, moods and reproductive hormones. Psychosom Med. 1979;41:605–15.

Adkins-Regan E. Sex hormones and sexual orientation in animals. Psychobiology. 1988;16:335–47.

Aizenberg D, Zemishlany Z, Dorfman-Etrog P, Weizman A. Sexual dysfunction in male schizophrenic patients. J Clin Psychiatry. 1995;56:137–41.

Aleksidze N. Fundamentals of Psychobiology. Tblisi: The Publishing House of Georgian National Academy of Sciences; 2014.

Alexander GM, Swerdloff RS, Wang C, Davidson T, McDonald V, Steiner B, et al. Androgen-behavior correlations in hypogonadal men and eugonadal men, I: mood and response to auditory sexual stimuli. Horm Behav. 1997;31:110–19.

Aron A, Fisher H, Mashek DJ, Strong G, Li H, Brown LL. Reward, motivation and emotion systems associated with early-stage intense romantic love: an fMRI study. J Neurophysiol. 2005;94:327–37.

Ashton AK. Sildenafil treatment of paroxetine-induced anorgasmia in woman. Am J Psychiatry. 1999;156:800.

Ashton AK, Bennett RG. Sildenafil treatment of serotonin reuptake inhibitor-induced sexual dysfunction. J Clin Psychiatry. 1999;60:194–5.

Bakke J. A double-blind study of a progestin estrogen combination in the management of the menopause. Pacific Med Surg. 1965;73:200–5.

Bales KL, Arias Del Razo R, Conklin QA, Hartman S, Mayer HS, Rogers FD, et al. Titi monkeys as a novel non-human primate model for the neurobiology of pair bonding. Yale J Biol Med. 2017;90:373–87.

Balthazart J. Sex differences in partner preferences in humans and animals. Phil Trans R Soc B. 2016;371:20150118.

Bancroft J, Sanders D, Davidson D, Warner P. Mood, sexuality, hormones, and the menstrual cycle; III: sexuality and the role of androgens. Psychosom Med. 1983;45:509–16.

Bancroft J, Tennent G, Loucas K, Cass J. The control of deviant sexual behavior by drugs, I: behavioral changes following oestrogens and antiandrogens. Br J Psychiatry. 1974;125:310–15.

Bancroft J, Wu FC. Changes in erectile responsiveness during androgen replacement therapy. Arch Sex Behav. 1983;12:59–66.

Barrett LF. The future of psychology: connecting mind to brain. Perspect Psychol Sci. 2009;4:326–39.

Baum MJ. Mammalian animal models of psychosexual differentiation: When is translation to the human situation possible? Horm Behav. 2006;50:579–88.

Benedek T, Rubenstein BB. The correlations between ovarian activity and psychodynamic processes, II: the menstrual phase. Psychosom Med. 1939;1:461–85.

Besser GM, Thorner MO. Prolactin and gonadal function. Pathol Biol (Paris). 1975;23:779–82.

Bitran D, Hull EM. Pharmacological analysis of male rat sexual behavior. Neurosci Biobehav Rev. 1987;11:365–89.

Blaicher W, Gruber D, Bieglmayer C, Blaicher AM, Knogler W, Huber JC. The role of oxytocin in relation to female sexual arousal. Gynecol Obstet Invest. 1999;47:125–6.

Bockstaele EJV, Pieribone VA, Aston-Jones G. Diverse afferents converge on the nucleus paragigantocellularis in the rat ventrolateral medulla: retrograde and anterograde tracing studies. J Comp Neurol. 1989;290:561–84.

Bowers MB Jr, Van Woert M, Davis L. Sexual behavior during L-dopa treatment for parkinsonism. Am J Psychiatry. 1971;127:1691–3.

Brown WA, Heninger G. Cortisol, growth hormone, free fatty acids, and experimentally evoked affective arousal. Am J Psychol. 1975;132:1172–6.

Brunetti M, Babiloni C, Ferretti A, Del Gratta C, Merla A, Olivetti Belardinelli M, et al. Hypothalamus, sexual arousal and psychosexual identity in human males: a functional magnetic resonance imaging study. Eur J Neurosci. 2008;27:2922–7.

Buffum J, Moser C, Smith D. Street drugs and sexual function. In: Sitsen JMA, editor. Handbook of Sexology. Vol. 6. The Pharmacology and Endocrinology of Sexual Function. New York: Elsevier Science; 1988.

Burger HG, Hailes J, Menelaus M, Nelson J, Hudson B, Balazs N. The management of persistent menopausal symptoms with oestradiol-testosterone implants: clinical, lipid and hormonal results. Maturitas. 1984;6:351–8.

Buvat J, Ashour M, Buvat-Heraut M, Fossati P. Prolactin and human sexual behavior. In: Rubin C, Harter M, editors. Progress in Prolactin Physiology and Pathology. Amsterdam: Elsevier/ North Holland Biomedical Press; 1978.

Campbell S. Double-blind psychometric studies on the effects of natural estrogens on postmenopausal women. In: Campbell S, editor. The Management of the Menopausal and Post-Menopausal Years. Baltimore, MD: University Park Press; 1976.

Campbell S, Whitehead M. Oestrogen therapy and the menopausal syndrome. Clin Obstet Gynecol. 1977;4:31–47.

Carani C, Bancroft J, Del Rio G, Granata ARM, Facchinetti F, Marrama P. The endocrine effects of visual erotic stimuli in normal men. Psychoneuroendocrinology. 1990;15:207–16.

Cardoso C, Ellenbogen MA, Serravalle L, Linnen AM. Stress-induced negative mood moderates the relation between oxytocin administration and trust: evidence for the tend-and-befriend response to stress? Psychoneurondocrinology. 2013;38 :2800–4.

Carmichael MS, Humbert R, Dixen J, Palmisano G, Greenleaf W, Davidson JM. Plasma oxytocin increases in the human sexual response. J Endocrinol Metab. 1987;64:27–31.

Carmichael MS, Warburton VL, Dixen J, Davidson JM. Relationships among cardiovascular, muscular and oxytocin responses during human sexual activity. Arch Sex Behav. 1994;23:59–77.

Carter CS. Neuroendocrine perspectives on social attachment and love. Psychoneuroendocrinology. 1998;23:779–818.

Carter CS. Oxytocin pathways and the evolution of human behavior. Annu Rev Psychol. 2014;65:17–39.

Carter CS. The oxytocin-vasopressin pathway in the context of love and fear. Front Endocrinol (Lausanne). 2017;8:356.

Carter CS, Porges SW. The biochemistry of love: an oxytocin hypothesis. EMBO Rep. 2013;14 :12–16.

Chan J, Alldredge BK, Baskin LS. Perphenazine induced priapism. DICP. 1990;24:246–9.

Chandrasekar SB, Pawar AT, Lokesh Prasad MS, Hemavati G. Anti-oxidants against stress-induced memory dysfunction. In: Uddin MS, Upaganlawar AB, editors. Oxidative Stress and Antioxidant Defence: Biomedical Value in Health and Diseases. Hauppauge, NY: Nova Science Publishers; 2019.

Chen KK, Chan JY, Chang LS, Chen MT, Chan SH. Elicitation of penile erection following activation of the hippocampal formation in the rat. Neurosci Lett. 1992;141:218–22.

Clark R, Hatfield E. Gender differences in receptivity to sexual offers. J Psychol Human Sex. 1989;2:39–55.

Clayton AH, Pradko JF, Croft HA, Montano CB, Leadbetter RA, Watson CB, et al. Prevalence of sexual dysfunction among newer antidepressants. J Clin Psychiatry. 2002;63:357–66.

Coope J. Double-blind cross-over study of estrogen replacement. In: Campbell S editor. The Management of the Menopausal and Post-Menopausal Years Baltimore, MD: University Park Press; 1976.

Cooper AJ. Progestogen in the treatment of male sex offenders: a review. Can J Psychiatry. 1986; 31 :73–9.

Crosson B, Ford A, McGregor KM, Meinzer M, Cheshkov S, Li X, et al. Functional imaging and related techniques: an introduction for rehabilitation researchers. J Rehabil Res Dev. 2010;47:vii–xxxiv.

Damasio A. Feelings of emotion and the self. Ann N Y Acad Sci. 2003;1001:253–61.

Danjou P, Alexandre L, Warot D, Lacomblez L, Puech AJ. Assessment of erectogenic properties of apomorphine and yohimbine in man. Br J Clin Pharmacol. 1988;26:733–9.

Danjou P, Lacomblez L, Warot D, Peuch AJ. Assessment of erectogenic drugs by numeric plethysmography. J Pharmacol Methods. 1989;21:61–9.

Davidson JM, Kwan M, Greenleaf WJ. Hormonal replacement and sexuality. Clin Endocrinol Metab. 1982;11:599–623.

de Boer A, van Buel EM, Ter Horst GJ. Love is more than just a kiss: a neurobiological perspective on love and affection. Neuroscience. 2012;201;114–24.

Dennerstein L, Gotts G, Brown JB, Morse CA, Farley TM, Pinol A. The relationship between the menstrual cycle and female sexual interest in women with PMS complaints and volunteers. Psychoneuroendocrinology. 1994;19:293–304.

Dfarhud D, Malmir M, Khanahmadi M. Happiness & health: the biological factors- systematic review article. Iran J Public Health. 2014;43:1468–77.

Dluzen DE, Ramirez VD, Carter CS, Getz LL. Male vole urine changes luteinizing hormone releasing hormone and norepinephrine in female olfactory bulb. Science. 1981;212:573–5.

Dulac C, O'Connell LA, Wu Z. Neural control of maternal and paternal behaviors. Science. 2014;345:765–70.

Dunbar RIM, Kaskatis K, MacDonald I, Barra V. Performance of music elevates pain threshold and positive affect: implications for the evolutionary function of music. Evol Psychol. 2012;10:688–702.

Earp BD, Wudarczyk OA, Foddy B, Savulescu J. Addicted to love: what is love addiction and when should it be treated? Philos Psychiatr Psychol. 2017;24:77–92.

Emanuele E. NGF and romantic love. Arch Ital Biol. 2011;149:265–8.

Emanuele E, Politi P, Bianchi M, Minoretti P, Bertona M, Geroldi D. Raised plasma nerve growth factor levels associated with early-stage romantic love. Psychoneuroendocrinology. 2006;31:288–94.

Erskine MS. Mating-induced increases in Fos protein in preoptic area and medial amygdala of cycling female rats. Brain Res Bull. 1993;32:447–51.

Etgen AM, Chu HP, Fiber JM, Karkanias GB, Morales JM. Hormonal integration of neurochemical and sensory signals governing female reproductive behavior. Behav Brain Res. 1999;105:93–103.

Etgen AM, Morales JC. Somatosensory stimuli evoke norepinephrine release in the anterior ventromedial hypothalamus of sexually receptive female rats. J Neuroendocrinol. 2002;14:213–18.

Everitt BJ, Cador M, Robbins TW. Interactions between the amygdala and ventral striatum in stimulus-reward associations: studies using a second-order schedule of sexual reinforcement. Neuroscience. 1989;30:63–75.

Exton MS, Bindert A, Krüger T, Scheller F, Hartmann U, Schedlowski M. Cardiovascular and endocrine alterations after masturbation-induced orgasm in women. Psychosom Med. 1999;61:280–9.

Exton NG, Truong TC, Exton MS, Wingenfeld SA, Leygraf N, Saller B, et al. Neuroendocrine response to film-induced sexual arousal in men and women. Psychoneuroendocrinology. 2000;25:187–99.

Fabre-Nys C. Steroid control of monoamines in relation to sexual behaviour. Rev Reprod. 1998;3:31–41.

Fabre-Nys C, Ohkura S, Kendrick KM. Male faces and odors evoke differential patterns of neurochemical release in the mediobasal hypothalamus of the ewe during estrus: an insight into sexual motivation. Eur J Neurosci. 1997;9:1666–77.

Ferla J, Anderson D, Schalch D. Psychoendocrine response to sexual arousal in human males. Psychosom Med. 1978;40:166–72.

Ferrari F, Giuliani D. Sexual attraction and copulation in male rats: effects of the dopamine agonist SND 919. Pharmacol Biochem Behav. 1995;50:29–34.

Field LH, Williams M. The hormonal treatment of sexual offenders. Med Sci Law. 1970;10:27–34.

Fisher E. Anatomy of Love. Columbine, NY: Fawcett; 1992.

Fisher H, Aron A, Brown LL. Romantic love: an fMRI study of aneural mechanism for mate choice. J Comp Neurol. 2005;493:58–62.

Fisher H, Aron A, Mashek D, Li H, Strong G, Brown LL. The neural mechanisms of mate choice: a hypothesis. Neuro Endocrinol Lett. 2002;4:92–7.

Fisher HE, Aron A, Brown LL. Romantic love: a mammalian brain system for mate choice. Philos Trans R Soc Lond B Biol Sci. 2006;361:2173–86.

Fisher HE, Brown LL, Aron A, Strone G, Mashek D. Reward, addiction, and emotion regulation systems associated with rejection in love. J Neurophysiol. 2010;104:51–60.

Fraley GS. Immunolesion of hindbrain catecholaminergic projections to the medial hypothalamus attenuates penile reflexive erections and alters hypothalamic peptide mRNA. J Neuroendocrinol. 2002;14:345–8.

Frye CA. Neurosteroids' effects and mechanisms for social, cognitive, emotional, and physical functions. Psychoneuroendocrinology. 2009;34:S143–61.

Furuhjelm M, Karlgren E, Carlstrom K. The effect of estrogen therapy on somatic and psychical symptoms in postmenopausal women. Acta Obstet Gynecol Scand. 1984;63:655–61.

Geary DC. The evolution and development of the human mind. In: Geary DC, editor. Male, Female: The Evolution of Human Sex Differences. Washington, DC: American Psychological Association; 2005.

Gingrich B, Liu Y, Cascio C, Wang Z, Insel TR. Dopamine D2 receptors in the nucleus accumbens are important for social attachment in female prairie voles (*Microtus ochrogaster*). Behav Neurosci. 2000;114:173–83.

Giuliano FA, Rampin O, Benoit G, Jardin A. Neurological control of penile erection. Urol Clin North Am. 1995;22:747–66.

Gooren L. Hypogonadotropic hypogonadal men respond less well to androgen substitution treatment than hypergonadotropic hypogonadal men. Arch Sex Behav. 1988;17:265–70.

Gower AJ, Berendsen HHG, Princen MM, Broekkamp CLE. The yawning-penile erection syndrome as a model for putative dopamine autoreceptor activity. Eur J Pharmacol. 1984;103:81–9.

Granata A, Bancroft J, Del Rio G. Stress and the erectile response to intracavernosal prostaglandin E1 in men with erectile dysfunction. Psychosom Med. 1995;57:336–44.

Greenblatt RB, Mortara F, Torpin R. Sexual libido in the female. Am J Obstet Gynecol. 1942;44 :658–63.

Griffin MG, Taylor GT. Norepinephrine modulation of social memory: evidence for a time-dependent functional recovery of behavior. Behav Neurosci. 1995;109:466–73.

Haber SN, Knutson B. The reward circuit: linking primate anatomy and human imaging. Neuropsychopharmacology. 2010;35:4–26.

Haber SN. Neuroanatomy of reward: a view from the ventral striatum. In Gottfried J, editor. Neurobiology of Sensation and Reward. Boca Raton, FL: CRC Press/Taylor & Francis; 2011.

Halpern CR, Udry JR, Suchindran C. Monthly measures of salivary testosterone predict sexual activity in adolescent males. Arch Sex Behav. 1998;27:445–65.

Halpern CT, Udry JR, Campbell B, Suchindran C, Mason GA. Testosterone and religiosity as predictors of sexual attitudes and activity among adolescent males: a biosocial model. J Biosoc Sci. 1994;26:217–34.

Halpern CT, Udry JR, Suchindran C. Testosterone predicts initiation of coitus in adolescent females. Psychosom Med. 1997;59:161–71.

Harvey RA, Clark MA, Finkel R. Lippincott's Illustrated Reviews Pharmacology. 5th edn. New York: Lippincott Williams & Wilkins; 2012.

Hatfield E, Rapson RL. Love and Sex: Cross-cultural perspectives. Needham Heights, MA: Allyn and Bacon; 1996.

Herbert J. Sexuality, stress and the chemical architecture of the brain. Annu Rev Sex Res. 1996;7:1–43.

Hiramoto K, Kobayashi H, Sekiyama A, F Sato E, Tsuruta D, Ishii M. Mild exercise suppresses exacerbation of dermatitis by increasing cleavage of the β-endorphin from proopiomelanocortin in NC/Nga mice. J Clin Biochem Nutr. 2013;52:58–63.

Horrobin DF. Prolactin and mental illness. Br J Psychol. 1974;124:456–7.

Huffer V, Levin L, Aronson H. Oral contraceptives: depression and frigidity. J Nerv Ment Dis. 1970;151:35–41.

Hull EM, Du J, Lorrain DS, Matuszewich L. Extracellular dopamine in the medial preoptic area: implications for sexual motivation and hormonal control of copulation. J Neurosci. 1995;15:7465–71.

Hull EM, Du J, Lorrain DS, Matuszewick L. Testosterone, preoptic dopamine, and copulation in male rats. Brain Res Bull. 1997;44:327–33.

Hull EM, Meisel RL, Sachs BD. Male sexual behavior. In: Pfaff DW, Arnold AP, Etgen AM, Fahrbach SE, Rubin RT, editors. Hormones, Brain, and Behavior. Vol. 2. New York: Academic Press; 2002.

Hunt MA, Dennerstein L. Increased female sexual response after oxytocin. BMJ. 1994;309:929.

Hunt MA, Dennerstein L. Oxytocin and female sexuality. Gynecol Obstet Invest. 1995;40:217–21.

Hyyppa M, Rinne UK, Sonninen V. The activating effect of L-dopa treatment on sexual functions and its experimental background. Acta Neurol Scand. 1970;43(S46):223–4.

Ismail AA, Davidson DW, Loraine JA, Fox CA. Relationship between plasma cortisol and human sexual activity. Nature. 1972;237:288–9.

Jackson SC, Walker JS. Self-administered intraurethral chlorpromazine: an unusual cause of priapism. Am J Emerg Med. 1991;9:171–5.

Jankowiak WR, Fischer EF. A cross-cultural perspective on romantic love. Ethnology 1992;31:149–55.

Johansen JA, Clemens LG, Nunez AA. Characterization of copulatory behavior in female mice: evidence for paced mating. Physiol Behav. 2008;95:425–9.

Kandel ER. The Disordered Mind: What Unusual Brains Tell Us About Ourselves. New York: Farrar, Straus, and Giroux; 2018.

Kane FJ, Lipton MA, Ewing JA. Hormonal influences in female sexual response. Arch Gen Psychiatry. 1969;20:202–9.

Kawashima S, Takagi K. Role of sex steroids on the survival, neuritic outgrowth of neurons, and dopamine neurons in cultured preoptic area and hypothalamus. Horm Behav. 1994;28:305–12.

Kendrick KM. Oxytocin, motherhood and bonding. Exp Physiol. 2000;85(s1):111s–24s.

Kenkel WM, Paredes J, Lewis GF, Yee JR, Pournajafi-Nazarloo H, Grippo AJ, et al. Autonomic substrates of the response to pups in male prairie voles. PLoS One. 2013;8:e69965.

Kruger T, Exton MS, Pawlak C, von zur Muhlen A, Hartmann U, Schedlowski M. Neuroendocrine and cardiovascular response to sexual arousal and orgasm in men. Psychoneuroendocrinology. 1998;23:401–11.

Kwan M, Greenleaf WJ, Mann J, Crapo L, Davidson JM. The nature of androgen action on male sexuality: a combined laboratory/selfreport study on hypogonadal men. J Clin Endocrinol Metab. 1983;57:557–62.

Lal S, Ackman D, Thavundayil JX, Kiely ME, Etienne P. Effect of apomorphine, a dopamine receptor agonist, on penile tumescence in normal subjects. Prog Neuropsychopharmacol Biol Psychiatry. 1984;8:695–9.

Lal S, Laryea E, Thavundayil JX, Nair NP, Negrete J, Ackman D, et al. Apomorphine-induced penile tumescence in impotent patients-preliminary findings. Prog Neuropsychopharmacol Biol Psychiatry. 1987;11:235–42.

Lal S, Tesfaye Y, Thavundayil JX, Thompson TR, Kiely ME, Nair NP, et al. Apomorphine: clinical studies on erectile impotence and yawning. Prog Neuropsychopharmacol Biol Psychiatry. 1989;13:329–39.

Landgraf R, Neumann ID. Vasopressin and oxytocin release within the brain: a dynamic concept of multiple and variable modes of neuropeptide communication. Front Neuroendocrinol. 2004;25:150–76.

Le Moene O, Agmo A. Modeling human sexual motivation in rodents: some caveats. Front Behav Neurosci. 2019;13:187.

Lee PA, Jaffe RB, Midgley AR Jr. Lack of alteration of serum gonadotropins in men and women following sexual intercourse. Am J Obstet Gynecol. 1974;120:985–7.

Leiblum S, Bachmann G, Kemmann E, Colburn D, Swartzman L. Vaginal atrophy in the postmenopausal woman: the importance of sexual activity and hormones. JAMA. 1983;249:2195–8.

Leonti M, Casu L. Ethnopharmacology of love. Front Pharmacol. 2018;9:567.

LeVay S. The Sexual Brain. Cambridge, MA: MIT Press; 1993.

Lieberwirth C, Wang Z. Social bonding: regulation by neuropeptides. Front Neurosci. 2014;8:171.

Liebowitz MR. The Chemistry of Love. Boston, MA: Little, Brown & Co; 1983.

Lim MM, Young LJ. Vasopressin-dependent neural circuits underlying pair bond formation in the monogamous prairie vole. Neuroscience. 2004;125:35–45.

Lincoln GA. Luteinizing hormone and testosterone in man. Nature. 1974;252:232–3.

Lo Monte G, Graziano A, Piva I, Marci R. Women taking the "blue pill" (sildenafil citrate): such a big deal? Drug Des Devel Ther. 2014;8:2251–4.

Love TM. Oxytocin, motivation and the role of dopamine. Pharmacol Biochem Behav. 2014;119:49–60.

MacLean PD. Brain mechanisms of primal sexual function. In: Sandler M, Gessa, GL, editors. Sexual Behavior, Pharmacology and Biochemistry. New York: Raven Press; 1975.

Manji HK, Quiroz JA, Payne JL, Singh J, Lopes BP, Viegas JS, et al. The underlying neurobiology of bipolar disorder. World Psychiatry. 2003;2:136–46.

Marazziti D, Akiskal HS, Rossi A, Cassano GB. Alteration of the platelet serotonin transporter in romantic love. Psychol Med. 1999;29:741–5.

Marazziti D, Cassano GB. The neurobiology of attraction. J Endocrinol Invest. 2003;26(3 Suppl):58–60.

Marazziti D, Roncaglia I, Del Debbio A, Bianchi C, Massimetti G, Origlia N, et al. Brain-derived neurotrophic factor in romantic attachment. Psychol Med. 2009;39:1927–30.

Marder SR, Meibach RC. Risperidone in the treatment of schizophrenia. Am J Psychiatry. 1994;151:825–35.

Meisell RL, Sachs BD. The physiology of male sexual behavior. In: Knobil E, Neill JD, editors. The Physiology of Reproduction. New York: Raven Press; 1994.

Merkin TE. Priapism as a sequela of chlorpromazine therapy. JACEP. 1977;6:367–8.

Mirin SM, Meyer RE, Mendelson JH, Ellingboe J. Opiate use and sexual function. Am J Psychiatry. 1980;137:909–15.

Montague PR, McClure SM, Baldwin RP, Phillips PE, Budygin EA, Stuber GD, et al., Dynamic gain control of dopamine delivery in freely moving animals. J Neurosci. 2004;24:1754–9.

Muller MN, Wrangham RW. Dominance, aggression and testosterone in wild chimpanzees: a test of the 'challenge hypothesis. Anim Behav. 2004;67:113–23.

Muller P, Musch K, Wolf AS. Prolactin variables in personality and sexual behavior. In: Zichella L, Pancheri P, editors. Psychoneuroendocrinology in Reproduction. Amsterdam: Elsevier/North-Holland Biomedical Press; 1979.

Murphy MR, Seckl JR, Burton S, Checkley SA, Lightman SL. Chances in oxytocin and vasopressin secretion during sexual activity in men. J Clin Endocrinol Metab. 1987;65:738–41.

Nathorst-Boos J, von Schoultz B, Carlström K. Elective ovarian removal and estrogen replacement therapy: effects on sexual life, psychological well-being and androgen status. J Psychosom Obstet Gynaecol. 1993;14:283–93.

Navneet M, Kalra S. The orgasmic history of oxytocin: love, lust, and labor. Indian J Endocrinol Metab. 2011;15:156–61.

Neumann ID. Oxytocin: the neuropeptide of love reveals some of its secrets. Cell Metab. 2007;5:231–3.

Ngun TC, Ghahramani N, Sánchez FJ, Bocklandt S, Vilain E. The genetics of sex differences in brain and behavior. Front Neuroendocrinol. 2011;32:227–46.

Nurnberg HG, Hensley PL, Lauriello J, Parker LM, Keith SJ. Sildenafil for women patients with antidepressant- induced sexual dysfunction. Psychiatr Serv. 1999;50:1076–8.

O'Brien CP, DiGiacomo JN, Fahn S, Schwarz GA. Mental effects of high-dosage levodopa. Arch Gen Psychiatry. 1971;24:61–4.

O'Carroll R, Bancroft J. Testosterone therapy for low sexual interest and erectile dysfunction in men: a controlled study. Br J Psychiatry. 1984;145:146–51.

O'Carroll R, Shapiro C, Bancroft J. Androgens, behaviour and nocturnal erection in hypogonadal men: the effects of varying the replacement dose. Clin Endocrinol (Oxf). 1985;23:527–38.

O'Sullivan JD, Hughes AJ. Apomorphine-induced penile erections in Parkinson's disease. Mov Disord. 1998;13:536–9.

Oei NY, Rombouts SA, Soeter RP, van Gerven JM, Both S. Dopamine modulates reward system activity during subconscious processing of sexual stimuli. Neuropsychopharmacology. 2012;37:1729–37.

Pake JA, Penn RD, Ryan WG. Altered sexual function and decreased testosterone in patients receiving intraspinal opioids. J Pain Symptom Manage. 1994;9:126–31.

Persky H, Charney N, Lief HI, O'Brien CP, Miller WR, Strauss D. The relationship of plasma estradiol level to sexual behavior in young women. Psychosom Med. 1978a;40:523–35.

Persky H, Lief HI, Strauss D, Miller WR, O'Brien CP. Plasma testosterone level and sexual behavior in couples. Arch Sex Behav. 1978b;7:157–73.

Persky J, O'Brien CP, Kahn MA. Reproductive hormone levels, sexual activity and moods during the menstrual cycle. Psychosom Med. 1976;38 :62–3.

Petersen JL, Hyde JS. A meta-analytic review of research on gender differences in sexuality. Psychol Bull. 2010;136:21–38.

Pfaff DW. Brain Arousal and Information Theory. Cambridge, MA: Harvard University Press; 2005.

Pfaus JG, Gorzalka BB. Opioids and sexual behavior. Neurosci Biobehav Rev. 1987;11:1–34.

Puigvert L, Gelsthorpe L, Soler-Gallart M, Flecha R. Girls' perceptions of boys with violent attitudes and behaviours, and of sexual attraction. Palgrave Commun. 2019;5:56.

Purvis K, Landgren BM, Cekan Z, Diczfalusy E. Endocrine effects of masturbation in men. J Endocrinol. 1976;70:439–44.

Reis SL, Abdo CH. Benefits and risks of testosterone treatment for hypoactive sexual desire disorder in women: a critical review of studies published in the decades preceding and succeeding the advent of phosphodiesterase type 5 inhibitors. Clinics (Sao Paulo). 2014;69:294–303.

Robinson DL, Heien ML, Wightman RM. Frequency of dopamine concentration transients increases in dorsal and ventral striatum of male rats during introduction of conspecifics. J Neurosci. 2002;22:10477–86.

Roelofs K. Freeze for action: neurobiological mechanisms in animal and human freezing. Philos Trans R Soc Lond B Biol Sci. 2017;372:20160206.

Roney JR, Mahler SV, Maestripieri D. Behavioral and hormonal responses of men to brief interactions with women. Evol Hum Behav. 2003;24:365–75.

Rowland DL, Heiman JR, Gladue BA, Hatch JP, Doering CH, Weiler SJ. Endocrine, psychological, and genital response to sexual arousal in men. Psychoneuroendocrinology. 1987;12:149–58.

Saad M, Rab LB, Bujaki B, Parvaresh A, Palamarchuk I, Koninck JD, et al. Using heart rate profiles during sleep as a biomarker of depression. BMC Psychiatry. 2019;19:168.

Saenz de Tejada I, Goldstein I, Azadzoi K, Krane RJ, Cohen RA. Impaired neurogenic and endothelium-mediated relaxation of penile smooth muscle from diabetic men with impotence. N Engl J Med. 1989;320:1025–30.

Salmon UJ, Geist SH. Effect of androgens upon libido in women. J Clin Endocrinol. 1943;3:235–8.

Scheele D, Striepens N, Güntürkün O, Deutschländer S, Maier W, Kendrick KM, et al. Oxytocin modulates social distance between males and females. J Neurosci. 2012;32:16074–9.

Schiavi RC, White D. Androgen and male sexual function: a review of human studies. J Sex Marital Ther. 1976;2:214–28

Schreiner-Engel P, Schiavi RC, Smith H, White D. Sexual arousability and the menstrual cycle. Psychosom Med. 1981;43:199–214.

Schreiner-Engel P, Schiavi RC, White D, Ghizzani A. Low sexual desire in women: the role of reproductive hormones. Horm Behav. 1989;23:221–34.

Segraves RT. Bethanechol reversal of imipramine-induced ejaculatory dysfunction. Am J Psychiatry. 1987;144:1243.

Segraves RT, Bari M, Segraves K, Spirnak P. Effect of apomorphine on penile tumescence in men with psychogenic impotence. J Urol. 1991;145:1174–5.

Seshadri KG. The neuroendocrinology of love. Indian J Endocrinol Metab. 2016;20:558–63.

Sherwin BB. Changes in sexual behavioral as a function of plasma sex steroid levels in postmenopausal women. Maturitas. 1985;7:225–33.

Skakkebaek NE, Bancroft J, Davidson DW, Warner P. Androgen replacement with oral testosterone undecanoate in hypogonodal men: a double-blind controlled study. Clin Endocrinol (Oxf). 1981;14:49–61.

Spinella, M. Clinical case report: hypersexuality and dysexecutive syndrome after a thalamic infarct. Int J Neurosci. 2004;114:1581–90.

Stanworth RD, Jones TH. Testosterone for the aging male; current evidence and recommended practice. Clin Interv Aging. 2008;3:25–44.

Steimer T. The biology of fear- and anxiety-related behaviors. Dialogues Clin Neurosci. 2002;4(3): 231–49.

Stoleru SG, Ennaji A, Counot A, Spira A. LH pulsatile secretion and testosterone blood levels are influenced by sexual arousal in human males. Psychoneuroendocrinology. 1993;18:205–18.

Stuart FM, Hammond DC, Pett MA. Inhibited sexual desire in women. Arch Sex Behav. 1987;16:91–106.

Studd JWW, Collins WP, Chakravarti S, Newton JR, Oram D, Parsons A. Oestradiol and testosterone implants in the treatment of psychosexual problems in the postmenopausal woman. Br J Obstet Gynaecol. 1977;84:314–15.

Stulberg D, Ewigman B. Antidepressants causing sexual problems? Give her Viagra. J Fam Pract. 2008;57:793–6.

Tetel MJ, Getzinger MJ, Blaustein JD. Fos expression in the rat brain following vaginal-cervical stimulation by mating and manual probing. J Neuroendocrinol. 1993;5:397–404.

Turner RA, Altemus M, Enos T, Cooper B, McGuinnesss T. Preliminary research on plasma oxytocin in normal cycling women: investigating emotion and interpersonal distress. Psychiatry. 1999;62:97–112.

Udry JR, Billy JO, Morris NM, Groff TR, Raj MH. Serum androgenic hormones motivate sexual behavior in adolescent boys. Fertil Steril. 1985;43:90–4.

Udry JR, Talbert LM, Morris NM. Biosocial foundations for adolescent female sexuality. Demography. 1986;23:217–30.

Utian WH. The true clinical features of postmenopause and oophorectomy and their response to oestrogen therapy. S Afr Med J. 1972;46:732–7.

Verberne AJ, Korim WS, Sabetghadam A, Llewellyn-Smith IJ. Adrenaline: insights into its metabolic roles in hypoglycaemia and diabetes. Br J Pharmacol. 2016;173:1425–37.

Verhaeghe J, Gheysen R, Enzlin P. Pheromones and their effect on women's mood and sexuality. Facts Views Vis Obgyn. 2013;5:189–95.

Wallen K, Baum MJ. Masculinization and defeminization in altricial and precocial mammals: comparative aspects of steroid hormone action. In: Pfaff DW, Arnold AP, Etgen AM, Fahrbach SE, Rubin RT, editors. Hormones, Brain and Behavior. San Diego, CA: Elsevier Science; 2002.

Wang Z, Yu G, Cascio C, Liu Y, Gingrich B, Insel TR. Dopamine D2 receptor-mediated regulation of partner preferences in female prairie voles (Microtus ochrogaster): a mechanism for pair bonding? Behav Neurosci. 1999;113:602–11.

Warner P, Bancroft J. Mood, sexuality, oral contraceptives and the menstrual cycle. J Psychosom Res. 1988;32:417–27.

Wenkstern D, Pfaus JG, Fibiger HC. Dopamine transmission increases in the nucleus accumbens of male rats during their first exposure to sexually receptive female rats. Brain Res. 1993;618:41–6.

Wiedeking C, Lake CR, Ziegler M, Kowarski AA, Money J. Plasma noradrenaline and dopamine– beta-hydroxylase during sexual activity. Psychosom Med. 1977;39:143–8.

Wiedeking C, Ziegler MG, Lake CR. Plasma noradrenaline and dopamine – beta-hydroxylase during human sexual activity. J Psychiatr Res. 1979;15:139–45.

Yager J. Bethanechol chloride can reverse erectile and ejaculatory dysfunction induced by tricyclic antidepressants and mazindol: case report. J Clin Psychiatry. 1986;47:210–11.

Zeitlin SI, Rajfer J. Hyperprolactinemia and erectile dysfunction. Rev Urol. 2000;2:39–42.

Zeki S. The neurobiology of love. FEBS Lett. 2007;581:2575–9.

16 Neurobiology of Love

A Comprehensive Analysis

Simrat Kaur, Neha Goyal, Ashok Tiwari, and Rachana
Department of Biotechnology, Jaypee Institute of Information Technology, Noida, Uttar Pradesh, India

CONTENTS

DOI: 10.1201/9780429265198-19

16.1 INTRODUCTION

Everyone can relate to the feeling of being in love. Love is an emotion which may be linked with consensual sexual activity, and the willingness or eagerness of individuals to be involved in developing a relationship that builds a space for two individuals to grow together (Crews, 1998). The aforementioned feeling of love has been described in numerous terms of commitment, compassion, attachment, grief on separation, intimacy, passion, and jealousy portrayed by social beings (Carter, 1998). The Oxford Dictionary describes love as an intense feeling of deep affection or fondness for a person or thing, a sexual passion, or sexual relations (Crews, 1998). This type of love is generally categorized as romantic love. In this case, a different treatment is meted out to the feelings that develop since they color our perception of the world in a different way, influence our thoughts, and all of this has mostly been described to be purely spiritual. However, scientists have reported that romantic love is also an interplay between different chemicals in one's body. Research in the area of neurobiology has examined the mechanisms and basis of love, its consequences for the individuals involved, and associated ontogenetic health outcomes and benefits (Stefano and Esch, 2005). Scientists have discovered some of the molecular processes involved in the nervous system that drives the behavior of individuals and autoregulate the feeling/process of love (Colon-Urban and Stefano, 1985).

Along with the above descriptions of love, there is another kind of love that exists between parents and their offspring, between two friends, between siblings, between a teacher and a student, and many more. In this kind of relationship, romance or sexual desire is not involved. The love which one feels for his or her family and friends is altogether different from romantic, intoxicating love, and is one of the biggest reasons that separates us (humans) from other beings on this planet. Various studies have shown the involvement of different, but overlapping, areas of the brain and sets of biomolecules in raising such a type of love between two or more individuals (Bartels and Zeki, 2004). These chemicals react in different ways in the case of parental love, driving feelings, emotions, and thoughts to initiate and to maintain love between and among individuals (parents, children, friends, and siblings).

The objective of this chapter is to identify the involvement of different parts of the brain, hormones, and other biochemical pathways that initiate, continue, or break up a relationship. Therefore, this chapter revolves around the neurobiology and physiology of love and attachment, hormonal changes in the brain, mechanisms, and steroids that interact to suspend the condition of being in love, with a special focus on romantic and parental attachment and behavior. It covers common principles and other aspects related to different kinds of love initially, followed by a categorical description of parental love and romantic love.

16.2 EVOLUTION OF LOVE

Charles Darwin and other biologists studied the evolution of living beings and found various mechanisms that drive the behavior required for their biological development (Crews, 1998). In other words, they studied the motivation behind the behavior that drives and evolves them in such a way that it ends with rewards and helps them to find a mutual support system to survive and thrive on this planet. Theodosius Dobzhansky's famous dictum holds true to explain the evolution of love: 'Nothing in biology makes sense except in the light of evolution.' Life on Earth is fundamentally social and there is the ability to interact dynamically with other living organisms and support mutual homeostasis, growth, and reproduction (Carter and Porges, 2013). The pathway of evolution occurred from reptiles to mammals and has led to the advent of anatomic systems and biochemical mechanisms that enable social engagement and selectively reciprocal sociality (Carter and Porges, 2013). Human love is, however, more complex than a simple feedback mechanism. The biology of love originates in the primitive parts of the brain – the emotional core of the human nervous system –and evolved long before the cerebral cortex (Carter and Porges, 2013). A human's brain in love is mainly flooded with sensations transmitted by the vagus nerve, forming the experience of falling in love, and building a narrative around incoming visceral experiences, reacting to that narrative rather than the reality (Carter and Porges, 2013).

Emotions and feelings, for example attachment, parental bonding, and love, are typical in advanced mammals (Esch and Stefano, 2005). The psychological interpretation of love can be described as the satisfaction of a craving associated with obtaining sensory stimulation (Komisaruk and Whipple,1998). Thus, it can be inferred that love has a close relation with reward and pleasure phenomena, and is an appetitive and addictive behavior as well (Esch and Stefano, 2004). Love and social attachments function to facilitate reproduction, provide a sense of security, and reduce feelings of stress or anxiety (Carter, 1998). The neurobiology of love is enmeshed in adulthood with reproduction and homeostasis, phylogenetically and ontogenetically (Uvnäs-Moberg, 1997).

16.3 THE SCIENTIFIC PERSPECTIVE OF LOVE

Scientists have always been curious to know what exactly makes individuals dive into this feeling of love and how they stay in love for years or separate after some time. Several neurobiological changes that take place in the brain and the entire body during the course of 'being in love' have already been discovered. Our knowledge of the physiology of the brain and its various neurobiological mechanisms helps us to understand why humans and animals behave in a certain way when it comes to love and bonding. Earlier in science, love was thought to be hypothetical and multidimensional, and had a number of interpretations and implications (Esch and Stefano, 2005). It has always been amazing to notice that how birds or any other animals take care of their offspring, how they take the responsibility for the tiny living being, and then allow them or make them leave the nest or home. In a similar way, it is amazing to see how a mother or a father goes beyond limits to take care of their human offspring as well. On the one hand, two friends can change their lives for each other; on the other hand, they may break up the friendship, become distant, or lose the feeling of attachment in their relationships after a certain period of time. As mentioned above, it has been observed by scientists that it all has a neurological and biochemical basis, and there are areas allocated in the brain for certain activities at a particular point in time, regulating emotions and other biologic events through certain biochemicals (Carter and Porges, 2013)

16.4 PARTS OF THE BRAIN INVOLVED IN AND RELEVANT BIOMOLECULES

16.4.1 Parts of the Brain Involved in Love

Various parts of the brain are responsible for a variety of events during the initiation and continuation of a relationship. From a behavioral point of view, the areas of the limbic brain (i.e., hypothalamus and medial preoptic area) are important for primarily motivated behavior such as maternal care, feeding, and sexual and aggressive behavior (Magon and Kalra, 2011). In animals with relatively smaller brains, these areas of the brain are under potent hormonal and visceral influences, and are activated primarily by olfactory cues. (Bales and Carter, 2003).

Several studies have revealed the role of hypothalamic, limbic, and brainstem structures, as well as biochemicals such as neuropeptides and brain monoamines, in partner preference, sexual desire, erection, and copulation (Magon and Kalra, 2011). Males and females experience differing pleasure experiences while in love (Magon and Kalra, 2011). Modulation of testosterone concentration causes a male to become sensitive and, modifying estrogen levels causes a female to differ from stereotypical female behavior (Gillies and McArthur, 2010).

The amygdala is highly involved in the sex and sexuality of an individual, and the male amygdala is slightly bigger than that of a female (Pfau et al., 2016). The medial part of the female amygdala plays an important role in pregnancy and appropriate coordination of the endocrine system (Pfau et al., 2016). Stimulation of the amygdala produces penile erection, sexual sensation, and orgasm (Gillies and McArthur, 2010). Precortical region epilepsy has been shown to elicit spontaneous sexual arousal and orgasm, clearly demonstrating the role of the amygdala in sexual pleasure (Pfau et al., 2016). There is a close relationship between reward and pleasure (Nestler, 2004). Anatomically, the reward pathways are linked to the brain's limbic system (Esch and Stefano, 2004). All these factors depend on brain activity for their effective functioning, which warrant a higher depth of understanding and research initiatives.

Multilevel kernel density meta-analysis (MKDA) of functional magnetic resonance imaging (fMRI) in love and sexual desire have provided evidence that both phenomena activate a very specific and similar set of neural networks. The networks related to sexual desires involve the hippocampus, ventral striatum, insula, amygdala, the hypothalamus and ventral tegmental area (VTA), and cortical and limbic areas

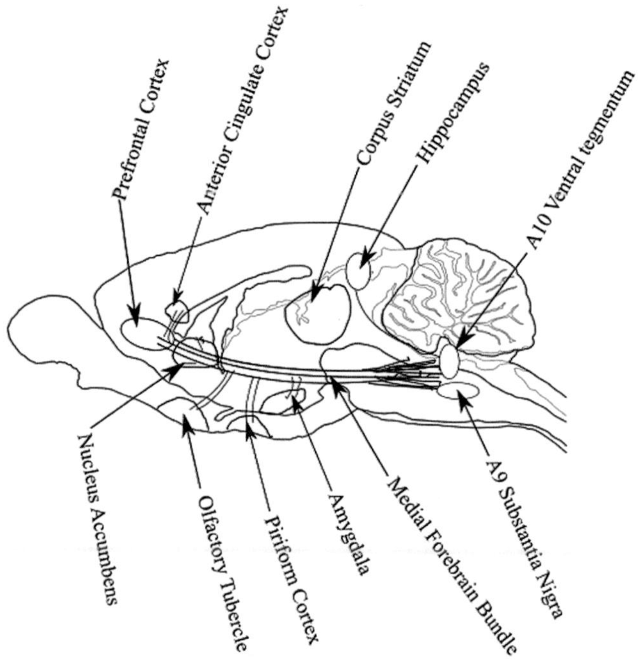

FIGURE 16.1 Different areas of the rat brain involved in the mechanism of love.

[e.g., inferior parietal lobule specific regions of the middle frontal gyrus, temporal cortex, middle frontal gyrus, superior temporal and precentral gyrus, anterior cingulate cortex (ACC), temporoparietal junction, somatosensory cortex and inferior parietal lobule] (Cacioppo et al., 2012) (Figures 16.1 and 16.2). MKDA studies have also revealed that the intrinsic differences between love and desire are related to the anterior insula (Cacioppo et al., 2012).

In 2015, fMRI of the brains of 2500 college scholars who viewed images of someone special to them was compared to the imaging of students who looked at pictures of acquaintances, and revealed that looking at a loved person had romantically activated regions abundant in dopamine. The two primary regions of the brain that showed activity on fMRI were the VTA and the caudate nucleus area (Schwartz and Olds, 2019). The caudate nucleus is a region related to expectation and reward detection plus the integration of sensory experiences into social behavior (Schwartz and Olds, 2019).

The VTA is linked to motivation, focused attention, and the pleasure in pursuing and obtaining rewards. This is why that part is known as the reward circuit of the brain. It is reviewed as being a primitive neural network complex, meaning it is evolutionarily aged; and it connects to the nucleus accumbens. Some other structures that contribute to the rewards circuit are the hippocampus, the amygdala, and the prefrontal cortex. These areas are very sensitive to the behaviors that generate pleasure, such as drug use, food consumption, and – more relevantly – sex (Schwartz and Olds, 2019).

Another area of the brain, the cingulate cortex, is one of the essential cortical areas jumbled with both negative (fear and pain) and positive (sexual arousal and attraction) affective

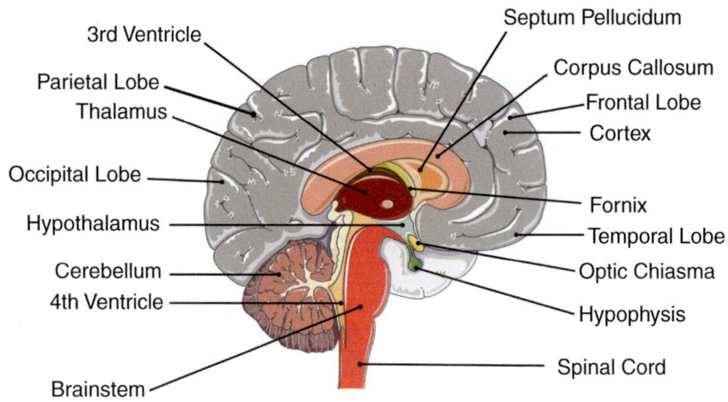

FIGURE 16.2 A mapping of different parts of the brain. The primary parts involved in romantic attraction range from the hippocampus to the somatosensory cortex (image procured from Shutterstock).

states (Chen et al., 2014; Devinsky et al., 1995; Leknes and Tracey, 2008; Hu et al., 2008). Neurons in rodents have been shown to become activated after the immediate use of early gene *c-Fos* as a marker (Moncho-Bogani et al., 2005; Pfaus and Heeb, 1997). Sexual attraction initiates a functional rise in presynaptic glutamate release to trigger neurons in ACC slices while using transgenic Fos green fluorescent protein mice. Animal studies have confirmed that the central supraspinal arrangements controlling arousal are confined mostly to the limbic system. Inactivation of the ACC by a local infusion of muscimol lowered the interest of male mice in female mice, and confirmed the role of ACC in sexual attraction (Wu et al., 2009).

16.4.2 Biochemicals Involved in Communication in Various Parts of the Brain

Various factors initiate the mechanism of love and are discussed in other sections, specifically for romantic love. Briefly, they involve sight, listening, and smells (pheromones), and so on (Doty, 2014.). Once, the signal is received by the brain and by the sensory organs, various biochemicals are secreted in different areas of the brain. The following subsections provide a description of such chemicals with their mechanisms of action.

16.4.2.1 Role of Oxytocin and Vasopressin

16.4.2.1.1 Importance of Oxytocin and Vasopressin

Two primary molecules are majorly responsible for various reward factors for love, especially in compassionate love between a parent and child (Esch and Stefano, 2011), and are described in detail here. These molecules are both relatively smaller nine-amino acid peptides with a 6:3 body-to-tail split. Oxytocin (OXT) differs from vasopressin (VSP) by the presence of isoleucine and leucine at positions 3 and 8, respectively, where VSP contains phenylalanine and arginine instead (Carter, 1998) (Figure 16.3a, b). The gene for these two peptides occupies the same chromosome.

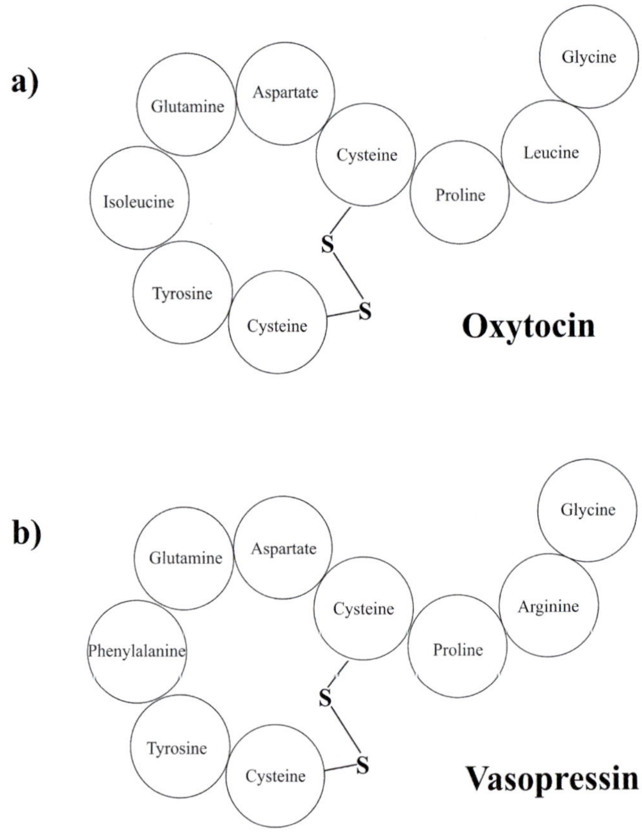

FIGURE 16.3 Structure of (a) oxytocin and (b) vasopressin.

OXT and VSP are characteristic to mammals alone (Moore et al., 1992). Their circulation is overseen by the larger-sized magnocellular neurons located in the supraoptic nucleus (SON) and paraventricular nucleus (PVN; Brown et al., 2013). The responsibility of storage and secretion of these hormones into the bloodstream lies with the posterior pituitary gland, which, in turn, receives them from the aforementioned magnocellular neurons. The PVN also contains parvocellular neurons which are responsible for injecting the hormones directly into the central nervous system (CNS; Brown et al.,

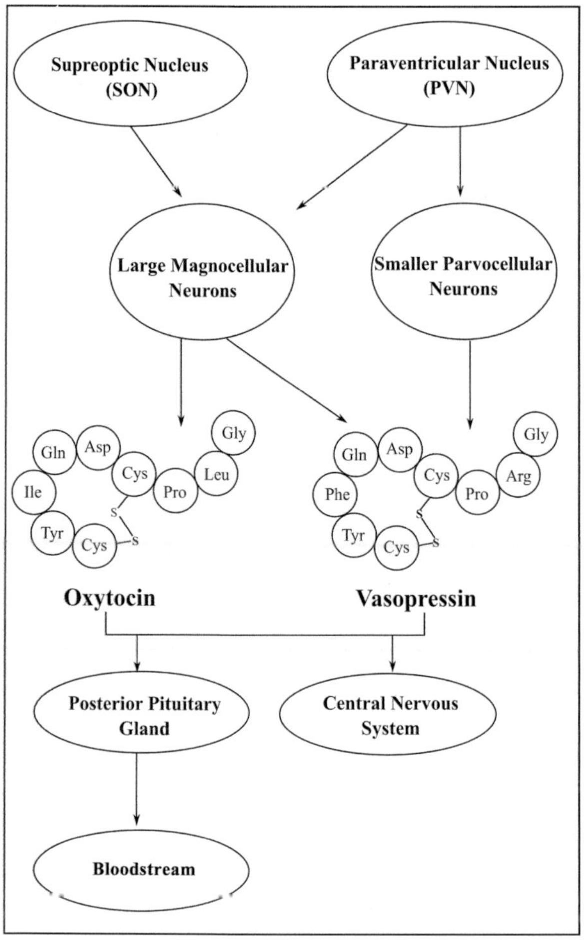

FIGURE 16.4 Mechanism of the release of oxytocin and vasopressin.

2013). The two release mechanisms are exclusive in terms of dependency (Insel and Young, 2001) (Figure 16.4).

This peptidergic system involving OXT and VSP inhibits defensive behaviors associated with stress and anxiety, and allows positive social interactions and the development of social bonds (Carter, 1998). Attachment formation among individuals occurs as a result of behavioral changes such as mammalian birth, lactation, and sexual interactions. These physiological changes, along with the neural changes, form the basis of building a relationship between a mother and an infant (Carter, 1998).

16.4.2.1.2 Receptors for Oxytocin and Vasopressin

The method of functioning for peptides is particularly dependent on their affinity for different receptors. OXT and VSP command one and three binding receptors (for VSP: V1a, V1b, and V2), respectively (Ostrowski et al., 1998). While the first kind of the latter receptors, V1a, can be observed prominently throughout the cardiovascular and CNS (Magon and Kalra, 2011), V1b receptors are more sparsely located in the CNS in parallel with OXT receptors and are instead housed by the pituitary gland. V2 receptors, however, reside in the kidneys. In the nervous system, OXT and VSP receptors are present in the smell-and-sense factory, brainstem, and regions of the spinal cord that dictate autonomy and procreation (Carter, 1998). OXT and VSP further challenge our understanding of their circuits, through which they affect behavior by having the capability to bind to other receptors as well (Barberis and Tribollet, 1996).

16.4.2.1.3 The Function of Oxytocin and Vasopressin

OXT governs various sociosexual behaviors in mammals (Witt et al., 1992), such as maternal care and the formation of mother–infant bonds (Kendrick et al., 1997). Cells produce oxytocinergic fibers in three different areas: the bed nucleus of stria terminalis (BNST), the anterior commissural nucleus, and the spinal cord (Dong and Swanson, 2006). Fibers within the spinal cord are responsible for OXT transmission into the cerebrospinal fluid, if they do not end in the CNS. OXT may even be created in tissues other than the neural framework. Childbirth forms a critical juncture at which OXT levels may spike within the PVN and SON, and hormonal situations that recreate similar conditions to childbirth may also recreate this increase (Magon and Kalra, 2011). VSP, just like OXT, is sourced most prominently by the PVN and SON regions (Brown, 2013). The parvocellular neurons present in the PVN are also important sources for the same (Dong and Swanson, 2006).

Vasopressinergic entities are found in the suprachiasmatic nucleus, medial amygdala (MAMY), BNST, and other areas of the caudal brainstem. The creation of VSP is handled by androgens present in the MAMY and lateral septum (DeVries and Villalba, 1997), which helps us obtain a clearer picture about why VSP is more prevalent in the female CNS than that of males.

As stated previously, OXT plays a primary role in individuals who recognize each other socially; the absence of a functional *Oxt* gene in experimentally prepared mice resulted in the subjects being continually unable to discern individuals who had been repeatedly introduced into the system (Insel and Young, 2001). This tendency to not maintain a grasp on social recognition does not seem to hold a link with sensory or memorial difficulties, which was indicated by the performance of the same *Oxt* knockout mice in habituating toward socio-irrelevant odors and the Morris water maze (Insel and Young, 2001). Prairie voles have a similar recognition correlation with OXT, and OXT antagonists have been shown to discourage the bonding process of individuals via the introduction of social memory loss of the partner (Insel and Shapiro, 1992). The nucleus accumbens and prelimbic cortex of these rodents are densely populated with OXT receptors (Insel and Shapiro, 1992). Stimulation and instant gratification reward cycles are controlled by these parts (McBride et al, 1999), but, more importantly, conditional learning is also dependent on them. In a similar process, OXT diffusion facilitates effective use of the reward circuitry and conditional reinforcement by enabling the mutual bonding syntax in the female sex. This highlights how the mutual attachment between individuals uses the same mental mechanism that dictates reinforcement and reward (Insel and Young, 2001).

OXT release is a pulsatile motion carried out by the posterior pituitary gland, which helps create uterine contractions during childbirth (Magon and Kalra, 2011). OXT also affects the breast by contracting the myoepithelial tissue and reducing milk production (Magon and Kalra, 2011). OXT is involved in a plethora of other neural functions like retention, sensory imaging, procreating, and regulation of body temperature, but the primary function remains stimulation of the parasympathetic nervous system (Uvnäs-Moberg, 1994). It also plays an important role in the regulation of the hypothalamic–pituitary–adrenal (HPA) axis through its ability to create 'immobility without fear' – the field of 'relaxed mind' activities and positions that require sustained stillness such as breastfeeding, childbirth, and even intercourse. OXT and VSP levels co-exist in humans for care-taking and protective aggression (Levy and Tasker, 2012). OXT regulates the HPA axis in responding to differences in body and environment temperatures, and external demands that require a trade-off between immunity and energy. Its activity is tuned down through the emotional comfort of a long-term relationship, but high-stress activities like childbirth stimulate its activity along with the sympathetic nervous system (Levy and Tasker, 2012).

The range of control that OXT exerts is indicative of women's ability to manage the intricacies of childbirth. Apart from this, it is interesting to note that our capacity for love and social interaction is linked to our health and prosperity by the same hormone. The same molecule is credited with the ability to heal, especially concerning the cardiac region and large arteries, where there is an abundance of receptors. The expression of these receptors occurs in these arteries and ventricles, and this process is especially critical through the fetal maturity of a mammal's heart (Gillies and McAurthe, 2010). Stressful experiences can be followed up with reparation and recreation of neural and tissue material in the body alongside anti-inflammation and reduction processes, which is why the feeling of 'healing a broken heart' is also attributed to these molecules (Berman et al., 2014).

As mentioned above, VSP plays an important neurobiological role in the reward and reinforcement circuitries related to external and psychotic stimulation (McBride et al., 1999). VSP is a central component of the sympathetic nervous system and may also regulate the parasympathetic nervous system function (Porges, 1998). Arginine VSP is a peptide hormone released during mating, which activates VSP receptor 1A (V1aR). In the single-spouse prairie vole, V1aR is at the start of the creation of a condition of partiality while choosing a mate, thus facilitating pair bond formation via a suitable substrate. Conversely, the absence of this receptor in non-monogamous species shows their inability to form partner preferences after mating. VSP is known to cause blood pressure spikes amongst other changes in the cardiovascular system (Berecek et al., 1991). Comparison of these two important biochemicals, with respect to their functions, is provided in Table 16.1.

16.4.2.1.4 Dopamine

Dopamine was described as a CNS neurotransmitter in the 1950s. The primary factories for dopamine include the hypothalamus, the VTA, and the substantia nigra. The level of dopamine transmission has been reported to increase in response to any reward and addictive drugs. However, dopamine's adverse role in maintaining the health of a person due to oxidative stress has also been documented (Juárez et al., 2016). Understandably, the mitigation of social attachment disorders is attempted with medicines that focus on the dopamine circuit (Nakamura et al., 2010). Dopamine has significant applications in learning and memory retention, where it becomes important in pair bonding (Love, 2014). The latter was initially observed in feminine prairie voles (Aragona et al., 2003), where peripherally administered apomorphine (dopamine agonist) was able to support mating and bonding, while haloperidol (antagonist) had the opposite effect and ruled out bond formation via intercourse (Aragona et al., 2003). The above experiment also unearthed the role of D2 receptors in pair bond formation (Aragona et al., 2003). In

TABLE 16.1
Comparison of Oxytocin and Vasopressin

Physiological Features	Oxytocin	Vasopressin
Aggression and territorial behavior	Decrease in general; women show a sharper decline	Increase in general; men show a sharper increase
Sensory processing	General increase	Increase in reward and limbic processing
Relaxation and secure well-being around partner	Increase	Increase
Social attachment and bonding	Increase	Increase
Blood pressure and stress	Decrease (HPA axis regulation)	Increase in blood pressure
Partner selection tendencies	Increase	Increase
Memory function	Increase in memory processing	Increase; attention and learning capacity also increase
Sexual and reproductive behavior	Increase	Increase
Anxiety	Decrease	Decrease
Parasympathetic autonomic functions	Increase	Sympathetic and parasympathetic regulation
Stress response through corticoid release	Agonist (positive stress response); reduced glucocorticoid release	Partial antagonist

female prairie voles, D2 receptor enabling or disabling had the same effect as adding a dopamine agonist or antagonist did, respectively (Gingrich et al., 2000).

16.4.2.1.5 Corticosterone

Cortisol is shown to influence preferences in prairie voles. It is the primary endogenous adrenal steroid in mammals, while corticosterone is the primary adrenal corticosteroid in rodents. Humans, however, produce both cortisol and corticosterone (Raff, 2016). In one study, it was seen that male prairie voles would react to induced stress or corticosterone injections by attempting to form renewed bonds with female mates. Surprisingly, female voles developed novel predilections for other females instead of forming different bonds with males as they had been interested in. This can be explained by a tendency for females to physiologically inspire others to retreat to the family nest under external duress since they may still find procreation opportunities, unlike their masculine counterparts. It was hence perceived that the presence of stressors may cause swift developments in sexual and social predilections for both kinds of rodents, be they of the opposite sex or not (Carter, 1998).

It has been found that cortisol levels are higher in individuals who are in love than those who are not (Marazziti and Canale, 2004). The initial 'honeymoon' phase of a relationship, including stressful and arousing associations, that come with the pair bond formation must overcome the initial fear of the unknown, and it is accentuated by changes characterized by hypercortisolemia (Marazziti and Canale, 2004). These stressful conditions have been demonstrated to endorse and encourage social attachment in mammals. It would be remiss to rule out HPA axis activation by the conjoining of stress and attachment, resulting in anxiety-free physiological make-ups and positive feelings associated with that partner (Legros, 2001).

16.4.2.2 Role of Luteinizing Hormone, Nitric Oxide, and Other Chemicals

Various chemicals are directly or indirectly involved in triggering and maintaining various events related to sexual behavior, reproduction, and love. (Pardo-Bellver et al., 2012; Luef, 2008; Salamon et al., 2005). Luteinizing hormone (LH) is an important hormone produced by the pituitary gland and regulates sexual development in vertebrates. Nitric oxide (NO) is found to be important for the co-occurrence of basal LH release in males and for the LH surge in ovariectomized females treated with estradiol plus progesterone (Figure 16.5). A rise in LH is induced by NO in estradiol-treated ovariectomized females and has a progesterone-like effect (Luef, 2008) (Figure 16.3). nNOS expressions stimulated by estradiol in the preoptic area also exerts a helping influence on NO-producing neurons (Salamon et al., 2005). Gonadotropin-releasing neurons (GnRH) neuron activity is modulated by the released NO. This implicates neural NO in the regulation of GnRH cell activity in the preoptic area. Therefore, a more robust signaling system including both NO from amygdala and hypothalamus can be hypothesized as important for maintaining hormonal relationships (Pardo-Bellver et al., 2012).

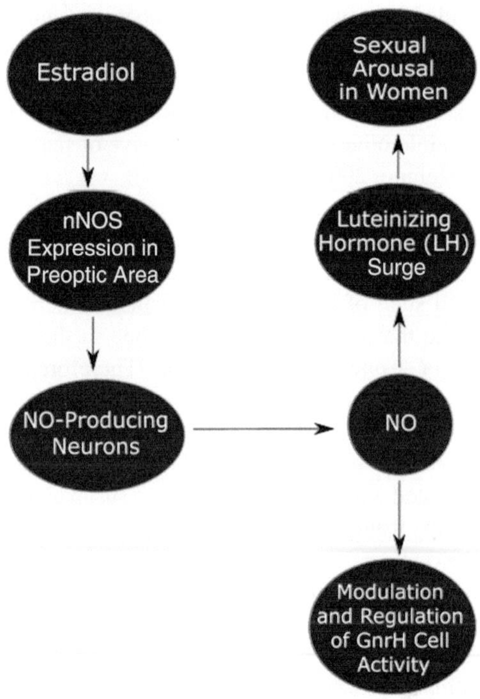

FIGURE 16.5 Role of estradiol and nitrous oxide (NO) in maintaining hormonal relationships. nNOS, ??; GnRH, gonadotropin-releasing hormone.

Further, levels of LH, estrogen steroid hormone, and a female sex hormone – estradiol, progesterone, dehydroepiandrosterone, and androstenedione – are a few more chemicals reported to play a role in love. The levels of these chemicals do not differ between men and women in love; however, testosterone levels show sex-related peculiarities in both men and women who were at the early stage of a relationship and they were lower and higher, respectively, than those who are not in a relationship (Marazziti and Canale, 2004). It was hypothesized that individuals falling in love tend to temporarily eliminate differences between the sexes or soften masculine features in males, while increasing them in females. This can be deduced in linking the changes in testosterone levels to the changes in behaviors, sexual attitudes, and aggressive traits, which move in different directions for the two sexes (Zitzmann and Nieschlag, 2001). It has been reported that levels of follicle-stimulating hormone in males are decreased, possibly another marker of hypothalamic involvement in the process of falling in love. However, these hormonal changes have been found to be reversible, state-dependent, and could be related to physiological features associated with falling in love (Marazziti and Canale, 2004).

16.5 THE PHENOMENON OF 'REWARDS VS PUNISHMENT' IN STRENGTHENING ATTACHMENTS

Most behaviors in animals can be explained by the process of approaching pleasant and avoiding painful stimuli (i.e., motivational behaviors; Carter, 1998). Thus, motivation can be divided into two categories – appetitive and aversive motivation. The tendency to approach or avoid a set of social stimuli is fundamental to social and attachment behaviors as some stimuli are inherently positive or elicit positive responses, while others are innately aversive or fear-inducing (Esch and Stefano, 2004).

Pleasure is associated with motivating behavior called reward (Esch and Stefano, 2004). Normal behavior is governed through pleasurable experiences (Bozarth, 1994). Pleasure is a state of happiness or satisfaction derived from an experience that one enjoys, as described in the *Longman Dictionary of Contemporary English*. Therefore, a close relationship between reward and pleasure exists (Nestler, 2004). Anatomically, reward pathways are linked to the brain's limbic system (Esch and Stefano, 2004). As pleasure can be associated with events that facilitate survival, it benefits the organism from an evolutionary perspective (Uvnäs-Moberg, 1997).

Pain describes conditions that would have undesirable biological consequences for an organism (Uvnäs-Moberg, 1997). With regard to segregated brain compartments involved in motivational processes, the physiological substrate for appetitive or aversive motivation lies within the limbic system (Hui et al., 2000).

Love has the capacity to influence the autonomic–emotional integration system (i.e., the limbic system) (Esch and Stefano, 2004; Salamon et al., 2003). The autonomic nervous system and emotions are wired together. Sympathetic activity and stress hormone production are embedded in underlying autoregulatory circuits (Esch et al., 2003). The influence of love on vital functions such as breathing, respiratory rate, blood pressure, and cardiac output, due to autonomic–emotional integration, often lead to a new consciousness or change of mind (Esch and Stefano, 2004). Thus, activation of the brain's reward system results in changes from a slight mood elevation to passionate pleasure and euphoria, directing behavior toward natural reward – love (Rossetti et al., 1992). These reward systems are formed of complex circuits of nerve cells that have evolved to produce a 'flying feeling' after pleasurable activities (Esch and Stefano, 2004).

The brain's circuitry for motivation and rewards are responsible for placebo physiology, which includes different CNS regions that serve separate functions but overlap in their reward signaling pathways (Esch and Stefano, 2005). Reward pathways originate from the limbic system. The prefrontal cortex, cingulate gyrus, amygdala, hippocampus, and nucleus accumbens participate in reward physiology (Stefano et al., 2001). Activation of the lateral orbitofrontal cortex is achieved via tactile, olfactory stimuli, or pleasant visuals, with its response depending on the pleasantness rather than the intensity of stimulation (Bartels and Zeki, 2004). Memories of the pleasure of wellness are accessible to this system through hippocampal mechanisms. With regard to frequent CNS reward tracks, activation of the medial forebrain bundle (MFB), as it courses through the lateral hypothalamus to the VTA, has been shown to produce robust rewarding effects (Bozarth, 1994).

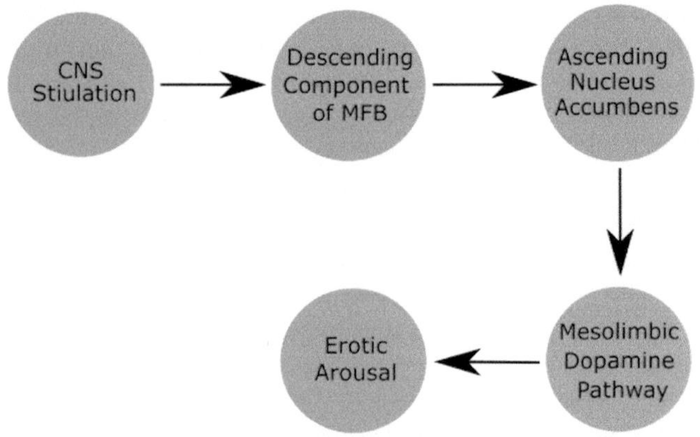

FIGURE 16.6 Flow diagram showing the pathway leading to erotic arousal. CNS, central nervous system; MFB, medial forebrain bundle.

An important component of CNS reward and motivation circuitries are nerve cells originating in the VTA, closer to the base of the brain (Goode and Maren, 2017). These cells send projections to target regions in the frontal brain, to a structure beneath the frontal cortex (i.e., nucleus accumbens; Nestler, 2001). The vital neurotransmitter of this connection is dopamine (Goode and Maren, 2017). The VTA system represents an old but effective part of motivational physiology and behavior. So, in mammals, the neurobiology of behavior, including reward circuitry involvement, is complex and integrated with several other brain regions that serve to enrich an experience with emotion (Esch and Stefano, 2004).

Electrophysiological and neurochemical techniques have shown that the descending component of the MFB can be activated by the CNS, which is synaptically coupled at the VTA to the ascending nucleus accumbens (Nestler, 2004). Thus, pleasure is induced by a circuitry rewarding pathway, first activating a descending MFB component and then the ascending mesolimbic dopamine pathway (Figure 16.6).

16.6 MECHANISM OF LOVE: AT A GLANCE

We are now aware of the structure of the brain, its parts, and the biochemicals involved in the various processes of love. Now, we present a brief explanation of the mechanism of love in general. A number of specific things regarding parental love and romantic love will also be discussed.

In a state of love, when an individual who is suffering and undergoing through external stress, she or he is supported by an empathetic behavior along with compassion from their partner. This leads to the activation of the limbic system, which involves the autonomic and peripheral nervous systems (Decety et al., 2016). This system helps to produce an emotional perspective in the individual, along with a cognitive perspective where the temporal lobe is activated. CNS activation helps in stress reduction, reduces suffering, and builds coping mechanisms (Panossian and Wikman, 2010). These mechanisms are autoregulated by a low production of VSP and a simultaneously high production of OXT

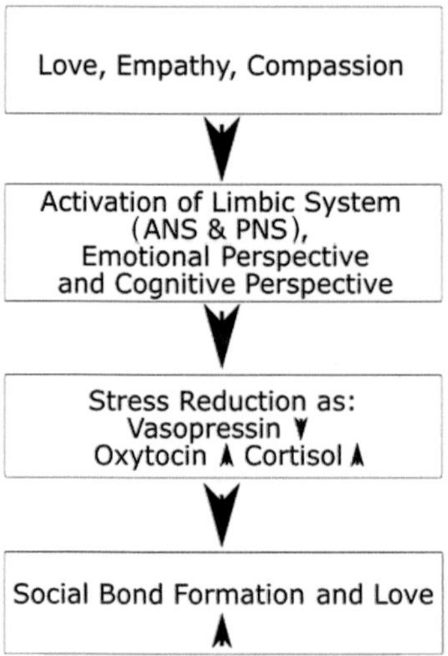

FIGURE 16.7 General mechanism of love. ANS, autonomic nervous system; PNS, peripheral nervous system.

(Panossian and Wikman, 2010). Further, stress hormones such as cortisol are increased, which further increases the social bond formation and feeling of love. This seems to be similar in the case of a parent and their offspring (Decety et al., 2016) (Figure 16.7).

16.6.1 A Detailed Mechanism of Parental Love

It has been discussed previously that there are several distinctions and overlaps between romantic love and maternal love. It has been observed for animals with small brains that olfactory recognition determines their behavior, and is dependent on pregnancy and parturition events, which simultaneously activate limbic mechanisms for bonding (Broad et al., 2006). In primates with large executive brains, the relative contribution of olfaction and physiology in determining behavioral output has diminished (Broad et al., 2006). There are biological constraints on maternalism in large-brained primates through lactation, but bonding and parental care can be activated by cognitive mechanisms. Cognitive capacity increases with larger executive brains, the development and growth of which extends substantially into the postpartum period, and necessitates extended parental care.

Maternal attachment is the most accepted form of enduring social bond. The concept of a mother's love implies a selective behavioral response by the parent to its offspring. Hormones that regulate birth and lactation are important in caregiver–child attachments (Kendrick et al., 1997). OXT and VSP are shown in both maternal behavior and maternal attachment. Catecholamines are involved in response to novelty, arousal, selective attention, certain kinds of learning, play a role in the development of attachments, and may reinforce or reward the expression of these attachments. Catecholamines, endogenous opioids, and prolactin affect parental behavior by modulating the rewarding aspects of this behavior (Panksepp and Bernatzky, 2002), and by the pacing of mother–infant interactions, and through their documented abilities to influence the release of other peptides.

16.6.1.1 Parental Attachment

In sheep, maternal behavior is directed only toward the lamb that the ewe birthed; a tangible rejection of other lambs is easily seen. Maternal bonding is explained by the release of OXT and endogenous opioids during mother–child activities such as breastfeeding and uterine stimulation (Kendrick et al., 1997). Injecting OXT in ewes can cause them to adopt or mistake an unrelated lamb for their own, provided the lamb is introduced at the time OXT is circulating through the ewe's body.

Hormones during pregnancy and birth facilitate maternal responses and lactation, including prolonged exposure to high levels of progesterone and estrogen, leading to a subsequent decline in progesterone, and increase in OXT and prolactin (Pedersen, 1999). Maternal behavior shares behavioral features with attachment. Maternal behavior is facilitated when the tendency to approach infants is stronger than the tendency to avoid infants. It is also suggested that reproductively naïve rats of both sexes have an inherent tendency to avoid potentially aversive novel stimuli. Hormones associated with birth and habituation to pup odors inhibit the fear-based system and permit the expression of maternal responses.

16.6.1.2 Specific Role of Biomolecules Involved in Parental Behavior

As we now know, various hormones and other molecules are involved in parental love. The sequence of events is shown in Figure 16.8.

(1) **Adrenal steroids:** Glucocorticoid levels rise in late pregnancy, and hormones of the HPA axis have been implicated in parturition. In human females, high levels of cortisol on postpartum days 2 or 3 have been correlated with positive maternal behaviors and attitudes (Fleming et al., 1997). Cortisol levels are also correlated with positive responses to infant odors, suggesting a role of cortisol in sensory processes (Stallings et al., 2001).

(2) **VSP:** Intracerebroventricular (ICV) VSP can facilitate maternal behavior but acts more slowly than OXT. Because the direct effects of peptides on behavior are often rapid, the slower actions of VSP in the induction of maternal behavior suggest that the observed effects are mediated by intermediary processes (Esch et al., 2003). Injecting selective V1aR antagonists can inhibit maternal behavior (Pedersen, 1999). VSPergic activity increases in the hypothalamus in late pregnancy. In male prairie voles, central VSP levels are found to be elevated, followed by mating, and under conditions when male parental behavior is likely (DeVries and Villalba, 1997).

(3) **OXT:** Since OXT is a mammalian hormone which plays an important role in birth and lactation, studies on its involvement in maternal behavior and

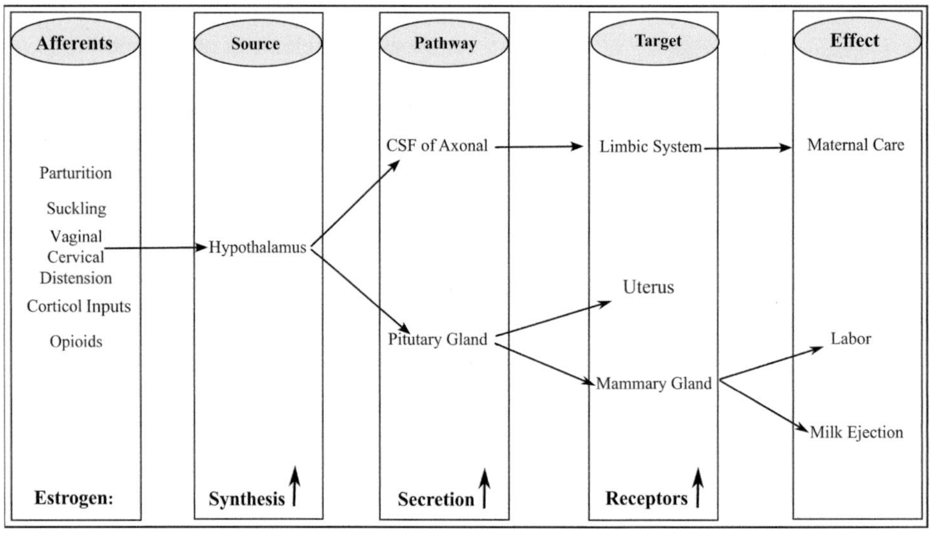

FIGURE 16.8 Mechanism of parental love (Insel, 1997). CSF, cerebrospinal fluid.

attachment have been carried out. ICV treatment with OXT has the potential to facilitate the onset of maternal responsiveness in virgin females (Pedersen, 1999). When treated with OXT antagonists, antibodies to OXT or lesions of OXT-producing neurons inhibit the induction of maternal behavior. OXT is released in or acts on brain areas associated with maternal behavior such as parturition (Pedersen, 1999; Liberzon et al., 1997; Uvnäs-Moberg; 1997, 1998).

(4) **Opioids:** Activation of the endogenous opioid system at parturition and during sucking promotes the positive effect arising from maternal behavior (Keverne, 2006). Naloxone treatment in monkeys has been shown to reduce mothers' caregiving and protective behaviour toward their infants (Martel et al., 1993). In sheep and rodents, interference of opioid blockade severely impairs maternal behavior (Keverne, 2006). The brain's endogenous opioid system that acts on naloxone-binding receptors in the ventral striatum dictates 'reward' alongside the mesolimbic dopamine system which manages the sensing of feelings that result in a reward (Merrer et al., 2009). Primates show extended postpartum care which is required for the development of infants, both behaviorally and morphologically, engaging prefrontal–ventral striatal reward pathways and intimate connections within the emotion-governing portion of the mind via parts such as the amygdala. Feminine members of the species are able to provide parental care constantly without the need for further procreative stages or childbirth because of their charges' emotional activity (Merrer et al., 2009).

16.6.1.3 Evolutionary Genetics of Caregiving

Studies have identified a category of mammalian genes that are autosomal haploid but expressed according to the parent of origin. Many paternally expressed genes have an important function in ensuring maternal care and provisioning for offspring, both pre- and postnatally (Keverne, 2001). A number of paternally expressed genes promote growth and enlargement of the placenta (*peg3, pegl, igf2, dlkl, insl, pllv*), nutrient transfer through the placenta (*igf2*), milk letdown (*peg3*), maternal care (*pegl, peg3*), placentophagia (*pegl*), suckling (*peg3, gnas*), and pre-weaning offspring growth (*peg3, pegl, gnas, igf2, grfl*). The parental genome ensures good maternal care across the biologically relevant parameters. Females govern sociability and the cohesiveness of social cliques over sustained time periods, while males deploy aggressive strategies and show mobility from the group. Females are responsible for providing care and nutrition, which is why the social standing of girls over boys becomes a part of the matriline. Asymmetries between the sexes in the context of parental care and in the context of social behavior (Keverne, 1993) have provided differential selection pressures through which imprinted genes might operate (Haig, 1997).

The knowledge set of offspring and the expression and self-regulation of emotions increases on the basis of interactions with their parents, primarily the mother/caretaker, and the constantly evolving action–reaction set between the two (Glaser, 2000). The evolution of a large neocortex that develops postnatally would not have been possible if parental and alloparental care had not extended beyond the biological dictation of hormonal factors. In the early years of neocortex formation and connection and association, one part that already remains in an advanced stage of development is the limbic emotional brain. The development of the mind into an adult specimen still entails the comprehension and channeling of emotions as a complex, yet necessary, portion. Cognitive capability has detailed interactions with the limbic brain in the context of the emotional interplay between mother and child. The child's need for an attachment figure and the mother's hormonal tendency to establish bonds results in an optimized socio-emotional setup for the development of the human adult brain. With the evolutionary increase in executive brainpower, maternalism has shifted from being primarily a component of physiological determinism perfected to match the biology of pregnancy and parturition to being heavily dependent on cognitive control. A human mother has both biology and emotional cognition to facilitate her bonding, which makes her the primary figure for infant attachment (Carter, 2005).

16.7 ROMANTIC LOVE

In this section, the primary focus is on romantic love, its types, and mechanisms by which it is distinct from parental love.

16.7.1 Classification of Romantic Love

The difference between romantic love and parental love has been explained earlier in this chapter. Romantic love has been classified further. Dr Helen Fisher and her team from Rutgers have classified 'romantic love' into three components. These components are divided on the basis of the neurochemical secretions involved, as follows.

16.7.1.1 Lust

Lust is induced by the essential sex hormones (estrogen and testosterone); the parts of the brain involved in lust are the hypothalamus, amygdala, posterior part of the insula, and prefrontal cortex (Dwivedi, 2012). Lust is one of the most basic human urges. It is also known as carnal desire or libido, and is inspired by the need for carnal gratification. Lust evolved to push people to seek a sexual union with an appropriate partner. The upswing in quantities of estrogen and testosterone creates the urge for gratification, and is physically manifested as lust (Fisher, 1998). Our so-called 'animal tendencies' are also influenced by the amygdala and, in particular, the hypothalamus (Vigué-Guix, 2017). Lust activates the posterior region of the insula which is generally activated through immediate bodily sensations, feelings, and reactions to external stimuli. It does not have a similar effect in the insula's anterior part, which is more active in relatively intangible representations that we perceive as complex thought processes (Cacioppo et al., 2012; Craig and Craig, 2009), explaining why lust remains an in-the-moment, relatively fleeting form of love.

16.7.1.2 Attraction

The hormones involved in attraction are happy hormones – dopamine, norepinephrine and serotonin – and the parts of the brain involved are the hypothalamus (Dwvedi, 2012) and prefrontal cortex. Another strong indicator of romantic attraction is a continuous decrease in serotonin activity around the brain (Vigué-Guix, 2017). As mentioned above, high levels of dopamine and norepinephrine are hypothesized to be associated with attraction (Vigué-Guix, 2017). It is the most vaunted manifestation of the three types, and is commonly referred to as passion for or obsession/infatuation with another human being. It is an emotional setup where the person chooses to fixate on a specific partner to mate with over a longer period of time. Selectively sorting out a partner saves on various costs associated with pursuing multiple partners. The attraction system is a system in humans that displays an increase in energy level and focusing of attention on a preferred partner. Feelings associated with this are exhilaration, intrusive thoughts about the beloved, and desires for emotional joining. This process also provides an advantage in natural selection, since animals have a higher probability of being attracted to fitter partners and thus increase their chances of successful procreation (Earp et al., 2012).

16.7.1.3 Attachment

Attachment is governed by the long-standing hormones OXT and VSP, and the portions of the brain involved in this case are the piriform cortex (Sullivan, 2012) and amygdala. In humans, attachment is known as compassionate love and is generally recognized as the epitome of romantic love from a social and physiological standpoint. Neurochemically, OXT and VSP play key roles in establishing the order of attachment within the brain (Vigué-Guix, 2017). It is usually characterized by a sense of security, feeling of calmness, social comfort, and emotional amalgamation. This system has evolved to motivate people to sustain their relationships for longer periods of time.

These three forms of love form important and mutually independent (but not exclusive) emotion systems which conciliate mating, procreation, and the raising of offspring (Fisher, 2000). Each of these emotional systems relates to overlapping but specific neurobiology in the brain. Different neurotransmitters produce different reactions upon reaching different parts of the brain. Figure 16.9 describes the involvement of different sets of hormones in the three different types of romantic love, while Table 16.1 provides a more in-depth analysis.

The neural makeup for each system can be often observed to be different at the highest level such as between species, at lower levels such as individualistically, and even at different

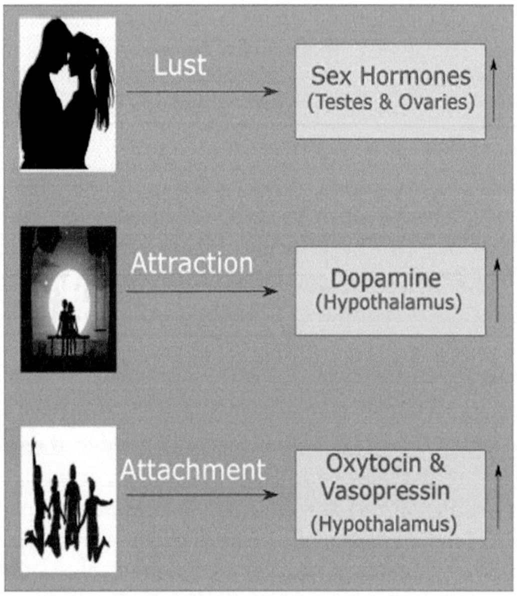

FIGURE 16.9 Hormonal changes signifying different types of love.

stages of an individual's lifetime. An increase in VSP levels in men and OXT in women causes attachment and may contribute to the feeling of closeness after coitus (Fisher, 2000). The three emotional systems can act independently in any individual as well, which means humans are capable of 'loving' more than one person at a time.

16.7.2 Principles of Attraction

Many variables are responsible for causing the phenomenon of attraction in an individual like similarity, dissimilarity, and individual traits. Here, we discuss the primary factors observed to be responsible for bringing individuals together.

16.7.2.1 The Similarity Principle

The first principle follows from an intrinsic, embedded value within us: 'like seeks like'. As Heider illustrated in 1958 with her theory of balance, individuals with interests and life experiences in common have more chances of being attracted to one another as they create an emotional stasis characterized by harmonious and constructive sensations and feelings. An imbalance, however, increases the risk of psychological disorders and uncertainty in the relationship.

16.7.2.2 Reciprocity

Newcomb called the phenomenon of reciprocity 'attraction breeds attraction', in 1956. Numerous experiments have been conducted where participants were provided with random feedback apparently left by others showed that just the existence of information about a person's preference with respect to them made them more susceptible to being attracted back (Luo and Zhang, 2009). Reciprocity was divided into two types by Kenny et al.:

(1) General reciprocal liking – if one usually likes other people for their qualities, their chances of looking attractive to people increases.
(2) Dyadic reciprocal liking – a more intimate form of reciprocity that exists between two interested individuals.

16.7.2.3 Beauty (Physical Attractiveness)

Evidence suggests that, especially during the first few encounters, attraction can be predicted to occur on the basis of physical beauty. This principle directs the male's attention more than their female counterpart's, as proved by Buss in 1989 and Buss & Schmitt in 1993 (Luo and Zhang, 2009).

16.7.2.4 Security

Security is a less widely recognized principle which hypothesizes that secure individuals with a reduced tendency to anxiety and avoidance form the most attractive partners due to their willingness to engage with their respective partners in stronger, more emotionally assured relationships. Research and empirical evidence suggest that certain characteristics tend to serve as relationship 'buffers'. Hence, a frank, outgoing, conscientious person who projects a 'positive' image in society would also project a sense of security with their concept of love (Luo and Zhang, 2009).

16.7.3 Initiators of Attraction

Attraction is a phenomenon of evoking interest or simply liking someone. We have discussed the principles that generally govern attraction, but these, in turn, come about via the underlying biochemical processes triggered in the body. More specifically, the following factors are responsible for the feelings one experiences and qualify as an attraction, either individually or in combination (Hilhorst, 2002).

16.7.3.1 Sight

It is said that beauty lies in the eyes of the beholder, yet there are numerous ways in which appearance has a subconscious effect on whom one is attracted to. For instance, having a symmetrical face is generally considered as a hallmark of conventional attractiveness. Similarly, a few more characteristics such as masculinity, shape, fashion sense/clothing, make-up, perfume, or simply artistic looks and personality further impress the other person before starting a relationship (Hilhorst, 2002).

16.7.3.2 Masculinity

In general, a woman tends to prefer men with more conventionally masculine features. Universally vaunted features such as broad shoulders, shaped jaw lines, and so on, play a role here, since they subconsciously project a positive outlook with regard to health and chivalry (DeBruine et al., 2006).

16.7.3.3 Positive Personality

Apart from an individual's chemical makeup and looks, emotional features can be defined in terms of how a potential partner is perceived. For instance, kindness in a human is an instant allure. These positive features can even combine with the aforementioned superficial attractiveness and magnify it (Mehrabian and Blum, 1997).

16.7.3.4 Smell

Here, a typical type of smell due to pheromones has been described, although all smells are not pheromones. Pheromones are chemicals present in bodily fluids such as sweat, which release odorous signatures upon secretion. Previous experiments have illustrated their roles in sexual attraction in the animal kingdom, and these chemicals may be important in human attraction as well. Our bodies' most active pheromone agents come from skin's sebaceous glands and the two types of sweat gland – apocrine and eccrine. These are all capable of producing chemicals with distinctive smells, and both men and women prefer smells that are MHC-dissimilar to themselves (Bui, 2009).

Several categories of pheromones such as releaser, signaler, modulator, and primer have effects on recipients via olfaction, ingestion, or absorption (Bruce, 1970; Wysocki and Preti,

2004). Releasers are the only type of pheromones involved in romantic attraction. These categories have different roles to play in eliciting a behavioral response, affecting mood and emotions, and the secretion of hormones. Other categories such as signaler pheromones send signals to members of groups of the same species to indicate the identity, emotional state, reproductive status, and even genetic information of the sender. (Wysocki and Preti, 2004). The category of the modulators is to adjust the ongoing behavior for physical and social stimuli, and to regulate psychological reactions toward a particular context, without triggering specific behaviors or thoughts. Primers are responsible for long-term neuroendocrine changes, which are further responsible for regulating the endocrine system and development of the recipient. (Wysocki and Preti, 2004)

16.7.3.5 Fertility

Men can sense fertility periods in women. A study has proved that during the peak fertile time in their menstrual cycle, a woman releases a distinct scent that makes her extra appealing and desired. A woman's face may also look attractive to men during this period. A woman's voice also plays a major role as her voice is most seductive at the most fertile point of her menstrual cycle (Nathan Pipitone, 2012).

16.7.3.6 Stress

Stress, which is usually said to be an emotional or physical strain on the body, is also one of the main components that activates or initiates a longing for proximity, closeness, privacy, and desire. Stress-related physiological processes and hormones are activated in the HPA axis. These factors have the ability to develop and promote social bonding. The subjects of experiments in such cases show high concentrations of cortisol when compared to control groups (Marazziti and Canale, 2004). The term 'love-induced hypercortisolemia' represents a type of stress response observed in any kind of change in a person's life or mind (Marazziti and Canale, 2004). This change and related stress response illustrate the early phase of the physiological state of sudden alertness and building up of relationships, which in some cases helps people to overcome the condition of neophobia. Naturally, positive social relations are seen to increase stress levels, which are further balanced by the production of OXT, which, in turn, assists a sense of security and support (Mottolese et al., 2014).

16.8 SEXUAL BEHAVIORAL CHANGES

16.8.1 Sex Behavior in Males

In men, OXT receptors may be located in the epididymis and corpus cavernosum of the penis; and binding to the OXT receptor in this region can affect subsequent ejaculation (Filippi et al., 2003) and contractility (Vignozzi et al., 2004). Plasma OXT levels rise during orgasm and sexual arousal, and significantly increases in males (Carmichael et al., 1987). In men, intranasal inhalation of OXT significantly increases epinephrine and plasma OXT levels for at least 60 minutes and surges self-perception of arousal during masturbation (Burri et al., 2008).

16.8.2 Sex Behavior in Females

Apart from being used in medicine to induce labor, OXT has also been used to facilitate breastfeeding. Plasma OXT levels can be supremely correlated with towering levels of arousal and lubrication, as measured by the Female Sexual Function Index (Salonia et al., 2005). Plasma OXT levels rise in females during sexual arousal and are raised further by orgasm (Blaicher et al., 1999; Carmichael et al., 1987).

The neuropeptides OXT and arginine vasopressin (AVP) serve as both peripheral hormones and as neurotransmitters or neuromodulators. The PVN and SON both contain OXT and AVP, producing neurons that release projections to the neurohypophysis where they are stored and sent to the bloodstream (Ishunina and Swaab,1999).

With these important factors, initiators and balancing strategies are covered for both romantic and parental love.

16.9 OPPORTUNITIES AND CHALLENGES

There are great opportunities in this field to understand the neurobiology of love. If we were able identify the exact molecules and their combinations along with their mechanisms of action, it would help us understand the relationships, emotional status, psychological behaviors, and reproduction preferences of human beings and other animals. These kinds of studies will definitely be fruitful in the field of psychology and behavioral science, as well as in reproduction biology with regard to mental and physiological abnormalities. For example, knowledge on the use of OXT has been used for years (Akingba, 1962) to induce labor, and it is now very clear that it also plays an important role in maintaining relationships, sustaining them for longer and forming the basis of the purest form of love between parents and children, between two friends, or among siblings (Vigué-Guix, 2017). OXT is also found to be important for maintaining relationships between two life partners, as this hormone can be stabilizing (Vigué-Guix, 2017). Similarly, other hormones like sex hormones initiate and trigger a relationship (Dwivedi, 2012), but, later – when it becomes a compassionate love – OXT and VSP play a major role. Therefore, knowledge of the interactions among hormones and other molecules, changes in brain endocrinology, and the sequence of events can help us to maintain a better family and societal structures. These kinds of studies will help us to understand the evolution of animal species and the development of the brain and complexity of human behavior.

The happy hormones – dopamine and serotonin – have been shown to be released during physical activities (Basso and Suzuki, 2017) and also while in certain relationships that make a person feel happy and enthusiastic about themselves and their life. A better understanding of such phenomena may also help people to overcome certain mental health issues such as depression and anxiety, which affect a large number of people worldwide.

There is a great possibility that a drug or hormonal modulator may balance out the levels of OXT and VSP during anger and unusual sexual behavior among individuals, or positively affect the ACC of sexually incompetent males. Products like pheromone-based perfumes are currently on the market, but nothing concrete can so far be said about their use in improving romantic partnerships (Wysocki and Preti, 2004). The challenge for this field is that human pheromones cannot so far be isolated as they are produced in very small quantities from various sources, though insect-based pheromones have been isolated and studied with regard to their impacts on social and individual behaviors, among or between partners. If we look at these studies from a different angle, insects are equally important for our sustainability on Earth as multiple natural tasks are fulfilled by them (Patel et al., 2021). Similarly, understanding the role of these molecules and mechanisms in other animals would also help us in understanding their physiology and neurobiology along with improving our understanding of how to use this knowledge for the benefit of humans.

So, we can conclude here that the major challenge is that pheromones are produced in very small quantities and their isolation is very difficult. A larger amount would be required to further explore their mechanism and applications. But, this challenge may be overcome very soon by advances in molecular biology and other techniques. Also, we are at a very nascent stage of understanding and deeper research has to be performed in this field to further elucidate important facts. Another big challenge in understanding the neurobiology of love is that it is not easy to work with humans as their perceptions and reactions might not be the same as they can be in a natural environment when in such relationships. Also, most people will not allow any kind of interference in a relationship for the sake of science and knowledge.

16.10 CONCLUSION

The chapter has established the relationship between love and several factors, including hormone levels, stress, health, and various other external stimuli. It gives deep insight into the physiological changes occurring in the body during the course of all types of affections. All the studied physiological and hormonal changes occurring during the course of love and maternal care may prove to be the milestone discovery for countering various human sociological problems, including the swiftly increasing arrogance in the modern man, higher rates of divorces, alarming cases of rape and physical harassment, and sexual problems in both males and females. Research has been performed in this area but not to that extent that concrete conclusions can be made and applied to improving relationships or treating medical or psychological problems. In short, in-depth research is still required to completely understand neuroscience in directing romantic attachment and parental love using advanced technology with which future research opportunities will shine, with the prospect of advances in social neuroscience in combination with attachment theory.

ACKNOWLEDGEMENT

The authors would like to acknowledge Shreya Bhargava for her valuable contribution in gathering data and partially formatting the article.

REFERENCES

Akingba JB. Induction of labour with intramuscular pitocin (oxytocin): a review of 223 consecutive cases. An International Journal of Obstetrics and Gynecology. 1962;69:924–928.

Albers HE. Species, sex and individual differences in the vasotocin/vasopressin system: relationship to neurochemical signaling in the social behavior neural network. Frontiers in Neuroendocrinology. 2015;36:49–71.

Aragona BJ, Liu Y, Curtis JT, Stephan FK, Wang Z. A critical role for nucleus accumbens dopamine in partner-preference formation in male prairie voles. Journal of Neuroscience. 2003;23:3483–3490.

Bales KL, Carter CS. Sex differences and developmental effects of oxytocin on aggression and social behavior in prairie voles (*Microtus ochrogaster*). Hormones and Behavior. 2003;44:178–184.

Barberis C, Tribollet E. Vasopressin and oxytocin receptors in the central nervous system. Critical Reviews in Neurobiology. 1996;10:119–154.

Bartels A, Zeki S. The neural correlates of maternal and romantic love. Neuroimage. 2004;21:1155–1166.

Basso JC, Suzuki WA. The effects of acute exercise on mood, cognition, neurophysiology, and neurochemical pathways: a review. Brain Plasticity. 2017;2:127–152.

Berecek KH, Wyss JM, Swords BH. Alterations in vasopressin mechanisms in captopril-treated spontaneously hypertensive rats. Clinical and Experimental Hypertension Part A: Theory and Practice. 1991;13:1019–1031

Berman H, Mason R, Hall J, Rodger S, Classen CC, Evans MK, et al. Laboring to mother in the context of past trauma: the transition to motherhood. Qualitative Health Research. 2014;24:1253–1264.

Blaicher W, Gruber D, Bieglmayer C, Blaicher AM, Knogler W, Huber JC. The role of oxytocin in relation to female sexual arousal. Gynecologic and Obstetric Investigation. 1999;47:125–126.

Bozarth MA. Pleasure systems in the brain. In: Warburton DM (ed.) Pleasure: The Politics and the Reality. New York: Wiley;1994:5–14.

Broad KD, Curley JP, Keverne EB. Mother–infant bonding and the evolution of mammalian social relationships. Philosophical Transactions of the Royal Society B: Biological Sciences. 2006;361:2199–2214.

Brown CH, Bains JS, Ludwig M, Stern JE. Physiological regulation of magnocellular neurosecretory cell activity: integration of intrinsic, local and afferent mechanisms. Journal of Neuroendocrinology. 2013;25:678–710.

Bruce HM. Pheromones. British Medical Bulletin. 1970;26:10–13.

Bui, S. Discovering the genetic magic behind love. Daily Bruin. 2009. Accessed: 19 October 2020. Available at: https://dailybruin.com/2009/02/03/discovering-genetic-magic-behind-love

Burri A, Heinrichs M, Schedlowski M, Kruger TH. The acute effects of intranasal oxytocin administration on endocrine and sexual function in males. Psychoneuroendocrinology. 2008;33:591–600.

Cacioppo S, Bianchi-Demicheli F, Frum C, Pfaus JG, Lewis JW. The common neural bases between sexual desire and love: a multilevel kernel density fMRI analysis. The Journal of Sexual Medicine. 2012;9:1048–1054.

Carmichael MS, Humbert R, Dixen J, Palmisano G, Greenleaf W, Davidson JM. Plasma oxytocin increases in the human sexual response. The Journal of Clinical Endocrinology & Metabolism. 1987;64:27–31.

Carmichael MS, Warburton VL, Dixen J, Davidson JM. Relationships among cardiovascular, muscular, and oxytocin responses during human sexual activity. Archives of Sexual Behavior. 1994;23:59–79.

Carter CS. Neuroendocrine perspectives on social attachment and love. Psychoneuroendocrinology. 1998;23:779–818.

Carter CS. The chemistry of child neglect: Do oxytocin and vasopressin mediate the effects of early experience? Proceedings of the National Academy of Sciences U S A. 2005;102:18247–18248.

Carter CS, Grippo AJ, Pournajafi-Nazarloo H, Ruscio MG, Porges SW. Oxytocin, vasopressin and sociality. Progress in Brain Research. 2008;170:331–336.

Carter CS, Lederhendler II, Kirkpatrick B (eds). The Integrative Neurobiology of Affiliation. Cambridge, MA: MIT Press;1999.

Carter CS, Porges SW. The biochemistry of love: an oxytocin hypothesis: Science & Society Series on Sex and Science. EMBO Reports. 2013;14:12–16.

Chen Q, Schreiber SS, Brinton RD. Vasopressin and oxytocin receptor mRNA expression during rat telencephalon development. Neuropeptides. 2000;34:173–180.

Chen T, O'Den G, Song Q, Koga K, Zhang MM, Zhuo M. Adenylyl cyclase subtype 1 is essential for late-phase long term potentiation and spatial propagation of synaptic responses in the anterior cingulate cortex of adult mice. Molecular Pain. 2014;10:65.

Cho MM, DeVries AC, Williams JR, Carter CS. The effects of oxytocin and vasopressin on partner preferences in male and female prairie voles (Microtus ochrogaster). Behavioral Neuroscience. 1999;113:1071–1079.

Colon-Urban R, Stefano GB. Behavioral effects of morphine on several species of bryozoans. In: Nielsen C, Larwood GP (eds.) Bryozoa: Ordovician to Recent. Denmark: Olsen and Olsen;1985:67–71.

Craig AD, Craig AD. How do you feel—now? The anterior insula and human awareness. Nature Reviews Neuroscience. 2009;10:59–70.

Crews D. The evolutionary antecedents to love. Psychoneuroendocrinology. 1998;23:751–764.

DeBruine LM, Jones BC, Little AC, Boothroyd LG, Perrett DI, Penton-Voak IS, et al. Correlated preferences for facial masculinity and ideal or actual partner's masculinity. Proceedings of the Royal Society B: Biological Sciences. 2006;273:1355–1360.

Decety J, Bartal IB, Uzefovsky F, Knafo-Noam A. Empathy as a driver of prosocial behaviour: highly conserved neurobehavioural mechanisms across species. Philosophical Transactions of the Royal Society B: Biological Sciences. 2016;371:20150077.

Devinsky O, Morrell MJ, Vogt BA. Contributions of anterior cingulate cortex to behaviour. Brain. 1995;118:279–306.

DeVries AC, DeVries MB, Taymans SE, Carter CS. The effects of stress on social preferences are sexually dimorphic in prairie voles. Proceedings of the National Academy of Sciences U S A. 1996;93:11980–11984.

Devries GJ, Villalba C. Brain sexual dimorphism and sex differences in parental and other social behaviors. Annals of the New York Academy of Sciences. 1997;807:273–286.

Domes G, Heinrichs M, Michel A, Berger C, Herpertz SC. Oxytocin improves "mind-reading" in humans. Biological Psychiatry. 2007;61:731–733.

Dong HW, Swanson LW. Projections from bed nuclei of the stria terminalis, anteromedial area: cerebral hemisphere integration of neuroendocrine, autonomic, and behavioral aspects of energy balance. Journal of Comparative Neurology. 2006;494:142–178.

Doty RL. Human pheromones: do they exist? In: Doty RL, Mucignat-Caretta C (eds.) Neurobiology of Chemical Communication. Boca Raton, FL: CRC Press/Taylor & Francis;2014:535–560.

Du Vigneaud V, Ressler C, Trippett S. The sequence of amino acids in oxytocin, with a proposal for the structure of oxytocin. Journal of Biological Chemistry. 1953;205:949–957.

Dunbar RI. Neocortex size as a constraint on group size in primates. Journal of Human Evolution. 1992;22:469–493.

Dwivedi M. Love—role of neurochemicals (hormones) behind this. Journal of Advancement in Pharmacology. 2012;1:1–5.

Earp BD, Sandberg A, Savulescu J. Natural selection, childrearing, and the ethics of marriage (and divorce): building a case for the neuroenhancement of human relationships. Philosophy & Technology. 2012;25:561–587.

Esch T, Fricchione GL, Stefano GB. The therapeutic use of the relaxation response in stress-related diseases. Medical Science Monitor. 2003;9:RA23–34.

Esch T, Stefano GB. The neurobiology of pleasure, reward processes, addiction and their health implications. Neuroendocrinology Letters. 2004;25:235–251.

Esch T, Stefano GB. The neurobiology of love. Neuroendocrinology Letters. 2005;26:175–192.

Esch T, Stefano GB. The neurobiological link between compassion and love. Medical Science Monitor. 2011;17:RA65–75.

Filippi S, Vignozzi L, Vannelli GB, Ledda F, Forti G, Maggi M. Role of oxytocin in the ejaculatory process. Journal of Endocrinological Investigation. 2003;26(3 Suppl):82–86.

Fisher HE. Lust, attraction, and attachment in mammalian reproduction. Human Nature. 1998;9:23–52.

Fisher HE. Lust, attraction, attachment: biology and evolution of the three primary emotion systems for mating, reproduction, and parenting. Journal of Sex Education and Therapy. 2000;25:96–104.

Fleming AS, Steiner M, Corter C. Cortisol, hedonics, and maternal responsiveness in human mothers. Hormones and Behavior. 1997 Oct 1;32(2):85–98.

Gillies GE, McArthur S. Estrogen actions in the brain and the basis for differential action in men and women: a case for sex-specific medicines. Pharmacological Reviews. 2010;62:155–198.

Gingrich B, Liu Y, Cascio C, Wang Z, Insel TR. Dopamine D2 receptors in the nucleus accumbens are important for social attachment in female prairie voles (*Microtus ochrogaster*). Behavioral Neuroscience. 2000;114:173–183.

Glaser D. Child abuse and neglect and the brain—a review. Journal of Child Psychology and Psychiatry. 2000;41:97–116.

Gonzaga GC, Turner RA, Keltner D, Campos B, Altemus M. Romantic love and sexual desire in close relationships. Emotion. 2006;6:163–179.

Goode TD, Maren S. Role of the bed nucleus of the stria terminalis in aversive learning and memory. Learning & Memory. 2017;24:480–491.

Haig D. Parental antagonism, relatedness asymmetries, and genomic imprinting. Proceedings of the Royal Society of London. Series B: Biological Sciences. 1997;264:1657–1662.

Heider F. The Psychology of Interpersonal Relations. Hove: Psychology Press;1982.

Hilhorst MT. Physical beauty: only skin deep? Medicine, Health Care and Philosophy. 2002;5:11–21.

Hu SH, Wei N, Wang QD, Yan LQ, Wei EQ, Zhang MM, et al. Patterns of brain activation during visually evoked sexual arousal differ between homosexual and heterosexual men. American Journal of Neuroradiology. 2008;29:1890–1896.

Hui KK, Liu J, Makris N, Gollub RL, Chen AJ, Moore CI, et al. Acupuncture modulates the limbic system and subcortical gray structures of the human brain: evidence from fMRI studies in normal subjects. Human Brain Mapping. 2000;9:13–25.

Insel TR. A neurobiological basis of social attachment. American Journal of Psychiatry. 1997;154:726–735.

Insel TR, Fernald RD. How the brain processes social information: searching for the social brain. Annual Review of Neuroscience. 2004;27:697–722.

Insel TR, Gelhard R, Shapiro LE. The comparative distribution of forebrain receptors for neurohypophyseal peptides in monogamous and polygamous mice. Neuroscience. 1991;43:623–630.

Insel TR, Hulihan TJ. A gender-specific mechanism for pair bonding: oxytocin and partner preference formation in monogamous voles. Behavioral Neuroscience. 1995;109:782–789.

Insel TR, Shapiro LE. Oxytocin receptor distribution reflects social organization in monogamous and polygamous voles. Proceedings of the National Academy of Sciences U S A. 1992;89:5981–5985.

Insel TR, Young LJ. The neurobiology of attachment. Nature Reviews Neuroscience. 2001;2:129–136.

Ishunina TA, Swaab DF. Vasopressin and oxytocin neurons of the human supraoptic and paraventricular nucleus, size changes in relation to age and sex. The Journal of Clinical Endocrinology & Metabolism. 1999;84:4637–4644.

Jern P, Westberg L, Johansson A, Jonsson L, Corander J, Sandnabba NK, Santtila P. Are single nucleotide polymorphisms in the oxytocin and vasopressin 1A/1B receptor genes likely candidates for variation in ejaculatory function? BJU International. 2012;110:E1173–E1180.

Jonas W, Mileva-Seitz V, Girard AW, Bisceglia R, Kennedy JL, Sokolowski M, et al. Genetic variation in oxytocin rs2740210 and early adversity associated with postpartum depression and breastfeeding duration. Genes, Brain and Behavior. 2013;12:681–694.

Juárez OH, Calderon GD, Hernandez GE, Barragán Mejía G. The role of dopamine and its dysfunction as a consequence of oxidative stress. Oxidative Medicine and Cellular Longevity. 2016;2016:9730467.

Kendrick KM. Introduction. The neurobiology of social recognition, attraction and bonding. Philosophical Transactions of the Royal Society of London. Series B, Biological Sciences. 2006;B361:2057–2059.

Kendrick KM, Da Costa AP, Broad KD, Ohkura S, Guevara R, Lévy F, Keverne EB. Neural control of maternal behaviour and olfactory recognition of offspring. Brain Research Bulletin. 1997;44:383–395.

Keverne EB. Sex differences in primate social behavior. In: The Development of Sex Differences and Similarities in Behavior. Dordrecht: Springer;1993:227–240.

Keverne EB. Genomic imprinting, maternal care, and brain evolution. Hormones and Behavior. 2001;40:146–155

Keverne EB. Trophoblast regulation of maternal endocrine function and behaviour. In: Moffett A, Loke C, McLaren A (eds.) Biology and Pathology of Trophoblast. Cambridge: Cambridge University Press;2006:148–163.

Kleiman DG. Monogamy in mammals. The Quarterly Review of Biology. 1977;52:39–69.

Komisaruk BR, Whipple B. Love as sensory stimulation: physiological consequences of its deprivation and expression. Psychoneuroendocrinology. 1998;23:927–44.

Lee HJ, Macbeth AH, Pagani JH, Young 3rd WS. Oxytocin: the great facilitator of life. Progress in Neurobiology. 2009;88:127–151.

Legros JJ. Inhibitory effect of oxytocin on corticotrope function in humans: are vasopressin and oxytocin ying–yang neurohormones? Psychoneuroendocrinology. 2001;26:649–655.

Leknes S, Tracey I. A common neurobiology for pain and pleasure. Nature Reviews Neuroscience. 2008;9:314–320.

Levy B, Tasker J. Synaptic regulation of the hypothalamic–pituitary–adrenal axis and its modulation by glucocorticoids and stress. Frontiers in Cellular Neuroscience. 2012;6:24.

Liberzon I, Trujillo KA, Akil H, Young EA. Motivational properties of oxytocin in the conditioned place preference paradigm. Neuropsychopharmacology. 1997;17:353–359.

Lim MM, Murphy AZ, Young LJ. Ventral striatopallidal oxytocin and vasopressin V1a receptors in the monogamous prairie vole (*Microtus ochrogaster*). Journal of Comparative Neurology. 2004;468:555–570.

Liu Y, Young KA, Curtis JT, Aragona BJ, Wang Z. Social bonding decreases the rewarding properties of amphetamine through a dopamine D1 receptor-mediated mechanism. Journal of Neuroscience. 2011;31:7960–7966.

Love TM. Oxytocin, motivation and the role of dopamine. Pharmacology Biochemistry and Behavior. 2014;119:49–60.

Love TM, Enoch MA, Hodgkinson CA, Peciña M, Mickey B, Koeppe RA, et al. Oxytocin gene polymorphisms influence human dopaminergic function in a sex-dependent manner. Biological Psychiatry. 2012;72:198–206.

Luef GJ. Epilepsy and sexuality. Seizure. 2008;17:127–130.

Luo S, Zhang G. What leads to romantic attraction: Similarity, reciprocity, security, or beauty? Evidence from a speed-dating study. Journal of Personality. 2009;77:933–964.

McBride SM, Giuliani G, Choi C, Krause P, Correale D, Watson K, et al. Mushroom body ablation impairs short-term memory and long-term memory of courtship conditioning in *Drosophila melanogaster*. Neuron. 1999;24:967–977.

Magon N, Kalra S. The orgasmic history of oxytocin: Love, lust, and labor. Indian Journal of Endocrinology and Metabolism. 2011;15(Suppl3):S156–S161.

Marazziti D, Canale D. Hormonal changes when falling in love. Psychoneuroendocrinology. 2004;29:931–936.

Martel FL, Nevison CM, Rayment FD, Simpson MJ, Keverne EB. Opioid receptor blockade reduces maternal affect and social

grooming in rhesus monkeys. Psychoneuroendocrinology. 1993;18:307–321.
Mehrabian A, Blum JS. Physical appearance, attractiveness, and the mediating role of emotions. Current Psychology. 1997;16:20–42.
Merrer JL, Becker JA, Befort K, Kieffer BL. Reward processing by the opioid system in the brain. Physiological Reviews. 2009;89:1379–1412.
Moncho-Bogani J, Martinez-Garcia F, Novejarque A, Lanuza E. Attraction to sexual pheromones and associated odorants in female mice involves activation of the reward system and basolateral amygdala. European Journal of Neuroscience. 2005;2:2186–2198.
Moore FA, Feliciano DV, Andrassy RJ, McArdle AH, Booth FV, Morgenstein-Wagner TB, et al. Early enteral feeding, compared with parenteral, reduces postoperative septic complications. The results of a meta-analysis. Annals of Surgery. 1992;216:172–183.
Mottolese R, Redouté J, Costes N, Le Bars D, Sirigu A. Switching brain serotonin with oxytocin. Proceedings of the National Academy of Sciences. 2014;111:8637–8642.
Munesue T, Yokoyama S, Nakamura K, Anitha A, Yamada K, Hayashi K, et al. Two genetic variants of CD38 in subjects with autism spectrum disorder and controls. Neuroscience Research. 2010;67:181–191.
Nakamura K, Sekine Y, Ouchi Y, Tsujii M, Yoshikawa E, Futatsubashi M, et al. Brain serotonin and dopamine transporter bindings in adults with high-functioning autism. Archives of General Psychiatry. 2010;67:59–68.
Nathan Pipitone R, Gallup GG, Jr. The unique impact of menstruation on the female voice: Implications for the evolution of menstrual cycle cues. Ethology. 2012;118:281–291.
Nestler EJ. Molecular basis of long-term plasticity underlying addiction. Nature Reviews Neuroscience. 2001;2:119–128.
Nestler EJ. Molecular mechanisms of drug addiction. Neuropharmacology 2004;47:24–32.
Ortigue S, Bianchi-Demicheli F, Patel N, Frum C, Lewis JW. Neuroimaging of love: fMRI meta-analysis evidence toward new perspectives in sexual medicine. The Journal of Sexual Medicine. 2010;7:3541–3552.
Ostrowski K, Hermann C, Bangash A, Schjerling P, Nielsen JN, Pedersen BK. A trauma-like elevation of plasma cytokines in humans in response to treadmill running. The Journal of Physiology. 1998;513:889–894.
Panksepp J, Bernatzky G. Emotional sounds and the brain: the neuro-affective foundations of musical appreciation. Behavioural Processes. 2002;60:133–155.
Panossian A, Wikman G. Effects of adaptogens on the central nervous system and the molecular mechanisms associated with their stress–protective activity. Pharmaceuticals. 2010;3:188–224.
Pardo-Bellver C, Cádiz-Moretti B, Novejarque A, Martínez-García F, Lanuza E. Differential efferent projections of the anterior, posteroventral, and posterodorsal subdivisions of the medial amygdala in mice. Frontiers in Neuroanatomy. 2012;6:33.
Patel V, Pauli N, Biggs E, Barbour L, Boruff B. Why bees are critical for achieving sustainable development. Ambio. 2021;50:49–59.
Pedersen CA. Oxytocin control of maternal behavior: regulation by sex steroids and offspring stimuli. In: Carter CS, Lederhendler II, Kirkpatrick B (eds.) The Integrative Neurobiology of Affiliation. Cambridge, MA: MIT Press;1999:301–320.
Pfau DR, Hobbs NJ, Breedlove SM, Jordan CL. Sex and laterality differences in medial amygdala neurons and astrocytes of adult mice. Journal of Comparative Neurology. 2016;524:2492–2502.
Pfaus JG, Heeb MM. Implications of immediate-early gene induction in the brain following sexual stimulation of female and male rodents. Brain Research Bulletin. 1997;44:397–407.
Popenoe EA, du Vigneaud V. A partial sequence of amino acids in performic acid-oxidized vasopressin. Journal of Biological Chemistry. 1954;206:353–360.
Porges SW. Love: an emergent property of the mammalian autonomic nervous system. Psychoneuroendocrinology. 1998;23:837–861.
Raff H. CORT, Cort, B, corticosterone, and now cortistatin: enough already! Endocrinology. 2016;157:3307–3308.
Rossetti ZL, Melis F, Carboni S, Gessa GL. Dramatic depletion of mesolimbic extracellular dopamine after withdrawal from morphine, alcohol or cocaine: a common neurochemical substrate for drug dependence. Annals of the New York Academy of Sciences. 1992;654:513–516.
Salamon E, Esch T, Stefano GB. Role of amygdala in mediating sexual and emotional behavior via coupled nitric oxide release 1. Acta Pharmacologica Sinica. 2005;26:389–395.
Salamon E, Kim M, Beaulieu J, Stefano GB. Sound therapy induced relaxation: down regulating stress processes and pathologies. Medical Science Monitor. 2003;9:RA96.
Salonia A, Nappi RE, Pontillo M, Daverio R, Smeraldi A, Briganti A, et al. Menstrual cycle-related changes in plasma oxytocin are relevant to normal sexual function in healthy women. Hormones and Behavior. 2005;47:164–169.
Schwartz R, Olds J. Love and the Brain. On the brain. 2019. Accessed: 20 October 2020. Available at: https://neuro.hms.harvard.edu/harvard-mahoney-neuroscience-institute/brain-newsletter/and-brain/love-and-brain
Sheldrick EL, Flint AP. Endocrine control of uterine oxytocin receptors in the ewe. Journal of Endocrinology. 1985;106:249–258.
Stallings J, Fleming AS, Corter C, Worthman C, Steiner M. The effects of infant cries and odors on sympathy, cortisol, and autonomic responses in new mothers and nonpostpartum women. Parenting. 2001;1:71–100.
Stefano GB, Esch T. Love and stress. Neuroendocrinology Letters. 2005;26:173–174.
Stefano GB, Fricchione GL, Slingsby BT, Benson H. The placebo effect and relaxation response: neural processes and their coupling to constitutive nitric oxide. Brain Research Reviews. 2001;35:1–9.
Uvnäs-Moberg K. Role of efferent and afferent vagal nerve activity during reproduction: integrating function of oxytocin on metabolism and behaviour. Psychoneuroendocrinology. 1994;19:687–695.
Uvnäs-Moberg K. Neuroendocrinology of the mother–child interaction. Trends in Endocrinology & Metabolism. 1996;7:126–131.
Uvnäs-Moberg K. Physiological and endocrine effects of social contact. Annals of the New York Academy of Sciences. 1997;807:146–163.
Uvnäs-Moberg K. Oxytocin may mediate the benefits of positive social interaction and emotions. Psychoneuroendocrinology. 1998;23:819–835.
Vignozzi L, Filippi S, Luconi M, Morelli A, Mancina R, Marini M, et al. Oxytocin receptor is expressed in the penis and mediates an estrogen-dependent smooth muscle contractility. Endocrinology. 2004;145:1823–1834.

Vigué-Guix I. Lust, romance, attachment—what brain scanning says about the drive to love. Accessed: 20 October 2020. Available at: https://irenevigueguix.wordpress.com/2017/02/07/lust-romance-attachment-what-brain-scanning-says-about-the-drive-to-love/

Witt DM. Oxytocin and rodent sociosexual responses: from behavior to gene expression. Neuroscience & Biobehavioral Reviews. 1995;19:315–324.

Witt DM, Winslow JT, Insel TR. Enhanced social interactions in rats following chronic, centrally infused oxytocin. Pharmacology Biochemistry and Behavior. 1992;43:855–861.

Wu LJ, Kim SS, Li X, Zhang F, Zhuo M. Sexual attraction enhances glutamate transmission in mammalian anterior cingulate cortex. Molecular Brain. 2009;2:1–4.

Wysocki CJ, Preti G. Facts, fallacies, fears, and frustrations with human pheromones. The Anatomical Record Part A: Discoveries in Molecular, Cellular, and Evolutionary Biology. 2004;281:1201–1211.

Zhuo M. Neuronal mechanism for neuropathic pain. Molecular Pain. 2007;3:1744–8069.

Zitzmann M, Nieschlag E. Testosterone levels in healthy men and the relation to behavioural and physical characteristics: facts and constructs. European Journal of Endocrinology. 2001;144:183–197.

Index

For Product Safety Concerns and Information please contact our EU representative GPSR@taylorandfrancis.com Taylor & Francis Verlag GmbH, Kaufingerstraße 24, 80331 München, Germany